Lecture Notes in Physics

Volume 859

For further volumes:
http://www.springer.com/series/5304

The Lecture Notes in Physics

The series Lecture Notes in Physics (LNP), founded in 1969, reports new developments in physics research and teaching—quickly and informally, but with a high quality and the explicit aim to summarize and communicate current knowledge in an accessible way. Books published in this series are conceived as bridging material between advanced graduate textbooks and the forefront of research and to serve three purposes:

- to be a compact and modern up-to-date source of reference on a well-defined topic
- to serve as an accessible introduction to the field to postgraduate students and nonspecialist researchers from related areas
- to be a source of advanced teaching material for specialized seminars, courses and schools

Both monographs and multi-author volumes will be considered for publication. Edited volumes should, however, consist of a very limited number of contributions only. Proceedings will not be considered for LNP.

Volumes published in LNP are disseminated both in print and in electronic formats, the electronic archive being available at springerlink.com. The series content is indexed, abstracted and referenced by many abstracting and information services, bibliographic networks, subscription agencies, library networks, and consortia.

Proposals should be sent to a member of the Editorial Board, or directly to the managing editor at Springer:

Christian Caron
Springer Heidelberg
Physics Editorial Department I
Tiergartenstrasse 17
69121 Heidelberg/Germany
christian.caron@springer.com

Sandibek B. Nurushev · Mikhail F. Runtso ·
Mikhail N. Strikhanov

Introduction to Polarization Physics

Sandibek B. Nurushev
Experimental Physics Department
Institute for High-Energy Physics
Protvino, Russia

Mikhail N. Strikhanov
Nat. Research Nuclear Univ. "MEPhI"
Moscow, Russia

Mikhail F. Runtso
Exp. Methods of Nuclear Physics
Nat. Research Nuclear Univ. "MEPhI"
Moscow, Russia

Translated from Russian by R. Tjapaev. The original Russian edition was published by the Moscow Engineering Physics Institute in 2007

ISSN 0075-8450
Lecture Notes in Physics
ISBN 978-3-642-32162-7
DOI 10.1007/978-3-642-32163-4
Springer Heidelberg New York Dordrecht London

ISSN 1616-6361 (electronic)

ISBN 978-3-642-32163-4 (eBook)

Library of Congress Control Number: 2012951700

Printed on acid-free paper

Springer is part of Springer Science+Business Media (www.springer.com)

Preface

Polarization physics represents the section of physics devoted to investigate the statistical and dynamical characteristics of processes associated with spin, which is one of the fundamental characteristics of elementary particles and nuclei. The spin is a tool to investigate and test fundamental questions in all type of well known interactions like the electroweak, strong and gravitation fields. This occurs for reason that all elementary particles involved in those interactions have the integer or half-integer spins.

During more than century development, polarization physics has created serious theoretical and methodical foundations leading to many discoveries. Its development became particularly fast in last two decades in connection with the "spin crisis" phenomenon discovered in 1987 by European Muon Collaboration (EMC) at CERN.

Large polarization effects have been revealed in the production of hyperons and strong spin effects have been found in the exclusive and inclusive production of hadrons. Investigations on polarization physics are conducted at the largest proton and electron accelerators and colliders. The great successes are reached in engineering methods of polarization physics. These successes concern the development of the methods for obtaining and accelerating polarized particles and the polarimetry methods for such beams. Impressive results have been obtained in the development of high-current and highly polarized ion sources for accelerators, as well as polarized targets.

To summarize the results of polarization physics the polarization community organizes the biannual High Energy Spin Physics Symposia, started in 1974 at the Argonne National Laboratory (Chicago); follow-up meetings were at Argonne 1976 and 1978, Lausanne 1980, Brookhaven 1982, Marseille 1984, Protvino 1986, Minneapolis 1988, Bonn 1990, Nagoya 1992, Bloomington 1994, Amsterdam 1996 and Protvino 1998. In 2000 in Osaka this Symposium was united with another polarization meeting under the title "The Symposia on Polarization Phenomena in Nuclear Physics", started at Basel in 1960 and conducted every five years at Karlsruhe 1965, Madison 1970, Zürich 1975, Santa Fe 1980, Osaka 1985, Paris

1990 and Bloomington 1994, the last one being in parallel to the High Energy Spin Physics meeting. The new Conference was named as The 14th International Spin Physics Symposium, followed by meetings at Brookhaven 2002, Trieste 2004, Kyoto 2006, Charlottesville (Virginia) 2008, and in 2010 19th International Spin Physics Symposium was held in Juelich (Germany). Between symposia, workshops on various important sections of polarization physics were organized.

The time has come to gather the basic results of polarization physics, to systematize them so that they were accessible to a wide audience.

A number of monographs were devoted to this field, but they have an appreciable theoretical bias. In addition into those monographs were not included many important works on the theoretical and experimental polarization physics done by Soviet and later by Russian physicists. We aim to fill this gap.

In this book, we aimed to compile and systematize theoretical, experimental, and particularly methodical aspects of polarization physics and to present them in the accessible form. We include in book the experimental data starting approximately around the middle of 1970s when the polarization of $\Lambda-$ hyperons was discovered at Fermilab and a lot of polarization results became available from high energy accelerators over the world.

This book is primarily designed for final-year and post-graduate students of faculties of physics of technical universities and assumes the corresponding basic knowledge. At the same time, to assist the study of the material, we tried to present the basic terms and definitions with the corresponding explanations. The most recent results are taken for the presentation of the corresponding fields. When writing the book, we use our lectures for students and wider audience, as well as our original papers, review articles and presentations at conferences.

The book consists of three parts. The parts are divided into chapters (numbered sequentially as 1, 2, etc. throughout the book) and the chapters are divided into sections (having common numbering throughout the book: Section 1.1, Section 1.2, first number denoting to which section chapter belong). Some sections are divided into subsections (numbered as Section 1.1.1, Section 1.1.2, etc.). The numbering of the formulas, tables, and figures is the same as numbering of sections in each chapter. The lists of references are given after each chapter. A reference mentioned in the text is indicated by the surname of the first author and the surname of the second author, if there are two authors only, or et al., if there are more then two authors, and the publication year. If a list of references contains several works of one author in one year, these works are marked by letters (a, b, c, etc.) after the publication year.

In order to unify the notations of the spin observables we introduce Sect. 13.4, under the title "The Ann Arbor Convention" adopted in 1977. It might be useful in identifications of the different labeling of polarization observables used in various scientific papers.

We are deeply grateful to PhDs A.A. Bogdanov and V.J. Khodyrev for continuous assistance in the preparation and editing of the book and to Drs. L.S. Azhgirei (deceased) and M.G. Ryskin for advices and stimulating discussions of many sections of the book.

S.B. Nurushev
M.F. Runtso
M.N. Strikhanov

Contents

Introduction

By the end of the XIX century and the beginning of the XX century, physicists have collected extensive experimental data on the spectral lines of hydrogen-like atoms; a consistent theoretical interpretation of these data was absent at that time. This particularly concerned the fine and hyperfine splittings of spectral lines, the so-called Zeeman effects. In 1911, Rutherford proposed a model of the atom with a central nucleus. In 1913, Bohr developed a model of the atom with electrons revolving around the nucleus and theoretically deduced a radiation formula for hydrogen atoms (Balmer series). The Bohr model included the relativistic relation between the energy and momentum of the electron. However, the *Zeeman effects* remained unexplained. The Bohr model of the atom was noticeably improved by Sommerfeld who expanded the Bohr postulates by allowing elliptic electron orbits (Bohr allowed only circular orbits) and quantized the action integral for any pair of conjugate variables. As a result, he arrived at the magnetic quantum number m, which was already the third quantum number (in addition to the principal quantum number n and orbital quantum number l). The quantum number m, which is defined as the projection of the orbital angular momentum vector $\vec{l}$ on any axis z, can have the integer values from $-l$ to $+l$. This number explained the *fine splitting* of levels. However, the anomalous Zeeman effect (hyperfine splitting of levels) remained unexplained. Pauli approached close to the understanding of this effect in 1924. He wrote that this phenomenon "is explained by the characteristic two-valuedness of the quantum properties of valence electrons, which cannot be described in classical mechanics" (Pauli 1925a). There is an opinion that Pauli easily could make the next step and discovered spin (Fidecaro 1998). However, he focused on another problem and soon arrived at the well-known "Pauli forbidden principle" (Pauli 1925b). The discovery of spin did not occur.

In 1925, Uhlenbeck and Goudsmit published an article where the spin concept was introduced for the first time as a quantum operator having two eigenvalues (Uhlenbeck and Goudsmit 1925). Following Pauli's proposal, they supplemented three quantum numbers (n, l, m), which describe the motion of electrons in the atom, by the fourth degree of freedom characterized by the quantum number m_s having two possible values. Similar to Pauli, they attributed this quantum number directly to the electron and treated it as the internal quantum number. However, in contrast to Pauli, they interpreted it as the quantum parameter appearing as a result of the rotation of the electron about its axis. For this reason, their next article appearing approximately in three months after the first one had the title "Spinning electrons and the structure of spectra" (Uhlenbeck and Goudsmit 1926) from which the term spin appeared. This spin degree of freedom was attributed directly to the electron. The authors could explain the Landé factor equal two (the Landé factor appears in the proportionality factor between the magnetic field and the energy difference of the split levels).

The final word in scientific investigations obviously belongs to theory. In 1927, Pauli wrote a nonrelativistic equation (Pauli equation) with the introduction of the sigma matrix and two-component spinors; this equation explained many spin effects in nuclear physics, including the anomalous Zeeman effects. In 1928, Dirac wrote a relativistic equation for the electron; unexpectedly for him, the existence of the electron (and the positron) with a half-integer spin and the required magnetic moment follows from this equation. By the end of 1928, spin was commonly accepted as the fundamental characteristic of elementary particles in addition to the charge and mass. This was the formation of spin physics.

The majority of known elementary particles have spin and are classified into two groups: bosons and fermions, having integer and half-integer spins, respectively. These groups are described by the Bose–Einstein and Fermi–Dirac statistics, respectively. Spin plays a very noticeable role in the dynamics of the strong, weak, and electromagnetic interactions of particles. However, spin in electroweak interactions is described completely, whereas its description for strong interactions remains a serious problem. Spin is directly involved in many important discoveries in modern physics such as the hyperfine splitting of atomic lines, shell models of atoms and nuclei, discovery of parity violation in weak and electromagnetic processes, the spin crisis phenomenon, the existence of the magnetic moments of particles.

According to modern representations, spin appears as a consequence of the symmetry of Minkowski space with respect to the space-time translations and rotations of four-dimensional space. The Poincaré group describing these transformations has two Casimir operators, which are responsible for two universal observables for any physical system. The first of these operators leads to the definition of the mass of a system and the second, to the spin concept (Ji 2002).

This book consists of three parts presenting the foundations of polarization physics. The first part contains theoretical introduction and covers a wide range of problems. Some of them are as follows. The technique for creating polarized ion sources, polarized solid and gas targets is based on the hyperfine splitting of hydrogen levels. These levels in the region of very low energies are calculated using the nonrelativistic Schrödinger equation. For this reason, the nonrelativistic Schrödinger equation and its solution with the inclusion of Pauli matrices and an external magnetic field are presented. For the region of intermediate energies, elastic scattering processes are primarily discussed. In the absence of the quantitative theory of strong interaction, the concept of complete experiment plays a special role in this energy range. This concept is based on the construction of the scattering matrix of any process using the general principles of the invariance of interactions. In this approach, spin and its transformation properties (under continuous and discrete transformations of space and time) are of particular importance. In the theoretical part, the general principles of construction of the density matrix, reaction matrix, and invariants are discussed and their applications to particular reactions of pion–nucleon and nucleon–nucleon scattering are illustrated. Examples of complete experiments are given.

Relativistic quantum mechanics is required to describe processes at high energies. In this case, particular difficulty is associated with the necessity of the relativistic description of spin for both a free particle and a particle in a magnetic field. The correct description of spin dynamics is especially important for the calculation of depolarizing effects in the acceleration of protons in accelerators. These problems are also briefly presented in the first part of the book. Finally, the interpretation of the experimental results presented in the third part of the book requires theoretical models of the dynamics of the interactions of particles with spin. A number of such models are also described in this part of the book.

Experimental investigations of spin phenomena in high energy physics require a beam of particles with a known degree of the orientation of their spin in a certain direction. Such prevailing orientation is called polarization; for this reason, experimental spin physics is also called polarization physics.

Experiments on deep inelastic scattering are now leading among polarization experiments. In the first part of the book, the status of these investigations is briefly reviewed on the basis of the latest experimental results.

The polarization technique is described in the second part of the book. It consists of four chapters. Any experiment in polarization physics involves the scattering of polarized particle beams on a polarized target (experiments with a fixed target) or a polarized beam (colliders). Chapter 5 describes the production and acceleration of polarized protons and electrons. The unique polarized muon beam is also separately considered. Chapter 6 is devoted to polarized targets. The polarized solid targets of certain large experimental setups are described so that they can be compared in many parameters. At accelerators, polarized jet and gas targets with and without storage cells are most successfully used. In Chap. 7, two polarized ion sources with the best parameters are described. Chapter 8 is devoted to polarimetry, i.e., the measurement of the polarization of beams and targets.

The third part of the book is called "Polarization experiments and their results". The latest results obtained at the largest setups are compiled. It is emphasized that polarization is the most sensitive tool for testing predictions of the Standard Model (SM) and quantum chromodynamics (QCD). By present time, the distribution functions of polarized valence quarks have been quite accurately determined, the distribution functions of sea quarks have been less precisely determined, and it is yet almost impossible to determine accurately the spin distribution functions of gluons. Single-spin asymmetries, i.e., non uniformities of the azimuthal distribution of the secondary particles appearing in the interactions of polarized beams and targets, are measured with good accuracy. Contrary to the theoretical expectations, spin effects survive in the inclusive production of pions in the polarized particle fragmentation region to a center-of-mass energy of 200 GeV. A similar large asymmetry is for the first time observed for the neutrons inclusively produced in the same kinematic region. The effects of spin information transfer in interactions are also discussed.

We have concluded that numerous unsolved problems still remain in polarization physics. For this reason, physicists working in this field develop new programs of polarization investigations.

References

Fidecaro, G.: In: Proceedings of the 13th International Symposium on High Energy Spin Physics, p. 50. Protvino, Russia (1998)

Ji, X.: In: Proceedings of the 15th International Symposium on High Energy Spin Physics, Upton, New York, p. 3 (2002)

Pauli, W.: Z. Phys. **31**, 373 (1925a)

Pauli, W.: Z. Phys. **31**, 765 (1925b)

Uhlenbeck, G.E., Goudsmit, S.: Nature **113**, 953 (1925)

Uhlenbeck, G.E., Goudsmit, S.: Nature **117**, 264 (1926)

Part I
The Theoretical Bases of Polarization

In this part of the manual, we briefly recall the basic elements of nonrelativistic and relativistic quantum mechanics that are useful for understanding of the subsequent material. We also try to present the basic conservation laws following from the Lorentz and generalized Lorentz transformations, as well as from discrete transformations. The aim of any experimental study, particularly in hadron physics, is to collect maximally complete information on a reaction under investigation. In this approach, we aim at quite completely presenting the determination of the complete set of experiments in the pion–nucleon and nucleon–nucleon interactions. Taking into account that the spin is a complicated subject (Tomonaga 1997) having no classical analogue, we present various definitions of the spin and its transformation properties.

We consider the theoretical part as a necessary tool for understanding the foundations of the technical methods of polarization physics and a basis for analyzing the results of polarization experiments (Feynman et al. 1963; Fermi 1961). Another aim of this compilation of the materials is to avoid diverting readers to the search for additional sources of information.

References

Feynman, R.P., Leighton, R.B., Sands, M.: The Feynman Lectures on Physics, vol. 3. Addison-Wesley, Reading (1963)
Fermi, E.: Notes on Quantum Mechanics. Univ. of Chicago Press, Chicago (1961)
Tomonaga, S.: The Story of Spin. Univ. of Chicago Press, Chicago (1997)

Chapter 1
Spin and Its Properties

Before describing the spin, we should obviously define the notion of spin. Uhlenbeck and Goudsmit (1925, 1926) proposed a hypothesis that, in addition to the mass and charge, the electron has an intrinsic angular momentum and magnetic moment. This intrinsic angular momentum was called spin and denoted by the symbol s as the first latter of this word. This angular momentum is not associated with the orbital motion of the particle. It is difficult to realize the notion of intrinsic angular momentum in application to an elementary particle such as the electron. Let us extend the hypothesis of spin to a nucleus and use the consideration from book (Landau and Lifshitz 1963). Since the nucleus is a complex system consisting of nucleons, its state should be specified not only by the internal energy, but also by the intrinsic angular momenta of the nucleons $\vec{s}_i$ and their orbital momenta $\vec{l}_i$. The total angular momentum of the nucleus $\vec{S}$ (that is, its spin) is determined as the sum $\vec{S} = \sum_i \vec{s}_i + \vec{l}_i$ where the sum runs over the nucleons in the nucleus and $\vec{S}$ can have $2S + 1$ values. Thus, the angular momentum distribution of the nucleons in the nucleus determines its spin. A similar consideration is also applicable to the nucleon, since it is also complicated system. In the quark model, a nucleon consists of three quarks mediated by gluons. The angular momentum distribution of quarks and gluons in the nucleon determines the nucleon spin. According to the naïve quark model, the nucleon spin should be completely determined by the spins of valence quarks. However, in 1984, the European Muon Collaboration (EMC) revealed that the valence quarks carry only 25 %, rather than 100 %, of the nucleon spin. This fact was called "spin crisis". Thus, the problem of the origin of the proton spin has not yet been quantitatively explained in the parton model.

For the electron (point particle), we cannot find such a simple explanation for the origin of spin, because the spin is a quantum-mechanical operator and has no analog in classical physics.

The hypothesis of spin opened possibilities for the simple explanation of a huge number of experimental facts.

The problem of the possibility of a direct experimental determination of the magnetic moment (correspondingly, spin) of the electron was formulated by Mott (1929). He showed that the uncertainty principle excludes the direct measurement of

S.B. Nurushev et al., *Introduction to Polarization Physics*,
Lecture Notes in Physics 859, DOI 10.1007/978-3-642-32163-4_1,
© Moskovski Inzhenerno-Fisitscheski Institute, Moscow, Russia 2013

the electron spin in the experiments, e.g., in the Stern–Gerlach experiment (Dehmet 1990). At the same time, he proposed an experiment that makes it possible to determine the average spin value, i.e., polarization, from the double scattering of electrons (Mott 1932). The essence of the experiment is as follows. An unpolarized low-energy electron beam is scattered from a highly charged target at large angles. The scattered electrons should be polarized due to the spin–orbit interaction. These polarized electrons are scattered in the same plane by the second identical target. The left–right asymmetry at the second target is measured. This asymmetry is the product of the polarization of the electrons after the first scattering act by the analyzing power of the second scattering act. The presence of nonzero asymmetry obviously confirms the presence of the polarization, i.e., spin of the electron. There were several attempts to observe this effect in the experiments; however, they were unsuccessful owing to various problems. Only in 1943, the first such experiment was successfully performed and its results completely confirmed Mott's predictions (Schull et al. 1943). In those measurements, a Mott polarimeter was used; for more details, see review of Gay (1992).

1.1 Elements of Nonrelativistic Quantum Mechanics

Spin is a purely quantum characteristic of objects of the microcosm, and its description requires the technique of quantum mechanics both nonrelativistic (Bethe and Salpeter 1957) and relativistic (Dirac 1958). In this section, we briefly consider the basic elements of nonrelativistic quantum mechanics that are required for the presentation of the materials on polarization physics (Shpol'skii 1984a, 1984b). Below, in Sect. 1.7, we present the necessary elements of relativistic quantum mechanics. We use the monographs given in the below list of references.

Recall the basic notions used in quantum mechanics.

Linear operators are very often used in quantum mechanics. We present the information on the operators following Fermi (1961). The operators act on functions specified on a certain domain such as the numerical x axis (one-dimensional or linear space), a set of points, points on the sphere, and three-dimensional space of numbers x, y, and z. The functions can be considered as vectors in a space, finite- or infinite-dimensional. An operator is generally a rule (mathematical operation) according to which the function f is transferred to the function g:

$$g = \hat{O} f. \tag{1.1}$$

The operators are denoted by a letter with a "hat". The functions and operators in quantum mechanics are generally complex. Operators $\hat{O}$ should include the identity operator $\hat{I}$ reproducing the initial function:

$$g = \hat{O} f = \hat{I} f = f. \tag{1.2}$$

Almost any mathematical operation can be associated with the corresponding operator.

Linear operators are important in quantum mechanics. They satisfy the requirement

$$\hat{O}(\alpha f + \beta g) = \alpha \hat{O} f + \beta \hat{O} g \tag{1.3}$$

for any pair of functions f and g and any complex constants α and β. The multiplication by numerical factors and functions, differentiation and integration operations, etc. are linear operators.

The sum and difference of the linear operators $\hat{C}_\pm = \hat{A} \pm \hat{B}$ are also linear operators:

$$\hat{C}_\pm f = \hat{A} f \pm \hat{B} f. \tag{1.4}$$

The summation (subtraction) is commutative:

$$\hat{C}_\pm f = \pm \hat{B} f + \hat{A} f. \tag{1.5}$$

The linear operators have the associativity property:

$$\hat{A} + (\hat{B} + \hat{C}) = (\hat{A} + \hat{B}) + \hat{C}. \tag{1.6}$$

The product of two linear operators also has the associativity property:

$$(\hat{A}\hat{B}) f = \hat{A}(\hat{B} f). \tag{1.7}$$

The multiplication of an operator by a number is equivalent to the multiplication of this number by the result of the action of the operator on a function.

The product of two linear operators is generally noncommutative, i.e.,

$$\hat{A}\hat{B} \neq \hat{B}\hat{A}. \tag{1.8}$$

In order to illustrate this statement, we consider the case where $\hat{A} = x$, $\hat{B} = d/dx$:

$$(\hat{A}\hat{B}) f = \left(x \frac{d}{dx} \right) f = x \frac{df}{dx}, \qquad (\hat{B}\hat{A}) f = \frac{d}{dx}(xf) = f + x \frac{df}{dx}. \tag{1.9}$$

The commutator of two operators $\hat{A}$ and $\hat{B}$ is defined as

$$[\hat{A}, \hat{B}] = -[\hat{B}, \hat{A}] = \hat{A}\hat{B} - \hat{B}\hat{A}. \tag{1.10}$$

If $[\hat{A}, \hat{B}] = 0$, the operators commute.

Let us also define the anticommutator as

$$\{\hat{A}, \hat{B}\} = \{\hat{B}, \hat{A}\} = \hat{A}\hat{B} + \hat{B}\hat{A}. \tag{1.11}$$

According to relation (1.9),

$$\left[\frac{d}{dx}, x \right] = 1. \tag{1.12}$$

The power of an operator specifies the multiplicity of the action of the operator, for example, for $\hat{A} = \frac{d}{dx}$, $\hat{A}^n = \frac{d^n}{dx^n}$ or $\hat{A}^{n+m} = \hat{A}^n \hat{A}^m$. The commutation relation $[\hat{A}^n, \hat{A}^m] = 0$ is valid for any operator. The inverse operator (its action cancels the action of the initial operator) $\hat{A}^{-1}$ also commutes with the initial operator $[\hat{A}^{-1}, \hat{A}] = 0$. An operator function $F(\hat{A})$ is useful in applications. By analogy with an ordinary function, this function can be expanded in the Taylor series:

$$F(\hat{A}) = \sum_{n=0}^{\infty} \frac{F^{(n)}(0)}{n!} \hat{A}^n. \tag{1.13}$$

Let us consider an example of the function $F(\hat{A}) = e^{\alpha \hat{A}}$, where $\hat{A} = d/dx$. In this case, the expansion has the form

$$e^{\alpha \hat{A}} = 1 + \alpha \hat{A} + \frac{\alpha^2}{2!} \hat{A}^2 + \cdots + \frac{\alpha^n}{n!} \hat{A}^n + \cdots = \sum_{n=0}^{\infty} \frac{\alpha^n}{n!} \hat{A}^n. \tag{1.14}$$

The substitution of the operator $\hat{A} = \frac{d}{dx}$ yields

$$e^{\alpha \hat{A}} = 1 + \alpha \frac{d}{dx} + \frac{\alpha^2}{2!} \frac{d^2}{dx^2} + \cdots + \frac{\alpha^n}{n!} \frac{d^n}{dx^n} + \cdots = \sum_{n=0}^{\infty} \frac{\alpha^n}{n!} \frac{d^n}{dx^n}. \tag{1.15}$$

Then, the action of the operator $F(\hat{A})$ on the function f provides

$$F(\hat{A}) f(x) = e^{\alpha \frac{d}{dx}} f(x) = \sum_{n=0}^{\infty} \frac{\alpha^n}{n!} \frac{d^n f(x)}{dx^n} = f(x + \alpha). \tag{1.16}$$

The last relation corresponds to the power-series expansion of the function $f(x + \alpha)$ in the variable α near the point $\alpha = 0$. As seen, the action of this operator reduces to the shift of the argument of the function by α.

Let us introduce the wave function $\psi(x)$ in the form of the column with n elements

$$\psi(x) = \begin{pmatrix} \psi_1 \\ \cdots \\ \psi_m \\ \cdots \\ \psi_n \end{pmatrix}, \tag{1.17}$$

where $x = x_1, x_2, \ldots$ is the set of all continuous arguments, e.g., coordinates and m is the discrete variable changing from 1 to n.

The action of the operator $\hat{F}$ on the function ψ_m sometimes yields the same function multiplied by a certain number λ_m : $\hat{F} \psi_m = \lambda_m \psi_m$. If the function ψ_m satisfies the so-called "standard conditions" (the requirements of its finiteness, continuity, and single-valuedness in the entire region of its independent arguments) and the square integrability condition (the integral of the squared absolute value of the function is finite), the function ψ_m is called an eigenfunction of the operator $\hat{F}$ and λ_m is its eigenvalue corresponding to the eigenfunction ψ_m.

The matrix element of an operator is defined as

$$F_{kl} = \int \psi_k^*(x) \hat{F} \psi_l(x) dx. \tag{1.18}$$

Let us introduce the notion of the Hermitian (self-adjoint) operator. For each linear operator $\hat{F}$, the adjoint linear operator $\hat{F}^+$ can be obtained from the initial operator $\hat{F}$ by exchanging the columns and rows and taking complex conjugation. The matrix element of the Hermitian conjugate operator satisfies the condition $F_{kl} = \int \psi_k^* \hat{F} \psi_l dX = \int (\hat{F}^+ \psi_k)^* \psi_l dX$, where $dX = dx_1 \cdot dx_2 \cdots$ are independent continuous variables, integration is performed over the entire region of independent variables (phase space), and the asterisk, as usual, denotes the complex conjugation. If the adjoint operator coincides with the initial operator, the operator is called self-adjoint or Hermitian. In this case, $\int \psi^* \hat{F} \varphi dX = \int (\hat{F} \psi)^* \varphi dX$. There is an important theorem according to which the eigenvalues of a self-adjoint operator are real. Let us prove it. We have

$$\hat{F} \psi_m = \lambda_m \psi_m. \tag{1.19}$$

The Hermitian conjugation gives

$$\psi_m^+ \hat{F}^+ = \lambda_m^* \psi_m^+. \tag{1.20}$$

Let us multiply Eq. (1.19) by ψ_m^+ from the left and Eq. (1.20), by ψ_m from the right. Subtracting one resulting relation from the other and taking into account that $\hat{F} = \hat{F}^+$ according to the definition of the Hermitian operator, we obtain

$$\lambda_m = \lambda_m^*, \tag{1.21}$$

quod erat demonstrandum.

According to this theorem, all observables (energy, momentum, angular momentum, spin moment, etc.) are represented in quantum mechanics by Hermitian operators.

Nonrelativistic quantum mechanics is based on the following six principles (Bjorken and Drell 1964, 1965):

1. A given physical system is described by the vector of state Φ, which contains all information on the system. In application to the single-particle system, the vector of state in the coordinate representation is called wave function. This wave function is denoted as ψ and is a complex function of the entire set of the arguments describing the given physical system. This set of arguments can include coordinates, momenta, time, spin, isospin, etc. These parameters should describe all degrees of freedom of the particle. We denote this set of independent variables except for time t as q. Then, the wave function is written as $\psi(t, q)$. The wave function $\psi(t, q)$ has no direct physical interpretation. However, the square of its absolute

value, $|\psi(t, q)|^2 \geq 0$, is treated as the probability of finding the particle at time t at the multidimensional-space point q. According to the probability interpretation, $|\psi(t, q)|^2$ should be finite in the entire physical region of the variables q.

2. Any physical observable corresponds to a linear Hermitian operator. In particular, the momentum p_i corresponds to the following operator in the q_i coordinate representation:

$$p_i \rightarrow \frac{\hbar}{i} \frac{\partial}{\partial q_i}. \tag{1.22}$$

3. The state of the physical system is the eigenfunction Φ of an arbitrary operator $\hat{O}$ if the following equality is valid:

$$\hat{O}\Phi_n(q, t) = O_n \cdot \Phi_n(q, t), \tag{1.23}$$

where Φ_n is the nth eigenstate (or eigenfunction of $\hat{O}$) corresponding to the eigenvalue O_n. If $\hat{O}$ is a Hermitian operator, the eigenvalue O_n is real.

4. An arbitrary wave function or vector of state of the physical system can be represented in terms of the complete orthonormalized set of wave functions ψ_n of the complete set of the operators commuting with the Hamiltonian and with each other. The completeness and orthonormalization of the system of the wave functions $\psi_n(q, s)$ (q means all continuous variables and s denotes all discrete variables) is expressed by the relation (Schiff 1968):

$$\sum_s \int dq \psi_n^*(q, s) \psi_m(q, s) = \delta_{nm}. \tag{1.24}$$

Therefore, an arbitrary wave function ψ of the physical system can be expanded in terms of this complete set as follows:

$$\psi = \sum_n a_n \psi_n. \tag{1.25}$$

The quantity $|a_n|^2$ is the probability that the physical system is in the nth eigenstate.

5. The experimental measurement of an observable provides one of its eigenvalues. For example, if the physical system is described by wave function ψ (1.25) and the function ψ_n is the eigenfunction of the operator $\hat{O}$ corresponding to the eigenvalue O_n, i.e., $\hat{O}\psi_n = O_n \cdot \psi_n$, then the measurement of the physical observable O provides the eigenvalue O_n with the probability $|a_n|^2$. The mean value of the operator $\hat{O}$ (taking into account the orthogonality of the eigenfunctions) is defined as

$$\langle \hat{O} \rangle = \sum_{n,s} \int dq \psi_n^*(q, s) \hat{O} \psi_m(q, s) = \sum_n |a_n|^2 O_n. \tag{1.26}$$

6. The following Schrödinger equation describes the time evolution of the physical system:

$$i\hbar\frac{\partial}{\partial t}\psi = \hat{H}\psi. \tag{1.27}$$

Here, the Hamiltonian $\hat{H}$ (operator corresponding to the energy of the system) is a linear Hermitian operator. The Hamiltonian of a closed (isolated) physical system does not explicitly depend on time; hence,

$$\frac{\partial H}{\partial t} = 0. \tag{1.28}$$

The solutions of the equation of motion with such a Hamiltonian specify possible stationary states of the physical system. The superposition principle (the fourth of the above principles) follows from the linearity of the Hamiltonian operator. Since the Hamiltonian is Hermitian, the probability of finding the particle at the point with the coordinates q is conserved as seen from the following relation obtained using formula (1.27):

$$\frac{\partial}{\partial t}\sum_s \int dq\psi^*\psi = \frac{i}{\hbar}\sum_s \int dq[(\hat{H}\psi)^*\psi - \psi^*(\hat{H}\psi)] = 0. \tag{1.29}$$

This relation expresses the conservation of the probability density.

Let us consider the simplest Hamiltonian of a free isolated particle moving with the momentum $\vec{p}$. This Hamiltonian is equal to the kinetic energy of the particle

$$H = \frac{p^2}{2m}. \tag{1.30}$$

For the passage from classical mechanics to quantum mechanics, each dynamical variable of classical mechanics is associated with a linear Hermitian operator in quantum mechanics and the following change is made:

$$H \to i\hbar\frac{\partial}{\partial t}, \qquad \vec{p} \to -i\hbar\nabla. \tag{1.31}$$

As a result, we arrive at the nonrelativistic Schrödinger equation for a free particle:

$$i\hbar\frac{\partial\psi(q,t)}{\partial t} = -\frac{\hbar^2}{2m}\nabla^2\psi(q,t). \tag{1.32}$$

In the presence of interaction, the Hamiltonian contains not only the kinetic energy K, but also the potential energy V of interaction and has form

$$H = K + V. \tag{1.33}$$

Therefore, the Schrödinger equation with allowance for the interaction between particles is written in the general form

$$i\hbar\frac{\partial\psi(q,t)}{\partial t} = \hat{H}\psi(q,t) = \left[-\frac{\hbar^2}{2m}\nabla^2 + V(q,t)\right]\psi(q,t). \qquad (1.34)$$

A general definite recipe for finding the Hamiltonian is absent. The Hamiltonian for a particular problem is constructed in terms of the basic independent kinematic parameters (momenta, orbital angular momenta, spins, external electromagnetic fields, magnetic moments, etc.), and the requirements of invariance under coordinate transformations (translations, rotations, space inversions, and time reversal) are imposed in order to obtain the scalar (or pseudoscalar) Hamiltonian. The correctness of the chosen Hamiltonian is determined by comparing the calculation results with experimental data.

The examples of the Hamiltonians for particular cases are considered in the following sections.

1.2   Angular Momentum Operator

At the beginning of this section, we present the basic information on the Dirac notation which is widely used below (Shpol'skii 1984a, 1984b).

Any vector ψ in the n-dimensional Euclidean space is unambiguously specified by the set of its components in a fixed basis; this representation can be written in the form of a column consisting of the components of this vector $\psi_1, \ldots, \psi_n$. Let us formally introduce the space of adjoint vectors ψ^+, which are obtained from ψ by Hermitian conjugation (transposition into row and complex conjugation of the components). The dot product of the vectors φ and ψ (the corresponding scalar product of two functions of x in the interval $a < x < b$ is usually defined as $(\varphi, \psi) = \int_a^b \varphi(x)\psi(x)dx$) can be written in the short matrix form $(\varphi, \psi) = \varphi^+\psi$. Following Dirac, we denote $\psi \equiv |\psi\rangle$, $\psi^+ \equiv \langle\psi|$, where $|\psi\rangle$ and $\langle\psi|$ are ket and bra vectors, respectively (named after word "bracket"). These two types of vectors are formally related by the Hermitian conjugation: $\langle\psi| \equiv |\psi\rangle^+$, $|\psi\rangle \equiv \langle\psi|^+$. As mentioned above, the symbol $\langle\varphi|\psi\rangle$ (the second vertical dash is omitted) means the dot product of vectors φ and ψ, which is a number. Let us introduce complete orthonormalized bases in the space of ket and bra vectors; these bases are the sets of the basis vectors that are obtained from each other by Hermitian conjugation. These basis vectors are denoted as $|1\rangle, |2\rangle, \ldots, |n\rangle$ and $\langle 1|, \langle 2|, \ldots, \langle n|$, respectively. In this notation, the condition of the orthonormalization of the basis is represented in the form

$$\langle j|k\rangle = \delta_{jk}, \qquad (1.35)$$

where δ_{jk} is the Kronecker delta function ($\delta_{jk} = 1$ for $j = k$ and $\delta_{jk} = 0$ for $j \neq k$).

The vectors $|\psi\rangle$ and $\langle\psi|$ can be expressed in terms of the respective basis vectors of the complete set of operators (tilde stands for transposition):

$$|\psi\rangle = \sum_{j=1}^{n} \psi_j |j\rangle; \qquad \langle\psi| = \sum_{j=1}^{n} \tilde{\psi}_j \langle j|. \tag{1.36}$$

Multiplying the first relation by $\langle j|$ from the left and the second relation by $|k\rangle$ from the right, we obtain

$$\psi_j = \langle j|\psi\rangle, \tag{1.37}$$

$$\tilde{\psi}_k = \langle\psi|k\rangle = \langle k|\psi\rangle^* = \psi_k^*, \quad \text{i.e. } \tilde{\psi}_j = \psi_j^*. \tag{1.38}$$

The sets of the numbers ψ_j and $\tilde{\psi}_j$ are the sets of the components of the vectors $|\psi\rangle$ and $\langle\psi|$, respectively, and unambiguously specify them.

Using the above expressions for the components of the vectors $|\psi\rangle$ and $\langle\psi|$, we can represent them in the form

$$|\psi\rangle = \sum_{j=1}^{n} |j\rangle\langle j|\psi\rangle; \qquad \langle\psi| = \sum_{j=1}^{n} \langle\psi|j\rangle\langle j|. \tag{1.39}$$

The action of the operator $\hat{F}$ on the vector ψ in the n-dimensional Euclidean space is written in the Dirac notation as $|\varphi\rangle = \hat{F}|\psi\rangle$.

The expression $\langle\varphi|\hat{F}|\psi\rangle$ means that the operator $\hat{F}$ acts on the vector ψ from the left and, then, the dot product of the resulting vector and the left vector φ is taken. In other words, the operator $\hat{F}$ acts on the initial state ψ and transfers it to the final state φ.

The vector $|f\rangle$ satisfying the equation

$$\hat{F}|f\rangle = f|f\rangle \tag{1.40}$$

is called an eigenvector of the operator $\hat{F}$, whereas the number f is the eigenvalue of this operator and corresponds to this eigenvector. The vector (or spinor, this notion will be introduced in the next sections) character of the wave function is represented in the form of Dirac brackets.

Then, we determine the matrix of the operator $\hat{F}$ in its own basis, i.e., in the basis of its eigenvectors $|f\rangle$. An element of the matrix of the operator is generally defined as

$$(F)_{f'f''} = \langle f'|\hat{F}|f''\rangle. \tag{1.41}$$

Taking into account Eq. (1.40) and the orthonormalization condition, we have

$$(F)_{f'f''} = \langle f'|\hat{F}|f''\rangle = \langle f'|f''|f''\rangle = f''\langle f'|f''\rangle = f''_{f'f''} = \delta_{f'f''}f'; \tag{1.42}$$

i.e., the matrix of the Hermitian operator $\hat{F}$ in its own basis is diagonal.

Its elements on the main diagonal are the eigenvalues of the operator $\hat{F}$ (some of them can be coinciding, so-called degenerate elements), whereas all off-diagonal elements are zero.

Thus, the pure algebraic problem of the diagonalization of the matrix of a given Hermitian operator (i.e., the determination of a basis in which this matrix is diagonal) is solved simultaneously with the determination of the eigenvalues of this operator.

Let us now present the basic content of this section concerning the angular momentum operator.

In quantum mechanics, the Hamiltonian is an operator determining the time evolution of the state of a quantum system. The basic conservation laws in physics are due to the requirement that space for a closed system be uniform and isotropic. The first requirement leads to the momentum conservation law (three conserved momentum components). The second requirement leads to the angular momentum conservation law (six invariant quantities: three angular-momentum components and three rotations involving the time axis of the four-dimensional space).

In this section, following books of Landau and Lifshitz (1963) and Schiff (1968), we present the properties of the angular momentum. They are also useful for the consideration of the "own (or intrinsic) angular momentum" of a particle, i.e., spin.

Let us consider a closed physical system with the Hamiltonian H. In view of the isotropy of the space, the Hamiltonian of the system should remain unchanged under the rotation of this system by an arbitrary angle about an arbitrary axis. It is sufficient to apply this condition to an infinitesimal rotation; in this case, it is also valid for finite rotations.

Let the physical system consists of n particles and be described by the wave function $\psi(\vec{r}_i)$, where $i = 1, 2 \ldots n$. The increment of the vector $\vec{r}$ under the infinitesimal rotation describing by the vector $\delta\vec{\varphi}$ that has the length $\delta\phi$ and is aligned with the rotation axis can be represented in the form

$$\delta\vec{r}_i = \delta\vec{\varphi} \times \vec{r}_i. \tag{1.43}$$

Here, the symbol $\times$ stands for the cross (vector) product. An arbitrary wave function under this transformation is transformed as follows (we take the first two terms of the expansion and, in the third transformation, use the commutativity of the dot product of the vectors and the property of the scalar triple product of vectors):

$$\psi(\vec{r}_i + \delta\vec{r}_i) = \psi(\vec{r}_i) + \sum_i \delta\vec{r}_i \cdot \vec{\nabla}_i \psi(\vec{r}_i) = \psi(\vec{r}_i) + \sum_i \delta\vec{\varphi} \times \vec{r}_i \cdot \vec{\nabla}_i \psi(\vec{r}_i)$$

$$= \left(1 + \delta\vec{\varphi} \cdot \sum_i \vec{r}_i \times \vec{\nabla}_i\right)\psi(\vec{r}_i).$$

The gradient operator $\vec{\nabla}$ (it is also denoted as grad) acting on a scalar function yields a vector function and, in the Cartesian coordinates, has the form $\vec{\nabla}_i = \frac{\partial}{\partial x_i}\vec{i} + \frac{\partial}{\partial y_i}\vec{j} + \frac{\partial}{\partial z_i}\vec{k}$, where the arrow denoting the vector is often omitted over ∇. The dot product of the vectors is denoted by the symbol $\cdot$.

The operator $1 + \delta\vec{\varphi} \cdot \sum_i \vec{r}_i \times \vec{\nabla}_i$ is the infinitesimal rotation operator; since space is homogeneous, it conserves the total energy of the system and should commute with the Hamiltonian $\hat{H}$ (Landau and Lifshitz 1963). Excluding the first term (unity commutes with any operator) and introducing the notation

$$\hat{L} = \sum_i \vec{r}_i \times \vec{\nabla}_i, \tag{1.44}$$

we write the condition of the commutativity of the operator $\hat{L}$ with the Hamiltonian

$$\lfloor \hat{L}, \hat{H} \rfloor = \hat{L}\hat{H} - \hat{H}\hat{L} = 0. \tag{1.45}$$

As known, any operator commuting with the Hamiltonian is an operator of a conserving quantity. Therefore, the operator $\hat{L}$ appearing from the requirement of the isotropy of space for a closed system corresponds to a conserving quantity. This operator is called the space moment of momentum operator according to its definition as the cross product of the coordinate vector operator by the momentum vector operator. It is also called the orbital angular momentum operator. Here, we deviate from the main theme of this section and point to a number of the properties of the operator $\hat{L}$ that are useful for considering the spin operator.

According to the following classical definition of the orbital angular momentum for one particle:

$$\vec{l} = \vec{r} \times \vec{p}, \tag{1.46}$$

where $\vec{r}$ and $\vec{p}$ are the radius vector and momentum of the particle, respectively, $\vec{l}$ is a pseudovector (or axial vector); i.e., under space inversion, $\vec{l}$ does not change sign in contrast to $\vec{r}$ and $\vec{p}$ that change signs (such vectors are called polar). Another important property of $\vec{l}$ is associated with the time reversal operation. Since the radius vector under this operation does not change sign and the momentum changes sign, the orbital angular momentum changes sign. Since the spin is not a classical object such as the orbital angular momentum, an analog of relation (1.46) is absent for the spin. Therefore, there is no a similar simple illustrative way to derive the same properties for the spin operator, as was done for $\vec{l}$. At the same time, we can extend these properties of the orbital angular momentum to the spin moment; otherwise, it would be impossible to make the summation operation providing the total angular momentum $\vec{j} = \vec{l} + \vec{s}$, where $\vec{s}$ is the spin vector. Now, we return to the main theme.

Taking into account the relation $\hat{p} = -i\hbar\nabla$ between the momentum operator and gradient operator, the quantum-mechanical representation of the angular momentum operator of the particle can be written by representing the cross product in the form of the determinant

$$\hbar\vec{l} = \begin{vmatrix} \vec{i} & \vec{j} & \vec{k} \\ x & y & z \\ \hat{p}_x & \hat{p}_y & \hat{p}_z \end{vmatrix}. \tag{1.47}$$

In the shorter representation,

$$\hbar l_m = x_i \hat{p}_k \varepsilon_{ikm}, \tag{1.48}$$

where $\hat{p}_k = -i\hbar\frac{\partial}{\partial x^k}$ (until the end of this section, we set Planck's constant $\hbar = 1$, as often make in theoretical works; in what follows, such cases will be mentioned), ε_{ikm} is the antisymmetric unit tensor of the third rank ($i = x, y, z = 1, 2, 3$), which is also called the unit axial tensor, and is defined as a tensor antisymmetric in pair of all three indices with the condition $\varepsilon_{123} = 1$. It is obvious that only 6 of its 27 components are nonzero; these are the components whose indices i, j, k constitute any permutation of $1, 2, 3$. These components are $+1$ and -1 if the set i, j, k is obtained from $1, 2, 3$ by means of even and odd numbers of pair permutations (transpositions), respectively. It is obvious that $\varepsilon_{ijk}\varepsilon_{ijl} = 2\delta_{kl}$ and $\varepsilon_{ijk}\varepsilon_{ijk} = 6$.

The commutation relations between $\hat{l}$ and coordinates x_i (operator $\hat{x}_i$ in the coordinate representation reduces to the multiplication by the coordinate; for this reason, it is written without hat) can be obtained by straightforward calculations and represented in the form

$$[\hat{l}_i, x_k] = i\varepsilon_{ikm}x_m. \tag{1.49}$$

The commutation relations between the orbital angular momentum operator $\hat{l}$ and the momentum operator $\hat{p}$ of the particle have the same form

$$[\hat{l}_i, \hat{p}_k] = i\varepsilon_{ikm}\hat{p}_m. \tag{1.50}$$

Similar commutation relations can be also obtained for the components of the orbital angular momentum operator $\hat{l}$:

$$[\hat{l}_i, \hat{l}_k] = i\varepsilon_{ikm}\hat{l}_m. \tag{1.51}$$

Let us define the square of the orbital angular momentum operator

$$\hat{l}^2 = \hat{l}_x^2 + \hat{l}_y^2 + \hat{l}_z^2. \tag{1.52}$$

This operator commutes with each of the components of the operator $\hat{l}_i$ ($i = x, y, z$). For example,

$$\begin{aligned}[\hat{l}_x^2, \hat{l}_z] &= \hat{l}_x^2\hat{l}_z - \hat{l}_z\hat{l}_x^2 = \hat{l}_x(-i\hat{l}_y + \hat{l}_z\hat{l}_x) - (i\hat{l}_y + \hat{l}_x\hat{l}_z)\hat{l}_x = -i(\hat{l}_x\hat{l}_y + \hat{l}_y\hat{l}_x), \\ [\hat{l}_y^2, \hat{l}_z] &= i(\hat{l}_x\hat{l}_y + \hat{l}_y\hat{l}_x), \qquad [\hat{l}_z^2, \hat{l}_z] = 0.\end{aligned} \tag{1.53}$$

Summing these relations, we obtain $[\hat{l}^2, \hat{l}_z] = 0$. As a result, we arrive at the relation

$$[\hat{l}^2, \hat{l}_i] = 0, \quad i = x, y, z. \tag{1.54}$$

The physical meaning of relation (1.54) is that the square of the orbital angular momentum can be accurately measured simultaneously with one of its components.

For applications, it is sometimes appropriate to change the operators $\hat{l}_x$ and $\hat{l}_y$ to their linear combinations

$$\hat{l}_\pm = \hat{l}_x \pm i\hat{l}_y. \tag{1.55}$$

According to relations (1.51),

$$[\hat{l}_+, \hat{l}_-] = 2\hat{l}_z, \qquad [\hat{l}_z, \hat{l}_+] = \hat{l}_+, \qquad [\hat{l}_z, \hat{l}_-] = -\hat{l}_-. \qquad (1.56)$$

The following relation can also be derived:

$$\hat{l}^2 = \hat{l}_+\hat{l}_- + \hat{l}_z^2 - \hat{l}_z = \hat{l}_-\hat{l}_+ + \hat{l}_z^2 + \hat{l}_z. \qquad (1.57)$$

Let us pass from the Cartesian coordinate system to the spherical coordinate system by means of the standard change of variables (as usual, the polar angle θ is measured from the positive z semiaxis in the clockwise direction and the angle φ, from the positive x semiaxis in the counterclockwise direction):

$$x = r \sin\theta \cos\varphi, \qquad y = r \sin\theta \sin\varphi, \qquad z = r \cos\theta.$$

In view of expressions (1.48) for the components of the orbital angular momentum operator, simple calculations provide the necessary expressions

$$\hat{l}_z = -i\frac{\partial}{\partial\varphi}, \qquad \hat{l}_\pm = e^{\pm i\varphi}\left(\pm\frac{\partial}{\partial\theta} + i\cot\theta\frac{\partial}{\partial\varphi}\right). \qquad (1.58)$$

The substitution of these expressions into Eq. (1.57) yields

$$\hat{l}^2 = -\left[\frac{1}{\sin^2\theta}\frac{\partial}{\partial\varphi^2} + \frac{1}{\sin\theta}\frac{\partial}{\partial\theta}\left(\sin\theta\frac{\partial}{\partial\theta}\right)\right]. \qquad (1.59)$$

This expression up to a constant factor is the angular part of the Laplace operator.

Now, we can determine the eigenvalues of the angular momentum projection on a certain direction. We use the above formulas for the spherical coordinate system. First, we consider the operator $\hat{l}_z$ defined by the first of formulas (1.58). In order to determine an eigenvalue of this operator, we should write the equation for its eigenfunction:

$$\hat{l}_z\psi = l_z\psi. \qquad (1.60)$$

Here, l_z without an operator symbol (hat) over it is eigenvalue of the operator $\hat{l}_z$. The substitution expression (1.58) for the operator $\hat{l}_z$ gives

$$-i\frac{\partial\psi}{\partial\varphi} = l_z\psi. \qquad (1.61)$$

The solution of this equation has the form

$$\psi = f(r, \theta)e^{il_z\varphi}. \qquad (1.62)$$

Here, $f(r, \theta)$ is an arbitrary function of its arguments. For the function ψ to be single-valued, it should be a periodic function of φ with a period of 2π. The condition of this periodicity has the form $e^{il_z\varphi} = e^{il_z(\varphi+2\pi)}$; therefore, $1 = e^{il_z(2\pi)}$. Hence,

$l_z = m$, where $m = 0, \pm 1, \pm 2, \ldots$; i.e., m takes integer values. Let us introduce the normalized eigenfunction of the operator $\hat{l}_z$:

$$\Phi_m(\varphi) = \frac{1}{\sqrt{2\pi}} e^{im\varphi}, \tag{1.63}$$

where the normalization is specified by the relation

$$\int_0^{2\pi} \Phi_v^*(\varphi)\Phi_{v'}(\varphi)d\varphi = \delta_{vv'}. \tag{1.64}$$

Therefore, the eigenfunction of the operator $\hat{l}_z$ can be written in the general form

$$\psi_m = f(r,\theta)e^{il_z\varphi}. \tag{1.65}$$

Let us determine the maximum and minimum values of l_z. Relation (1.52) can be represented in the form

$$\hat{l}^2 - \hat{l}_z^2 = \hat{l}_x^2 + \hat{l}_y^2. \tag{1.66}$$

Since the right-hand side contains the operators of positive quantities, the left-hand side should also be positive. Therefore,

$$-\sqrt{l^2} \leq l_z \leq +\sqrt{l^2}. \tag{1.67}$$

Thus, the absolute values of the upper and lower bounds of l_z values coincide with each other. Let us determine this boundary value l.

In view of relations (1.56) and (1.60), the action of the operator $\hat{l}_z\hat{l}_\pm$ on the wave function ψ_m yields

$$\hat{l}_z\hat{l}_\pm\psi_m = (m \pm 1)\hat{l}_\pm\psi_m. \tag{1.68}$$

Correspondingly, the function $\hat{l}_\pm\psi_m$ is an eigenfunction of the operator $\hat{l}_z$ with the eigenvalue $(m \pm 1)$ up to the normalization constant. Therefore,

$$\psi_{m+1} = N_1\hat{l}_+\psi_m, \qquad \psi_{m-1} = N_2\hat{l}_-\psi_m. \tag{1.69}$$

As seen, the operator $\hat{l}_+$ increases the eigenvalue m by unity, whereas the operator $\hat{l}_-$ reduces the eigenvalue m by unity. Taking the first of relations (1.69) with $m = l$, we obtain $\psi_{l+1} = 0$, because the maximum m value is l. Thus,

$$\hat{l}_+\psi_l = 0. \tag{1.70}$$

Applying the operator $\hat{l}_-$ to this equality and using relation (1.57), we obtain

$$\hat{l}_-\hat{l}_+\psi_l = (\hat{l}^2 - \hat{l}_z^2 - \hat{l}_z)\psi_l = 0. \tag{1.71}$$

Since ψ_l is an eigenfunction of all three operators in the parentheses,

$$\hat{l}^2\psi_l = l(l+1)\psi_l. \tag{1.72}$$

This formula specifies the eigenvalues of the square of the orbital angular momentum operator. The parameter l can be any nonnegative integer. For a given l values, the eigenvalues of the operator $\hat{l}_z$ are $m = -l, -(l-1), -(l-2)\ldots 0 \ldots (l-2)$, $(l-1), l$, i.e., $(2l+1)$ values. The parameter m is also called the magnetic quantum number or the projection of the orbital angular momentum $\hat{l}$ onto the z axis; it leads to "space quantization".

Let us calculate the matrix elements of the operators $\hat{l}_x$ and $\hat{l}_y$ in the representation where the matrices of energy, $\hat{l}_z$, and $\hat{l}^2$ are diagonal.

Since the operator is a rule according to which each vector ψ of the n-dimensional Euclidean space is transformed to a vector φ of the same space, the transformation of one vector to another can be represented in the form $\varphi_j = \sum_{k=1}^{n} a_{jk} \psi_k$; in the matrix notation, $\varphi = A\psi$. Thus, each $n \times n$ matrix specifies an operator in the n-dimensional Euclidean space.

Since the operators $\hat{l}_x$ and $\hat{l}_y$ commute with the Hamiltonian and $\hat{l}^2$ operator, their matrix elements are nonzero only for the transitions where the energy and l value remain unchanged. This means that it is sufficient to calculate only the matrix elements of the operators $\hat{l}_x$ and $\hat{l}_y$ between different m values.

According to formula (1.69), the operators $\hat{l}_-$ and $\hat{l}_+$ transfer the states $m+1$ and $m-1$, respectively, to the state m. Taking into account this property, using relations (1.49), and writing the matrix elements in the Dirac notation, we obtain

$$l(l+1) = \langle m|l_+|m-1\rangle\langle m-1|l_-|m\rangle + m^2 - m. \tag{1.73}$$

According to definition (1.55), the operators $\hat{l}_+$ and $\hat{l}_-$ are mutually Hermitian, because the operators $\hat{l}_x$ and $\hat{l}_y$ are Hermitian. Therefore,

$$\langle m-1|\hat{l}_-|m\rangle = \langle m|\hat{l}_+|m-1\rangle^*. \tag{1.74}$$

The substitution of this relation into relation (1.73) provides

$$\left|\langle m|\hat{l}_+|m-1\rangle\right|^2 = l(l+1) - m(m-1) = (l-m+1)(l+m). \tag{1.75}$$

Finally,

$$\langle m|\hat{l}_+|m-1\rangle = \langle m-1|\hat{l}_-|m\rangle = \sqrt{(l+m)(l-m+1)}. \tag{1.76}$$

From these relations, the nonzero matrix elements of the operators $\hat{l}_x$ and $\hat{l}_y$ are obtained in the form

$$\langle m|\hat{l}_x|m-1\rangle = \langle m-1|\hat{l}_x|m\rangle = \frac{1}{2}\sqrt{(l+m)(l-m+1)} \tag{1.77}$$

and

$$\langle m|\hat{l}_y|m-1\rangle = \langle m-1|\hat{l}_y|m\rangle = -\frac{i}{2}\sqrt{(l+m)(l-m+1)}. \tag{1.78}$$

These relations will be used in the following chapters (taking half-integer l values, we obtain the Pauli matrices, which will be considered in the next section).

1.3 Pauli Spin Operator

Spin is the intrinsic angular momentum of a particle and takes discrete values. Particles with integer spins are called bosons. Among them are photon, vector mesons, gluon, and intermediate bosons. Spin can also take half-integer values. Particles with half-integer spins are called fermions. Among them are nucleons, electrons, neutrinos, muons, and quarks. All elementary particles without exception can be classified in spin as bosons and fermions (Kane 1987), which are described by the Bose–Einstein and Fermi–Dirac statistics, respectively.

The hypothesis of the intrinsic angular momentum of the electron was proposed in different forms by many physicists (Fidecaro 1998). This hypothesis was most clearly formulated by Dutch scientists Uhlenbeck and Goudsmit (1925, 1926) in order to explain the presence of the hyperfine structures in the energy levels of hydrogen-like atoms. Spin appeared as an operator in quantum mechanics in 1927 owing to Pauli.

Spin is particularly important in weak decays of particles (Okun 1982). As an example, we point out that one of the largest discoveries in physics in the twentieth century, namely, the discovery of parity violation in weak interactions was made on polarized particles (beta decay of polarized nuclei), i.e., with spin (Lee and Wu 1965).

Let us consider a number of the properties of the spin operator $\hat{s}$ with a value of $1/2$. This operator $\hat{s}$ is related to the Pauli operator $\hat{\sigma}$ as

$$\hat{s} = \frac{1}{2}\hat{\sigma}. \tag{1.79}$$

Both operators acting in the spin space are axial vectors in the usual coordinate representation. The operator $\hat{\sigma}$ in the rest system of the particle has the form

$$\|\sigma_x\| = \|\sigma_1\| = \begin{Vmatrix} 0 & 1 \\ 1 & 0 \end{Vmatrix}, \qquad \|\sigma_y\| = \|\sigma_2\| = \begin{Vmatrix} 0 & -i \\ i & 0 \end{Vmatrix},$$

$$\|\sigma_z\| = \|\sigma_3\| = \begin{Vmatrix} 1 & 0 \\ 0 & -1 \end{Vmatrix}. \tag{1.80}$$

The commutation properties of the Pauli matrices σ can be expressed by the relation

$$\sigma_\alpha \sigma_\beta = \delta_{\alpha\beta} I + i\varepsilon_{\alpha\beta\gamma}\sigma_\gamma, \tag{1.81}$$

where $\delta_{\alpha\beta}$ is the unit symmetric tensor of the second rank and $\varepsilon_{\alpha\beta\gamma}$ is the unit anti-symmetric tensor of the third rank (both tensors are defined in the three-dimensional space).

Using relations (1.81), we can derive the following expressions for the commutators and anticommutators

$$[\hat{\sigma}_\alpha, \hat{\sigma}_\beta] = \hat{\sigma}_\alpha \hat{\sigma}_\beta - \hat{\sigma}_\beta \hat{\sigma}_\alpha = 2i\varepsilon_{\alpha\beta\gamma}\hat{\sigma}_\gamma, \tag{1.81a}$$

$$\{\hat{\sigma}_\alpha, \hat{\sigma}_\beta\} = \hat{\sigma}_\alpha \hat{\sigma}_\beta + \hat{\sigma}_\beta \hat{\sigma}_\alpha = 2\delta_{\alpha\beta}. \tag{1.81b}$$

A number of useful properties follow from these relations. First, according to Eq. (1.81a), the product of two different components of the spin operator is expressed in terms of the first power of the third component of this operator. This means that any matrix in the two-dimensional spin space cannot contain the powers of sigma-matrices higher than the first, i.e., can be represented as a linear expression of the Pauli matrices. Second, the spin operator components anticommute with each other. Third, the square of each spin component is the identity matrix I. From relations (1.81), (1.81a), and (1.81b), we can obtain the useful equality

$$(\hat{\vec{\sigma}} \cdot \vec{A})(\hat{\vec{\sigma}} \cdot \vec{B}) = (\vec{A} \cdot \vec{B}) + i\hat{\vec{\sigma}} \cdot (\vec{A} \times \vec{B}), \tag{1.82}$$

where the vectors $\vec{A}$ and $\vec{B}$ are independent of the spin variables. As can be verified by direct transformation of the Pauli matrices, the operator $\hat{\vec{\sigma}}$ is Hermitian; therefore, its eigenvalue is a real number. The mean value of the operator $\hat{\vec{\sigma}}$ between the spin states of the particle is called the polarization vector $\vec{P}$:

$$\vec{P} = \langle \hat{\vec{\sigma}} \rangle. \tag{1.83}$$

Since

$$\sigma^2 = \sigma_1^2 + \sigma_2^2 + \sigma_3^2 = 3, \tag{1.84}$$

the eigenvalue of the square of the spin operator is $\vec{s}^2 = \frac{1}{4} \cdot \vec{\sigma}^2 = \frac{3}{4}$.

At the same time, it can be shown that the length of the polarization vector P is always smaller than unity, $|P| \leq 1$ (Bilen'kii et al. 1964). This will be proved below.

The transformation properties (properties under the coordinate transformation) of the spin operator (and, correspondingly, the Pauli operator) can be defined by analogy with the orbital angular momentum operator of the particle, the sum of which with the spin is the total angular momentum of the particle,

$$\vec{j} = \vec{s} + \vec{l}.$$

In the general case, the requirement of the isotropy of space provides the following relations for the components of the total angular momentum $\vec{j}$ of the particle (Landau and Lifshitz 1963):

$$(j_x + ij_y)Y_{lm} = \sqrt{(j-m)(j+m+1)}Y_{jm+1}, \tag{1.85}$$

$$(j_x - ij_y)Y_{lm} = \sqrt{(j+m)(j-m+1)}Y_{jm-1}, \tag{1.86}$$

$$j_z Y_{lm} = mY_{lm}, \qquad j^2 Y_{lm} = j(j+1)Y_{lm}. \tag{1.87}$$

Changing the angular momentum j and spherical functions Y_{lm} in these equations to the spin operator $\hat{s}$ and spinors χ_{sm} (see the next section), respectively, we arrive at the equations for the eigenfunctions and eigenvalues of the spin operators. From these equations, we can also derive commutation relations (1.81), (1.81a), and (1.81b) for the spin operators, as well as explicit expressions (1.80) for the Pauli matrices. Moreover, relations (1.85)–(1.87) allow one to find explicit expressions for the spin operators of any rank including the deuteron spin ($s = 1$).

The spin operator $\hat{\vec{s}}$ is a pseudovector; i.e., it is transformed as a normal vector under the rotation of the coordinate system and, as well as the orbital angular momentum, does not change under space inversion, i.e., is an axial vector. Under time reversal, spin, as well as orbital angular momentum, changes sign. As mentioned above, the use of analogy with the orbital angular momentum for the definition of the transformation properties of the spin operator is simplest and quite convincing.

1.4 Spinors

Let $\psi(x, y, z; \sigma)$ be the wave function of the particle with the spin σ (we follow the notation from Landau and Lifshitz (1963); σ should not be confused with the Pauli matrix), σ in this case is the z component of the spin and ranges from $-s$ to $+s$. The functions $\psi(\sigma)$ with various σ values will be treated as the wave function "components".

In contrast to the usual variables (coordinates), the variable σ is discrete. The most general linear operator acting on functions of the discrete variable σ has the form

$$(\hat{f}\psi)(\sigma) = \sum_{\sigma'} f_{\sigma\sigma'}(\sigma'), \tag{1.88}$$

where $f_{\sigma\sigma'}$ are constants. The expression $(\hat{f}\psi)$ is written in the parentheses in order to show that the argument (σ) refers to the function appearing as a result of the action of the operator $\hat{f}$ on the function ψ, rather than to the function ψ itself. It can be shown that the quantities $f_{\sigma\sigma'}$ coincide with the matrix elements of the operator $\hat{f}$ that are defined in the ordinary way. Therefore, the operators acting on the functions of σ can be represented in the form of $2s + 1$-row matrices.

In the case of zero spin, the wave function has only one component $\psi(0)$. Since the spin operators are related to the rotation operators, this means that the wave function of a particle with spin 0 does not change under the rotations of the coordinate system, i.e., is a scalar or a pseudoscalar.

The wave functions of the particles with spin $1/2$ have two components $\psi(1/2)$ and $\psi(-1/2)$. We denote them as ψ^1 and ψ^2. Under an arbitrary rotation of the coordinate system, they undergo the linear transformation

$$\psi^{1'} = \alpha\psi^1 + \beta\psi^2, \qquad \psi^{2'} = \gamma\psi^1 + \delta\psi^2. \tag{1.89}$$

The coefficients $\alpha, \beta, \gamma, \delta$ are generally complex and are the functions of the rotation angles. Linear transformations (1.89) under which the bilinear form

$$\psi^1\psi^2 - \psi^2\psi^1, \tag{1.90}$$

is invariant are called binary. The two-component quantity (ψ^1, ψ^2) that is transformed according to a binary transformation under the rotation of the coordinate system is called the spinor.

Let us consider the spinors χ_{sm} (s is the spin value and m is its projection), which are the eigenfunctions of the square of the spin operator, $\hat{s}^2$, and the spin projection operator $\hat{s}_z$. Let us assume that they are defined in a given coordinate system K with the axes (x, y, z). Let a new coordinate system K' with the axes (x', y', z') be obtained from K by means of the rotation about the z axis by the angle ϕ. The rotation operator by the infinitesimal angle $\delta\varphi$ about the z axis is expressed in terms of the angular momentum operator (spin in this case) in the form $1 + i\delta\varphi \cdot \hat{s}_z$. Therefore, under the rotation, the wave function $\psi(\sigma)$ is transformed to $\psi(\sigma) + \delta\psi(\sigma)$, where $\delta\psi(\sigma) = i\delta\varphi \cdot \hat{s}_z\psi(\sigma)$. Since $\hat{s}_z\psi(\sigma) = s_z\psi(\sigma)$, we have $\delta\psi(\sigma) = is_z\psi(\sigma)\delta\varphi$. For the rotation by the finite angle φ, the finite spinor takes the form of the function $\psi(\sigma)' = e^{is_z\varphi}\psi(\sigma)$.

In this case, the finite spinor is defined by the expression

$$\psi_z(\varphi) = \hat{U}_z(\varphi)\chi_{sm} = e^{is_z\varphi}\chi_{sm} = e^{i\frac{1}{2}\sigma_z\varphi}\chi_{sm}$$

$$= \left[1 + i\left(\frac{\varphi}{2}\right)\sigma_z - \frac{1}{2!}\left(\frac{\varphi}{2}\right)^2 - i\frac{1}{3!}\left(\frac{\varphi}{2}\right)^3\sigma_z - \cdots\right]\chi_{sm}$$

$$= \left(\cos\frac{\varphi}{2} + i\sin\frac{\varphi}{2}\|\sigma_z\|\right)\chi_{sm} = \left\|\begin{matrix} e^{\frac{1}{2}i\varphi} & 0 \\ 0 & e^{-\frac{1}{2}i\varphi} \end{matrix}\right\|\chi_{sm}, \tag{1.91}$$

where the operator

$$\hat{U}_z(\varphi) = \left\|\begin{matrix} e^{\frac{1}{2}i\varphi} & 0 \\ 0 & e^{-\frac{1}{2}i\varphi} \end{matrix}\right\| \tag{1.92}$$

ensures the rotation of the coordinate system K about the z axis by the angle φ.

The rotation operator by the angle θ about the x axis can be expressed by the matrix (see Landau and Lifshitz 1963)

$$\hat{U}_x(\theta) = \left\|\begin{matrix} \cos\frac{\theta}{2} & i\sin\frac{\theta}{2} \\ i\sin\frac{\theta}{2} & \cos\frac{\theta}{2} \end{matrix}\right\|. \tag{1.93}$$

Let the quantization axis be specified by the Euler angles φ, θ, ψ (see Fig. 1.1). Performing the Euler transformations, we obtain the spinor with the new spin quantization axis,

$$\Psi = \left\|\begin{matrix} \cos\frac{\theta}{2}e^{\frac{1}{2}i(\varphi+\psi)} & i\sin\frac{\theta}{2}e^{-\frac{i}{2}(\varphi-\psi)} \\ i\sin\frac{\theta}{2}e^{\frac{i}{2}(\varphi-\psi)} & \cos\frac{\theta}{2}e^{-\frac{i}{2}(\varphi+\psi)} \end{matrix}\right\|\chi_{sm}. \tag{1.94}$$

Setting two arbitrary angles equal zero, we naturally obtain the rotation about the third direction. To clarify this item, we introduce the unit vector $\vec{n}(n_1, n_2, n_3)$ about which the rotation by the angle ε is performed. Then, the rotation operator is written in the form (see Eq. (1.92))

$$U_n(\varepsilon) = e^{i\vec{\sigma}\cdot\vec{n}\varepsilon} = \left\|\begin{matrix} \cos\varepsilon + in_3\sin\varepsilon & (in_1 + n_2)\sin\varepsilon \\ (in_1 - n_2)\sin\varepsilon & \cos\varepsilon - in_3\sin\varepsilon \end{matrix}\right\|. \tag{1.95}$$

Fig. 1.1 Euler
transformation

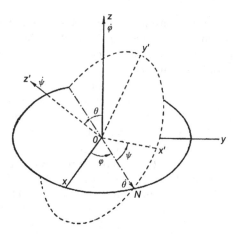

Under such rotations, the spin vector, as well as the orbital angular momentum, is transformed as an ordinary vector, namely (counterclockwise rotation about the z axis):

$$\sigma_x' = \cos\varphi \cdot \sigma_x + \sin\varphi \cdot \sigma_y, \qquad \sigma_y' = -\sin\varphi \cdot \sigma_x + \cos\varphi \cdot \sigma_y. \qquad (1.96)$$

According to formula (1.91), the spinor χ changes sign under the rotation of the coordinate system by the angle 2π. This is characteristic of almost all spinors describing the particles with half-integer spins. However, the spinor square $|\chi|^2$ is a positively defined function, as should be expected, because this quantity corresponds to the probability that the particle is in a certain spin state.

As an example, let us determine the explicit form of the Pauli operators in the rest frame of the particle.

Formulas (1.94)–(1.96) from the preceding section are applicable to the spin operators. In the case of the particle with the spin $s = 1/2$, we introduce the notation

$$\alpha = Y_{\frac{1}{2}\frac{1}{2}}, \qquad \beta = Y_{\frac{1}{2}-\frac{1}{2}}; \qquad \hat{s}_+ = \hat{s}_x + i\hat{s}_y, \qquad \hat{s}_- = \hat{s}_x - i\hat{s}_y. \qquad (1.97)$$

In the rest system of the particle, we take the z axis as the quantization axis and represent the spinor components α and β in the orthonormalized form

$$\alpha = \left\| \begin{matrix} 1 \\ 0 \end{matrix} \right\|, \qquad \beta = \left\| \begin{matrix} 0 \\ 1 \end{matrix} \right\|. \qquad (1.98)$$

Then, Eqs. (1.94) and (1.95) from the preceding section provide four equations in the matrix form

$$\hat{s}_+\alpha = 0, \qquad \hat{s}_+\beta = \alpha; \qquad \hat{s}_-\alpha = \beta, \qquad \hat{s}_-\beta = 0. \qquad (1.99)$$

All spin operators can be represented in the form of the rank-2 matrix with unknown elements $\hat{s} = \|a_{ij}\|$, where $i, j = 1, 2$. Substituting relations (1.98) into

Eq. (1.99) and solving them, we obtain

$$\hat{s}_+ = \frac{1}{2}\begin{Vmatrix} 0 & 1 \\ 0 & 0 \end{Vmatrix}, \qquad \hat{s}_- = \frac{1}{2}\begin{Vmatrix} 0 & 0 \\ 1 & 0 \end{Vmatrix};$$

$$\hat{s}_x = \frac{1}{2}\begin{Vmatrix} 0 & 1 \\ 1 & 0 \end{Vmatrix}, \qquad \hat{s}_y = \begin{Vmatrix} 0 & -i \\ i & 0 \end{Vmatrix}, \qquad \hat{s}_z = \frac{1}{2}\begin{Vmatrix} 1 & 0 \\ 0 & -1 \end{Vmatrix}.$$

(1.100)

The expression for the operator $\hat{s}_z$ is naturally obtained due to the condition that the spinor components α and β are the eigenfunctions of $\hat{s}_z$ with the eigenvalues $\pm\frac{1}{2}$.

Let us find the explicit matrix representation for spin-1 operators.

The difference from the preceding example is that the spinors are three-component; i.e., the spinors

$$\alpha = \begin{Vmatrix} 1 \\ 0 \\ 0 \end{Vmatrix}, \qquad \beta = \begin{Vmatrix} 0 \\ 1 \\ 0 \end{Vmatrix}$$

should be supplemented by the spinor

$$\gamma = \begin{Vmatrix} 0 \\ 0 \\ 1 \end{Vmatrix},$$

and the spin matrices are 3×3 matrices.

Each spin operator includes nine unknown coefficients $\hat{s} = \|a_{ij}\|$, where $i, j = 1, 2, 3$.

With the same representation for the spin operators, we write the equations following from Eqs. (1.94)–(1.96) from the preceding section for spin 1:

$$\hat{s}_+\alpha = 0, \qquad \hat{s}_+\beta = \sqrt{2}\alpha, \qquad \hat{s}_+\gamma = \sqrt{2}\beta;$$

$$\hat{s}_-\alpha = \sqrt{2}\beta, \qquad \hat{s}_-\beta = \sqrt{2}\gamma, \qquad \hat{s}_-\gamma = 0.$$

(1.101)

Solving these equations, we obtain

$$\hat{s}_+ = \begin{Vmatrix} 0 & \sqrt{2} & 0 \\ 0 & 0 & \sqrt{2} \\ 0 & 0 & 0 \end{Vmatrix}, \qquad \hat{s}_- = \begin{Vmatrix} 0 & 0 & 0 \\ \sqrt{2} & 0 & 0 \\ 0 & \sqrt{2} & 0 \end{Vmatrix};$$

$$\hat{s}_x = \frac{1}{\sqrt{2}}\begin{Vmatrix} 0 & 1 & 0 \\ 1 & 0 & 1 \\ 0 & 1 & 0 \end{Vmatrix}, \qquad \hat{s}_y = \frac{i}{\sqrt{2}}\begin{Vmatrix} 0 & -1 & 0 \\ 1 & 0 & -1 \\ 0 & 1 & 0 \end{Vmatrix}, \qquad \hat{s}_z = \begin{Vmatrix} 1 & 0 & 0 \\ 0 & 0 & 0 \\ 0 & 0 & -1 \end{Vmatrix}.$$

(1.102)

These are the explicit expressions for the spin operators of the particle with spin 1.

1.5 Schrödinger Equation

In many applications in this book, we will use the Schrödinger equation. As an example of the problems with a discrete spectrum, we consider below the hydrogen atom in the ground state. As an example of the scattering problem (the problem on the continuous spectrum), we consider the scattering of the nucleon on the nucleus in the Born approximation (Fermi model).

Another example of the application of the Schrödinger equation will be given in the section devoted to nucleon–nucleon scattering when the unitarity relation is derived. Specific applications of these and other formulas will be illustrated in the corresponding sections of the book.

Equation (1.34) in Sect. 1.1 is the Schrödinger equation in the presence of the interaction:

$$i\hbar \frac{\partial \psi(q,t)}{\partial t} = \hat{H}\psi(q,t) = \left[-\frac{\hbar^2}{2m}\nabla^2 + V(q,t) \right]\psi(q,t). \tag{1.103}$$

For stationary problems (when the Hamiltonian is time independent), this equation has the form

$$E\psi(q,t) = \left[-\frac{\hbar^2}{2m}\nabla^2 + V(q,t) \right]\psi(q,t). \tag{1.104}$$

Let us consider the application of this equation in the cases of the discrete and continuous spectra.

A. Hydrogen atom in the ground state and its energy levels

For this problem, the following interaction Hamiltonians are known and described in detail in the literature:

$$H = H_c + H_r + H_{sl} + H_{ss} + H_{sB}. \tag{1.105}$$

A.1. Here, the Coulomb interaction Hamiltonian has the form

$$H_c = \frac{Ze^2}{r}. \tag{1.106}$$

This Hamiltonian determines the Balmer terms (i.e., the energy levels in the spectroscopic terminology) of the hydrogen atom

$$E_n = -\frac{2\pi\hbar c Z^2 R}{n^2}. \tag{1.107}$$

Here, $n = 1, 2, 3, \ldots, \infty$ is called the principal quantum number and determines the energy levels of the hydrogen atom in the leading approximation; $R = \frac{\mu e^4}{4\pi\hbar^3 c}$ is the Rydberg constant ($2\pi\hbar c R = 13.6$ eV), where μ is the reduced mass of the electron–proton system ($\frac{1}{\mu} = \frac{1}{m_e} + \frac{1}{m_p}$); and Z is the charge number of the nucleus.

Term (1.106) provides the leading contribution to the level energy; the other terms (H_r, H_{sl}, H_{ss}, H_{sB}) can be treated as small perturbations.

A.2. The Hamiltonian H_r presents the relativistic corrections to the electron energy at high velocities. The perturbative calculations give (Shpol'skii 1984b)

$$\Delta E_r = \frac{\alpha^2 R Z^4}{n^3}\left(\frac{1}{l+\frac{1}{2}} - \frac{3}{4n}\right). \tag{1.108}$$

Here, $\alpha = \frac{e^2}{\hbar c} \approx 1/137$ is the fine structure constant introduced by Sommerfeld and l is the azimuthal quantum number. The comparison of Eq. (1.108) with Eq. (1.107) shows that the relativistic corrections to the electron energy are about α^2 of the main Balmer term.

A3. Spin–orbit interaction Hamiltonian H_{sl} and the remaining terms in expression (1.105) are directly responsible for polarization phenomena, because present the spin interaction. Let us consider them in more detail (Shpol'skii 1984b).

To clarify the physical picture, we use the classical model in which the hydrogen atom is a system of a proton and an electron revolving around it in an elliptic orbit. The electron has the spin s and, as was found experimentally, the magnetic moment $\vec{\mu}_e$. The proton also has spin I and the corresponding magnetic moment $\vec{\mu}_p$. The magnetic moment of the proton creates the magnetic field $\vec{H}_p$ at the electron location point. In order to estimate its magnitude, we pass from the reference frame with the origin at the proton to the frame with the origin at the electron. If the electron moves with the velocity $\vec{v}_e$, the proton moves in the opposite direction with the velocity $-\vec{v}_e$. This means that the current $\vec{j} = Ze\vec{v}_e$ appears in the rest frame of the electron. According to the Biot–Savart law, this current creates the following magnetic field at the electron location point:

$$\vec{H}_p = -\frac{Ze\vec{v} \times \vec{r}}{cr^3} = \frac{Ze\vec{r} \times \vec{v}}{cr^3}, \tag{1.109}$$

where $\vec{r}$ is the vector from the current element to the point where the field is determined and c is the speed of light. The orbital angular momentum of the electron, $\vec{l}_e$, is related to its velocity as

$$\vec{l}_e = \mu\vec{r} \times \vec{v}, \tag{1.110}$$

where μ is the reduced mass. The substitution of this relation into Eq. (1.109) yields

$$\vec{H}_p = \frac{Ze}{\mu cr^3}\vec{l}_e. \tag{1.111}$$

The additional energy due to the interaction of the magnetic dipole with the magnetic moment of the electron, $\vec{\mu}_e$, and magnetic field is given by the expression

$$H_{ls} = -\frac{Ze}{\mu cr^3}\vec{\mu}_e \cdot \vec{l}. \tag{1.112}$$

This expression was obtained in the rest frame of the electron (R frame). In order to return to the laboratory frame (L frame), where the hydrogen atom is at rest, it is necessary to perform a Lorentz transformation. As a result, as shown by Frenkel (1926) before the appearance of the Dirac theory, the energy H_{ls} is halved:

$$H_{ls} = -\frac{Ze}{2\mu cr^3}\vec{\mu}_e \cdot \vec{l}. \tag{1.113}$$

This factor $1/2$ is of significant importance for the consideration of the motion of the electron in the magnetic field, as well as for the explanation of the Landé factor and Thomas precession of the spin under relativistic spin transformations.

Considering this Hamiltonian as perturbation for the basic Hamiltonian H_c in relation (1.105), we can determine the corresponding addition to the energy in perturbation theory:

$$\Delta E_{ls} = \bar{H}_{ls} = -\left\langle \frac{Ze}{2\mu cr^3}\vec{\mu}_e \cdot \vec{l} \right\rangle. \tag{1.114}$$

The angular brackets $\langle \, \rangle$ stand for coordinate averaging. Coordinate averaging concerns only the factor $\frac{1}{r^3}$, because the other factors are independent of the coordinates:

$$\left\langle \frac{1}{r^3} \right\rangle = \int \psi_{nl}^* \frac{1}{r^3} \psi_{nl} d\tau.$$

Here, ψ_{nl} are the eigenfunctions of the basic Hamiltonian H_c with the quantum numbers n and l. The calculations give the expression

$$\left\langle \frac{1}{r^3} \right\rangle = \frac{Z^3}{a_1^3 n^3 l (l + \frac{1}{2})(l + 1)}. \tag{1.115}$$

Here, a_1 is the first Bohr radius given by the expression

$$a_1 = \frac{\hbar^2}{\mu e^2} \approx \frac{\hbar^2}{m_e e^2}. \tag{1.116}$$

Then, it is necessary to calculate the dot product $\vec{\mu}_e \cdot \vec{l}$ of the magnetic moment of the electron and its orbital angular momentum. The magnetic moment and spin moment are related as $\vec{\mu}_e = -g_e \mu_B \vec{s}_e$ (the minus sign appears because the electron charge is negative), where g_e is the Landé factor (for the electron, $g_e \cong 2$), $\mu_B = \frac{e}{2m_e c}$ is the Bohr magneton, e is the elementary charge, and m_e is the electron mass. Thus, the problem reduces to the calculation of the dot product $\vec{s} \cdot \vec{l}$, where the subscripts are omitted for brevity.

We introduce the total angular momentum operator as

$$\vec{j} = \vec{l} + \vec{s}. \tag{1.117}$$

Taking square of this equality, we find

$$ls\cos(\vec{l}\vec{s}) = \frac{1}{2}(j^2 - l^2 - s^2) = \frac{1}{2}[j(j+1) - l(l+1) - s(s+1)]. \qquad (1.118)$$

The eigenvalues for $\hat{l}^2$ are given by Eq. (1.72) from Sect. 1.2. Similar formulas are valid for the square of the spin operator $\hat{s}^2$ and the square of the total angular momentum operator $\hat{j}^2$.

The substitution of relations (1.115) and (1.118) into formula (1.114) gives

$$\Delta E_{ls} = \frac{\pi R \alpha^2 \hbar c Z^4}{n^3 l(l + \frac{1}{2})(l+1)}[j(j+1) - l(l+1) - s(s+1)]. \qquad (1.119)$$

The sum of relativistic (1.108) and spin–orbit (1.119) terms is

$$\Delta E(r, ls) = \Delta E_r + \Delta E_{ls} = \frac{-2\pi \hbar c R \alpha^2 Z^4}{n^3} \left(\frac{1}{j + \frac{1}{2}} - \frac{3}{4n} \right). \qquad (1.120)$$

Formula (1.120) for the fine structure of the atomic spectral lines that is derived from the Schrödinger equation coincides with the expression obtained from the solution of the Dirac equation.

A4. Let us consider the Hamiltonian H_{ss} in expression (1.105) (we follow Feynman et al. 1963). This Hamiltonian corresponds to the interaction between the spins of the electron and proton in the hydrogen atom and is responsible for the hyperfine splitting of the atomic spectral terms. This splitting in the absence of an external magnetic field occurs as follows. According to quantum mechanics, the vector sum of two spins corresponds to new states with the total spins $s = 0$ (singlet state) and $|\vec{s}| = 1$ (triplet state). The difference between the energies of these two states, singlet and triplet, is $\Delta e = \Delta E_{st} = h\nu$. For the hydrogen atom in the ground state, this energy corresponds to the magnetic field $H_c \sim 509$ Oe. This field is called the critical field. When creating polarized targets, the external magnetic field is always normalized to this critical field.

As known, the spins of both the electron and proton are $1/2$. Therefore, we have four configurations of their spins:

1. The spins of the electron and proton are parallel and directed upwards. We again use the Dirac brackets for the notation of these spin configurations. For example, $|m_e m_p\rangle$ denotes the state of the system of two spins with their projections m_e and m_p for the electron and proton, respectively. Let the upward direction be the positive direction of the z axis. Then, for the case under consideration where both spins are directed upwards, the state is denoted as $|++\rangle$. For brevity, we denote this state as the first state: $|1\rangle$.
2. The spins of the electron and proton are directed upwards and downwards, respectively: $|2\rangle = |+-\rangle$.
3. The spins of the electron and proton are directed downwards and upwards, respectively: $|3\rangle = |-+\rangle$.

4. The spins of the electron and proton are directed downwards: $|4\rangle = |--\rangle$.

The basis functions chosen above have the properties of the completeness and orthogonality. This means that these wave functions in the spin space completely describe the system of the spins of the electron and proton. We emphasize two circumstances. First, the coordinate and spin variables are independent and, second, the electron spin operators act only on the spin indices m_e, whereas the proton spin operators act only on the spin indices m_p.

The states of the system before and after interaction can be described in a certain basis. This means that the choice of the basis is independent of the Hamiltonian of the process under consideration.

Owing to the completeness and orthogonality of the basis vectors, any state of the physical system of two spins can be expanded in terms of the basis vectors:

$$|\psi\rangle = \sum_{i=1-4} C_i |i\rangle, \tag{1.121}$$

and the expansion coefficients are expressed as

$$C_i = \langle i|\psi\rangle. \tag{1.122}$$

The quantity $|C_i|^2$ determines the probability that the system of two spins is in the state i. The coefficients C_i depend on the Hamiltonian of the system and are determined from the solution of the Schrödinger or Dirac equation.

Let us find the Hamiltonian of the problem. For the description of the system of two spins at rest, there are two vectors $\vec{\sigma}^e$ and $\vec{\sigma}^p$ corresponding to the electron and proton spins. The Hamiltonian should be a scalar operator composed of these vectors and identity matrices in these two spin spaces. Thus, the Hamiltonian has the general form

$$H_{ss} = E_0 \cdot I + A\vec{\sigma}^e \cdot \vec{\sigma}^p, \tag{1.123}$$

where E_0 determines the zero energy, which is insignificant in this problem on the hyperfine splitting of the spectral terms of the hydrogen atom and, hence, $E_0 = 0$ can be set, and I is the identity matrix.

Applying the operator $H_{ss} = A\vec{\sigma}^e \cdot \vec{\sigma}^p$ to the basis wave functions specified by Eq. (1.121), we obtain the matrix elements of the Hamiltonian in the form (this procedure was described in detail in Feynman et al. 1963)

$$(H_{ss})_{ij} = \begin{Vmatrix} A & 0 & 0 & 0 \\ 0 & -A & 2A & 0 \\ 0 & 2A & -A & 0 \\ 0 & 0 & 0 & A \end{Vmatrix}. \tag{1.124}$$

For the Hamiltonian H_{ss}, the Schrödinger equation has the form

$$i\hbar\frac{\partial \psi(q,t)}{\partial t} = H_{ss}\psi(q,t). \tag{1.125}$$

Since the problem is stationary, the expansion coefficients of the wave function can be represented in the form

$$\psi(t) = C(t) = ae^{-iEt/\hbar}. \tag{1.126}$$

Here, C is the four-component time-dependent amplitude from Eq. (1.121). The new four-component amplitudes a are time independent. Substituting Eq. (1.126) into Eq. (1.125) and using the Hamiltonian in matrix representation (1.124), we arrive at the system of equations for the coefficients a_i (where E is the eigenvalue of the Hamiltonian)

$$Ea_1 = Aa_1, \; Ea_2 = -Aa_2 + 2Aa_3,$$
$$Ea_3 = 2Aa_2 - Aa_3, \tag{1.127}$$
$$Ea_4 = Aa_4.$$

This system has two simple solutions. First, $E = A$, $a_1 = 1$, $a_2 = a_3 = a_4 = 0$ and the wave function has the form (we denote this solution as the state $|I\rangle$)

$$|I\rangle = |1\rangle = |++\rangle. \tag{1.128}$$

Second, $E = A$, $a_4 = 1$, $a_2 = a_3 = a_1 = 0$ and the state $|II\rangle$ has the form

$$|II\rangle = |4\rangle = |--\rangle. \tag{1.129}$$

Two remaining equations contain mixed amplitudes a_2 and a_3. The sum and difference of the second and third of Eq. (1.127) have the form

$$E(a_2 + a_3) = A(a_2 + a_3) \tag{1.130}$$

and

$$E(a_2 - a_3) = -3A(a_2 - a_3). \tag{1.131}$$

These two equations have two solutions

$$a_2 = a_3, \quad E = A \quad \text{and} \quad a_2 = -a_3, \quad E = -3A. \tag{1.132}$$

The respective normalized states and corresponding energies can be represented in the form

$$|III\rangle = \frac{1}{\sqrt{2}}(|2\rangle + |3\rangle) = \frac{1}{\sqrt{2}}(|+-\rangle + |-+\rangle), \qquad E_{III} = A, \tag{1.133}$$

$$|IV\rangle = \frac{1}{\sqrt{2}}(|2\rangle - |3\rangle) = \frac{1}{\sqrt{2}}(|+-\rangle - |-+\rangle), \qquad E_{IV} = -3A. \tag{1.134}$$

As a result, four states with definite energies are found; three of these states have the same energy, i.e., are degenerate. The state $|IV\rangle$ has the energy $-3A$, and, as a result, the sum of the energies of four states is zero. This corresponds to our choice $E_0 = 0$ in the definition of the Hamiltonian H_{ss}.

The energy A remains indefinite. It can be calculated theoretically. We present the value measured in Crampton et al. (1963). According to those measurements, $A = h\nu$, where

$$\nu = (1\,420\,405\,751.800 \pm 0.026) \text{ Hz.}$$

The states presented above are also orthogonal and normalized and can be treated as new basis vectors.

A5. The Hamiltonian H_{sB} is responsible for the Zeeman splitting of the spectral terms of the hydrogen atom in the external magnetic field.

For the hydrogen atom in the external magnetic field, the Hamiltonian H_{ss} is supplemented by the new Hamiltonian

$$H_{sB} = -\mu_e \vec{\sigma}^e \cdot \vec{B} - \mu_p \vec{\sigma}^p \cdot \vec{B}. \tag{1.135}$$

Since both Hamiltonians H_{ss} and H_{sB} contribute to the Zeeman splitting of the spectral terms of the hydrogen atom, we consider their sum

$$H_Z = H_{ss} + H_{sB} = A\vec{\sigma}^e \cdot \vec{\sigma}^p - \mu_e \vec{\sigma}^e \cdot \vec{B} - \mu_p \vec{\sigma}^p \cdot \vec{B}. \tag{1.136}$$

If the external field $\vec{B}$ is aligned with the z axis, Eq. (1.127) reduce to the form

$$\begin{aligned}
Ea_1 &= \left[A - (\mu_e + \mu_p)B\right]a_1, \\
Ea_2 &= -\left[A + (\mu_e - \mu_p)B\right]a_2 + 2Aa_3, \\
Ea_3 &= 2Aa_2 - \left[A - (\mu_e - \mu_p)B\right]a_3, \\
Ea_4 &= \left[A + (\mu_e + \mu_p)B\right]a_4.
\end{aligned} \tag{1.137}$$

As above, the first and fourth equations have the solutions ($a_1 = 1$, $a_2 = a_3 = a_4 = 0$):

$$\begin{aligned}
|I\rangle &= |1\rangle = |++\rangle, & E_I &= A - (\mu_e + \mu_p)B, \\
|II\rangle &= |4\rangle = |--\rangle, & E_{II} &= A + (\mu_e + \mu_p)B.
\end{aligned} \tag{1.138}$$

For two remaining states, there are two homogeneous equations with zero determinant. The matrix elements have the form

$$\begin{aligned}
H_{11} &= -A - (\mu_e - \mu_p)B, & H_{12} &= 2A, \\
H_{21} &= 2A, & H_{22} &= -A + (\mu_e - \mu_p)B.
\end{aligned} \tag{1.139}$$

The level energy is expressed in terms of these matrix elements as follows:

$$\begin{aligned}
E &= \frac{1}{2}(H_{11} + H_{22}) \pm \sqrt{\frac{1}{4}(H_{11} - H_{22})^2 + H_{12}H_{21}} \\
&= -A \pm \sqrt{(\mu_e - \mu_p)^2 B^2 + 4A^2}.
\end{aligned} \tag{1.140}$$

Thus, two level energies are given by the expressions

$$E_{III} = A\left[-1 + 2\sqrt{1 + (\mu_e - \mu_p)^2 B^2/4A^2}\,\right],$$

$$E_{IV} = A\left[-1 - 2\sqrt{1 + (\mu_e - \mu_p)^2 B^2/4A^2}\,\right]. \tag{1.141}$$

The correctness of the resulting formulas can be verified taking zero magnetic field. In this case, the results should coincide with the solutions for the spin–spin Hamiltonian (see relations (1.128) and (1.129)).

In terms of the parameters $\mu = -(\mu_e + \mu_p)$ and $\mu' = -(\mu_e - \mu_p)$, the energy levels are expressed as

$$E_I = A + \mu B, \qquad E_{II} = A - \mu B, \qquad E_{III} = A\left(-1 + 2\sqrt{1 + \mu'^2 B^2/4A^2}\right),$$

$$E_{IV} = -A\left(1 + 2\sqrt{1 + \mu'^2 B^2/4A^2}\right). \tag{1.142}$$

As known, the magnetic moment of the electron is negative and its absolute value is approximately three orders of magnitude larger than the positive magnetic moment of the proton. Therefore, both introduced parameters μ and μ' are nearly μ_e, but are positive. As seen, in the absence of a magnetic field, the first three energies are A and the fourth energy is $-3A$. The behaviors of these energies with an increase in the magnetic field are different. Beginning with the value A, the energy E_I increases linearly and the energy E_{II} decreases linearly with the same rate with an increase in the field B. The energies E_{III} and E_{IV} beginning with the values A and $-3A$, respectively, increase first as a square function and then as a linear function of B.

We return to the calculation of the proton polarization as a function of the magnetic field. According to quantum mechanics, the squares of the absolute values of the coefficients a_2 and a_3 in the expansion of the wave function of state III,

$$\psi_{III} = a_2|+-\rangle + a_3|-+\rangle, \tag{1.143}$$

are the probabilities that the hydrogen atom is in the states where the proton is polarized against and along the field, respectively. If the basis wave functions are normalized to unity, then

$$|a_2|^2 + |a_3|^2 = 1. \tag{1.144}$$

Therefore, the polarization (with positive sign) of the protons in state III is given by the expression

$$P = |a_3|^2 - |a_2|^2. \tag{1.145}$$

We determine this polarization as follows. From the second of equations (1.137), we obtain

$$a_3/a_2 = \frac{E + A - \mu' B}{2A}, \tag{1.146}$$

where E is given by the first of expressions (1.141),

$$E_{III} = A\left(-1 + 2\sqrt{1 + \mu'^2 B^2/4A^2}\right) = A\left(-1 + 2\sqrt{1 + x^2}\right), \qquad (1.147)$$

where

$$x = \mu' B/2A. \qquad (1.148)$$

The substitution of (1.147) into (1.146) yields

$$a_3/a_2 = x + \sqrt{1 + x^2}. \qquad (1.149)$$

In order to satisfy the normalization condition, we set $a_3 = \cos\theta$, $a_2 = \sin\theta$. The substitution of these relations into Eq. (1.149) gives

$$\cot\theta = x + \sqrt{1 + x^2}. \qquad (1.150)$$

The condition $x = 1$ corresponds to a critical field of 509 Oe.

Simple algebra provides the relation $\tan 2\theta = 1/x$. Therefore, $\cos 2\theta = x/\sqrt{1 + x^2}$. Substituting this expression into (1.145), we determine the proton polarization in energy state *III*:

$$P_{III} = \cos^2\theta - \sin^2\theta = \cos 2\theta = x/\sqrt{1 + x^2}. \qquad (1.151)$$

The polarization for energy state *IV* is the same in absolute value as that for state *III*, but is opposite in sign. The polarizations of the first two states (*I* and *II*) are linear functions of the magnetic field.

1.6 Dirac Equation

The Dirac equation will be often used in this book. In particular, the analysis of the properties of relativistic spin involves the Lorentz transformation of spin, and this transformation can be understood only with the Dirac equation. It is known that the spin operator naturally appears from this equation. In addition, the Dirac equation provides the correct Landé g-factor of the electron.

The Schrödinger equation was considered in Sect. 1.5, where the nonrelativistic expression for the Hamiltonian was used. To derive the Dirac equation, it is necessary to satisfy three main requirements. The first requirement is the covariance condition; i.e., this equation should have the same form in any Lorentz frame. Second, the wave functions should be the eigenfunctions of the Hamiltonian, constitute a complete and orthonormalized system, and allow probability interpretation. Third, any physical observable should correspond to a linear Hermitian operator.

Let us write the relativistic Hamiltonian of a free particle in the form

$$H = \sqrt{p^2 c^2 + m^2 c^4} = -i\hbar c(\vec{\alpha} \cdot \vec{\nabla}) + \beta mc^2, \qquad (1.152)$$

where $\vec{\alpha}(\alpha_1, \alpha_2, \alpha_3)$ and β are the parameters to be determined. Here, we use the formal representation of the square root, because a Lorentz-invariant Hamiltonian should include a linear dependence on the momentum operator. The operator $\vec{\nabla}(\frac{\partial}{\partial x^1}, \frac{\partial}{\partial x^2}, \frac{\partial}{\partial x^3})$ presents the three-dimensional momentum of the particle. Using Eq. (1.152), we write the equation

$$i\hbar \frac{\partial \psi}{\partial t} = H\psi = \left[\frac{\hbar c}{i}(\vec{\alpha} \cdot \vec{\nabla}) + \beta mc^2\right] \cdot \psi. \tag{1.153}$$

Let us take the square of expression (1.152), apply the result to ψ, and require that the resulting equation coincides with the classic Klein–Gordon wave equation

$$\left[\frac{\partial}{\partial x_\mu} \frac{\partial}{\partial x^\mu} + \left(\frac{mc}{\hbar}\right)^2\right]\psi = 0. \tag{1.154}$$

This requirement leads to the following conditions on the parameters introduced above:

$$\{\alpha_i, \alpha_k\} = 2\delta_{ik}, \qquad \{\alpha_i, \beta\} = 0, \quad \alpha_i^2 = \beta^2 = 1, \tag{1.155}$$

where $i, k = 1, 2, 3$. For the Hamiltonian to be Hermitian, the matrices α_i and β should be Hermitian. According to the conditions $\alpha_i^2 = 1$ and $\beta^2 = 1$, the eigenvalues of the operators α_i and β are ± 1. Using the anticommutation conditions, we can show that the traces (the trace of a matrix is the sum of the elements of the main diagonal and is denoted as Tr) of these matrices are zero. For example,

$$Tr\,\alpha_i = -Tr\,\beta\alpha_i\beta = -Tr\,\beta^2\alpha_i = -Tr\,\alpha_i = 0. \tag{1.156}$$

Similarly, it can be proved that the trace of the β matrix is also zero. This means that the rank of the matrices α_i and β should be even. The minimum rank in this case is $n = 4$, because $n = 2$ corresponds to the spin space of the Pauli operators and identity matrix, which describe a nonrelativistic two-dimensional system. In one of particular choices of the matrices, α_i and β can be expressed in terms of the Pauli matrices σ_i:

$$\alpha_i = \begin{Vmatrix} 0 & \sigma_i \\ \sigma_i & 0 \end{Vmatrix}, \qquad \beta = \begin{Vmatrix} I & 0 \\ 0 & -I \end{Vmatrix}. \tag{1.157}$$

Here, I is the identity matrix in this representation. As seen, the Dirac equation directly leads to the appearance of the spin operator for a point particle (electron) with spin $s = 1/2$. As shown below, the Dirac equation predicts the correct gyromagnetic ratio equal to 2, which could not be obtained without this equation.

From Dirac equation (1.153) and the Hermitian-conjugate equation, one can obtain the current conservation law in the differential form

$$\frac{\partial}{\partial t}\rho + \mathrm{div}\,\vec{j} = 0. \tag{1.158}$$

The probability density ρ and three-dimensional vector of the probability current density $\vec{j}$ are specified by the respective formulas

$$\rho = \psi^+\psi = \sum_{\sigma=1}^{4} \psi_\sigma^* \psi_\sigma, \tag{1.159}$$

$$j^k = c\psi^+\alpha_k\psi. \tag{1.160}$$

Using the divergence theorem, we can show that

$$\frac{\partial}{\partial t}\int d^3x\, \psi^*\psi = -\int d^3x \cdot \mathrm{div}\, \vec{j} = -\oint_\Sigma \vec{j}\cdot d\vec{s} = 0.$$

Let us consider the Dirac equation in the presence of the external electromagnetic field $A^\mu(\Phi, \vec{A})$. This interaction can be introduced by means of the gauge-invariant change (invariant when the derivative of an arbitrary function with respect to the corresponding coordinates is added to the potential)

$$p^\mu \to \pi^\mu = p^\mu - \frac{e}{c}A^\mu. \tag{1.161}$$

In this case, the Dirac equation is transformed to the form

$$i\hbar\frac{\partial\psi}{\partial t} = H\psi = \left[c\vec{\alpha}\cdot\left(\hat{\vec{p}} - \frac{e}{c}\vec{A}\right) + \beta mc^2 + e\Phi\right]\psi. \tag{1.162}$$

We seek the solution of this equations in the form

$$\psi = \left\|\begin{array}{c}\bar{\varphi}\\\bar{\chi}\end{array}\right\|, \tag{1.163}$$

where $\bar{\varphi}$ and $\bar{\chi}$ are the two-component spinors. Then, the substitution of Eqs. (1.163) and (1.157) into Eq. (1.162) gives

$$i\hbar\frac{\partial}{\partial t}\left\|\begin{array}{c}\bar{\varphi}\\\bar{\chi}\end{array}\right\| = c\vec{\sigma}\cdot\left(\hat{\vec{p}} - \frac{e}{c}\vec{A}\right)\left\|\begin{array}{c}\bar{\chi}\\\bar{\varphi}\end{array}\right\| + mc^2\left\|\begin{array}{c}\bar{\varphi}\\-\bar{\chi}\end{array}\right\| + e\Phi\left\|\begin{array}{c}\bar{\varphi}\\\bar{\chi}\end{array}\right\|. \tag{1.164}$$

Let us consider the case where $E_{\mathrm{kin}} \ll mc^2$ and seek the solution in the form

$$\left\|\begin{array}{c}\bar{\varphi}\\\bar{\chi}\end{array}\right\| = \exp\left(-i\frac{mc^2}{\hbar}t\right)\left\|\begin{array}{c}\varphi\\\chi\end{array}\right\|. \tag{1.165}$$

The substitution of Eq. (1.165) into Eq. (1.164) yields

$$i\hbar\frac{\partial}{\partial t}\left\|\begin{array}{c}\varphi\\\chi\end{array}\right\| = c\vec{\sigma}\cdot\vec{\pi}\left\|\begin{array}{c}\chi\\\varphi\end{array}\right\| - 2mc^2\left\|\begin{array}{c}\varphi\\\chi\end{array}\right\| + e\Phi\left\|\begin{array}{c}\varphi\\\chi\end{array}\right\|. \tag{1.166}$$

This matrix relation presents two equations corresponding to the upper and lower elements of the matrices. Under the assumptions that χ is time independent and the term $e\Phi$ is negligible compared to the term containing the mass, the lower equation provides

$$\chi = \frac{\vec{\sigma} \cdot \vec{\pi}}{2mc} \varphi. \tag{1.167}$$

The component χ is much smaller than φ, and their ratio is much smaller than 1 in the nonrelativistic case. The substitution of expression (1.167) into the upper equation in matrix relation (1.166) gives

$$i\hbar \frac{\partial \varphi}{\partial t} = \left(\left(\frac{(\vec{\sigma} \cdot \vec{\pi})(\vec{\sigma} \cdot \vec{\pi})}{2mc} \right) + e\Phi \right) \varphi. \tag{1.168}$$

The Pauli operators satisfy the relation

$$(\vec{\sigma} \cdot \vec{a})(\vec{\sigma} \cdot \vec{b}) = \vec{a} \cdot \vec{b} + i\vec{\sigma} \cdot \vec{a} \times \vec{b}.$$

Applying this relation to Eq. (1.168), we obtain

$$i\hbar \frac{\partial \varphi}{\partial t} = \left[\frac{(\vec{p} - \frac{e}{c}\vec{A})^2}{2m} - \frac{e\hbar}{2mc} \vec{\sigma} \cdot \vec{B} + e\Phi \right] \varphi. \tag{1.169}$$

This equation coincides with the nonrelativistic Pauli equation, which was written one year before (1927) Dirac derived his equation in the relativistic form. This equation describes the electron with the spin 1/2, and two components of the function φ correspond to two orientations of the electron spin. In this equation, the correct value of the magnetic moment corresponding to the gyromagnetic ratio $g = 2$ is automatically obtained. This can be verified by retaining only terms linear in the external field in Eq. (1.169) with the following homogeneous magnetic field $\vec{B} = \operatorname{rot} \vec{A}$, $\vec{A} = \frac{1}{2}\vec{B} \times \vec{r}$:

$$i\hbar \frac{\partial \varphi}{\partial t} = \left[\frac{\vec{p}^2}{2m} - \frac{e}{2mc}(\vec{L} + 2\vec{s}) \cdot \vec{B} \right] \varphi. \tag{1.170}$$

Here, we introduce the orbital angular momentum $\vec{L} = \vec{r} \times \vec{p}$ and spin $\vec{s} = \frac{1}{2}\vec{\sigma}$. The coefficient 2 of the term with spin is the gyromagnetic ratio of the electron. Thus, the equality $g = 2$ following from the Dirac equation was one of the important achievements in theoretical physics. More recently, the experimental verification of the same relation for muons became one of the important directions of the search for the effects beyond the Standard Model.

Another important consequence of Eq. (1.170) is the prediction of the spin–orbit interaction of the particle. In particular, this interaction underlies the shell model of atoms and nuclei. The term containing the interaction of the magnetic moment of the Dirac particle with the external magnetic field is remarkable. This term explains the Zeeman splitting of the levels of the hydrogen atom. The comparison of this term with the first term in Eq. (1.170) shows that the spin interaction should weaken with an increase in the particle energy. As seen from the discussion of the polarization data in the third part of the book, the energy scale above which this statement is valid has not yet determined.

The use of the projection operators noticeably simplifies the calculations of the spin observables in relativistic theory. The specificity of the Dirac equation is that it has the solutions with positive and negative energies. In addition, the Dirac particle for each energy has two spin states. Four projection operators are necessary in order to separate one of these four states in the calculations. We present their explicit form (for details, see Bjorken and Drell 1964).

The operator separating the solution with positive energy and positive spin projection has the form (Λ is the energy projection operator, Σ is the spin projection operator, and u_z is the spin variable):

$$P_1(p) = \Lambda_+(p)\Sigma(u_z), \qquad P_2(p) = \Lambda_+(p)\Sigma(-u_z),$$
$$P_3(p) = \Lambda_-(p)\Sigma(u_z), \qquad P_4(p) = \Lambda_-(p)\Sigma(-u_z). \tag{1.171}$$

The energy projection operator is written in the explicit form

$$\Lambda_r = \frac{\varepsilon_r \hat{p} + mc}{2mc}, \tag{1.172}$$

where ε_r specifies the sign of the energy: the $+$ and $-$ signs are taken for the energy at $r = 1$ and 2, respectively. The projection operator has the properties

$$\Lambda_r(p)\Lambda_{r'}(p) = \left(\frac{1+\varepsilon_r\varepsilon_{r'}}{2}\right)\Lambda_r(p), \qquad \Lambda_+^2(p) = \Lambda_+(p), \tag{1.173}$$

$$\Lambda_+(p) + \Lambda_{r-}(p) = 1, \qquad \Lambda_+(p)\Lambda_-(p) = 0. \tag{1.174}$$

For the spin projection operator, the following formula is derived ($u_z = \hat{s}$):

$$\Sigma(s) = (1/2)(1 + \gamma_5\hat{s}), \tag{1.175}$$

where the four-spin operator satisfies the condition $(sp) = s^\mu p_\mu = 0$ and quantities γ_i will be specified in the next sections. In view of the covariance of the projection operator Σ, its action on the Dirac wave functions u and v with positive and negative energies, respectively, is specified as

$$\Sigma(s)u(p, s) = u(p, s), \qquad \Sigma(s)v(p, s) = v(p, s),$$
$$\Sigma(-s)u(p, s) = \Sigma(-s)v(p, s) = 0. \tag{1.176}$$

In the rest frame of the particle, the solution of the Dirac equation with positive energy, u, is transformed to the Pauli wave function. In this case, there is the complete correspondence in the transformation of the spin states; namely, the state with spin $+$ is transformed to plus and the state with spin $-$ is transformed to minus. However, under the same transformation from the state with negative energy to the rest frame, the signs of the spins change to opposite. This specificity is explained in a theory where the existence of a real antiparticle (positron in this case) is postulated.

1.7 Elements of Relativistic Quantum Mechanics

Some elements of relativistic quantum mechanics are often used in applications. For example, when constructing the reaction matrix, it is necessary to know the constraints implied by certain conservation laws or by the interaction symmetry properties (discrete and continuous) characteristic of the process under investigation. In view of this circumstance, as well as to avoid diverting readers to other sources of information, we present in detail the properties of the Dirac spinors under the transformations of the proper Lorentz group (the determinant of the transformations is $+1$) and extended Lorentz group (rotation group, space inversion, and time reversal). Here, we consider the transformation properties of the Dirac spinors under rotations of the coordinate system, inversion of the coordinate axes, and time reversal following Fermi (1961).

A. Transformations of Dirac spinors under rotations of the coordinate system

Let us represent the Dirac equation for the electron with charge e in the external electromagnetic field $\vec{A}$ in a certain basic Lorentz frame in the form

$$\left(\frac{mc}{\hbar} + \vec{\gamma} \cdot \vec{\nabla} - \frac{ie}{\hbar c} \vec{\gamma} \cdot \vec{A} \right) \psi = 0. \tag{1.177}$$

We introduce a new Lorentz frame where the coordinates are determined by the linear orthogonal transformation (the vector operators ∇_μ and A_μ are transformed similarly):

$$x'_\mu = a_{\mu\nu} x_\nu, \qquad \nabla'_\mu = a_{\mu\nu} \nabla_\nu, \qquad A'_\mu = a_{\mu\nu} A_\nu. \tag{1.178}$$

Hereinafter, the repeated indices imply summation.

Let us introduce the linear unitary operator $\hat{T}$ specifying the transformation of the Dirac spinors:

$$\psi' = \hat{T}^{-1} \psi. \tag{1.178a}$$

Assuming that the Dirac matrices remain unchanged in the new frame, we arrive at the following covariant form of the Dirac equation in this frame:

$$\left(\frac{mc}{\hbar} + \vec{\gamma} \cdot \vec{\nabla}' - \frac{ie}{\hbar c} \vec{\gamma} \cdot \vec{A}' \right) \psi' = 0. \tag{1.179}$$

In order to determine the properties of the operator $\hat{T}$, we multiply this equation by $\hat{T}$ from the left and substitute expression (1.178a) for ψ'. As a result, we obtain

$$\left(\frac{mc}{\hbar} + \hat{T} \vec{\gamma} \hat{T}^{-1} \cdot \vec{\nabla}' - \frac{ie}{\hbar c} \hat{T} \vec{\gamma} \cdot \vec{A}' \hat{T}^{-1} \right) \psi = 0.$$

Changing the primed vector operators ∇' and A' to unprimed by formulas (1.178), we arrive at the equation

$$\left(\frac{mc}{\hbar} + \hat{T} \gamma_\lambda \hat{T}^{-1} a_{\lambda\nu} \nabla_\nu - \frac{ie}{\hbar c} \hat{T} \gamma_\lambda a_{\lambda\nu} A_\nu \hat{T}^{-1} \right) \psi = 0. \tag{1.180}$$

The comparison of this equation with Eq. (1.177) shows that $\hat{T}\gamma_\lambda\hat{T}^{-1}a_{\lambda\nu} = \gamma_\nu$. Using the orthogonality relation

$$a_{\mu\nu}a_{\lambda\nu} = a_{\nu\mu}a_{\nu\lambda} = \delta_{\mu\lambda}, \tag{1.181}$$

we arrive at the final result

$$\hat{T}\gamma_\mu\hat{T}^{-1} = a_{\mu\nu}\gamma_\nu. \tag{1.182}$$

Let us consider the infinitesimal transformation of the matrix elements $a_{\mu\nu}$:

$$a_{\mu\nu} = \delta_{\mu\nu} + \varepsilon_{\mu\nu}. \tag{1.183}$$

Taking into account orthogonality condition (1.181), we can write

$$a_{\lambda\nu}a_{\mu\nu} = (\delta_{\lambda\nu} + \varepsilon_{\lambda\nu})(\delta_{\mu\nu} + \varepsilon_{\mu\nu}) = \delta_{\lambda\nu}\delta_{\mu\nu} + \delta_{\lambda\nu}\varepsilon_{\mu\nu} + \delta_{\mu\nu}\varepsilon_{\lambda\nu}$$

$$= \delta_{\lambda\mu} + \varepsilon_{\mu\lambda} + \varepsilon_{\lambda\mu} = \delta_{\lambda\mu}.$$

Therefore,

$$\varepsilon_{\mu\lambda} = -\varepsilon_{\lambda\mu}. \tag{1.184}$$

Since the coordinate variables x, y, and z and time t are real, the components ε_{ik} ($i, k = 1, 2, 3$) of the antisymmetric second-rank tensor $\varepsilon_{\mu\nu}$ are real and $\varepsilon_{4n} = -\varepsilon_{n4}$ ($n = 1, 2, 3$) are imaginary.

Under such small rotations of the Lorentz frame, the operator $\hat{T}$ assumingly also undergoes the infinitesimal transformation

$$\hat{T} = 1 + \hat{s}, \qquad T^{-1} = 1 - \hat{s}. \tag{1.185}$$

It is assumed that the operator $\hat{s}$ is of the same smallness order as ε. Here, the second equation is obtained from the first equation under the assumption that the square of the operator $\hat{s}$ can be neglected. The substitution of relations (1.185) into Eq. (1.182) yields

$$\hat{s}\gamma_\mu - \gamma_\mu\hat{s} = \varepsilon_{\mu\nu}\gamma_\nu. \tag{1.186}$$

The direct calculations show that the solution of this equation is given by the expression

$$\hat{s} = -\frac{1}{4}\varepsilon_{\mu\nu}\gamma_\mu\gamma_\nu. \tag{1.187}$$

As a result, the spinor transformation matrix $\hat{T}$ corresponding to Lorentz transformations (1.178) and (1.183) has the form

$$\hat{T} = 1 - \frac{1}{4}\varepsilon_{\mu\nu}\gamma_\mu\gamma_\nu. \tag{1.188}$$

As known, the Lorentz group is of fundamental importance in relativistic theory, in particular, in relativistic quantum mechanics. A Lorentz transformation consists of coordinate translations and rotations.

Let us consider several examples of the rotations of the reference frame. Six planes exist for the four coordinates x, y, z, and t (the number of two-term combinations from a set of four elements is six). Therefore, there are six Lorentz rotations.

Example 1.1 The infinitesimal rotation about the z axis is specified by the relations

$$x'_1 = x_1 - \varepsilon x_2, \qquad x'_2 = x_2 + \varepsilon x_1, \qquad x'_3 = x_3, \qquad x'_4 = x_4. \qquad (1.189)$$

In this case, all $\varepsilon_{\mu\nu}$ elements are zero except for $\varepsilon_{12} = -\varepsilon_{21} = \varepsilon$.

The four-row matrix $\hat{T}$ in this case has the form

$$\hat{T}_\varepsilon = 1 + \frac{1}{2}\varepsilon\gamma_1\gamma_2 = \begin{Vmatrix} 1+\frac{i}{2}\varepsilon & 0 & 0 & 0 \\ 0 & 1-\frac{i}{2}\varepsilon & 0 & 0 \\ 0 & 0 & 1+\frac{i}{2}\varepsilon & 0 \\ 0 & 0 & 0 & 1-\frac{i}{2}\varepsilon \end{Vmatrix}. \qquad (1.190)$$

For the finite rotations by the angle φ, taking the power φ/ε of the matrix $\hat{T}$ and passing to the limit $\varepsilon \to 0$, we obtain

$$\hat{T}_\varphi = \begin{Vmatrix} \exp(\frac{i}{2}\varphi) & 0 & 0 & 0 \\ 0 & \exp(-\frac{i}{2}\varphi) & 0 & 0 \\ 0 & 0 & \exp(\frac{i}{2}\varphi) & 0 \\ 0 & 0 & 0 & \exp(-\frac{i}{2}\varphi) \end{Vmatrix}. \qquad (1.191)$$

Under such a rotation of the reference frame about the z axis by the angle φ, the spinors in the new system, ψ', are expressed in terms of the spinors in the initial system, ψ, as

$$\psi'_1 = e^{\frac{i}{2}\varphi}\psi_1, \qquad \psi'_2 = e^{-\frac{i}{2}\varphi}\psi_2, \qquad \psi'_3 = e^{\frac{i}{2}\varphi}\psi_3, \qquad \psi'_4 = e^{-\frac{i}{2}\varphi}\psi_4. \qquad (1.192)$$

Note that the wave functions change signs under the rotation of the reference frame by the angle $\varphi = 2\pi$.

Example 1.2 Let us consider an infinitesimal Lorentz transformation involving the coordinates x and t. It has the form

$$x'_1 = x_1 - \varepsilon ct = x_1 + i\varepsilon x_4, \qquad x'_4 = x_4 - i\varepsilon x_1, \qquad x'_2 = x_2, x'_3 = x_3. \qquad (1.193)$$

The corresponding spinor transformation operator has the form

$$\hat{T}_\varepsilon = 1 - \frac{i}{2}\varepsilon\gamma_1\gamma_4 = 1 + \frac{\varepsilon}{2}\alpha_1 = \begin{Vmatrix} 1 & 0 & 0 & \varepsilon/2 \\ 0 & 1 & \varepsilon/2 & 0 \\ 0 & \varepsilon/2 & 1 & 0 \\ \varepsilon/2 & 0 & 0 & 1 \end{Vmatrix}. \qquad (1.193a)$$

In order to obtain the finite transformation, it is necessary to apply transformation (1.193a) n times, where $n = (1/\varepsilon)\,\text{Arcth}\,\beta$ and to pass to the limit $\varepsilon \to 0$ and

$n \to \infty$. Then, ($\gamma = 1/\sqrt{1 - \beta^2}$ is the Lorentz factor)

$$x_1' = \gamma(x_1 - \beta x_0), \qquad x_0' = \gamma(x_0 - \beta x_1), \qquad x_4 = ix_0 = ict. \qquad (1.194)$$

The finite transformation matrix $\hat{T}$ can be modified as

$$\hat{T}_\beta = (T_{14})^n = \left(1 + \frac{1}{2}\varepsilon\alpha_1\right)^n$$

$$\to \quad e^{\frac{1}{2}n\varepsilon\alpha_1} = \cosh\left(\frac{1}{2}n\varepsilon\right) + \alpha_1 \sinh\left(\frac{1}{2}n\varepsilon\right)$$

$$= \cosh\left(\frac{1}{2}\operatorname{Arcth}\beta\right) + \alpha_1 \sinh\left(\frac{1}{2}\operatorname{Arcth}\beta\right)$$

$$= \sqrt{(\gamma + 1)/2} + \alpha_1 \sqrt{(\gamma - 1)/2}. \qquad (1.194a)$$

Here, we take into account that $\alpha_1^2 = 1$. Thus, the spinor transformation operator takes the final form

$$\hat{T}_\beta = \sqrt{(\gamma + 1)/2} + \alpha_1 \sqrt{(\gamma - 1)/2}. \qquad (1.195)$$

B. Transformations of Dirac spinors under space inversion

Space inversion is described by the transformation

$$x_i' = -x_i, \quad i = 1, 2, 3; \qquad x_4' = x_4. \qquad (1.196)$$

Taking into account the relations

$$\hat{T}\gamma_\mu \hat{T}^{-1} = a_{\mu\nu}\gamma_\nu, \qquad (1.196a)$$

and denoting the spinor transformation operator as $\hat{T}_{rev}$, we obtain

$$\hat{T}_{rev}\gamma_i \hat{T}_{rev}^{-1} = -\gamma_i, \quad i = 1, 2, 3, \qquad \hat{T}_{rev}\gamma_4 \hat{T}_{rev}^{-1} = \gamma_4. \qquad (1.197)$$

The solution of these equations has the form

$$\hat{T}_{rev} = \gamma_4 = \beta. \qquad (1.198)$$

The properties of the matrix β ensure the following transformation characteristics of the matrix $\hat{T}_{rev}$:

$$\hat{T}_{rev} = \hat{T}_{rev}^{-1} = \hat{T}^+. \qquad (1.199)$$

For the operator $\hat{T}_{rev}$ in form (1.198), the wave functions in the inverse coordinate system are expressed in terms of the wave functions in the initial coordinate system as

$$\psi_1' = \psi_1, \qquad \psi_2' = \psi_2, \qquad \psi_3' = -\psi_3, \qquad \psi_4' = -\psi_4. \qquad (1.200)$$

According to these relations, two pairs of the wave functions (ψ_1, ψ_2) and (ψ_3, ψ_4) are transformed differently under space inversion: the first pair does not change sign, i.e., is an even function, whereas the second pair is an odd function. Since the first pair of the wave functions describes a particle (electron) and the second pair, an antiparticle (positron), the particle and antiparticle have opposite parities. As known, under space inversion, the wave functions acquire the factor $(-1)^l$, where l is the orbital angular momentum. As a result, the transformation for even l values under inversion has the form

$$\psi_1(x) = \psi_1(-x), \qquad \psi_2(x) = \psi_2(-x),$$
$$\psi_3(x) = -\psi_3(-x), \qquad \psi_4(x) = -\psi_4(x). \tag{1.201}$$

The transformation for odd l values has the form

$$\psi_1(x) = -\psi_1(-x), \qquad \psi_2(x) = -\psi_2(-x),$$
$$\psi_3(x) = \psi_3(-x), \qquad \psi_4(x) = \psi_4(x). \tag{1.202}$$

Some useful properties of the space inversion operator are as follows:

$$\hat{T}_{rev}\gamma_\mu\hat{T}_{rev} = \begin{cases} -\gamma_\mu, & \mu = 1, 2, 3; \\ +\gamma_\mu, & \mu = 4; \end{cases} \qquad \hat{T}_{rev}\beta\gamma_\mu\hat{T}_{rev} = \begin{cases} -\beta\gamma_\mu, & \mu = 1, 2, 3; \\ +\beta\gamma_\mu, & \mu = 4. \end{cases}$$
$$\tag{1.203}$$

C. Transformations of Dirac spinors under time reversal

The time reversal operation is specified by the transformations

$$\vec{x} \to \vec{x}, \qquad \nabla \to \nabla, \qquad \vec{A} \to -\vec{A},$$
$$x_4 \to -x_4, \qquad \nabla_4 \to -\nabla_4, \qquad A_4 \to A_4. \tag{1.204}$$

The Dirac equation for the wave function ψ in the presence of the external four-component electromagnetic field A_μ is written in the form

$$\left(\frac{mc}{\hbar} + \vec{\gamma} \cdot \vec{\nabla} - \frac{ie}{\hbar c}\vec{\gamma} \cdot \vec{A}\right)\psi + \gamma_4\left(\frac{\partial}{\partial x_4} - \frac{ie}{\hbar c}A_4\right)\psi = 0. \tag{1.205}$$

The Dirac equation in which time reversal (transformations (1.204)) is performed has the form

$$\left(\frac{mc}{\hbar} + \vec{\gamma} \cdot \vec{\nabla} + \frac{ie}{\hbar c}\vec{\gamma} \cdot \vec{A}\right)\psi' - \gamma_4\left(\frac{\partial}{\partial x_4} + \frac{ie}{\hbar c}A_4\right)\psi' = 0. \tag{1.206}$$

It is impossible to solve Eq. (1.206) for ψ' as above. The solution is sought through the transformation

$$\psi' = S\psi^*. \tag{1.207}$$

Let us take the complex conjugation of Eq. (1.205):

$$\left(\frac{mc}{\hbar} + \vec{\gamma}^* \cdot \vec{\nabla} + \frac{ie}{\hbar c}\vec{\gamma}^* \cdot \vec{A}\right)\psi^* - \gamma_4^*\left(\frac{\partial}{\partial x_4} + \frac{ie}{\hbar c}A_4\right)\psi^* = 0. \qquad (1.208)$$

Multiplying Eq. (1.208) by the operator S from the left and using relation (1.207), we obtain

$$\left(\frac{mc}{\hbar} + S\vec{\gamma}^*S^{-1} \cdot \vec{\nabla} + \frac{ie}{\hbar c}S\vec{\gamma}^*S^{-1} \cdot \vec{A}\right)\psi' - S\gamma_4^*S^{-1}\left(\frac{\partial}{\partial x_4} + \frac{ie}{\hbar c}A_4\right)\psi' = 0. \qquad (1.209)$$

The requirement that this equation coincides with Eq. (1.206) provides

$$S\vec{\gamma}^*S^{-1} = \vec{\gamma}, \qquad S\gamma_4^*S^{-1} = \gamma_4, \qquad \psi' = S\psi^*. \qquad (1.210)$$

These relations are satisfied by the S matrix given by the expression

$$S = i\gamma_1\gamma_3 = \begin{Vmatrix} 0 & -i & 0 & 0 \\ i & 0 & 0 & 0 \\ 0 & 0 & 0 & -i \\ 0 & 0 & i & 0 \end{Vmatrix}. \qquad (1.211)$$

Thus, we obtain the solution of the Dirac equation describing the time-reversed motion of the system.

D. Charge conjugation operation

The Dirac equation should describe both the electron and positron. These particles constitute a particle–antiparticle pair and differ from each other in the opposite electric charges. For this reason, it is natural to expect that the Dirac equation has a symmetric solution under change in the charge sign

$$e \leftrightarrow -e. \qquad (1.212)$$

The Dirac equation for the electron was written in form (1.205),

$$\left(\frac{mc}{\hbar} + \vec{\gamma} \cdot \vec{\nabla} - \frac{ie}{\hbar c}\vec{\gamma} \cdot \vec{A}\right)\psi + \gamma_4\left(\frac{\partial}{\partial x_4} - \frac{ie}{\hbar c}A_4\right)\psi = 0.$$

Let us change the charge sign, introduce the wave function of the positron ψ^C, and rewrite this equation in the form

$$\left(\frac{mc}{\hbar} + \vec{\gamma} \cdot \vec{\nabla} + \frac{ie}{\hbar c}\vec{\gamma} \cdot \vec{A}\right)\psi^C + \gamma_4\left(\frac{\partial}{\partial x_4} + \frac{ie}{\hbar c}A_4\right)\psi^C = 0. \qquad (1.213)$$

We define the charge conjugation operation C as

$$\psi^C = C\psi^*. \qquad (1.214)$$

As above, we take the complex conjugation of Dirac equation (1.205):

$$\left(\frac{mc}{\hbar} + \vec{\gamma}^* \cdot \vec{\nabla} + \frac{ie}{\hbar c}\vec{\gamma}^* \cdot \vec{A}\right)\psi^* - \gamma_4^*\left(\frac{\partial}{\partial x_4} + \frac{ie}{\hbar c}A_4\right)\psi^* = 0. \qquad (1.215)$$

Multiplying this equation by the operator C from the left, we arrive at the equation

$$\left(\frac{mc}{\hbar} + C\vec{\gamma}^*C^{-1} \cdot \vec{\nabla} + \frac{ie}{\hbar c}C\vec{\gamma}^*C^{-1} \cdot \vec{A}\right)\psi^C - C\gamma_4^*C^{-1}\left(\frac{\partial}{\partial x_4} + \frac{ie}{\hbar c}A_4\right)\psi^C = 0.$$
$$(1.216)$$

The requirement that Eqs. (1.213) and (1.216) coincide with each other leads to the conditions

$$\hat{C}\vec{\gamma}^*\hat{C}^{-1} = \vec{\gamma}^*, \qquad \hat{C}\gamma_4^*\hat{C}^{-1} = -\gamma_4. \qquad (1.217)$$

For the standard form of the Dirac matrices, relations (1.216) are satisfied for the choice

$$\hat{C} = \gamma_2. \qquad (1.218)$$

Thus, the solution of the charge-conjugate equation is related to the solution of the initial equation as

$$\psi^C = \gamma_2\psi^*. \qquad (1.219)$$

1.8 Tensors and Lorentz Transformations of Spinors

As known, under change in the reference frame, mathematical objects are transformed differently and classified according to their transformation properties. Quantities that remain unchanged under these transformations are called either scalars or pseudoscalars. Pseudoscalars differ from scalars in their behavior under space inversion. A quantity that does not change sign under this inversion is called scalar, whereas a quantity that changes sign under inversion is called pseudoscalar. Other classes of quantities called tensors of the first, second, etc. ranks. Among the first rank tensors are the momentum vector (it is a polar vector; i.e., it changes sign under space inversion) and angular momentum (it is an axial vector or a pseudovector; i.e., it does not change sign under space inversion). There are also spinors with two, four, etc. components. In relativistic quantum mechanics, the requirement of the invariance of the Dirac equation under the Lorentz transformations is of decisive importance. For this reason, we consider the operators appearing in this equation such as scalars and tensors, as well as their properties. Then, we formulate the conditions of the invariance of the Dirac equation and define the Lorentz transformations of the spinors and their bilinear combinations.

A. Time and space coordinates (t, x, y, z) in the Minkowski space constitute a four-vector with the following components, which are conditionally called contravariant and marked by superscripts:

$$x^\mu \equiv (x^0, x^1, x^2, x^3) \equiv (t, x, y, z). \tag{1.220}$$

The metric tensor g is defined as

$$g = g_{\mu\nu} = g^{\mu\nu} = \begin{Vmatrix} 1 & 0 & 0 & 0 \\ 0 & -1 & 0 & 0 \\ 0 & 0 & -1 & 0 \\ 0 & 0 & 0 & -1 \end{Vmatrix}. \tag{1.221}$$

In another possible definition of the metric tensor, the first diagonal element in matrix (1.221) is -1 and the other diagonal elements are 1. This variant is not considered below. This tensor allows one to raise and lower indices and, in particular, to obtain a vector with covariant components (they are marked by subscripts):

$$x_\mu \equiv (x_0, x_1, x_2, x_3) \equiv (t, -x, -y, -z) = g_{\mu\nu} x^\nu. \tag{1.222}$$

The scalar product of any two four-vectors x and p is defined as

$$px = p_\mu x^\mu = p^\nu g_{\nu\mu} x^\mu = x_\nu p^\nu = xp. \tag{1.223}$$

The indices of any tensor a can be similarly raised or lowered:

$$a^\nu_\mu = a_{\mu\rho} g^{\rho\nu} = g_{\mu\rho} a^{\rho\nu}; \quad a^{\mu\nu} = g^{\mu\rho} a_{\rho\lambda} g^{\lambda\nu}. \tag{1.224}$$

Applying this relation to the metric tensor, we obtain

$$g^\nu_\mu = g_{\mu\lambda} g^{\lambda\nu} = \delta_{\mu\nu} = \begin{pmatrix} 1 & 0 & 0 & 0 \\ 0 & 1 & 0 & 0 \\ 0 & 0 & 1 & 0 \\ 0 & 0 & 0 & 1 \end{pmatrix}. \tag{1.225}$$

The scalar quantity defined by expression (1.223) is invariant under the Lorentz transformations. Another example of a scalar is the square of the four-momentum of a particle with mass m:

$$p^2 = p_\mu p^\mu = \varepsilon^2 - \vec{p}^2 = m^2. \tag{1.226}$$

Similar invariants can be composed of combinations of tensors of various ranks. For example, using four-vectors a and b and tensors B and C, we can compose the invariant combinations

$$aBC = a_\mu B^{\mu\nu} C_\nu, \qquad BD = B^{\mu\nu} D_{\mu\nu}. \tag{1.227}$$

These examples imply the following rule for composing invariants: a quantity is invariant if and only if dummy indices appear pairwise so that one of them is a superscript and the other is a subscript. The absence of such correspondence can mean error in the calculations of invariants.

Let us consider examples of invariants with differentiation operators. Let F be a scalar, i.e., invariant function. Then, the differential of this operator should also be scalar:

$$dF = \frac{\partial F}{\partial x_\mu} dx_\mu = \frac{\partial F}{\partial x^\nu} dx^\nu = \text{invariant.} \tag{1.228}$$

For this quantity to be invariant, the derivatives should be transformed as follows:

$$\frac{\partial F}{\partial x_\mu} = \partial^\mu F\text{—contravariant,} \qquad \frac{\partial F}{\partial x^\mu} = \partial_\mu F\text{—covariant.} \tag{1.229}$$

Therefore, the gradient vector components are transformed oppositely to the coordinates with respect to which the derivatives are taken. Thus, the covariant derivative is specified by the formula

$$\frac{\partial}{\partial x^\mu} = \partial_\mu = \left(\frac{\partial}{\partial t}, \frac{\partial}{\partial x}, \frac{\partial}{\partial y}, \frac{\partial}{\partial z} \right). \tag{1.230}$$

The contravariant derivative is given by the formula

$$\frac{\partial}{\partial x_\mu} = \partial^\mu = \left(\frac{\partial}{\partial t}, -\frac{\partial}{\partial x}, -\frac{\partial}{\partial y}, -\frac{\partial}{\partial z} \right). \tag{1.231}$$

The Klein–Gordon operator is invariant and specified by the expression

$$\Box = \partial_\mu \partial^\mu = \frac{\partial^2}{\partial t^2} - \frac{\partial^2}{\partial x^2} - \frac{\partial^2}{\partial y^2} - \frac{\partial^2}{\partial z^2}. \tag{1.232}$$

Using this operator, we can write the invariant relativistic Klein–Gordon equation for a free spinless particle with positive energy:

$$\Box \psi = \partial_\mu \partial^\mu \psi = -p_\mu p^\mu \psi = -m^2 \psi. \tag{1.233}$$

This equation has a solution in the form of the plane wave

$$\psi_p(x) = (2\pi)^{-3/2} e^{-ipx}. \tag{1.234}$$

The form of Eq. (1.233) implies the definition of the covariance of an equation: the equation is covariant if its left- and right-hand sides are transformed identically. For example, if the left-hand side of the equation is scalar, then its right-hand side should also be scalar (see Eq. (1.233)); if the left-hand side is vector, then the right-hand side should also be vector with the same property under space inversion: if one side is covariant, the other side should also be covariant. If the left-hand side of the equation has a free index in a certain position (upper or lower), then the right-hand side of the equation should have the same index in the same position. Some examples illustrating these statements are given below.

The examples of correctly written equations are

$$a_\mu b^\mu = c, \qquad a_\mu b^\mu c_\lambda = d_\lambda, \qquad a_\mu b^\mu = D^{\lambda\rho} c_\lambda k_\rho. \tag{1.235}$$

The examples of incorrectly written equations are

$$a_\mu b^\mu = c_\lambda, \qquad a_\mu b^\mu c_\lambda = d^\lambda, \qquad a_\mu b^\mu = D^{\lambda\rho} c_\lambda k^\rho. \tag{1.236}$$

This section is based on Hagedorn (1963).

B. The transformation rules for the Dirac matrices and their mean values under the Lorentz transformations are important for, e.g., the construction of the scattering matrix, density matrix, and experimental observables. Using the results from the preceding section, we consider below some examples.

The operator of infinitesimal rotation of the reference frame is given by the expression (see Eq. (1.188) in Sect. 1.7)

$$\hat{T} = 1 - \frac{1}{4}\varepsilon_{\mu\nu}\gamma_\mu\gamma_\nu. \tag{1.237}$$

This expression can be represented by separating the terms associated with the time component:

$$\hat{T} = 1 - \frac{1}{4}\varepsilon_{\mu\nu}\gamma_\mu\gamma_\nu = 1 - \frac{1}{4}\varepsilon_{nm}\gamma_n\gamma_m - \frac{1}{2}\varepsilon_{4m}\beta\gamma_m. \tag{1.238}$$

The inverse matrix with an accuracy of ε^2 is given by the expression

$$\hat{T}^{-1} = 1 + \frac{1}{4}\varepsilon_{\mu\nu}\gamma_\mu\gamma_\nu = 1 + \frac{1}{4}\varepsilon_{nm}\gamma_n\gamma_m + \frac{1}{2}\varepsilon_{4m}\beta\gamma_m. \tag{1.239}$$

Here, $\varepsilon_{\mu\nu}$, ε_{nm} are the antisymmetric unit tensor of the second rank in the four- and three-dimensional spaces, respectively, with $\mu, \nu = 1, 2, 3, 4$ and $m, n = 1, 2, 3$.

The terms $\frac{1}{4}\varepsilon_{nm}\gamma_n\gamma_m$ and $\frac{1}{2}\varepsilon_{4m}\beta\gamma_m$ in the matrix $\hat{T}$ are real and imaginary, respectively. The Dirac spin matrices satisfy the relations

$$\{\gamma_\mu, \gamma_\nu\} = 2\delta_{\mu\nu}, \qquad \gamma_\mu\gamma_\nu + \gamma_\nu\gamma_\mu = 2\delta_{\mu\nu}, \qquad \beta = \gamma_4. \tag{1.240}$$

The Hermitian conjugation of operator equation (1.238) gives

$$\hat{T}^+ = 1 + \frac{1}{4}\varepsilon_{\mu\nu}^*(\gamma_\mu\gamma_\nu)^+ = 1 + \frac{1}{4}\varepsilon_{nm}\gamma_n\gamma_m - \frac{1}{2}\varepsilon_{4m}\beta\gamma_m. \tag{1.241}$$

The comparison of Eqs. (1.239) and (1.241) indicates that the operator $\hat{T}$ is not unitary; i.e., $\hat{T} \neq \hat{T}^+$ in the general case. It becomes unitary only for purely spatial rotations, when $\varepsilon_{4m} = 0$. In the general case, the following commutation relations are valid:

$$\beta\hat{T}^+\beta = \hat{T}^{-1}, \qquad \hat{T}^+\beta = \beta\hat{T}^{-1}, \qquad \beta T^+ = \hat{T}^{-1}\beta. \tag{1.242}$$

Let us now determine which scalars, vectors, and tensors of the second rank can be constructed in terms of the Dirac matrices. We begin with the construction of the scalar matrix. Under the Lorentz transformation

$$x'_\mu = a_{\mu\nu} x_\nu, \tag{1.243}$$

the wave function is transformed by the rule

$$\psi' = \hat{T}\psi. \tag{1.244}$$

The mean value of the scalar operator u should remain unchanged under the above Lorentz transformations

$$\psi'^+ u \psi' = \psi^+ u \psi. \tag{1.245}$$

The substitution of Eq. (1.244) into Eq. (1.245) provides

$$(\hat{T}\psi)^+ u (T\psi) = \psi^+ (\hat{T}^+ u T)\psi. \tag{1.246}$$

The comparison of this equation with Eq. (1.245) shows that

$$u = \hat{T}^+ u \hat{T} \tag{1.247}$$

for any operator $\hat{T}$. Using relations (1.242), we obtain

$$u = \hat{T}^+ u \hat{T} = \beta \hat{T}^{-1} \beta u \hat{T}. \tag{1.248}$$

Multiplying the two extreme terms by β and, then, by $\hat{T}$ and taking into account that $\beta^2 = 1$, we arrive at the equation

$$(\beta u)\hat{T} = \hat{T}(\beta u). \tag{1.249}$$

This equation has two solutions

$$\beta u = 1 \quad \text{or} \quad \beta u = \gamma_1 \gamma_2 \gamma_3 \gamma_4 = \gamma_5. \tag{1.250}$$

Therefore, two solutions, $u_1 = \beta$ and $u_2 = \beta \gamma_5$, exist for u. These two solutions are different. Indeed, the first solution commutes with the space inversion operator $\hat{T}_{rev} = \beta$; i.e.,

$$\hat{T}^+ u_1 \hat{T}^{-1} = \hat{T}^+ \beta \hat{T}^{-1} = \beta. \tag{1.251}$$

The second solution anticommutes with this operator $\hat{T}_{rev}$:

$$\hat{T}^+ u_2 \hat{T}^{-1} = \hat{T}^+ \beta \gamma_5 \hat{T}^{-1} = -\beta \gamma_5. \tag{1.252}$$

Thus, the matrix β corresponds to the true scalar $\psi^+ \beta \psi$ and the quantity $\beta \gamma_5$ is pseudoscalar $\psi^+ \beta \gamma_5 \psi$. The quantity $\bar{\psi}\psi$, where $\bar{\psi} = \psi^+ \beta$, is transformed as a scalar and the quantity $\bar{\psi}\gamma_5\psi$ is transformed as a pseudoscalar.

The behavior of the entire set of the independent Dirac matrices is determined similarly. They have the following properties:

1. $1 = S$ is a scalar,
2. $\gamma_5 = P$ is a pseudoscalar,
3. $\gamma_\mu = V$ is a vector $(\mu = 1, 2, 3, 4)$,
4. $\gamma_5\gamma_\mu = A$ is a pseudovector or axial four-vector,
5. $\frac{1}{2}(\gamma_\mu\gamma_\nu - \gamma_\nu\gamma_\mu) = \sigma_{\mu\nu}$ is an antisymmetric tensor of the second rank.

The 16 Dirac matrices presented above constitute the complete set; i.e., any 4×4 matrix acting on Dirac spinors can be expanded in terms of this set. This property is used, e.g., to construct the relativistic density matrix or reaction matrix.

1.9 Spin of a Relativistic Particle with Nonzero Mass

Let us consider a particle with the spin $s = 1/2$ and mass m. We define the polarization four-vector $S(s_0, \vec{s})$ as follows:

- In the rest frame of the particle (R frame), it coincides with the nonrelativistically defined spin $S(0, \vec{s}_R)$.
- It is an axial vector by analogy with the orbital angular momentum. As known (Hagedorn 1963), the four-vector $x(ct, \vec{x})$ under the Lorentz transformation is transformed as follows:

$$\vec{x} = \vec{x}' + \vec{\beta}\gamma\left(\frac{\gamma}{\gamma + 1}\vec{\beta} \cdot \vec{x}' + ct'\right); \qquad ct = \gamma(ct' + \vec{\beta} \cdot \vec{x}'). \qquad (1.253)$$

Here, the primed variables are defined in the rest frame of the particle (R frame), whereas the unprimed variables are defined in another Lorentz frame, e.g., in the laboratory frame (L frame); and β and γ are the velocity and Lorentz factor of the particle in the L frame, respectively. By analogy, the transformation of the spin under the transition from the R frame to the L frame is $S(s_{0L}, \vec{s}_L)$, where

$$\vec{s}_L = \vec{s}_R + \vec{\beta}\frac{\gamma^2}{\gamma + 1}\vec{\beta} \cdot \vec{s}_R, \qquad s_{0L} = \gamma\vec{\beta} \cdot \vec{s}_R. \qquad (1.254)$$

Here, it is taken into account that the fourth component of the spin s_{0R} is zero in the R frame.

Let us determine the length of the three-dimensional spin vector $\vec{s}_L$. The square of the polarization four-vector, S^2, is invariant

$$S^2 = s_{0L}^2 - \vec{s}_L^2 = -\vec{s}_R^2 = \text{Inv.} \qquad (1.255)$$

In view of Eq. (1.254), from this relation we obtain

$$\vec{s}_L^2 = s_{0L}^2 + \vec{s}_R^2 = \vec{s}_R^2\left[1 + \gamma^2\beta^2\cos^2(\theta_R)\right]. \qquad (1.256)$$

Here, θ_R is the angle between the polarization vector $\vec{s}_R$ and particle velocity $\vec{\beta}$. This relation shows that the magnitude of the spin and its direction in the laboratory frame depend on the particle velocity and angle θ_R, whereas these quantities in the R frame are constant. Moreover, the magnitude of the vector $\vec{s}_L$ can be as large as desired and is infinite for a massless particle. Thus, we arrive at two conclusions: first, the spin has no physical meaning in any frame other than the R frame, and, second, the spin for massless particles should be defined in another way (see below).

Let us determine the angle between the polarization vector and velocity in the L frame as a function of the angle θ_R. The dot product of S and four-velocity $V(\gamma, \gamma\vec{v})$ in the L and R frames is written as

$$S \cdot V = s_{0L}\gamma - \gamma\vec{s}_L \cdot \vec{v} = \text{Inv} = 0, \qquad (1.257)$$

taking into account Eqs. (1.254) and (1.256), from this relation we obtain

$$\cos(\theta_L) = \vec{s}_L \cdot \vec{v}/s_L v = s_{0L}/s_L v = \gamma \cos(\theta_R)/\left[1 + \gamma^2\beta^2\cos^2(\theta_R)\right]^{1/2}. \quad (1.258)$$

According to Eq. (1.258),

- if a beam is transversely polarized in the R frame, i.e., $\theta_R = 90°$, it is also transversely polarized in the L frame, i.e., $\theta_L = 90°$;
- if a beam is longitudinally polarized in the R frame, i.e., $\theta_R = 0^0$ (180°), it is also longitudinally polarized in the L frame only in two extreme cases: the nonrelativistic and relativistic ones.

For the case $\gamma \neq 1$ or intermediate angles, it is necessary to plot the dependences on the energy of the polarized beam for insight into the behavior of the polarization in the L frame when it changes in the R frame.

The same relation determines an expression for "helicity" that is the projection of the polarization vector onto the direction of the particle velocity:

$$h = \frac{\vec{s}_L \cdot \vec{\beta}}{\beta} = \gamma\frac{\vec{s}_R \cdot \vec{\beta}}{\beta} = \gamma s_R \cos\theta_R. \qquad (1.259)$$

This expression shows that the polarization in the L frame can exceed unity, which is physically infeasible. This means that the polarization is meaningful only in the R frame.

1.10 Spin of a Massless Particle

As seen from the relation for the polarization four-vector $S(s_{0L}, \vec{s}_L)$, where

$$\vec{s}_L = \vec{s}_R + \vec{\beta}\frac{\gamma^2}{\gamma+1}\vec{\beta} \cdot \vec{s}_R, \qquad s_{0L} = \gamma\vec{\beta} \cdot \vec{s}_R, \qquad (1.260)$$

when the particle velocity coincides with the speed of light c, the Lorentz factor γ becomes infinite and the notion of spin in the L frame becomes meaningless, because spin also becomes infinite. A similar situation arose previously with the four-velocity $V(\gamma, \gamma \vec{v})$. However, the four-momentum $p = mV$ is not divergent, because all particles moving with the speed of light are massless and the product $m\gamma$ coincides with the energy of a massless particle. By analogy, let us take the product of the four-vector S by mass m,

$$W = Sm = \left(m\gamma \vec{\beta} \cdot \vec{s}_R, m\vec{s}_R + m\frac{\gamma^2}{\gamma + 1}\vec{\beta} \cdot \vec{s}_R \vec{\beta} \right). \tag{1.261}$$

W is so-called Pauli–Lubanski operator. Let us denote the dot product $\vec{\beta} \cdot \vec{s}_R / \beta = |s_R| \cos\theta$ as s (helicity) and rewrite

$$W = Sm = \left(m\gamma\beta s, m\vec{s}_R + m\beta s \frac{\gamma^2}{\gamma + 1}\vec{\beta} \right). \tag{1.262}$$

This four-vector in the limit $m = 0$ (and, correspondingly, $\beta \to 1$, $\gamma \to \infty$) with $\varepsilon = m\gamma$, $\vec{p} = m\gamma\vec{\beta} = \varepsilon\vec{l}$ (where $\vec{l} = \vec{\beta}/\beta$) is given by the simple formula

$$W = s(\varepsilon, \vec{l}\varepsilon) = sp. \tag{1.263}$$

Here, p is the four-momentum of the massless particle with the helicity s. Since W is also a four-vector, the following scalars are invariant:

$$W^\mu W_\mu = W^\mu p_\mu = p^\mu p_\mu = 0. \tag{1.264}$$

The spin value s in the following expression for the spin vector is also invariant:

$$\vec{s} = s\vec{l}, \tag{1.265}$$

where the spin direction coincides with the particle velocity direction. In this case, the spin projection onto this direction (which is helicity by definition) can have only two values, $+$ or $-$. One of such particles is the neutrino whose spin is $1/2$ and the spin projection is always negative (left-handed neutrino), whereas the spin projection of the antineutrino is positive (right-handed antineutrino). The helicity of the massless particles is conserved in any interactions.

The above definition of spin is also applicable to photons, although it differs from the standard definition of the photon spin.

1.11 Motion of the Polarization Vector in an External Electromagnetic Field

Many applied problems of the acceleration and transport of polarized particles are solved using the relativistic equations of motion of the polarization vector in external electromagnetic fields. The solution of this problem was first given in Frenkel (1926) before the appearance of the Dirac equation and known article Thomas (1927) concerning relativistic spin precession. Frenkel (1926) has not been cited for a long time, although the reference to it was given in often cited work of Bargmann et al. (1959).

In this section, we derive such an equation from the general principles of relativistic mechanics (Hagedorn 1963).

The first principle is formulated as follows: "Expected values of observables in quantum mechanics are described by the equations of classical mechanics" (Ehrenfest). This means that the motion of the polarization vector, which is defined as the mean value of the spin vector (e.g., Pauli operator), in the electromagnetic field is described by a classical equation. This equation in the rest frame of the particle has the form

$$\frac{d\vec{s}}{dt} = \vec{\mu} \times \vec{H}, \quad \vec{\mu} = g\mu_0\vec{s}. \tag{1.266}$$

Here, $\vec{\mu}$ is the magnetic moment of the particle, g is the gyromagnetic ratio, $\vec{s}$ is the nonrelativistic spin vector of the particle, and μ_0 is the nuclear magneton. The aim is to write this equation of motion in the four-dimensional covariant form. In this case, we use the rule: "If an equation specified in a particular Lorentz frame can be represented in the covariant form and this form can be reduced to the original form in the particular Lorentz frame, this generalization is single".

The notion of spin was generalized in the covariant form in Sect. 1.3. Therefore, the covariant form of the equation should be

$$\frac{dS}{d\tau} = Z, \tag{1.267}$$

where Z should be a covariant four-vector. Equation (1.266) suggests that the right-hand side should include only certain combinations of vectors. Namely, the spin S should appear linearly and homogeneously. The equation is linear in the electromagnetic tensor $F = F^{\mu\nu}$ (in order to ensure the covariant representation of the electromagnetic field). It should contain the parameters of the particle motion in the field F, i.e., four-velocity V and four-acceleration $\dot{V}$. Various four-vectors should be composed of four quantities S, F, V and $\dot{V}$. The combination $\dot{V}F$ is forbidden, because it leads to the quadratic function of F owing to the dependence of $\dot{V}$ on F.

Let us compose the dot product of the spin four-vector and the four-velocity $V(\gamma, \gamma\vec{v})$

$$SV = s_0v_0 - \vec{s} \cdot \vec{v} = 0. \tag{1.268}$$

This relation is valid in any reference frame, because it is valid in the rest frame of the particle. Taking the time derivative of this relation, we obtain

$$\frac{dS}{dt}V = -S\frac{dV}{dt}. \tag{1.269}$$

Taking into account that the four-velocity in the rest frame of the particle is

$$V(\gamma, \gamma\vec{v}) = V(1, 0), \tag{1.270}$$

we transform the left-hand side of relation (1.269) as

$$\left(\frac{dS}{dt}V\right)_R = \frac{ds_0}{dt} = -S\frac{dV}{dt}. \tag{1.271}$$

Thus, in the R frame, in view of the relation $(\vec{s} \times \vec{H}) = (\vec{S} \cdot \vec{F})_R$, we obtain

$$\left(\frac{dS}{dt}\right)_R \equiv \left(\frac{ds_0}{dt}, \frac{d\vec{s}}{dt}\right)_R = \left(-S\frac{dV}{dt}, g\mu_0(SF)_R\right) = Z_R. \tag{1.272}$$

To represent this relation in the covariant form, we perform the following transformations. First, we introduce the proper time τ (in the R frame) and denote the differentiation with respect to it by an overdot:

$$\frac{d}{d\tau} = \gamma\frac{d}{dt}. \tag{1.273}$$

In the R frame, $\gamma = 1$. Second, the term $\vec{\mu} \times \vec{H}$ should be rewritten in the covariant form. To this end, we represent the electromagnetic field tensor in the matrix form

$$F^{\mu\nu} = \begin{Vmatrix} 0 & E_1 & E_2 & E_3 \\ -E_1 & 0 & H_3 & -H_2 \\ -E_2 & -H_3 & 0 & H_1 \\ -E_3 & H_2 & -H_1 & 0 \end{Vmatrix}. \tag{1.274}$$

Here, the Greek superscripts are $\mu, \nu = 0, 1, 2, 3$ and E and H are the electric and magnetic fields, respectively. It is easy to see that only the magnetic field remains in this matrix if the first row and first column are excluded.

Analyzing combinations of four quantities and taking into account the above discussion, we obtain the following useful terms:

$$SF, \qquad V(S\dot{V}), \quad \text{and} \quad V(SFV). \tag{1.275}$$

Therefore, the vector $\dot{S}$ can be represented in the form of a linear combination of these three terms:

$$\dot{S} = aSF + bV(S\dot{V}) + cV(SFV). \tag{1.276}$$

In order to determine the parameters a, b, and c, we pass to the R frame, where the velocity has the components $V(1,0)$. Then, $\dot{S}_R$ can be represented in terms of the components as

$$\dot{S}_R = \left\{a(SF)_R + b(S\dot{V})_R + c(SF)_R, a(SF)_R\right\}. \tag{1.277}$$

Comparing this expression with expression (1.272), we obtain

$$a = g\mu_0 = -c, \qquad b = -1. \tag{1.278}$$

Finally, the generalization of Eq. (1.272) written in the R frame to the arbitrary Lorentz frame is derived in the form

$$\dot{S} = g\mu_0[SF - V(SFV)] - V(S\dot{V}). \tag{1.279}$$

Formula (1.279) is derived under two conditions: first, the magnetic moment of the particle is constant and, second, the particle has no electric moments of any orders and no magnetic moments higher than linear.

If these two conditions are not satisfied, the spin in the R frame is given by another expression. In this case, another formula for the covariant spin can be derived though with large complications. However, the covariant spin vector cannot be constructed for an electrically polarizable particle.

Formula (1.279) can be simplified for the case of the homogeneous field. For this simplification, we write the equation of motion of a charged particle in the homogeneous external field:

$$\dot{V} = -\frac{e}{m}FV. \tag{1.280}$$

The substitution of this expression into Eq. (1.279) yields

$$\dot{S} = g\mu_0 SF + \left(\frac{e}{m} - g\mu_0\right)V(SFV). \tag{1.281}$$

This equation with $e = 0$ is also applicable to neutral particles. Using the relation $g\mu_0 = g(e/2m)$, we obtain

$$\dot{S} = \frac{e}{2m}\big[gSF - (g-2)V(SFV)\big]. \tag{1.282}$$

This equation written in three-dimensional space is the known BMT (Bargmann, Michel, Telegdi) equation (Bargmann et al. 1959), which should be recalled to the FTBMT (Frenkel, Thomas, Bargmann, Michel, Telegdi) equation, because, as mentioned above, this equation was previously derived in Frenkel (1926) and Thomas (1927).

Another approach to the solution of the problem discussed above can be found in Leader (2001).

1.12 Thomas Spin Precession

The first measurements of the magnetic moment of the electron led to the contradiction with the expected theoretical result: the gyromagnetic ratio g (Landé factor) appeared to be 2 rather than an expected value of 1. Soviet physicist Frenkel (1926) and American physicist Thomas (1926, 1927) were among first scientists who explained this contradiction. They proposed and justified a hypothesis that a coefficient of 2 in the g-factor is due to the kinematic effect associated with the relativistic transformations of the polarization vector to various inertial reference frames. Let us consider the production of a particle with the polarization $\vec{P}$ in interaction (or in decay). The particle production plane is the plane containing the momenta of the initial and desired (final) particles. If the interaction is strong (or electromagnetic), the polarization vector is perpendicular to the particle production plane. Let the particle in the laboratory frame has the four-momentum $p(\varepsilon, \vec{p})$ and production angle θ, where $\vec{p} = \varepsilon\beta\vec{l}/c$, $\varepsilon = \gamma mc^2$, $\gamma = \frac{1}{\sqrt{1-\beta^2}}$, and $\vec{l}$ is the unit vector along the direction of the momentum $\vec{p}$. The same parameters in the center-of-mass frame are marked by asterisks. Below, we take into account that the transverse vector components remain unchanged under proper Lorentz transformations. For this reason, we assume that the polarization vector lies in the reaction plane. The angle between the polarization vector in the rest frame of the particle (R frame) and the particle momentum in the laboratory frame (continued to the R frame) is denoted as α, the angle between the polarization vector and the particle momentum in the center-of-mass frame (continued to the R frame) is denoted as α^*, and the difference between these two angles is denoted as ω. These angles satisfy the relation (Galbraith and Williams 1963)

$$\sin\omega = \frac{\beta_c\gamma_c}{\beta\gamma}\sin\theta^*. \tag{1.283}$$

Here, β_c, γ_c are the velocity and Lorentz factor of the center-of-mass frame (C frame), respectively, and β, γ are the velocity and Lorentz factor of the particle in the laboratory frame (L frame).

The physical meaning of the angle ω is as follows. An observer in the R frame sees that the C and L frames move at the angle ω to each other rather than in parallel. Therefore, if a polarization vector is transformed to the R frame, first, directly from the L frame and, second, from the L frame to the C frame and, then, to the R frame, the resulting vector is in two different angular positions. This is a consequence of the noncommutativity of the rotation and translation in the Lorentz transformations. For these two positions of the polarization vector to coincide, it is necessary to rotate one of the vectors $\vec{P}$ by the angle ω about the normal to the plane. This is the sense of the Thomas precession.

The same result was obtained in another way in Stapp (1956). According to this work, in terms of the spin transformation, the particle scattering process can be divided into three stages.

1. It is necessary to transform the spin vector of the initial particle from its rest frame to the center-of-mass frame of the colliding particles. In this frame, scattering occurs and the process is theoretically analyzed.
2. The spin vector is transformed from the center-of-mass frame to the laboratory frame, where the polarization is measured.
3. It is necessary to transform the spin vector to the rest frame of the final particle. These transformations are schematically shown in Fig. 1.2. The calculations provide the following formula determining the kinematic rotation of the spin in the

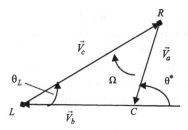

Fig. 1.2 Scheme of the rotation of the polarization vector under relativistic transformation: R, C, and L correspond to three reference frames; the relative velocities of the scattered particle in these reference frames are given on the sides of the triangle. The angles θ_L, θ^* correspond to the scattering angles of the particle in the laboratory and center-of-mass frames, respectively. The meanings of the other symbols are given in the main text

reaction plane:

$$\sin \Omega = |\vec{V}_a \times \vec{V}_b| \cdot \frac{1 + \gamma^a + \gamma^b + \gamma^c}{(1 + \gamma^a)(1 + \gamma^b)(1 + \gamma^c)}. \tag{1.284}$$

Here, $\gamma^a, \gamma^b, \gamma^c$ are the Lorentz factors of the particle in the center-of-mass frame of the colliding particles, in the laboratory frame, and in the new c.m. frame, respectively; and $\vec{V}_a, \vec{V}_b$, and $\vec{V}_c$ are the spatial components of the four-vector of the relative relativistic velocity of the particle in the center-of-mass frame, in the laboratory frame, and in the new c.m. frame, respectively.

The derivation of the above complicated formulas will be considered elsewhere.

One of the important applications of the relativistic transformation of the polarization vector is the creation of a polarized proton (antiproton) beam at the Tevatron (FNAL) due to the parity-violating decays of $\Lambda^0(\bar{\Lambda})$ (Grosnick et al. 1990). Protons from the decays of Λ^0 particles are polarized along their momenta in the rest frame of the Λ^0 particles. Under the transformation from the rest frame of Λ^0 (C frame) to the laboratory frame, the angle of the proton polarization vector is transformed as (Overseth 1969; Dalpiaz and Jansen 1972):

$$\tan \varepsilon = \frac{\sin \theta^*}{\gamma_0(\cos \theta^* + \beta_0/\beta_\Lambda)}, \tag{1.285}$$

where ε is the angle between the proton polarization vector and its momentum in the laboratory frame; $\beta_0, \gamma_0, \theta^*$ are the velocity, Lorentz factor, and the proton emission angle in the rest frame of Λ^0, respectively; and β_Λ is the velocity of Λ^0 in the laboratory frame.

The expression (1.285) was not derived but only postulated. Recently this expression was checked by applying the direct Lorentz transformation in Chetvertkova and Nurushev (2007) and looks like

$$\tan \varepsilon = \frac{\sin \theta^*}{\gamma_\Lambda \gamma_0(\cos \theta^* + \beta_0/\beta_\Lambda)}. \tag{1.286}$$

So comparing expressions (1.285) and (1.286) we see their difference by the relativistic factor γ_Λ, which is very essential in magnitude.

According to expressions (1.285) and (1.286), if the proton emission angle in c.m.s. is zero, we obtain an almost longitudinally polarized proton beam. For instance, in the case if the proton emission angle in c.m.s. is 90° the tan ε derived from (1.285) and (1.286) differs by factor γ_Λ. So the correct formula is (1.286).

References

Bargmann, V., Michel, L., Telegdi, V.A.: Phys. Rev. Lett. **2**, 435 (1959)

Bethe, H.A., Salpeter, E.E.: Quantum Mechanics of One- and Two-Electron Atoms. Academic Press, New York (1957)

Bilen'kii, S.M., Lapidus, L.I., Ryndin, R.M.: Usp. Fiz. Nauk **84**, 241 (1964) [Sov. Phys. Uspekhi **7**, 721 (1965)]

Bjorken, J.D., Drell, S.D.: Relativistic Quantum Mechanics. McGraw-Hill, New York (1964)

Bjorken, J.D., Drell, S.D.: Relativistic Quantum Fields. McGraw-Hill, New York (1965)

Chetvertkova, V.A., Nurushev, S.B.: In: Proceedings of the XII Advanced Research Workshop on High Energy Spin Physics (DSPIN07), Dubna, 3–7 September, p. 37 (2007)

Crampton, S.B., et al.: Phys. Rev. Lett. **11**, 338 (1963)

Dalpiaz, P., Jansen, J.A.: CERN/ECFA/72/4 **1**, 284 (1972)

Dehmet, H.: Science **247**, 539 (1990)

Dirac, P.A.M.: The Principles of Quantum Mechanics, 4th edn. Clarendon Press, Oxford (1958)

Fermi, E.: Notes on Quantum Mechanics. Univ. of Chicago Press, Chicago (1961)

Feynman, R.P., Leighton, R.B., Sands, M.: The Feynman Lectures on Physics, vol. 3. Addison-Wesley, Reading (1963)

Fidecaro, G.: In: Proc. 13th Int. Symp. on High Energy Spin Physics, Protvino, Russia, p. 50 (1998)

Frenkel, J.: Z. Phys. **37**, 243 (1926)

Galbraith, W., Williams, W.S.C. (eds.): High Energy and Nuclear Physics Data Handbook. Rutherford High Energy Laboratory, Chilton (1963)

Gay, T.J.: Rev. Sci. Instrum. **63**, 1635 (1992)

Grosnick, D.P., et al.: Nucl. Instrum. Methods Phys. Res. A **290**, 269 (1990)

Hagedorn, R.: Relativistic Kinematics. W.A. Benjamin, New York (1963), Sect. 8

Kane, G.: Modern Elementary Particles Physics. Addison-Wesley, New York (1987) (Mir, Moscow, 1990)

Landau, L.D., Lifshitz, E.M.: Quantum Mechanics: Non-Relativistic Theory. Nauka, Moscow (1963) (Pergamon Press, Oxford, 1977, 3rd edn.)

Leader, E.: Spin in Particle Physics. Cambridge Univ. Press, London (2001)

Lee, T.D., Wu, C.S.: Annu. Rev. Nucl. Sci. **15**, 381 (1965)

Mott, N.F.: Proc. R. Soc. A **124**, 425 (1929)

Mott, N.F.: Proc. R. Soc. A **135**, 429 (1932)

Okun, L.B.: Leptons and Quarks, 2nd edn. Nauka, Moscow (1982) (North-Holland, Amsterdam, 1984)

Overseth, O.E.: Polarized protons at the 200-GeV accelerator. National Accelerator Laboratory, 1969 Summer Study Report, SS-118 2250, vol. 1, p. 19 (1969)

Schiff, L.I.: Quantum Mechanics, 3rd edn. McGraw-Hill, New York (1968)

Schull, C.G., et al.: Phys. Rev. **63**, 29 (1943)

Shpol'skii, E.V.: Atomic Physics, vol. 1, 7th edn. Nauka, Moscow (1984a) [in Russian]

Shpol'skii, E.V.: Atomic Physics, vol. 2, 7th edn. Nauka, Moscow (1984b) [in Russian]

Stapp, H.P.: Phys. Rev. **103**, 425 (1956)
Thomas, L.H.: Nature **117**, 514 (1926)
Thomas, L.H.: Philos. Mag. **3**, 1 (1927)
Uhlenbeck, G.E., Goudsmit, S.: Nature **113**, 953 (1925)
Uhlenbeck, G.E., Goudsmit, S.: Nature **117**, 264 (1926)

Chapter 2
Spin in Strong Interactions

In this chapter, the basic theoretical relations underlying the design and analysis of
polarization experiments involving strong interactions are reported on classical examples of pion–nucleon and nucleon–nucleon scattering. After the introduction of
the density matrix and reaction matrix, we discuss such notions as the complete set
of experiments and the equality of the polarization $\vec{P}$ in the direct reaction and the
asymmetry $\vec{A}$ in the inverse reaction on the example of the simple pion–nucleon
system. On the example of the nucleon–nucleon system, we present the method for
explicitly constructing the reaction matrix, formulate the unitarity condition, and
point to the possibilities of seeking the effects of parity and time reversal violation.
The presentation is primarily based on the technique of nonrelativistic quantum mechanics. Relativistic pion–nucleon and nucleon–nucleon elastic scattering matrices
are discussed in the concluding sections. The inclusion of the relativistic effects
insignificantly changes the nonrelativistic results.

2.1 Density Matrix

As known, a particle with spin $\vec{s}$ is described by the wave function Ψ having $2s + 1$
components Ψ_α, where $\alpha = s, s - 1, \ldots, 0, \ldots, -(s - 1), -s$. If only one of these
components with a certain α value is nonzero, the particle is in a pure spin state. In
this case, the mean value of any operator $\hat{O}$ is given by the expression

$$\langle \hat{O} \rangle = \langle \Psi_\alpha^* | \hat{O} | \Psi_\alpha \rangle.$$

However, components for several α values are most often nonzero in reality. In
this case, the state is mixed. A simple example is a polarized proton beam. If its
polarization is 100 %, this is a pure spin state: the spins of all protons are oriented
identically. When the beam is partially polarized, the protons are in a mixed spin
state. This means that the spins of some protons are directed upwards and the spins

S.B. Nurushev et al., *Introduction to Polarization Physics*,
Lecture Notes in Physics 859, DOI 10.1007/978-3-642-32163-4_2,
© Moskovski Inzhenerno-Fisitscheski Institute, Moscow, Russia 2013

of other protons are directed downwards. In this case, the mean value of an arbitrary spin operator $\hat{O}$ in the mixed spin state Ψ is given by the expression

$$\langle \hat{O} \rangle = \sum_\alpha W_\alpha \langle \Psi_\alpha^* | \hat{O} | \Psi_\alpha \rangle, \tag{2.1}$$

where W_α is the weight of the pure state α. The Dirac brackets mean integration (summation) with respect to continuous (discrete) variables.

Let us consider the set of other $(2s + 1)$ components $\{\chi_m\}$, which are the orthogonal eigenfunctions of a certain spin operator as the basis functions. If this set is the complete orthogonal set, it can be used for the expansion

$$\Psi_\alpha = \sum_{m=-s}^{s} C_m^\alpha \chi_m. \tag{2.2}$$

The substitution of this expansion into Eq. (2.1) yields

$$\langle \hat{O} \rangle = \sum_{\alpha mn} W_\alpha C_m^{\alpha *} C_n^\alpha \left(\chi_m^* \hat{O} \chi_n \right) = \sum_{mn} \left(\sum_\alpha W_\alpha C_n^\alpha C_m^{\alpha *} \right) \cdot \left(\chi_m^* \hat{O} \chi_n \right)$$

$$= \sum_{mn} \rho_{nm} O_{mn}, \tag{2.3}$$

where O_{mn} is the matrix element of the operator $\hat{O}$ and

$$\rho_{nm} = \sum_\alpha W_\alpha \left(C_m^{\alpha *} C_n^\alpha \right) \tag{2.4}$$

is the matrix element of a certain operator $\hat{\rho}$, which is called the density matrix. The operator $\hat{\rho}$ can be represented in the matrix form

$$\hat{\rho} = \sum_\alpha W_\alpha \Psi_\alpha \Psi_\alpha^+. \tag{2.5}$$

Indeed, let us represent the matrix element of the operator $\hat{\rho}$ in the form

$$\rho_{mn} = \left(\chi_m^+ \hat{\rho} \chi_n \right) = \sum_\alpha W_\alpha \left(\chi_m^+ \Psi_\alpha \Psi_\alpha^+ \chi_n \right). \tag{2.6}$$

Since the eigenfunctions χ are orthogonal, expansion (2.2) provides

$$\chi_m^+ \Psi_\alpha = \sum_n C_n^\alpha \chi_m^+ \chi_n = C_m^\alpha, \qquad \Psi_\alpha^+ \chi_m^+ = \sum_n C_n^{\alpha *} \chi_n^+ \chi_m = C_m^{\alpha *}. \tag{2.7}$$

The substitution into Eq. (2.6) finally yields the expression

$$\rho_{mn} = \sum_\alpha W_\alpha C_m^\alpha C_n^{\alpha *}, \tag{2.8}$$

which coincides with Eq. (2.4).

Using Eqs. (2.4) or (2.5), one can show that the matrix ρ is Hermitian:

$$\hat{\rho}^+ = \rho; \tag{2.9}$$

i.e., the mean value of ρ is a real number. Let us define the sum of the diagonal elements of an arbitrary matrix C (its trace) as

$$Tr\,C = \sum_i C_{ii}. \tag{2.10}$$

If the operator $\hat{C}$ is the product of two operators $\hat{A}$ and $\hat{B}$, its matrix element is given by the expression

$$C_{ij} = \sum_m A_{im} B_{mj}. \tag{2.11}$$

In this case, the sum of the diagonal elements is

$$Tr\,\hat{C} = \sum_{mn} A_{nm} B_{mn} = Tr\,\hat{A}\hat{B}. \tag{2.12}$$

Comparison with Eq. (2.3) provides

$$\langle\hat{O}\rangle = Tr\,\hat{O}\hat{\rho} = Tr\,\hat{\rho}\hat{O}. \tag{2.13}$$

The latter equality shows that two operators can be transposed in the trace even if they do not commute. However, in the general case, only a clockwise or counterclockwise cyclic permutation without the transposition of the operators is possible (if the operators do not commute).

Thus, to determine the mean value of the operator $\hat{O}$, it is necessary to multiply this operator by the density matrix and to calculate the sum of the diagonal elements of the resulting matrix.

According to the theory of matrices, any matrix of rank $m = n = (2s+1)$ can be expanded in a complete set of $(2s+1)^2$ matrices $\{s_\nu\}$ of the same rank satisfying the orthogonality condition

$$Tr\,s_\nu s_\mu = \delta_{\nu\mu}(2s+1). \tag{2.14}$$

Therefore, we can write the expansion

$$\hat{\rho} = \sum_{\mu=1}^{(2s+1)^2} C_\mu s_\mu. \tag{2.15}$$

Multiplying this relation by s_ν from the right and calculating the trace, we obtain

$$Tr\,\hat{\rho}s_\nu = \sum_\mu C_\mu Tr\,s_\mu s_\nu = (2s+1)\sum_\mu C_\mu \delta_{\mu\nu} = (2s+1)C_\nu. \tag{2.16}$$

The substitution of the coefficients C_μ determined from Eq. (2.16) into Eq. (2.15) finally gives

$$\hat{\rho} = \frac{1}{2s+1} \sum_{\mu=1}^{(2s+1)^2} Tr\,(\hat{\rho}\hat{s}_\mu)\hat{s}_\mu. \tag{2.17}$$

Substituting $\hat{O} = \hat{s}_\mu$ (spin operator) into Eq. (2.13), we obtain the polarization vector $\vec{P}$:

$$\frac{1}{2}\vec{P} = \langle \hat{s}_\mu \rangle = Tr\,(\hat{\rho}\hat{s}_\mu). \tag{2.18}$$

Hence, the final expression for the density matrix in terms of observable $\vec{P} = \langle \hat{s} \rangle$ has the form

$$\hat{\rho} = \frac{1}{2s+1}\sum_{\mu=1}^{(2s+1)^2} \langle \hat{s}_\mu \rangle \hat{s}_\mu. \tag{2.19}$$

Thus, the density matrix is completely determined by the mean value of the operators $\hat{s}_\mu$. With the normalization of the mean value of the density matrix to unity, we obtain

$$\langle \hat{O} \rangle = Tr\,\hat{\rho}\hat{O}/Tr\,\hat{\rho}. \tag{2.20}$$

With this normalization condition, the final expression for the density matrix has the form

$$\hat{\rho} = \frac{1}{2s+1}\sum_{\mu=1}^{(2s+1)^2} Tr\,\hat{\rho}\langle \hat{s}_\mu \rangle \hat{s}_\mu. \tag{2.21}$$

The above presentation was based on works Martin and Spearman (1970) and Nurushev (1983).

2.2 Reaction Matrix

When considering reactions involving particles with spin, we use the wave functions Ψ and Φ of the initial and final states, respectively. The transition matrix from the initial to final state, M, is defined as follows (Nurushev 1983):

$$\Phi = M\Psi. \tag{2.22}$$

Then, we have two density matrices:

$$\rho_i = \sum_\alpha W_\alpha \Psi_\alpha \Psi_\alpha^+ \tag{2.23}$$

for the initial state and

$$\rho_f = \sum_\alpha W_\alpha \Phi_\alpha \Phi_\alpha^+ \tag{2.24}$$

for the final state. Here, $\sum_\alpha$ stands for averaging over the initial spin states and summation over the final spin states. In view of relation (2.22), the relation between these two matrices is obtained from Eq. (2.24) in the form

$$\rho_f = \sum_\alpha W_\alpha M \Psi_\alpha \Psi_\alpha^+ M^+ = M\left(\sum_\alpha W_\alpha \Psi_\alpha \Psi_\alpha^+\right)M^+ \tag{2.24a}$$

or

$$\rho_f = M\rho_i M^+. \tag{2.25}$$

Hence, the solution of the problem of the interaction between particles is reduced to the determination of ρ_i in terms of the mean values of the complete set of the spin matrices s_ν (the mean values $\langle s_\nu \rangle$ are called the observables) and the operator $\hat{M}$. Then, the density matrix of the final state is unambiguously determined by Eq. (2.25), and a necessary observable in the final state can be calculated.

Let us find the operator $\hat{\rho}_f$ as a function of s_ν and M. To this end, we multiply Eq. (2.25) by s_μ from the right and calculate the trace of the product of the matrices:

$$Tr\,\hat{\rho}_f s_\mu = Tr\left(M\hat{\rho}_i M^+ s_\mu\right)$$

$$= Tr\left[M\left(\frac{1}{2s+1}\sum_{\nu=1}^{(2s+1)^2} Tr\,\hat{\rho}_i\,\langle s_\nu\rangle_i s_\nu\right)M^+ s_\mu\right]$$

$$= \frac{1}{2s+1}Tr\,\hat{\rho}_i\sum_{\nu=1}^{(2s+1)^2} \langle s_\nu\rangle_i \cdot Tr\left(Ms_\nu M^+ s_\mu\right). \tag{2.26}$$

In view of the relation

$$Tr\left(\hat{\rho}_f s_\mu\right) = \langle s_\mu\rangle_f Tr\,\hat{\rho}_f, \tag{2.27}$$

we obtain

$$\langle s_\mu\rangle_f \cdot Tr\,\hat{\rho}_f = \frac{1}{2s+1}Tr\,\hat{\rho}_i\sum_{\nu=1}^{(2s+1)^2} \langle s_\nu\rangle_i \cdot Tr\left(Ms_\nu M^+ s_\mu\right). \tag{2.28}$$

Denoting the differential cross section as

$$I = \frac{Tr\,\hat{\rho}_f}{Tr\,\hat{\rho}_i}, \tag{2.29}$$

we arrive at the following expression for the mean value $\langle s_\mu\rangle_f$ of the spin operator s_μ final state:

$$\langle s_\mu\rangle_f \cdot I = \frac{1}{2s+1}\cdot\sum_{\nu=1}^{(2s+1)^2} \langle s_\nu\rangle_i \cdot Tr\left(Ms_\nu M^+ s_\mu\right). \tag{2.30}$$

This expression allows one to calculate the mean value of any spin operator s_μ in the final state using the known parameters of the initial state $\langle s_\nu\rangle_i$ and the scattering matrix M.

Let us consider reactions of the type

$$\pi + N = \pi + N, \tag{2.31}$$

or, in the spin notation, $0 + 1/2 \to 0 + 1/2$ (spins of the pion and nucleon are 0 and $1/2$, respectively). In the corresponding two-dimensional spin space, the Pauli matrices $\vec{\sigma}$ $(\sigma_x, \sigma_y, \sigma_z)$ together with the identity matrix 1 can be used as a complete set of spin operators. In this case, the density matrix of the initial state can be written in the form

$$\hat{\rho}_i = C_0 \cdot 1 + \vec{C}_1 \cdot \vec{\sigma}. \tag{2.32}$$

Let us determine the coefficients C_0 and $\vec{C}_1$. From the normalization condition of the trace of the density matrix $\hat{\rho}_i$ to unity, we obtain

$$Tr\,\hat{\rho}_i = C_o \cdot Tr\,1 + \vec{C}_1 \cdot Tr\,\vec{\sigma} = 2C_0 = 1, \tag{2.33}$$

because

$$Tr\,\vec{\sigma} = 0. \tag{2.34}$$

Relation (2.34) is easily verified by writing the Pauli matrices in the explicit form

$$\sigma_x = \begin{pmatrix} 0 & 1 \\ 1 & 0 \end{pmatrix}, \qquad \sigma_y = \begin{pmatrix} 0 & -i \\ i & 0 \end{pmatrix}, \qquad \sigma_z = \begin{pmatrix} 1 & 0 \\ 0 & -1 \end{pmatrix}. \tag{2.35}$$

Let the nucleon in the initial state be polarized and have polarization vector $\vec{P}_t$ (the subscript t means the target). For the initial state,

$$\vec{P}_t = \langle \vec{\sigma}_t \rangle = \frac{Tr\,\hat{\rho}\vec{\sigma}_t}{Tr\,\hat{\rho}} = Tr\,\vec{\sigma}_t \cdot \left(\frac{1}{2} + \vec{C}_1 \cdot \vec{\sigma} \right) = Tr\,\vec{\sigma}_t(\vec{C}_1 \cdot \vec{\sigma})$$

$$= C_{1k} \cdot \vec{e}_k \cdot Tr\,\sigma_k\sigma_i = C_{1k} \cdot \vec{e}_k \cdot 2\delta_{ik} = 2\vec{C}_{1t}. \tag{2.36}$$

Here, $\vec{e}_k$ are the unit coordinate vectors in the Cartesian coordinate system and we use the relation

$$(\vec{\sigma} \cdot \vec{A})(\vec{\sigma} \cdot \vec{B}) = \vec{A} \cdot \vec{B} + i\vec{\sigma} \cdot (\vec{A} \times \vec{B}), \tag{2.37}$$

which is easily proved using relations (2.35). Thus, we obtain

$$C_0 = \frac{1}{2}, \qquad \vec{C}_1 = \frac{1}{2}\vec{P}_t. \tag{2.38}$$

Then, the density matrix of the initial state $\hat{\rho}_i$, is represented in the form

$$\hat{\rho}_i = \frac{1}{2}(1 + \vec{P}_t \cdot \vec{\sigma}). \tag{2.39}$$

Hence, the density matrix $\hat{\rho}$ is completely determined by the target polarization vector $\vec{P}_t$ (or the beam polarization vector $\vec{P}_B$ in the case of the $1/2 + 0 \rightarrow 1/2 + 0$ reaction).

Note that, in view of the properties of the Pauli matrices, expression (2.39) cannot include operators above the first order, because all such operators are reduced to an operator maximally of the first order.

2.3 R, P, and T Transformations

In the next section, we consider nucleon–nucleon elastic scattering and, following Wolfenstein and Ashkin (1952), construct the elastic scattering matrix. Here, as a preparation to this consideration, we discuss the constraints on this matrix that follow from the physical requirements of the isotropy of space (R operation), space inversion (P operation), and time reversal (T operation). Below in this section, we follow Bilenky et al. (1964). In the interaction (or Heisenberg) representation, the S-matrix is defined in terms of the Dirac brackets as

$$|\psi(+\infty)\rangle = S|\psi(-\infty)\rangle. \tag{2.40}$$

Here, $|\psi(-\infty)\rangle$ is the wave function of the system in the initial state at $t \to -\infty$ and $|\psi(+\infty)\rangle$ is the wave function of the system in the final state. According to this definition, the S-matrix transforms the initial state of two free nucleons to the final state with allowance for their interactions. Thus, the S-matrix contains all information on the interaction between the nucleons. If the nucleons do not interact, it is reasonable to set $S = 1$. Then, we formulate the necessary physical requirements.

1. R operation. Let $|\psi(t)\rangle$ be the wave function of the system at time t in an arbitrary reference frame called base. We introduce the second reference frame R rotated by a certain angle and denote the wave function of the system in this reference frame as $|\psi(R,t)\rangle$. The wave functions in two reference frames should be related by a unitary transformation (according to the requirement that the numbers of particles in both reference frames should be the same). The unitarity of the operator implies the equality $U^+(R) = U^{-1}(R)$.

Hence,

$$|\psi(R,t)\rangle = U(R)|\psi(t)\rangle. \tag{2.41}$$

The matrix $U(R)$ is obviously a function of the rotation angles of the R frame with respect to the base frame. The multiplication of Eq. (2.40) by $U(R)$ from the left gives

$$|\psi(R,+\infty)\rangle = U(R)SU^{-1}(R)|\psi(R,-\infty)\rangle. \tag{2.42}$$

The wave functions $|\psi(R,-\infty)\rangle$ and $|\psi(R,+\infty)\rangle$ describe the initial and final states of the nucleons, respectively, in the rotated reference frame R. Hence, in this reference frame, by the definition of the S-matrix, which is determined only by the interaction dynamics but is independent of the choice of the reference frame, the wave functions should be related by the same S-matrix as in Eq. (2.40):

$$|\psi(R,+\infty)\rangle = S|\psi(R,-\infty)\rangle. \tag{2.43}$$

Comparison of Eqs. (2.42) and (2.43) shows that

$$U(R)SU^{-1}(R) = S. \tag{2.44}$$

Since the matrix $U(R)$ is unitary, this equality can be represented in the other form

$$U^{-1}(R)SU(R) = S. \tag{2.45}$$

Relation (2.44) (or (2.45)) expresses the invariance of strong interactions under rotations of the reference frame in physical space. An application of this relation will be considered elsewhere.

2. P operation. It is postulated that strong interactions are invariant under space inversion. This property is also called the parity conservation law. Let us consider the constraints imposed by this postulate on the S-matrix. As in the above consideration, we take the initial reference frame as base. As above, the wave function in this frame is denoted as $|\psi(t)\rangle$. We introduce the reference frame I, in which all coordinate axes are inverted, i.e., $x \rightarrow -x$, $y \rightarrow -y$, and $z \rightarrow -z$. If the base frame is left-handed, the frame I is right-handed. Let $U(I)$ be a unitary operator transforming the function $|\psi(t)\rangle$ in the base frame to the wave function in the I frame:

$$|\psi(I, t)\rangle = U(I)|\psi(t)\rangle. \tag{2.46}$$

In this case, both functions describe the same physical state, but in different coordinates. Hence, the S-matrices in two frames should be the same. Let us determine the matrix $S(I)$. By the definition,

$$|\psi(I, +\infty)\rangle = S(I)|\psi(I, -\infty)\rangle. \tag{2.47}$$

At the same time, from Eq. (2.46) we obtain

$$|\psi(I, +\infty)\rangle = U(I)|\psi(+\infty)\rangle = U(I)S|\psi(-\infty)\rangle = U(I)SU^{-1}|\psi(I, -\infty)\rangle. \tag{2.48}$$

Comparison of Eqs. (2.47) and (2.48) shows that

$$S(I) = U(I)SU^{-1}(I). \tag{2.49}$$

Thus, the postulate of the invariance of strong interactions under space inversion leads to the relation

$$S = U^{-1}(I)SU(I). \tag{2.50}$$

These two cases imply that the invariance under the R and P transformations reduces to the commutativity of the S-matrix and the corresponding transformation U matrices.

3. T operation. The principle of the invariance of the strong interaction under time reversal is formulated as follows. Let us consider the Schrödinger equation in the interaction representation

$$i\frac{\partial|\psi(t)\rangle}{\partial t} = H(t)|\psi(t)\rangle. \tag{2.51}$$

Changing $t \rightarrow -t$ and taking complex conjugation in this equation, we arrive at the equation

$$i\frac{\partial|\psi(-t)\rangle^*}{\partial t} = H^*(-t)|\psi(-t)\rangle^*. \tag{2.52}$$

Since $H^*(-t) \neq H(t)$ in the general case, this new equation is not a Schrödinger equation. However, let us assume that there is a unitary operator $U(T)$ providing the transformation

$$|\psi(T, t)\rangle = U(T)|\psi(-t)\rangle^*. \tag{2.53}$$

The substitution of this relation into Eq. (2.52) yields

$$i\frac{\partial|\psi(T,t)\rangle}{\partial t} = U(T)H^*(-t)U^{-1}(T)|\psi(T,t)\rangle. \tag{2.54}$$

Setting

$$H(t) = U(T)H^*(-t)U^{-1}(T), \tag{2.55}$$

we arrive at the Schrödinger equation

$$i\frac{\partial|\psi(T,t)\rangle}{\partial t} = H(t)|\psi(T,t)\rangle. \tag{2.56}$$

Thus, relation (2.55) presents a mathematical formulation of the physical postulate of the invariance of the interaction under time reversal.

According to Eq. (2.55), for each wave function $|\psi(t)\rangle$ that is a solution of Schrödinger equation (2.51), there is another function $|\psi(T,t)\rangle$ that satisfies Eq. (2.56) and describes the motion of the system in the reverse time direction. Let us obtain the requirement imposed on the S-matrix by the time reversibility condition.

We have two Schrödinger equations with two wave functions, but with the same S-matrix, because the S-matrix is independent of the initial state of the system, but is completely determined by the dynamics of interaction.

According to the above consideration and by analogy with relation (2.40), we write

$$|\psi(T,+\infty)\rangle = S|\psi(T,-\infty)\rangle.$$

Using relation (2.53), we represent this equality in the form

$$U(T)|\psi(-\infty)\rangle^* = SU(T)|\psi(+\infty)\rangle^*.$$

The multiplication of this relation by $U^{-1}(T)$ from the left yields

$$|\psi(-\infty)\rangle^* = U^{-1}(T)SU(T)|\psi(+\infty)\rangle^*. \tag{2.57}$$

Let us take into account that the S-matrix is unitary, i.e., $S^+S = 1$ and that $S^+ = \tilde{S}^*$ by the definition, where the asterisk and tilde mean the complex conjugation and transposition of the matrix, respectively. Multiplying Eq. (2.40) by S^+ from the left, taking complex conjugation, and taking into account the properties of the S-matrix, we obtain

$$|\psi(-\infty)\rangle^* = \tilde{S}|\psi(+\infty)\rangle^*. \tag{2.58}$$

The comparison of Eqs. (2.57) and (2.58) shows that

$$U^{-1}(T)SU(T) = \tilde{S}. \tag{2.59}$$

This is the requirement imposed on the S-matrix by the time-reversal invariance of the interaction.

The results are summarized as follows:

- the invariance of the interaction under rotation of the reference frame, the corresponding R operation is given by relation (2.45): $U^{-1}(R)SU(R) = S$;

- the invariance of the interaction under space inversion, the corresponding P operation is given by relation (2.50): $U^{-1}(I)SU(I) = S$;
- the invariance of the interaction under time reversal, the corresponding T operation is given by expression (2.59): $U^{-1}(T)SU(T) = \tilde{S}$.

In applications, another matrix M determined only by the interaction between the nucleons is used. The initial state of the particles at $t \to -\infty$ is denoted as $|i\rangle$, and the final state at $t \to +\infty$ is denoted as $|f\rangle$. The particles in the initial and final states do not interact and have the relative momenta $\vec{p}$ and $\vec{p}'$, total momenta $\vec{Q}$ and $\vec{Q}'$, and total energies E and E', respectively. The matrix M is expressed in terms of the S matrix as

$$S - 1 = M, \tag{2.60}$$

or in terms of the matrix elements, taking into account the conservation of the energy and momentum:

$$\langle f|S|i\rangle = \langle f|i\rangle - 2\pi i \delta(\vec{Q}' - \vec{Q})\delta(E' - E)\langle f|M|i\rangle. \tag{2.61}$$

Here, the δ functions ensure the conservation of the momentum and energy. The matrix M transforms the initial state $|i\rangle$ to the final state $|f\rangle$ and acts only in the spin space. By the definition of the matrix elements,

$$\langle f|M|i\rangle = (\chi'^{+}M(\vec{p}', \vec{p})\chi). \tag{2.62}$$

Here, χ and χ' are the spin wave functions of the initial and final states of the nucleons, respectively. The matrix M, as well as the S matrix, is determined by the dynamics of interaction and satisfies the requirements of the R, P, and T invariances. Let us consider their in more detail.

From the postulate of the invariance of the interaction under rotation of the reference frame, the following constraint on the S-matrix was obtained (see (2.45)):

$$U^{-1}(R)SU(R) = S.$$

Let us represent this relation in terms of the matrix elements in the base and rotated R frames

$$\left(f\left|U^{-1}(R)SU(R)\right|i\right) = \langle f, R|S|R, i\rangle = \langle f|S|i\rangle. \tag{2.63}$$

Here, $|R, i\rangle$ and $|R, f\rangle$ are the wave functions in the rotated R frame. The total, Q, and relative, p, momenta in the base reference frame are related to the respective momenta Q_R and p_R in the rotated reference frame as

$$(Q_R)_i = a_{il}Q_i, \qquad (p_R)_i = a_{il}p_i. \tag{2.64}$$

Here, a is the rotation matrix from the base frame to the R frame and a_{li} are its matrix elements (in this case, the cosines and sines of the rotation angle from the old to new reference frame).

Relation (2.63) indicates that the respective elements of the S matrix in different frames are the same. According to M-matrix definition (2.61), this is valid for its matrix elements:

$$\left(\chi'^{+}(R)M(\vec{p}'_R, \vec{p}_R)\chi(R)\right) = \left(\chi'^{+}(p)M(\vec{p}', \vec{p})\chi(p)\right), \tag{2.65}$$

where $\chi(R)$ and $\chi'(R)$ are the spin wave functions in the rotated frame and χ and χ', in the base frame. Since these sets of functions describe the same spin state, but in different frames, they should be related by a unitary transformation:

$$\chi(R) = U(R)\chi, \qquad \chi'(R) = U'(R)\chi'. \tag{2.66}$$

The nucleon–nucleon scattering under consideration involves two spin particles in the initial and final states of the reaction. This means that the spin functions in the initial and final states of the reaction are the products of the functions of individual nucleons. As a result, the unitary operators $U(R)$ and $U'(R)$ are the direct products of the matrices acting on the spin functions of the individual particles.

The mean value of the spin operator in quantum mechanics is an observable, namely, the polarization vector (more precisely, the polarization vector is $\vec{P} = \vec{\sigma} = 2\vec{s}$, where $\vec{\sigma}$ is the Pauli operator and $\vec{s}$ is the spin vector). The mean value of the spin operator should be transformed as a vector:

$$\chi^+(R)s_l\chi(R) = a_{li}\chi^+s_i\chi.$$

Here, s_l is the spin operator of one of the initial nucleons. Therefore, the matrix $U(R)$ should satisfy the condition

$$U^{-1}(R)s_lU(R) = a_{li}s_i. \tag{2.67}$$

The same condition is obviously imposed on the matrix $U'(R)$. For a given rotation angle of the frame R, the matrix U can be reconstructed from these conditions.

From Eqs. (2.65) and (2.66), it follows that

$$U'^{-1}(R)M(\vec{p}'_R, \vec{p}_R)U'(R) = M(\vec{p}', \vec{p}). \tag{2.68}$$

This is the mathematical expression of the postulate of the invariance of the interaction under space rotation.

Then, we consider the P operation, i.e., space inversion. Under this operation, the momenta, being polar vectors, change signs, whereas the spin vector, being an axial vector, does not change sign. Therefore,

$$s_l = U^{-1}(I)s_lU(I), \qquad s'_l = U'^{-1}(I)s'_lU'(I), \tag{2.69}$$

where the unitary matrix $U(I)(U'(I))$ ensures the transformation of the wave function of the initial (final) state from the base frame to the inverted frame I. This transformation is written as follows:

$$\chi(I) = U(I)\chi, \qquad \chi'(I) = U'(I)\chi'.$$

The condition on the S-matrix provides

$$\left(\chi'^+M(\vec{p}', \vec{p})\chi\right) = I_iI_f^*\left(\chi'^+(I)M(-\vec{p}', -\vec{p})\chi(I)\right). \tag{2.70}$$

Here, I_i and I_f are the internal parities of two initial and two final nucleons, respectively. Using the relation between the wave functions χ and χ', we obtain

$$M(\vec{p}', \vec{p}) = I_iI_f^*U^{-1}(I)M(-\vec{p}', -\vec{p})U(I). \tag{2.71}$$

2.4 Unitarity Condition

The Schrödinger equation for the wave function Ψ_k has the form

$$\nabla^2 \Psi_k + \frac{2\mu}{\hbar^2}(E - \hat{u})\Psi_k = 0, \tag{2.72}$$

where μ is the reduced mass of colliding particles and $\hat{u}$ is the potential energy of their interaction in the operator form (can include the spin and isospin operators). For the case of elastic scattering, Ψ_k can be represented in the form of the sum of two terms:

$$\Psi_k = e^{i\vec{k}\cdot\vec{r}}\chi + \frac{1}{r}e^{ikr}M(\vec{k}'',\vec{k})\chi. \tag{2.73}$$

The first term is an incident plane wave; the second term is the divergent scattering waves; $\vec{k}$ and $\vec{k}''$ are the wave vectors of the initial (before the interaction) and final states, respectively; χ is the spinor function of the initial state; and M is the reaction matrix. Since the elastic scattering is considered in the center-of-mass frame,

$$|\vec{k}''| = |\vec{k}|, \qquad \hat{u}^+ = \hat{u}. \tag{2.74}$$

The second condition is the Hermitian condition imposed on the interaction potential, because its eigenvalues should be real.

Then, the interaction in the intermediate state of the colliding particles leads to the transition of the wave vector $\vec{k}''$ to the wave vector $\vec{k}'$, determining a given direction (e.g., the direction to a recording detector).

At large distances from the collision center, the following expansion can be used (to define the absolute phases, it is necessary to include a known, for example, electromagnetic or weak interaction in addition to strong interactions):

$$\Psi_k^0 = e^{i\vec{k}\cdot\vec{r}} = e^{ikz\cos\theta} = \sum_{l=0}^{\infty} i^l (2l+1) P_l(\cos\theta)\frac{\sin(kr - \frac{1}{2}l\pi)}{kr}$$

$$= \frac{1}{ikr}\sum_{l=0}^{\infty}\frac{(2l+1)}{2}P_l(\cos\theta)e^{ikr}$$

$$- \frac{1}{ikr}\sum_{l=0}^{\infty}(-1)^l\frac{(2l+1)}{2}P_l(\cos\theta)e^{-ikr}. \tag{2.75}$$

It can be shown that

$$\sum_{l=0}^{\infty}\frac{(2l+1)}{2}P_l(\cos\theta) = \delta(1-\cos\theta),$$

$$\sum_{l=0}^{\infty}(-1)^l\frac{(2l+1)}{2}P_l(\cos\theta) = \delta(1+\cos\theta). \tag{2.76}$$

These formulas are verified by multiplying the both sides by $P_l(\cos\theta)$ and integrating with respect to $\cos\theta$. The validity of relations (2.76) is obvious in view of the orthogonality condition of the Legendre polynomials,

$$\int_{-1}^{1} P_{l'}(\cos\theta) P_l(\cos\theta) d\cos\theta = \frac{2}{2l+1}\delta_{ll'} \tag{2.77}$$

and the definition of the Dirac δ functions

$$\int f(x)\delta(x-a)dx = f(a). \tag{2.78}$$

Taking into account Eq. (2.76), from Eq. (2.75) we obtain

$$e^{i\vec{k}\cdot\vec{r}} = \frac{1}{ikr}e^{ikr}\delta(1-\cos\theta) - \frac{1}{ikr}e^{-ikr}\delta(1-\cos\theta). \tag{2.79}$$

Therefore,

$$\Psi_k(\vec{r}) = \frac{1}{r}e^{ikr} \cdot \chi\left(\frac{1}{ik}\delta(1-\cos\theta) + M(\vec{k}'',\vec{k})\right) - \frac{1}{ikr}e^{-ikr}\delta(1+\cos\theta) \cdot \chi. \tag{2.80}$$

Similarly,

$$\Psi_{k'}(\vec{r}) = \frac{1}{r}e^{-ikr} \cdot \chi^+\left(M(\vec{k}'',\vec{k}') - \frac{1}{ik}\delta(1-\cos\theta)\right) + \frac{1}{ikr}e^{ikr}\delta(1+\cos\theta) \cdot \chi^+. \tag{2.81}$$

From the Schrödinger equation

$$\nabla^2\Psi_k + \frac{2\mu}{\hbar^2}(E-\hat{u})\Psi_k = 0, \qquad \nabla^2\Psi_{k'}^+ + \frac{2\mu}{\hbar^2}\cdot\Psi_{k'}^+\cdot(E-\hat{u}) = 0, \tag{2.82}$$

we can obtain the relation

$$\Psi_{k'}^+\nabla^2\Psi_k - \nabla^2\Psi_{k'}^+\Psi_k = 0, \tag{2.83}$$

which can be modified to the form

$$\nabla\left(\Psi_{k'}^+\nabla\Psi_k - \nabla\Psi_{k'}^+\Psi_k\right) = 0. \tag{2.84}$$

The integration over the volume V provides

$$\int \nabla\left(\Psi_{k'}^+\nabla\Psi_k - \nabla\Psi_{k'}^+\Psi_k\right)dv = \int \nabla\left(\Psi_{k'}^+\nabla\Psi_k - \nabla\Psi_{k'}^+\Psi_k\right)ds = 0. \tag{2.85}$$

Using Eqs. (2.80) and (2.81), as well as the operator $\nabla = \frac{\partial}{\partial r}$, we obtain

$$\int\left\{\frac{1}{kr^2}\delta(1+\cos\theta)\delta(1+\cos\theta') - \frac{1}{r^2}\left[M^+(\vec{k}'',\vec{k}') + \frac{i}{k}\delta(1-\cos\theta')\right]\right.$$
$$\left.\cdot\left[M(\vec{k}'',\vec{k}') - \frac{1}{k}\delta(1-\cos\theta)\right]\right\}ds = 0, \tag{2.86}$$

where $ds = r^2 d\omega_{k''}$. After the integration, we arrive at the formula

$$\frac{1}{2i}[M(\vec{k}',\vec{k}) - M^+(\vec{k},\vec{k}')] = \frac{k}{4\pi}\int M^+(\vec{k}'',\vec{k}')\cdot M(\vec{k}'',\vec{k})d\omega_{k''}. \tag{2.87}$$

This condition can be generalized to the case of inelastic reactions:

$$\frac{1}{2i}\left[\frac{1}{k_a}M_{ab}(\vec{k}',\vec{k}) - \frac{1}{k_b}M_{ba}^+(\vec{k},\vec{k}')\right] = \sum_C \frac{1}{4\pi}\int M_{bc}^+(\vec{k}'',\vec{k}) \cdot M_{ac}(\vec{k}'',\vec{k}')d\omega_{k''},$$

(2.88)

where the summation over all possible reaction channels is implied.

Relation (2.87) provides a number of important consequences. At $\vec{k}' = \vec{k}$ (elastic scattering at zero angle), this relation gives the so-called optical theorem

$$\mathrm{Im}\,a(0) = \frac{k}{4\pi}\sigma_{\mathrm{TOT}},$$

(2.89)

which is the relation between the imaginary part of the forward elastic scattering amplitude $a(0)$ and the total cross section σ_{TOT}.

The application of relation (2.88) to the pion–nucleon scattering matrix gives two relations

$$\mathrm{Im}\,a(\vec{k}'',\vec{k}') = \frac{k}{4\pi}\int \left[a^*(\vec{k}'',\vec{k}') \cdot a(\vec{k}'',\vec{k}')\right.$$
$$\left. + i\vec{b}^*(\vec{k}'',\vec{k}') \times \vec{b}(\vec{k}'',\vec{k}') \cdot (\vec{n}'' \times \vec{n}')\right]d\omega_{k''},$$

(2.90)

$$\mathrm{Re}\,\vec{b}(\vec{k}',\vec{k}) = \frac{k}{4\pi}\int \left[a^*(\vec{k}'',\vec{k}') \cdot \vec{b}(\vec{k}'',\vec{k}') + \vec{b}^*(\vec{k}'',\vec{k}') \cdot a(\vec{k}'',\vec{k}')\right.$$
$$\left. + \mathrm{Re}[\vec{b}^*(\vec{k}'',\vec{k}') \times \vec{b}(\vec{k}'',\vec{k}') \cdot (\vec{n}'' \times \vec{n}') \cdot \vec{n}]\right]d\omega_{k''}.$$

(2.91)

Similar relations are also valid for nucleon–nucleon scattering. These relations are particularly suitable at low energies, when only the elastic channel is opened. As a result, the number of necessary experiments is halved (by two and five experiments for the πN and nucleon–nucleon scatterings, respectively). However, these statements should be considered carefully, because they are valid only under ideal conditions, which are almost inaccessible in actual experiments.

The presentation in this section follows Nurushev (1983).

2.5 Pion–Nucleon Scattering

Let us consider reactions of the type

$$\pi + N = \pi + N,$$

(2.92)

or, in the spin notation, $0 + 1/2 \rightarrow 0 + 1/2$. In the corresponding two-dimensional spin space, the Pauli matrices $\vec{\sigma}\,(\sigma_x, \sigma_y, \sigma_z)$ together with the identity matrix 1 can be used as a complete set of spin operators. In this case, the density matrix of the initial state can be written in the form (see Eq. (2.32) in Sect. 2.2)

$$\hat{\rho}_i = C_0 \cdot 1 + \vec{C}_1 \cdot \vec{\sigma}.$$

(2.93)

Hence, the density matrix $\hat{\rho}$ is completely determined by the target, $\vec{P}_t$, or beam, $\vec{P}_B$, polarization vector.

Then, it is necessary to determine the reaction matrix M. In the general case, it should be, first, a function of two variables, for example, momenta $\vec{k}_i$ and $\vec{k}_f$ before and after reactions, respectively, and, second, a two-row matrix in spin space. As the matrix, it can be expanded in the complete set consisting of the Pauli matrices and identity matrix:

$$M\left(\vec{k}_i, \vec{k}_f\right) = a\left(\vec{k}_i, \vec{k}_f\right) \cdot I + \vec{b}\left(\vec{k}_i, \vec{k}_f\right) \cdot \vec{\sigma}. \tag{2.94}$$

According to the experimental data, we require that strong interactions satisfy the following conditions (Nurushev 1983):

1. Parity conservation law. Since the parities of the initial and final systems are identical, the M-matrix should be a scalar function of the initial energy and scattering angle of the particles. This means that the vector $\vec{b}$, as well as $\vec{\sigma}$, should also be axial. The only axial vector that can be composed of two vectors $\vec{k}_i$ and $\vec{k}_f$ has the form

$$\vec{b}'(\vec{k}_i, \vec{k}_f) = b\vec{n}, \tag{2.95}$$

   where $\vec{n} = \vec{k}_i \times \vec{k}_f / |\vec{k}_i \times \vec{k}_f|$ is the unit vector perpendicular to the reaction plane. The quantity $a(\vec{k}_i, \vec{k}_f)$ should obviously be a scalar function.
2. Time reversibility. Under the time reversal operation, $\vec{k}_i \to -\vec{k}_f$ and $\vec{k}_f \to -\vec{k}_i$, so that $\vec{n}$ changes sign. Under this operation, $\vec{\sigma}$ also changes sign, so that the quantity $\vec{\sigma} \cdot \vec{n}$ is a scalar. For the $(0 + 1/2)$ system, this requirement is satisfied simultaneously with requirement 1. However, for more complex systems (e.g., $(1/2 + 1/2)$), time reversibility gives rise to additional constraints.

Thus, the πN elastic scattering matrix $M(\vec{k}_i, \vec{k}_f)$ has the form

$$M(\vec{k}_i, \vec{k}_f) = a(\vec{k}_i, \vec{k}_f) + b(\vec{k}_i, \vec{k}_f) \cdot \vec{\sigma} \cdot \vec{n}. \tag{2.96}$$

Here, b and a are called the amplitudes with and without spin flop, respectively, are complex quantities; therefore, four real functions (for a given initial energy and a given scattering angle) should be determined from experiments. A set of independent experiments necessary for the unambiguous determination of all reaction amplitudes is called the complete set of experiments. Hence, the complete set of experiments for the πN system should include at least four independent experiments.

However, reality is much more complicated that the above description. First, since the experimentally measured quantities are quadratic combinations of the amplitudes a and b, only the difference of the phases of the amplitudes a and b for strong interactions, rather than their absolute values, can be determined from an experiment. To determine the absolute values of the phases, it is necessary to include a known (for example, electromagnetic or weak) interaction in addition to strong interactions. Thus, the complete set of experiments should include more than four experiments. Second, if elastic scattering is the single allowed reaction channel, the unitarity condition gives rise to two additional relations between the amplitudes a and b; for this reason, to determine these amplitudes, it is sufficient to perform two independent experiments.

However, at energies above the pion production threshold, the number of experiments in the complete set is larger than four.

In reality, there are three reactions induced by charged pions:

$$\pi^+ + p \rightarrow \pi^+ + p(a), \qquad \pi^- + p \rightarrow \pi^- + p(b),$$
$$\pi^- + p \rightarrow \pi^0 + n(c). \tag{2.97}$$

These three reactions are related by the requirement of the isotopic invariance of strong interactions. As a result, the matrices of these reactions are related as

$$M(a) = M_1, \qquad M(b) = \frac{1}{2}(M_1 + M_0), \qquad M(c) = \frac{1}{2}(M_1 - M_0). \tag{2.98}$$

Here, the matrices M_1 and M_0 correspond to the isotopic states of the pion–nucleon system with $T = 3/2$ and $1/2$, respectively. These matrices are reconstructed similarly to the matrix M. Disregarding isotopic invariance, three reactions (2.97) are described by the set of 12 experiments. Allowance for isotopic invariance reduces this number to 8.

For the system of two particles with spin $1/2$ (example is pion + nucleon), we present the proof of the equality $P = A$, where P and A are the polarization of the particle and its asymmetry in the binary reactions, respectively; this relation is very important in applications (Bilenky et al. 1964).

Let us consider the wave function of the system with the inverted direction of the wave vector that satisfies the Schrödinger equation with reversed time ($t \rightarrow -t$). We represent it in the form

$$\Psi_{-k'} = e^{-i\vec{k}'\vec{r}} + \frac{1}{r}e^{ikr}M(\vec{k}'', -\vec{k}')$$
$$= \frac{i}{kr}e^{-ikr} \cdot \delta(1 - \cos\theta')$$
$$+ \frac{1}{r}e^{ikr}\left[M(\vec{k}'', -\vec{k}') - \frac{i}{k} \cdot \delta(1 - \cos\theta')\right]. \tag{2.99}$$

The substitution of this wave function into the following Eq. (2.85) from Sect. 2.4:

$$\int \nabla(\Psi_{k'}^+ \nabla\Psi_k - \nabla\Psi_{k'}^+ \Psi_k)dv = \int \nabla(\Psi_{k'}^+ \nabla\Psi_k - \nabla\Psi_{k'}^+ \Psi_k)ds = 0,$$

gives

$$\int \left[M(\vec{k}'', \vec{k}') \cdot \delta(1 - \cos\theta') + \frac{1}{ik} \cdot \delta(1 - \cos\theta') \cdot \delta(1 - \cos\theta)\right.$$
$$\left. - M(\vec{k}'', -\vec{k}') \cdot \delta(1 + \cos\theta)\right]d\omega_{k''} = 0. \tag{2.100}$$

After the integration, we obtain

$$M(\vec{k}', \vec{k}) = M(-\vec{k}, -\vec{k}'). \tag{2.101}$$

This relation is the condition of the time reversibility of the process.

Scattering matrix (2.96) constructed above satisfies this condition. Using the explicit form of the scattering matrix, we prove the following statement widely used in applications.

Let the polarization of the particle d be measured in the reaction

$$a(0) + b(1/2) \rightarrow c(0) + d(1/2) \tag{2.102}$$

(the spins of the particles are given in the parentheses). This is usually achieved by the rescattering of the particle d on a certain nucleus with a known analyzing power. To explain new terms, we make a brief digression.

Let a beam with a given energy and unit polarization be scattered on a nuclear target at a given angle. The number of the scattered particles per unit flux of the incident beam whose polarization is directed upward from the scattering plane is denoted as N_1. Under the same conditions, this number for the beam polarization downwards to the scattering plane is denoted as N_2. In this notation, the analyzing power of the target is defined by the formula

$$A_N = \frac{N_1 - N_2}{N_1 + N_2}.$$

If the beam is partially polarized, i.e., $P \neq 1$, raw asymmetry (directly measured in experiment), can be defined as

$$\varepsilon = P \cdot A.$$

The quantity ε is also called left–right asymmetry or, sometimes, "raw" asymmetry. According to the definition, the asymmetry ε coincides with the analyzing power A_N at the 100 % polarization of the beam.

We denote the polarization of the particle d as P. Let us consider the reverse reaction

$$c(0) + d_\uparrow(1/2) \rightarrow a(0) + b(1/2), \tag{2.103}$$

where the particle d is polarized. Let the left–right asymmetry A_N be measured in the formation of the particle a (or b). Theorem: the polarization P of the particle d in reaction (2.102) is equal to the asymmetry A of the particle b (or a) in reaction (2.103). This statement is expressed by the equality

$$P = A_N. \tag{2.104}$$

Let us prove this statement. Indeed, by the definition of polarization (Bilenky et al. 1964).

$$\vec{P} = \frac{Tr\,(M\rho_i M^+ \vec{\sigma})}{Tr\,(M\rho_i M^+)} \tag{2.105}$$

since the particle b in initial system (2.102) is unpolarized, $\rho_i = 1/2$ and, calculating P, we obtain

$$P \cdot I_0 = 2\mathrm{Re}\,a^* b = Tr\left(MM^+\vec{\sigma}\right), \tag{2.106}$$

where

$$I_0 = |a|^2 + |b|^2 \tag{2.107}$$

is the differential cross section for reaction (2.102). Since the particle d in reaction (2.103) is polarized, $\rho_i = \frac{1}{2}(1 + \vec{P}_0 \cdot \vec{\sigma})$ and the scattering cross section is given by the expression

$$I_f = Tr\left(M\hat{\rho}_i M^+\right) = \frac{1}{2}Tr\left(MM^+\right) + \frac{1}{2}\vec{P}_0 Tr\left(M\vec{\sigma}M^+\right). \qquad (2.108)$$

The direct calculation shows that $MM^+ = M^+M$ and

$$Tr\left(M\vec{\sigma}M^+\right) = Tr\left(MM^+\sigma\right) = 2I_0 P, \qquad (2.109)$$

where P is determined by expression (2.106). Then,

$$I_f = I_0(1 + \vec{P}_0 \cdot \vec{P}). \qquad (2.110)$$

By the definition of left–right asymmetry

$$A_N = \frac{1}{P_0} \frac{I_f(+) - I_f(-)}{I_f(+) + I_f(-)} = P, \qquad (2.111)$$

quod erat demonstrandum.

When deriving relation (2.111), we explicitly take into account the detailed balance principle, i.e., the equality of the cross sections for the direct and inverse reactions.

Note that the positive sign in relation (2.111) appears for a reaction in which the initial and final states have the same parity. For the case of different parities, the sign in relation (2.111) is negative.

Theorem (2.104) is also proved for the case of the binary reaction when both initial (and final) particles have spin $1/2$. This theorem is invalid for the inclusive reactions.

A test of the relation $P = A$ is simultaneously a test of the T invariance in strong interactions. At present, this problem is of very great interest in view of the creation of an absolute polarimeter for the RHIC collider (see Chap. 8 in the second part of the book, which is devoted to the polarimetry of beams).

It is known that a reaction with the production of hyperons is a convenient reaction for verifying the relation $P = A$. If a target is unpolarized, the polarization of a hyperon is determined from its decay. If the target is polarized, left–right asymmetry can be measured with the same instruments. The same is true for the beam. An example of such interactions is the reaction

$$\pi^- + p \rightarrow K^0 + \Lambda(\uparrow)(a), \qquad \pi^- + p(\uparrow) \rightarrow K^0 + \Lambda(b). \qquad (2.112)$$

This reaction is very convenient because channels (a) and (b) can be measured simultaneously if a polarized target is used. In this case, it is very important to detect both K mesons and Λ hyperons. Averaging the experimental results over the target polarizations (as if the target polarization vanishes), one can determine the polarization of the Λ hyperons (channel (a)). Averaging over the polarization of the Λ hyperons for the polarized target (channel (b)), we determine asymmetry. Comparison of these two observables ensures the direct test of the equality $P = A_N$. Such an experiment has not yet been carried out.

An important application of the relation $P = A_N$ is the measurement of asymmetry in the reaction

$$\pi^- + p(\uparrow) \rightarrow \pi^0 + n. \tag{2.113}$$

For the direct measurement of the neutron polarization on an unpolarized target, it is necessary to scatter the neutron by another target and to detect the scattered neutron. This is a difficult experimental problem owing to loss in the yields for the second scattering process and low neutron detection efficiency. Therefore, the use of the polarized proton target made it possible to measure the neutron polarization in reaction (2.113). These measurements provided first doubts in the Regge pole model very popular in the 1960s.

2.6 Nucleon–Nucleon Scattering

In this section, we present the method for determining the reaction matrix in the nonrelativistic case proposed in Wolfenstein and Ashkin (1952). We will see later that the relativistic approach does not change the results, but leads to the kinematic rotations of the observables lying in the reaction plane. The observables perpendicular to the reaction plane remain unchanged in this case.

2.6.1 Construction of the Reaction Matrix

The system of two nucleons is described by two spin operators $\vec{\sigma}_1$ and $\vec{\sigma}_2$ and two identity operators I_1 and I_2 acting in the spin spaces of the first and second particles, respectively. As a result, the scattering matrix is a four-dimensional matrix depending on the physical vectors $\vec{\sigma}_1$ and $\vec{\sigma}_2$, $\vec{k}_i$ and $\vec{k}_f$ (relative momenta of two nucleons in the initial and final states, respectively). When constructing the NN-scattering matrix, we follow Wolfenstein and Ashkin (1952). Since the initial (two-nucleon) and final (also two-nucleon) systems in our case have the same internal parity, the scattering matrix $M(\vec{\sigma}_1, \vec{\sigma}_2; \vec{k}_i, \vec{k}_f)$ should be a scalar function composed of the combination of spin operators and momenta. Using the spin operators, we can generally compose 16 combinations (complete set):

$$
\begin{array}{lll}
1 & \text{(scalar)} & \\
(\vec{\sigma}_1 \cdot \vec{\sigma}_2 - 1) & \text{(scalar)} & \\
(\vec{\sigma}_1 + \vec{\sigma}_2) & \text{(axial vector)} & \\
(\vec{\sigma}_1 - \vec{\sigma}_2) & \text{(axial vector)} & \text{(A)} \\
(\vec{\sigma}_1 \times \vec{\sigma}_2) & \text{(axial vector)} & \\
l_{\alpha\beta} = (\sigma_{1\alpha}\sigma_{2\beta} + \sigma_{1\beta}\sigma_{2\alpha}) & \text{(symmetric tensor).} &
\end{array}
$$

In view of the properties of the sigma operators, these combinations cannot include terms of the orders higher than the first order.

The combinations that can be composed of the momenta $\vec{k}_i$ and $\vec{k}_f$ are as follows:

$$
\begin{array}{ll}
1 & \text{(scalar)} \\
\vec{k}_f - \vec{k}_i = \vec{K} & \text{(polar vector)} \\
\vec{k}_f \times \vec{k}_i = \vec{n} & \text{(axial vector)} \\
\vec{n} \times \vec{K} = \vec{P} & \text{(polar vector)} \\
K_\alpha K_\beta, n_\alpha n_\beta & \text{(symmetric tensors)} \\
P_\alpha P_\beta, K_\alpha P_\beta + K_\beta P_\alpha & \text{(symmetric tensors)}.
\end{array} \tag{B}
$$

Multiplying the quantities from sets (A) and (B), we take into account the requirement of the invariance of the matrix $M(\vec{\sigma}_1, \vec{\sigma}_2; \vec{k}_i, \vec{k}_f)$ under rotation and inversion of space. Thus, the following combinations can be included in the amplitudes of the scattering matrix:

$$
1, \quad (\vec{\sigma}_1 \cdot \vec{\sigma}_2 - 1), \quad (\vec{\sigma}_1 + \vec{\sigma}_2) \cdot \vec{n}, \quad (\vec{\sigma}_1 - \vec{\sigma}_2) \cdot \vec{n}, \tag{2.114}
$$

$$
(\vec{\sigma}_1 \times \vec{\sigma}_2) \cdot \vec{n}, \tag{2.115}
$$

$$
\sum_{\alpha\beta} l_{\alpha\beta} K_\alpha K_\beta, \quad \sum_{\alpha\beta} l_{\alpha\beta} n_\alpha n_\beta, \quad \sum_{\alpha\beta} l_{\alpha\beta} P_\alpha P_\beta, \quad \sum_{\alpha\beta} l_{\alpha\beta} (K_\alpha P_\alpha + K_\beta P_\alpha). \tag{2.116}
$$

Combinations (2.116) can be represented in the form

$$
\vec{\sigma}_1 \cdot \vec{K} \vec{\sigma}_2 \cdot \vec{K}, \quad \vec{\sigma}_1 \cdot \vec{n} \vec{\sigma}_2 \cdot \vec{n}, \quad \vec{\sigma}_1 \cdot \vec{P} \vec{\sigma}_2 \cdot \vec{P}; \tag{2.117}
$$

$$
\vec{\sigma}_1 \cdot \vec{K} \vec{\sigma}_2 \cdot \vec{P} + \vec{\sigma}_1 \cdot \vec{P} \vec{\sigma}_2 \cdot \vec{K}. \tag{2.118}
$$

Three vectors $\vec{K}, \vec{n},$ and $\vec{P}$ are mutually orthogonal. As a result, the sum of three terms in (2.117) is equal to the dot product $\vec{\sigma}_1 \cdot \vec{\sigma}_2$, as can be verified by direct calculations. Hence, only two of three terms in (2.117) are independent.

Now, we require the time-reversal invariance of these terms. Under time reversal $t \to -t$, the spin operator and momentum are transformed as follows (prime marks the quantities with reversed time):

$$
\vec{\sigma}' = -\vec{\sigma}, \quad \vec{k}_i' = -\vec{k}_f, \quad \vec{k}_f' = -\vec{k}_i. \tag{2.119}
$$

Using Eq. (2.119) and the definition of vectors $\vec{K}, \vec{n},$ and $\vec{P}$ (see (B)), we can show that

$$
\vec{K}' = \vec{K}, \quad \vec{n}' = -\vec{n}, \quad \text{and} \quad \vec{P}' = -\vec{P}. \tag{2.120}
$$

Under this transformation, terms (2.115) and (2.118) change sign and, correspondingly, are rejected. Finally, the nucleon–nucleon elastic scattering matrix is expressed in the form

$$
\begin{aligned}
M(\vec{\sigma}_1, \vec{\sigma}_2; \vec{k}_i, \vec{k}_f) = {} & A + B(\vec{\sigma}_1 \cdot \vec{\sigma}_2 - 1) + C(\vec{\sigma}_1 + \vec{\sigma}_2) \cdot \vec{n} \\
& + D(\vec{\sigma}_1 - \vec{\sigma}_2) \cdot \vec{n} + E(\vec{\sigma}_1 \cdot \vec{K})(\vec{\sigma}_2 \cdot \vec{K}) \\
& + F(\vec{\sigma}_1 \cdot \vec{P})(\vec{\sigma}_2 \cdot \vec{P}).
\end{aligned} \tag{2.121}
$$

Let us introduce the triple of orthogonal unit vectors in the center-of-mass frame:

$$\vec{n} = \frac{\vec{k} \times \vec{k}'}{|\vec{k} \times \vec{k}'|}, \qquad \vec{m} = \frac{\vec{k}' - \vec{k}}{|\vec{k}' - \vec{k}|}, \qquad \vec{l} = \frac{\vec{k} + \vec{k}'}{|\vec{k} + \vec{k}'|}, \tag{2.122}$$

where $\vec{k} = \frac{\vec{k}_i}{|\vec{k}_i|}$ and $\vec{k}' = \frac{\vec{k}_f}{|\vec{k}_f|}$ are the unit vectors.

It is convenient to introduce these unit vectors, because the vectors $\vec{l}$ and $\vec{m}$ in the nonrelativistic approximation coincide with the directions of the momenta of the scattered and recoil particles in the laboratory frame, respectively. The scattering matrix is rewritten in the new notation as

$$M(\vec{\sigma}_1, \vec{\sigma}_2; \vec{k}_i, \vec{k}_f) = a + b(\vec{\sigma}_1 \cdot \vec{n})(\vec{\sigma}_2 \cdot \vec{n}) + c(\vec{\sigma}_1 + \vec{\sigma}_2) \cdot \vec{n}$$
$$+ d(\vec{\sigma}_1 - \vec{\sigma}_2) \cdot \vec{n} + e(\vec{\sigma}_1 \cdot \vec{m})(\vec{\sigma}_2 \cdot \vec{m})$$
$$+ f(\vec{\sigma}_1 \cdot \vec{l})(\vec{\sigma}_2 \cdot \vec{l}). \tag{2.123}$$

The amplitudes a, b, c, d, e, and f are complex functions of the energy and scattering angle $(\vec{k}\vec{k}') = \cos\theta$.

The term with the amplitude d should be absent for nucleon–nucleon scattering. This is proved as follows (Bilenky et al. 1964). Two nucleons in the initial state have the internal parity $(-1)^l$, total spin S, and total isospin T. According to the Pauli exclusion principle, the wave function of two nucleons should be antisymmetric under permutation, i.e., should change sign:

$$P_i = (-1)^l (-1)^{S+1} (-1)^{T+1} = -1. \tag{2.124}$$

A similar relation can also be obtained for the final nucleons with the orbital angular momentum l', spin S', and isospin T':

$$P_f = (-1)^{l'} (-1)^{S'+1} (-1)^{T'+1} = -1. \tag{2.125}$$

To satisfy the condition $P_i = P_f$, we take into account that the parities of the nucleons in the interaction remain unchanged and the terms containing orbital angular momenta can be canceled. We also accept the hypothesis of the isotopic invariance of the strong interaction; for this reason, the terms with isotopic spin T can also be canceled. As a result, we arrive at the condition

$$(-1)^S = (-1)^{S'}. \tag{2.126}$$

Since the possible values of S and S' are 0 (singlet state) and 1 (triplet state), then $S = S'$. This means that transitions only within triplets and singlets separately are allowed in nucleon–nucleon scattering, whereas mixed singlet–triplet and inverse transitions are forbidden. This leads to the exclusion of the term with d in the scattering matrix. The nucleon–nucleon scattering matrix has the final form

$$M(\vec{\sigma}_1, \vec{\sigma}_2; \vec{k}_i, \vec{k}_f) = a + b(\vec{\sigma}_1 \cdot \vec{n})(\vec{\sigma}_2 \cdot \vec{n}) + c(\vec{\sigma}_1 + \vec{\sigma}_2) \cdot \vec{n}$$
$$+ e(\vec{\sigma}_1 \cdot \vec{m})(\vec{\sigma}_2 \cdot \vec{m}) + f(\vec{\sigma}_1 \cdot \vec{l})(\vec{\sigma}_2 \cdot \vec{l}). \tag{2.127}$$

The singlet and triplet projection operators are introduced as

$$\hat{S} = \frac{1}{4}[1 - (\vec{\sigma}_1 \cdot \vec{\sigma}_2)], \qquad \hat{T} = \frac{1}{4}[3 + (\vec{\sigma}_1 \cdot \vec{\sigma}_2)]. \qquad (2.128)$$

Then, expression (2.127) can be represented in the form

$$M(\vec{\sigma}_1, \vec{\sigma}_2; \vec{k}_i, \vec{k}_f) = B\hat{S} + \left[C(\vec{\sigma}_1 + \vec{\sigma}_2) \cdot \vec{n} + \frac{1}{2} G(\vec{\sigma}_1 \cdot \vec{m})(\vec{\sigma}_2 \cdot \vec{m}) \right.$$

$$+ (\vec{\sigma}_1 \cdot \vec{l})(\vec{\sigma}_2 \cdot \vec{l}) + \frac{1}{2} H(\vec{\sigma}_1 \cdot \vec{m})(\vec{\sigma}_2 \cdot \vec{m}) - (\vec{\sigma}_1 \cdot \vec{l})(\vec{\sigma}_2 \cdot \vec{l})$$

$$+ N(\vec{\sigma}_1 \cdot \vec{n})(\vec{\sigma}_2 \cdot \vec{n}) \right] \hat{T}. \qquad (2.129)$$

The amplitude B corresponds to singlet scattering, whereas the remaining four amplitudes describe triplet scattering.

The amplitudes in expressions (2.127) and (2.129) are related as:

$$\begin{array}{ccc} B = a - b - e - f, & C = c, & G = 2a + e + f, \\ H = e - f, & N = a + b. \end{array} \qquad (2.130)$$

For joint description of all possible types of nucleon–nucleon scattering (pp, nn, and np), the general matrix can be written taking into account isotopic invariance:

$$M(\vec{\sigma}_1, \vec{\sigma}_2; \vec{k}_i, \vec{k}_f) = M_0 \hat{T}_0 + M_1 \hat{T}_1. \qquad (2.131)$$

Here,

$$\hat{T}_0 = \frac{1}{4}(1 - \vec{\tau}_1 \cdot \vec{\tau}_2), \qquad \hat{T}_1 = \frac{1}{4}(3 + \vec{\tau}_1 \cdot \vec{\tau}_2) \qquad (2.132)$$

are the isosinglet and isotriplet projection operators, respectively; and $\vec{\tau}_1$ and $\vec{\tau}_2$ are the isospin operators of the first and second nucleons, respectively. Each of the matrices M_0 and M_1 is a scattering matrix of form (2.129).

The final wave function of the system of two nucleons can be written in the form

$$\chi_f = M(\vec{\sigma}_1, \vec{\sigma}_2; \vec{k}, \vec{k}') \chi_{iS} \chi_{iT}, \qquad (2.133)$$

where χ_{iS} and χ_{iT} are the spin and isospin wave functions of the initial system of two nucleons. In view of Eqs. (2.129) and (2.132), the requirement of the antisymmetry of this function provides the following conditions on the amplitude under the change $\theta \to \pi - \theta$:

(a) the isotriplet amplitudes B, C, and H do not change their signs, whereas G and N change their signs;

(b) on the contrary, the isosinglet amplitudes B, C, and H change their signs, whereas G and N do not change their signs.

These relations allow one to investigate pp and nn scatterings only in the angular range $0 \le \theta \le 90°$. Moreover, the amplitude analysis for angles $0°$, $90°$, and $180°$ can be performed only with three rather than five amplitudes; this significantly reduces the number of necessary experimental observables.

In the case of np scattering, where both isotopic matrices M_0 and M_1 are used, the measurements should be performed in a wider angular range, namely, $0 \le \theta \le 180°$.

2.6.2 Some Ways for Experimentally Seeking P- and T-noninvariant Terms in the Matrices of the Strong Interaction

The nucleon–nucleon scattering matrix given by Eq. (2.129) can be written in the form

$$M^{(0)} = (u + v) + (u - v)(\vec{\sigma}_1 \cdot \vec{n})(\vec{\sigma}_2 \cdot \vec{n}) + C\left[(\vec{\sigma}_1 \cdot \vec{n}) + (\vec{\sigma}_2 \cdot \vec{n})\right]$$
$$+ (g - h)(\vec{\sigma}_1 \cdot \vec{m})(\vec{\sigma}_2 \cdot \vec{m}) + (g + h)(\vec{\sigma}_1 \cdot \vec{l})(\vec{\sigma}_2 \cdot \vec{l}), \qquad (2.134)$$

where $\vec{l}, \vec{m}$, and $\vec{n}$ were defined in Eq. (2.122).

The aim of the complete set of experiments on nucleon–nucleon scattering is to reconstruct the amplitudes u, v, c, g, and h from the experimental data. In the case of parity violation or time reversal violation, the scattering matrix contains additional terms, which are considered below.

2.6.2.1 Parity Violation

The forward NN-scattering matrix in the case of parity violation can be written in the form

$$M = M^0 + (i/4)(M_{os} - M_{so})(\vec{\sigma}_1 \times \vec{\sigma}_2) \cdot \vec{k}$$
$$+ (i/4)(M_{os} + M_{so})(\vec{\sigma}_1 - \vec{\sigma}_2) \cdot \vec{k}. \qquad (2.135)$$

Here, M^0 is given by expression (2.134) and the amplitudes M_{os} and M_{so} determine P-violating triplet–singlet and singlet–triplet transitions, respectively. The total cross section for the interaction between polarized particles corresponding to the matrix M is written in the form (Bilenky and Ryndin 1963; Philips 1963):

$$\sigma_{P_1 P_2} = \sigma^{(0)}_{P_1 P_2} + (1/4)(\sigma_{os} - \sigma_{so})(\vec{P}_1 \times \vec{P}_2) \cdot \vec{k}$$
$$+ (1/4)(\sigma_{os} + \sigma_{so})(\vec{P}_1 - \vec{P}_2) \cdot \vec{k}, \qquad (2.136)$$

where $\vec{P}_1$ and $\vec{P}_2$ are the beam and target polarizations, respectively; and $\sigma_{os}(\sigma_{so})$ is the total cross section for the P-odd interaction with the triplet–singlet (singlet–triplet) transition. The cross section for the P-invariant interaction, $\sigma^{(0)}_{P_1 P_2}$, can be written in the form

$$\sigma^{(0)}_{P_1 P_2} = \sigma_0 + \sigma_1(\vec{P}_1 \cdot \vec{P}_2) + \sigma_2(\vec{P}_1 \cdot \vec{k})(\vec{P}_2 \cdot \vec{k})$$
$$= \sigma_0 + \Delta\sigma_T \vec{P}_{1T} \cdot \vec{P}_{2T} + \Delta\sigma_L \vec{P}_{1L} \cdot \vec{P}_{2L}, \qquad (2.137)$$

where the subscripts L and T mean the longitudinal and transverse polarization components of the beam. Formula (2.136) shows that, to detect a P-odd effect, it is necessary to measure the total cross section for the interaction of the longitudinally polarized beam with the unpolarized target or the unpolarized beam with the transversely polarized target (the third term). Such experiments have been performed, and they will be discussed in the next sections of this book.

The second term in the formula (2.136) corresponds to simultaneous parity and time-reversal violation. To measure it, the polarized beam and polarized target, whose polarization vectors are perpendicular to each other and to the beam momentum, should be used. Such experiments have not yet been carried out.

2.6.2.2 T-odd Terms

If the interaction is invariant under space inversion, the term corresponding to the T-odd effect has the form

$$M^{(1)} = M_T(\vec{\sigma}_1 \cdot \vec{l})(\vec{\sigma}_2 \cdot \vec{m}) + (\vec{\sigma}_1 \cdot \vec{m})(\vec{\sigma}_2 \cdot \vec{l}). \tag{2.138}$$

In the case of the matrix $M^{(0)}$, it can be shown that the polarization P of the final nucleon for the case of the unpolarized initial states is equal to left–right asymmetry A_N in the scattering of the polarized nucleon on the unpolarized nucleon: $P = A_N$.

Note that this equality holds for the more general case of the direct

$$a + b \rightarrow c + d \tag{2.139}$$

and inverse

$$c + d \rightarrow a + b \tag{2.140}$$

reactions. In this case, the polarization P refers to the particle c in the direct reaction with the unpolarized particles a and b, whereas the asymmetry A refers to the particle a in the inverse reaction with polarized particles (Baz 1957).

If the interaction contains the term $M^{(1)}$, the equality $P = A_N$ is violated and is replaced by the relation (for NN elastic scattering)

$$\sigma_0(P - A) = -8\,\mathrm{Im}\big(M_T^* \cdot h\big). \tag{2.141}$$

This relation should be tested at the angles at which h is noticeably nonzero. Such a test is simplified if the unambiguous phase or amplitude analysis has been performed.

2.7 Complete Set of Experiments

The idea of the complete set of experiments for the set of observables that completely and unambiguously determine the reaction matrix elements was first proposed in Puzikov et al. (1957) and Smorodinskii (1960). In application to nucleon–nucleon elastic scattering, the possible ways for reconstructing the matrix elements were first proposed in Schumacher and Bethe (1961). We use these works to reconstruct the nucleon–nucleon scattering amplitudes.

The nucleon–nucleon elastic scattering matrix was constructed in Sect. 2.6, where it was also shown that the total number of the independent complex amplitudes necessary for describing the $p + p \to p + p$ reaction at a fixed angle and a fixed energy is five. This means that ten real quantities, namely, five absolute values of the amplitudes and five their phases, should be measured at a fixed angle and a fixed initial energy. Hence, these ten observables constitute a minimum set for the complete experiment. Pion–nucleon scattering is described by two amplitudes, and the complete set should consist of no less than four observables. The complete set for pion–pion scattering consists of only two observables. The number of the components of the complete experiments generally depends on the spin of the interacting particles. The number of observables in the complete set increases with the spin. It is seen that most observables in the complete experiment are associated with spin, i.e., spin carries rich information on interaction.

Let us consider the examples of complete experiments at fixed angles and fixed initial energies.

2.7.1 Complete Experiment on pp Elastic Scattering at 0° in the Center-of-Mass Frame. The Total Cross Sections for Nucleon–Nucleon Interactions

First, the total cross section σ_T for the interaction between two particles with spin 1/2 should be a scalar. Second, it should be a linear function of the polarizations of the initial particles, $\vec{P}_1$ and $\vec{P}_2$. Third, it should be composed of the kinematic quantities determining the reaction (Bilenky and Ryndin 1963; Philips 1963). Thus,

$$\sigma = \sigma_0 + \sigma_1(\vec{P}_1 \cdot \vec{P}_2) + \sigma_2(\vec{P}_1 \cdot \vec{k})(\vec{P}_2 \cdot \vec{k}). \qquad (2.142)$$

Here, $\vec{k}$ is the unit vector in the incident beam direction and σ_0, σ_1, and σ_2 are the experimentally measured parameters depending only on the initial beam energy. Their meaning is as follows. By the definition, the polarization is the mean value of the Pauli operator; hence,

$$
\begin{aligned}
(\vec{P}_1 \cdot \vec{P}_2) &= \langle (\vec{\sigma}_1 \cdot \vec{\sigma}_2) \rangle = (2\vec{S}^2 - 3), \\
(\vec{P}_1 \cdot \vec{k})(\vec{P}_2 \cdot \vec{k}) &= \langle (\vec{\sigma}_1 \cdot \vec{k})(\vec{\sigma}_2 \cdot \vec{k}) \rangle = (2(\vec{S} \cdot \vec{k})^2 - 1).
\end{aligned}
\qquad (2.143)
$$

Here, $\vec{S} = \frac{1}{2}(\vec{\sigma}_1 + \vec{\sigma}_2)$ is the total spin of two initial interacting nucleons. According to Eqs. (2.142) and (2.143),

$$(\vec{P}_1 \cdot \vec{P}_2) = \sum_m w_m^t - 3w^s, \qquad (\vec{P}_1 \cdot \vec{k})(\vec{P}_2 \cdot \vec{k}) = \sum_m (-1)^{1+m} w_m^t - w^s, \quad (2.144)$$

where w^s and w^t are the probabilities of finding the system of two nucleons in the singlet and triplet states, respectively. Taking into account the normalization condition $w^s + \sum_m w_m^t = 1$, it follows from Eq. (2.145) that

$$w^s = \frac{1}{4}\left(1 - (\vec{P}_1 \cdot \vec{P}_2)\right),$$

$$w_0^t = \frac{1}{4}\left(1 + (\vec{P}_1 \cdot \vec{P}_2) - 2(\vec{P}_1 \cdot \vec{k})(\vec{P}_2 \cdot \vec{k})\right), \qquad (2.145)$$

$$w_+^t + w_-^t = \frac{1}{2}\left(1 + (\vec{P}_1 \cdot \vec{k})(\vec{P}_2 \cdot \vec{k})\right).$$

Using expression (2.142) and the definition of the triplet projection operator, we can show that $w_+^t = w_-^t$.

Representing the total cross section in the form of the sum of the weighted cross sections for the singlet and triplet states

$$\sigma = w^s \sigma^s + \sum w_m^t \sigma_m^t \qquad (2.146)$$

and substituting the expressions for w^s and w_m^t, we obtain

$$\sigma = \sigma_0 + \frac{1}{4}(\sigma_0^t - \sigma^s)(\vec{P}_1 \cdot \vec{P}_2) + \frac{1}{2}(\sigma_+^t - \sigma_0^t)(\vec{P}_1 \cdot \vec{k})(\vec{P}_2 \cdot \vec{k}). \qquad (2.147)$$

Comparison of Eqs. (2.142) and (2.147) provides the following relation between the coefficients:

$$\sigma_1 = \frac{1}{4}(\sigma_0^t - \sigma^s) \quad \text{and} \quad \sigma_2 = \frac{1}{2}(\sigma_+^t - \sigma_0^t). \qquad (2.148)$$

Thus, in an experiment with polarized nucleons, three total cross sections can be measured for the cases: (a) both nucleons are unpolarized, (b) both nucleons are polarized transversely to the beam, and (c) both nucleons are polarized along the beam. These three experiments constitute the complete set for determining the total cross sections for nucleon–nucleon interactions. As a result, σ_s, σ_0^t, and σ_+^t can be reconstructed and their individual contributions to the usual (unpolarized) total cross section σ_0 can be determined:

$$\sigma_0 = \frac{1}{4}\sigma^s + \frac{1}{4}\sigma_0^t + \frac{1}{2}\sigma_+^t. \qquad (2.149)$$

The applications of the above relations are discussed in the section "Polarization experiments and results."

2.7.2 Forward NN-Scattering Amplitudes

In view of the symmetry condition, the forward scattering amplitudes satisfy the conditions $c(0) = d(0) = 0$, $b(0) = e(0)$ and the forward scattering matrix has the form

$$M(\vec{\sigma}_1, \vec{\sigma}_2; \vec{k}_i, \vec{k}_f) = a(0) + e(0)(\vec{\sigma}_1 \cdot \vec{\sigma}_2) + [f(0) - e(0)](\vec{\sigma}_1 \cdot \vec{k})(\vec{\sigma}_2 \cdot \vec{k}). \qquad (2.150)$$

The above unitary condition (see matrix relation (2.87) in Sect. 2.4) applied to matrix (2.150) leads to the following relations between the imaginary parts of the amplitudes and total cross sections (optical theorem):

$$\text{Im}\,a(0) = \frac{k}{4\pi}\sigma_T, \qquad \text{Im}\,e(0) = \frac{k}{4\pi}\sigma_1, \qquad \text{Im}\big[f(0) - e(0)\big] = \frac{k}{4\pi}\sigma_2. \quad (2.151)$$

Here, k is the wave number in the center-of-mass frame. Thus, the measurements of three observables σ_0, σ_1, and σ_2 allow one to reconstruct the imaginary parts of three amplitudes a, e, and f of the forward pp elastic scattering.

The determination of three real parts of these amplitudes is necessary to complete the reconstruction of the scattering matrix. This requires the measurements of additional three parameters at very small angles (in the so-called Coulomb–nuclear interference region). One of them is the differential cross section. Such a measurement makes it possible to reconstruct the real part of the spin-independent amplitude $a(0)$. The measurements of two other parameters, for example, σ_1 and σ_2, allow one to reconstruct the real parts of the amplitudes e and f through the dispersion relations. Another way is to measure the spin–spin correlation parameters A_{ik} $(i, k = N, S, L)$ in the Coulomb–nuclear interference region. Here, N, S, and L denote the initial-proton polarizations that are (N) perpendicular to the reaction plane, (S) transverse, and longitudinal (L) with respect to the initial momentum in the reaction plane. Thus, the complete experiment on forward pp elastic scattering is finished. This approach is obviously universal and can be applied at any initial energy.

Unfortunately, such a complete experiment has not yet been performed at any energy (except for the low-energy region, where the phase analysis has been performed).

Relations (2.151) are useful in a number of cases. They impose additional conditions on the phases in the phase analysis, which can be substantial when choosing between several sets of phase solutions. The use of the dispersion relation allows the reconstruction of the real parts $\text{Re}\,a(0)$, $\text{Re}\,e(0)$, and $\text{Re}\,f(0)$, if the imaginary parts of these amplitudes are known from the experimental data (these imaginary parts appear in the integrands in the dispersion relations). The differential cross section for forward proton–proton scattering is of great interest for theoretical analysis. At the same time, a method for its direct measurement is absent and it is necessary to use an extrapolation method; i.e., the differential cross sections for elastic scattering are measured down to extremely small angles for which reliable data are yet obtained and, then, they are extrapolated to zero angle using a certain function, for example, an exponential. To test the resulting cross section, relations (2.151) are used as follows. The differential cross section for forward scattering can be written in the form

$$\frac{d\sigma}{d\Omega} = \left(\frac{k}{4\pi}\right)^2 \big(|a|^2 + 2|e|^2 + |f|^2\big). \quad (2.152)$$

This expression contains the squares of the real and imaginary parts of each of three amplitudes. The substitution of only imaginary parts from relations (2.151) gives the inequality

$$\frac{d\sigma}{d\Omega} \geq \left(\frac{k}{4\pi}\right)^2 \left[(\sigma_T)^2 + 2(\sigma_1)^2 + (\sigma_1 + \sigma_2)^2\right]. \tag{2.153}$$

This is the test relation that provides a lower limit and is widely used to measure the differential cross sections for forward scattering.

2.7.3 Complete Set of Experiments on pp Elastic Scattering at 90° in the Center-of-Mass Frame

One of the first attempts to solve this problem was made as early as in 1959 in Nurushev (1959). We briefly repeat the way proposed in that work.

The following five parameters of pp elastic scattering were measured at an energy of 660 MeV and an angle of 90° (Azhgirey et al. 1963): differential cross section $I = (2.07 \pm 0.03)$ mb/sr, spin correlation parameter $C_{nn} = (0.93 \pm 0.21)$, depolarization parameter $D = (0.93 \pm 0.17)$, transverse polarization rotation parameter $R = (0.26 \pm 0.07)$, and longitudinal polarization rotation parameter $A = (0.20 \pm 0.06)$. Using these five observables, one can reconstruct three absolute values and two relative phases of the nonzero amplitudes B, C and H (see Sect. 2.6 "Nucleon–nucleon scattering"). The phase of the amplitude B is taken to be zero. Thus, the amplitudes (dimensionless) normalized to the cross section are given by the formulas

$$|b|^2 = \frac{|B|^2}{4I} = \frac{1}{2}(1 - C_{nn}), \qquad |c|^2 = \frac{2|C|^2}{I} = \frac{1}{4}(1 + C_{nn} + 2D),$$

$$|h|^2 = \frac{|H|^2}{2I} = \frac{1}{4}(1 + C_{nn} - 2D), \qquad \sin(\delta_C - \delta_B) = -\frac{R + A}{2dc}, \tag{2.154}$$

$$\cos(\delta_H - \delta_B) = \frac{A - R}{2bh}.$$

The substitution of the numerical values of the observables gives (Kumekin et al. 1954)

$$|b|^2 = 0.35 \pm 0.11, \qquad |c|^2 = 1.00 \pm 0.10, \qquad |h|^2 = 0.02 \pm 0.10,$$

$$\sin \delta_c = 0.39 \pm 0.21, \qquad \cos \delta_h = -0.36 \pm 0.18.$$

Comparison with similar data for lower energies shows (Nurushev 1959) that the contributions from the triplet amplitudes c and h prevail in the energy range under consideration, whereas the contribution from the singlet term b is smaller. The terms h (tensor interaction) and c (spin–orbit interaction) dominate in the upper and lower energy ranges, respectively.

2.7.4 Complete Set of Experiments on pp Elastic Scattering at an Arbitrary Angle in the Center-of-Mass Frame

In this section, we try to answer the question: How many and what particular observables should be measured at a given initial energy in order to reconstruct the amplitude of nucleon–nucleon elastic scattering at an arbitrary angle θ in the center-of-mass frame? More briefly, how many measurements constitute the complete set of experiments? Unfortunately, the complete set of experiments has not yet been performed for any energy above 3 GeV. The direct reconstruction of the elements of the nucleon–nucleon scattering matrix from experimental data is the single method of analysis at energies above the meson production threshold. In such an approach, one common phase in the scattering matrix remains undetermined. It can be determined at energies below the meson production threshold by means of the unitarity relation. The possibility of directly reconstructing the scattering amplitudes was discussed in Puzikov et al. (1957), Smorodinskii (1960), and Schumacher and Bethe (1961). In the last work, 11 experimental observables (the differential cross section, polarization, and components of the depolarization, polarization transfer, and polarization correlation tensors) were used and the absolute values of five amplitudes, as well as four relative phases, were unambiguously reconstructed. In agreement with the expectations, one common phase was undetermined. According to Schumacher and Bethe (1961), the complete set includes tensors up to the second order. The possibilities of simplifying the procedure for reconstructing the amplitudes with the use of the polarization tensors of the third and fourth orders were considered in Bilenky et al. (1965) and Vinternitts et al. (1965).

It is not excluded that theoretical ideas can noticeably reduce the number of necessary experiments of the complete set at asymptotic energies.

Below, we discuss the method for reconstructing the scalar amplitudes of nucleon–nucleon scattering in the relativistic case and present the particular sets of the complete set of experiments. In this presentation, we follow Bilenky et al. (1966).

We write the nucleon–nucleon scattering matrix in the form (Wolfenstein and Ashkin 1952; Dalitz 1952)

$$M(\vec{p}', \vec{p}) = (u + v) + (u - v)(\vec{\sigma}_1 \cdot \vec{n})(\vec{\sigma}_2 \cdot \vec{n}) + c(\vec{\sigma}_1 + \vec{\sigma}_2) \cdot \vec{n}$$
$$+ (g - h)(\vec{\sigma}_1 \cdot \vec{m})(\vec{\sigma}_2 \cdot \vec{m}) + (g + h)(\vec{\sigma}_1 \cdot \vec{l})(\vec{\sigma}_2 \cdot \vec{l}). \quad (2.155)$$

Here, the complex scalar scattering amplitudes u, v, c, g, and h are functions of the energy and scattering angle θ. Our main aim is to express these amplitudes in terms of experimental quantities. The unit vectors $\vec{n}, \vec{l}, \vec{m}$ are given by the expressions

$$\vec{l} = \frac{\vec{p}' + \vec{p}}{|\vec{p}' + \vec{p}|}, \quad \vec{m} = \frac{\vec{p}' - \vec{p}}{|\vec{p}' - \vec{p}|}, \quad \vec{n} = \vec{l} \times \vec{m} = \frac{\vec{p} \times \vec{p}'}{|\vec{p} \times \vec{p}'|}. \quad (2.156)$$

These unit vectors defined in the center-of-mass frame are mutually orthogonal. In the nonrelativistic approximation, the vectors $\vec{l}$ and $\vec{m}$ are directed along the scattered and recoil particle momenta in the laboratory frame, respectively. In the relativistic case, this relation is invalid and an additional rotation angle appears.

The proton–proton scattering matrix should satisfy the Pauli exclusion principle:

$$M(\vec{p}', \vec{p}) = -\Pi(1, 2)M(-\vec{p}', \vec{p}) = -M(\vec{p}', -\vec{p})\Pi(1, 2). \qquad (2.157)$$

Here,

$$\Pi(1, 2) = \frac{1}{2}(1 + \vec{\sigma}_1 \cdot \vec{\sigma}_2) \qquad (2.158)$$

is the operator of the permutation of the spin variables. The substitution of pp-scattering matrix (2.155) into relation (2.157) indicates that the scalar scattering amplitudes satisfy the following symmetry conditions:

$$
\begin{aligned}
u(\pi - \theta) = -u(\theta), \qquad & h(\pi - \theta) = h(\theta), \\
c(\pi - \theta) = c(\theta), \qquad & v(\pi - \theta) = -g(\theta).
\end{aligned}
\qquad (2.159)
$$

It follows from these relations that, for example, three amplitudes, c, h and, e.g., g, are nonzero for an angle of 90°. Their reconstruction was considered above.

Neutron–proton elastic scattering was not discussed above. Such a discussion is most appropriate with the isotopic invariance hypothesis. In this case, two isotopic scattering matrices $M_1(\vec{p}', \vec{p})$ and $M_0(\vec{p}', \vec{p})$ with isospins 1 and 0, respectively, can be introduced. Both matrices are written in the form, where the scalar amplitudes have subscripts 1 and 0, and the matrix $M_1(\vec{p}', \vec{p})$ coincides with the pp-scattering matrix. In this case, the np-scattering matrix is determined by the expression

$$M_{np}(\vec{p}', \vec{p}) = \frac{1}{2}\left[M_1(\vec{p}', \vec{p}) + M_0(\vec{p}', \vec{p})\right]. \qquad (2.160)$$

The generalized Pauli exclusion principle can be applied to the isotopic scattering matrices. Introducing the subscript $i = 0, 1$, we write the Pauli condition as follows:

$$M_i(\vec{p}', \vec{p}) = (-1)^i \Pi(1, 2)M_i(-\vec{p}', \vec{p}) = (-1)^i M_i(\vec{p}', -\vec{p})\Pi(1, 2). \qquad (2.161)$$

This condition provides the following constraints on the scalar scattering amplitude in different isotopic states:

$$
\begin{aligned}
u_i(\pi - \theta) = (-1)^i u_i(\theta), \qquad & h_i(\pi - \theta) = (-1)^{i+1} h_i(\theta), \\
c_i(\pi - \theta) = (-1)^{i+1} c_i(\theta), \qquad & v_i(\pi - \theta) = (-1)^i g_i(\theta).
\end{aligned}
\qquad (2.162)
$$

In the nonrelativistic approximation, the problem of the joint analysis of the pp- and np-scattering data for reconstructing the scattering matrix was considered in Kazarinov (1956) and Golovin et al. (1959).

A theoretical analysis of collisions between particles is usually performed in the center-of-mass frame. However, observables such as cross sections and polarizations are measured in the laboratory frame. Therefore, it is necessary to determine the rules for the transition from one frame to the other taking into account the relativistic kinematics and the specificity of the spin transformations. Let us consider particular examples.

According to the general rules, the mean value of any spin operator $\langle \vec{\sigma}_{1L} \rangle$ is expressed as

$$\langle \vec{\sigma}_{1L} \rangle \vec{a}'_L = Tr\left[\vec{\sigma}_1(\vec{a}'_L)_R \rho_f\right]/Tr\, \rho_f. \tag{2.163}$$

Here, ρ_f is the density matrix of the final state and $\vec{a}'_L$ is an arbitrary unit vector in the laboratory frame (L frame). It is intended to measure the projection of the polarization vector of the first particle on this direction. The vector $(\vec{a}'_L)_R = R_n(\Omega')\vec{a}'_L$ includes the relativistic transformation of the spin vector and is obtained from the vector $\vec{a}'_L$ by means of its rotation by the angle $\Omega' = \theta - 2\theta_L$ about a vector perpendicular to the scattering plane. In the nonrelativistic limit, the center-of-mass scattering angle θ is equal to $2\theta_L$, where θ_L is the laboratory scattering angle. Hence, $\Omega' = 0$ in this case.

For the measurement of the projection of the polarization vector of the recoil particle (particle 2) on the direction of the unit vector $\vec{b}''_L$ in the L frame, we have

$$\langle \vec{\sigma}_{2L} \rangle \vec{b}''_L = Tr\left[\vec{\sigma}_2(\vec{b}''_L)_R \rho_f\right]/Tr\, \rho_f. \tag{2.164}$$

It is intended to measure the projection of the polarization vector of the second particle on this direction. The vector $(\vec{b}''_L)_R = R_n(\Omega'')\vec{b}''_L$ includes the relativistic transformation of the spin vector and is obtained from the vector $\vec{b}''_L$ by means of its rotation by the angle $\Omega'' = 2\varphi_L - \varphi$ about a vector perpendicular to the scattering plane, where $\varphi = \pi - \theta$ and φ_L are the emission angles of the second particle in the center-of-mass frame and L frame, respectively.

The correlation of polarization projections on the same unit vectors $(\vec{a}'_L, \vec{b}''_L)$ can be measured in the experiment:

$$\langle (\vec{\sigma}_1 \cdot \vec{a}'_L)(\vec{\sigma}_2 \cdot \vec{b}''_L) \rangle = Tr\left[(\vec{\sigma}_1 \cdot (\vec{a}'_L)_R)(\vec{\sigma}_2 \cdot (\vec{b}''_L)_R)\rho_f\right]/Tr\, \rho_f. \tag{2.165}$$

Let us introduce the following three sets of the orthonormalized unit vectors in the laboratory frame:

$$\vec{n}_L, \quad \vec{k}_L, \quad \vec{s}_L = \vec{n}_L \times \vec{k}_L, \tag{2.166}$$

$$\vec{n}_L, \quad \vec{k}'_L, \quad \vec{s}'_L = \vec{n}_L \times \vec{k}'_L, \tag{2.167}$$

$$\vec{n}_L, \quad \vec{k}''_L, \quad \vec{s}''_L = \vec{n}_L \times \vec{k}''_L. \tag{2.168}$$

Here, $\vec{k}_L, \vec{k}'_L, \vec{k}''_L$ are the unit vectors in the directions of the momenta of the incident, scattered, and recoil nucleons, respectively, and $\vec{n}_L = \vec{k}_L \times \vec{k}'_L/|\vec{k}_L \times \vec{k}'_L|$ is the unit vector perpendicular to the scattering plane; $\vec{n}_L = \vec{n}$, where $\vec{n}$ is the unit vector perpendicular to the scattering plane in the center-of-mass frame. In the subsequent presentation, the initial, scattered, and recoil particles are described in coordinate systems (2.166), (2.167), and (2.168), respectively. When calculating observables, we use tensors up to the second rank.

1. The zero-rank tensor is the differential cross section. It is given by the expression

$$\sigma_0 = Tr\, \rho_f = \frac{1}{4}Tr\left(MM^+\right) = 2\left(|u|^2 + |v|^2 + |c|^2 + |g|^2 + |h|^2\right). \tag{2.169}$$

2. The first-rank tensors:

(a) The initial particles are unpolarized. The i-th component of the polarization of the scattered particle is measured. It is given by the expression

$$P_{1i}\sigma_0 = \frac{1}{4}Tr\left(\sigma_{1i}MM^+\right).$$
(2.170)

(b) The initial particles are unpolarized. The i-th component of the polarization of the recoil particle is measured. It is given by the expression

$$P_{2i}\sigma_0 = \frac{1}{4}Tr\left(\sigma_{2i}MM^+\right) = P_{1i}\sigma_0.$$
(2.171)

(c) The incident (first) particle is polarized in the i-th direction. The target (second) particle is unpolarized. The i-th component of the asymmetry is measured. It is given by the expression

$$A_{1i}\sigma_0 = \frac{1}{4}Tr\left(M\sigma_{1i}M^+\right).$$
(2.172)

(d) The incident (first) particle is unpolarized. The target (second) particle is polarized in the i-th direction. The i-th component of the asymmetry of the recoil particle is measured. It is given by the expression

$$A_{2i}\sigma_0 = \frac{1}{4}Tr\left(M\sigma_{2i}M^+\right) = A_{1i}\sigma_0.$$
(2.173)

3. The second-rank tensors:
   (a) The depolarization tensor D_{ik}. The first particle is polarized (component i) in the initial state, and its polarization (component k) after scattering is measured. The expression for the tensor D_{ik} has the form

$$D_{ik}\sigma_0 = \frac{1}{4}Tr\left(\sigma_{1i}M\sigma_{1k}M^+\right).$$
(2.174)

   (b) The depolarization tensor of the second particle $D_{ik}^{(2)}$. The second particle is polarized (component i, the first particle is unpolarized) in the initial state, and its polarization (component k) after scattering is measured. The expression for the tensor $D_{ik}^{(2)}$ has the form

$$D_{ik}^{(2)}\sigma_0 = \frac{1}{4}Tr\left(\sigma_{2i}M\sigma_{2k}M^+\right) = D_{ik}\sigma_0.$$
(2.175)

   (c) The polarization transfer tensor from the first particle to the second K_{ik}. The first particle is polarized (the second particle is unpolarized) in the initial state, and the polarization of the second particle after scattering is measured. The expression for the tensor K_{ik} has the form

$$K_{ik}\sigma_0 = \frac{1}{4}Tr\left(\sigma_{1i}M\sigma_{2k}M^+\right).$$
(2.176)

   (d) The polarization transfer tensor from the second particle to the first one $K_{ik}^{(2)}$. The second particle is polarized (the first particle is unpolarized) in the initial state, and the polarization of the first particle after scattering is measured. The expression for the tensor $K_{ik}^{(2)}$ has the form

$$K_{ik}^{(2)} \sigma_0 = \frac{1}{4} Tr \left(\sigma_{2i} M \sigma_{1k} M^+ \right) = K_{ik} \sigma_0. \tag{2.177}$$

(e) The polarization correlation tensor C_{ik}. Both particles in the initial state are unpolarized, and the correlation of their polarizations after scattering is measured. The expression for the tensor C_{ik} has the form

$$C_{ik} \sigma_0 = \frac{1}{4} Tr \left(\sigma_{1i} \sigma_{2k} M M^+ \right). \tag{2.178}$$

(f) The polarization correlation tensor $C_{ik}^{(2)}$. Both particles in the initial state are unpolarized, and the correlation of their polarizations after scattering is measured. The difference from (e) is that the Pauli operators are enclosed by M matrices. The expression for the tensor $C_{ik}^{(2)}$ has the form

$$C_{ik}^{(2)} \sigma_0 = \frac{1}{4} Tr \left(\sigma_{1i} \sigma_{2k} M M^+ \right). \tag{2.179}$$

(g) The two-spin asymmetry tensor or asymmetry correlation tensor A_{ik}. Both particles in the initial state are polarized: the first and second particles are polarized in the directions i and k, respectively. Asymmetry after scattering is measured. The expression for the tensor A_{ik} has the form

$$A_{ik} \sigma_0 = \frac{1}{4} Tr \left(M \sigma_{1i} \sigma_{2k} M^+ \right). \tag{2.180}$$

(h) The two-spin asymmetry tensor $A_{ik}^{(2)}$. Both particles in the initial state are polarized: the first and second particles are polarized in the directions k and i, respectively. Asymmetry after scattering is measured. The expression for the tensor $A_{ik}^{(2)}$ has the form

$$A_{ik}^{(2)} \sigma_0 = \frac{1}{4} Tr \left(M \sigma_{1k} \sigma_{2i} M^+ \right) = A_{ik} \sigma_0. \tag{2.181}$$

The requirements of the invariance of strong interactions under a number of continuous (isotropy and uniformity of space) and discrete (space inversion and time reversal) transformations lead to relations between measured quantities. For example,

$$P_i = A_i = P n_i \qquad C_{ik} \left(\vec{p}', \vec{p} \right) = A_{ik} \left(-\vec{p}, -\vec{p}' \right). \tag{2.182}$$

For the case of the beam with polarization $\vec{P}_1$ and the polarized target with polarization $\vec{P}_2$ (or two colliding polarized proton beams with the indicated polarizations), the general requirements of the invariance of the scattering matrix under the transformations listed above in the parentheses provide the following expression for the differential cross section

$$\begin{aligned}
\sigma(\vec{P}_1, \vec{P}_2) = \sigma_0 \{ & 1 + P(\vec{P}_1 + \vec{P}_2) \cdot \vec{n}_L + A_{nn} (\vec{P}_1 \cdot \vec{n}_L)(\vec{P}_2 \cdot \vec{n}_L) \\
& + A_{ss} (\vec{P}_1 \cdot \vec{s}_L)(\vec{P}_2 \cdot \vec{s}_L) + A_{kk} (\vec{P}_1 \cdot \vec{k}_L)(\vec{P}_2 \cdot \vec{k}_L) \\
& + A_{sk} [(\vec{P}_1 \cdot \vec{s}_L)(\vec{P}_2 \cdot \vec{k}_L) + (\vec{P}_1 \cdot \vec{k}_L)(\vec{P}_2 \cdot \vec{s}_L)] \}.
\end{aligned} \tag{2.183}$$

Here, P is the polarization appearing after the scattering of the unpolarized particles and

$$A_{ab} = (\vec{a}_L)_i A_{ik} (\vec{b}_L)_k. \tag{2.184}$$

For the measurements of the depolarization parameters D_{ik} or the polarization transfer tensor K_{ik}, additional scattering is necessary in order to determine the polarization of the scattered particles. Such experiments are very difficult because, first, luminosity in the second scattering process is low and, second, it is difficult to find analyzers with a high analyzing power at high energies. This is one of the main reasons why, for example, such experiments are not planned at the RHIC collider. In contrast to these parameters, the asymmetry correlation tensor A_{ik} is directly measured at RHIC. For the measurements of, for example, the parameter A_{ll}, the polarization of both initial particles should be oriented along their momentum; i.e., both beams should be longitudinally polarized. To measure, for example, A_{sl}, one beam of particles should be polarized along the vector $\vec{s}$, and the other, along the vector $\vec{l}$. All these possibilities can be implemented at the RHIC polarized particle collider.

As mentioned above, the measurements of observables are carried out in the laboratory frame, whereas all theoretical jobs with these quantities are performed in the center-of-mass frame. Let us determine the transformations between these two frames. First, we recall the relation between the triples of unit orthogonal vectors in the laboratory and center-of-mass frames:

$$\vec{n}_L = \vec{n}, \qquad \vec{k}_L \neq \vec{k}, \qquad \vec{s}_L \neq \vec{s}. \tag{2.185}$$

Here, $\vec{k} = \vec{p}/|\vec{p}|$ is the unit vector in the direction of the incident particle momentum in the center-of-mass frame, $\vec{n}$ is the unit vector perpendicular to the scattering plane in this frame, and $\vec{s} = \vec{n} \times \vec{k}$ is the vector that is perpendicular to the momentum of the initial particles and lies in the scattering plane. Thus, we obtain the relations

$$\begin{aligned} A_{ss} &= C_+ - C_{lm} \sin\theta - C_- \cos\theta, \\ A_{kk} &= C_+ + C_{lm} \sin\theta + C_- \cos\theta, \\ A_{sk} &= -C_{lm} \cos\theta + C_- \sin\theta. \end{aligned} \tag{2.186}$$

Here,

$$C_+ = \frac{1}{2}(C_{ll} + C_{mm}), \qquad C_- = \frac{1}{2}(C_{ll} - C_{mm}). \tag{2.187}$$

The components A_{nn} and C_{nn} perpendicular to the scattering plane are the same in both frames. The other components of the tensors satisfy the relations

$$C_+ = \frac{1}{2}(A_{ss} + A_{kk}), \tag{2.188}$$

$$C_{lm} = -A_{sk} \cos\theta + \frac{1}{2}(A_{kk} - A_{ss}) \sin\theta, \tag{2.189}$$

$$C_- = A_{sk} \sin\theta + \frac{1}{2}(A_{kk} - A_{ss}) \cos\theta. \tag{2.190}$$

Here, the polarization correlation tensors C_{ik} are expressed in terms of the asymmetry correlation tensors A_{nn}. This representation is reasonable. Indeed, the tensors C_{ik} can also be measured with the use of unpolarized initial particles. However, this measurement requires the analysis of the final polarization (double scattering) and, therefore, analyzing scattering. As mentioned above, such experiments are very difficult. At the same time, it is easier to measure the parameters A_{nn} (in single scattering) and to determine C_{ik} in terms of these parameters by the above formulas.

In the first experiments at low energies (beam kinetic energy 100–600 MeV), the spin correlation tensor C_{nn} was measured and other parameters were measured later. We consider these tensors:

$$C_{s's''} = \left(\vec{s}'_L\right)_{Ri} C_{ik} \left(\vec{s}''_L\right)_{Rk}, \qquad C_{s'k''} = \left(\vec{s}'_L\right)_{Ri} C_{ik} \left(\vec{k}''_L\right)_{Rk},$$
$$C_{k's''} = \left(\vec{k}'_L\right)_{Ri} C_{ik} \left(\vec{s}''_L\right)_{Rk}, \qquad C_{k'k''} = \left(\vec{k}'_L\right)_{Ri} C_{ik} \left(\vec{k}''_L\right)_{Rk}. \tag{2.191}$$

Let us apply the rotation operator about the vector $\vec{n}$ to an arbitrary vector $\vec{a}$ and represent the result in terms of three vectors:

$$R_n(\Omega)\vec{a} = (\vec{a} \cdot \vec{n})\vec{n}(1 - \cos\theta) + \vec{a}\cos\Omega + (\vec{n} \times \vec{a})\sin\Omega. \tag{2.192}$$

Using this relation and formulas (2.183), we can find that

$$\left(\vec{k}'_L\right)_R = R_n(\Omega')\vec{k}'_L = \vec{l}\cos\alpha + \vec{m}\sin\alpha, \tag{2.193}$$

$$\left(\vec{s}'_L\right)_R = R_n(\Omega')\vec{s}'_L = -\vec{l}\sin\alpha + \vec{m}\cos\alpha. \tag{2.194}$$

Here, the relativistic spin rotation angle is $\alpha = \theta/2 - \theta_L$, where θ and θ_L are the particle scattering angles in the center-of-mass and laboratory frames, respectively.

Similar transformation formulas can be obtained for the recoil particle:

$$\left(\vec{k}''_L\right)_R = R_n(\Omega'')\vec{k}''_L = -\vec{l}\sin\alpha' - \vec{m}\cos\alpha', \tag{2.195}$$

$$\left(\vec{s}''_L\right)_R = R_n(\Omega'')\vec{s}''_L = \vec{l}\cos\alpha' - \vec{m}\sin\alpha'. \tag{2.196}$$

Here, the relativistic spin rotation angle is $\alpha' = \varphi/2 - \varphi_L$, where φ and φ_L are the scattering angles of the recoil particle in the center-of-mass and laboratory frames, respectively.

We emphasize two features. First, relativistic spin precession (Thomas precession) concerns only the polarization components lying in the scattering plane and does not involve its normal component. Second, in the nonrelativistic limit, angles are $\alpha = \alpha' = 0$ and Thomas precession at low energies does not affect observables. In the nonrelativistic case, we have the equalities

$$\left(\vec{k}'_L\right)_R = \vec{l}, \qquad \left(\vec{s}'_L\right)_R = \vec{m}, \qquad \left(\vec{k}''_L\right)_R = -\vec{m}, \qquad \left(\vec{s}''_L\right)_R = \vec{l}. \tag{2.197}$$

Let us express experimentally measured quantities (2.189) in terms of the tensors in the center-of-mass frame using relations (2.191)–(2.194). The desired formulas have the form

$$C_{s's''} = -C_+ \sin(\alpha + \alpha') + C_{lm}\cos(\alpha - \alpha') - C_-\sin(\alpha - \alpha'). \tag{2.198}$$
$$C_{s'k''} = -C_+ \cos(\alpha + \alpha') + C_{lm}\sin(\alpha - \alpha') + C_-\cos(\alpha - \alpha'). \tag{2.199}$$

$$C_{k's''} = C_+ \cos(\alpha + \alpha') + C_{lm} \sin(\alpha - \alpha') + C_- \cos(\alpha - \alpha'). \quad (2.200)$$

$$C_{k'k''} = -C_+ \sin(\alpha + \alpha') - C_{lm} \cos(\alpha - \alpha') + C_- \sin(\alpha - \alpha'). \quad (2.201)$$

It is easy to verify that the observables are related as

$$(C_{s's''} + C_{k'k''})/(C_{s'k''} - C_{k's''}) = \tan(\alpha + \alpha'). \quad (2.202)$$

Hence, the number of the independent observables is three rather than four.

Let us solve the inverse problem, i.e., express three parameters C_+, C_-, C_{lm} in terms of the observables $C_{s's''}$, $C_{s'k''}$, and $C_{k's''}$. Using relations (2.196)–(2.199), we obtain

$$C_+ = (C_{k's''} - C_{s'k''})/2\cos(\alpha + \alpha'), \quad (2.203)$$

$$C_- = \frac{1}{2}(C_{s'k''} + C_{k's''})\cos(\alpha - \alpha')$$
$$- \left[C_{s's''} + \frac{1}{2}\tan(\alpha + \alpha')(C_{k's''} - C_{s'k''}) \right] \sin(\alpha - \alpha'), \quad (2.204)$$

$$C_{lm} = \frac{1}{2}(C_{s'k''} + C_{k's''})\sin(\alpha - \alpha')$$
$$+ \left[C_{s's''} + \frac{1}{2}\tan(\alpha + \alpha')(C_{k's''} - C_{s'k''}) \right] \cos(\alpha - \alpha'). \quad (2.205)$$

Now, we consider the case with the polarized beam and unpolarized target. The polarization components of the scattered particles are measured after collisions. In view of invariance, these components can be written in the form

$$\sigma(\vec{P}_1)\langle\vec{\sigma}_1\rangle_L \cdot \vec{n}_L = \sigma_0\big(P + D_{nn}(\vec{P}_1 \cdot \vec{n}_L)\big), \quad (2.206)$$

$$\sigma(\vec{P}_1)\langle\vec{\sigma}_1\rangle_L \cdot \vec{k}'_L = \sigma_0\big(D_{k'k}(\vec{P}_1 \cdot \vec{k}_L) + D_{k's}(\vec{P}_1 \cdot \vec{s}_L)\big), \quad (2.207)$$

$$\sigma(\vec{P}_1)\langle\vec{\sigma}_1\rangle_L \cdot \vec{s}'_L = \sigma_0\big(D_{s'k}(\vec{P}_1 \cdot \vec{k}_L) + D_{s's}(\vec{P}_1 \cdot \vec{s}_L)\big). \quad (2.208)$$

Here, $\sigma(\vec{P}_1)$ is the differential cross section for the scattering of particles with polarization $\vec{P}_1$ arbitrarily oriented in space on the unpolarized target; it is given by the expression

$$\sigma(\vec{P}_1) = \sigma_0\big(1 + P(\vec{P}_1 \cdot \vec{n})\big). \quad (2.209)$$

Taking certain components of the initial and final polarizations, we arrive at the known Wolfenstein parameters:

$$D_{nn} = D = (\vec{n}_L)_i \, D_{ik}(\vec{n}_L)_k, \qquad D_{s's} = R = (\vec{s}'_L)_{Ri} \, D_{ik}(\vec{s}_L)_k; \quad (2.210)$$

$$D_{s'k} = A = (\vec{s}'_L)_{Ri} \, D_{ik}(\vec{k}_L)_k, \qquad D_{k's} = R' = (\vec{k}'_L)_{Ri} \, D_{ik}(\vec{s}_L)_k; \quad (2.211)$$

$$D_{k'k} = A' = (\vec{k}'_L)_{Ri} \, D_{ik}(\vec{k}_L)_k. \quad (2.212)$$

The spin rotation parameters (A, R, A', R') contain the final unit vectors with the subscript R, which means the necessity of the inclusion of the relativistic spin rotation in the reaction plane.

Referring readers interested in the details of the derivation of the following formulas to Bilenky et al. (1965), we present the expressions of physical observables in terms of the scattering amplitude in the relativistic case:

$$\sigma_0 = 2(|u|^2 + |v|^2 + |c|^2 + |g|^2 + |h|^2), \tag{2.213}$$

$$\sigma_0 D_{nn} = 2(|u|^2 + |v|^2 + |c|^2 - |g|^2 - |h|^2), \tag{2.214}$$

$$\sigma_0 K_{nn} = 2(|u^2| - |v|^2 + |c|^2 + |g|^2 - |h|^2), \tag{2.215}$$

$$\sigma_0 C_{nn} = 2(|u^2| - |v|^2 + |c|^2 - |g|^2 + |h|^2), \tag{2.216}$$

$$\sigma_0 P = 4 \operatorname{Re} cu^*, \tag{2.217}$$

$$\sigma_0 D_+ = 4 \operatorname{Re} uv^*, \tag{2.218}$$

$$\sigma_0 D_- = 4 \operatorname{Re} gh^*, \tag{2.219}$$

$$\sigma_0 D_{lm} = 4 \operatorname{Im} cv^*, \tag{2.220}$$

$$\sigma_0 K_+ = 4 \operatorname{Re} ug^*, \tag{2.221}$$

$$\sigma_0 K_- = 4 \operatorname{Re} vh^*, \tag{2.222}$$

$$\sigma_0 K_{lm} = 4 \operatorname{Im} cg^*, \tag{2.223}$$

$$\sigma_0 C_+ = 4 \operatorname{Re} vg^*, \tag{2.224}$$

$$\sigma_0 C_- = 4 \operatorname{Re} uh^*, \tag{2.225}$$

$$\sigma_0 C_{lm} = -4 \operatorname{Im} ch^*. \tag{2.226}$$

These 14 experimental observables are used to reconstruct five amplitudes and their phases. In this case, one common phase remains undetermined. Fundamentally, it also can be reconstructed. However, this problem is not discussed here.

One of the variants of reconstruction of the amplitudes is as follows (for details, see Bilenky et al. 1966):

$$|g|^2 = \frac{1}{8}\sigma_0(1 + K_{nn} - D_{nn} - C_{nn}); \tag{2.227}$$

$$|h|^2 = \frac{1}{8}\sigma_0(1 - K_{nn} - D_{nn} + C_{nn}); \tag{2.228}$$

$$|v|^2 = \frac{1}{8}\sigma_0(1 - K_{nn} + D_{nn} - C_{nn}); \tag{2.229}$$

$$|u|^2 + |c|^2 = \frac{1}{8}\sigma_0(1 + K_{nn} + D_{nn} + C_{nn}). \tag{2.230}$$

For further analysis, it is necessary to fix the phase of any of five amplitudes. For example, let the amplitude c be real positive. This means that the scattering matrix is determined up to the phase of the amplitude c. Under this condition, we obtain

$$\operatorname{Re} u = \frac{1}{4c}\sigma_0 P, \qquad \operatorname{Im} h = \frac{1}{4c}\sigma_0 C_{lm}, \tag{2.231}$$

$$\operatorname{Im} v = -\frac{1}{4c}\sigma_0 D_{lm}, \qquad \operatorname{Im} g = -\frac{1}{4c}\sigma_0 K_{lm}. \tag{2.232}$$

For further calculations, we write the following identity for any two complex quantities:

$$|x|^2|y|^2 - (\text{Re}\,xy^*)^2 = |x|^2(\text{Im}\,y)^2 + |y|^2(\text{Im}\,x)^2 - 2\,\text{Re}\,xy^*\,\text{Im}\,x\,\text{Im}\,y. \quad (2.233)$$

Taking $x = g$ and $y = h$ and using Eqs. (2.231) and (2.232), we obtain

$$c^2 = \frac{|g|^2 M^2 - |h|^2 N^2 - 2\,\text{Re}\,gh^* M N}{|g|^2|h|^2 - (\text{Re}\,gh^*)^2}. \quad (2.234)$$

Here, the quantities $|g|^2$, $|h|^2$, and $\text{Re}\,gh^*$ were determined above and

$$M = \frac{1}{4}\sigma_0 C_{lm}, \qquad N = -\frac{1}{4}\sigma_0 K_{lm}. \quad (2.235)$$

It remains to determine the signs of the quantities

$$\text{Im}\,u, \qquad \text{Re}\,h, \qquad \text{Re}\,v, \qquad \text{Re}\,g. \quad (2.236)$$

One relation can be written immediately in the form

$$\text{Re}\,gh^* = \text{Re}\,g\,\text{Re}\,h + \text{Im}\,g\,\text{Im}\,h = \frac{1}{4}\sigma_0 D_-. \quad (2.237)$$

Using Eq. (2.218), we can determine the signs of $\text{Im}\,u$ and $\text{Re}\,v$. Any unused equation from the set of Eqs. (2.213)–(2.226) can be used to eliminate the remaining ambiguity.

Thus, the problem of the relativistic reconstruction of the nucleon–nucleon scattering matrix has been solved in the general form. Since the volume of the book is limited, we omit such interesting problems as the reconstruction of the amplitudes by means of the measurements of the polarization parameters of the recoil particles and the joint analysis of pp and np scatterings using isotopic invariance.

2.8 Partial Wave Analysis

The results of polarization experiments on the elastic scattering of nucleons in the low-energy range (0.1–10 GeV) were considered using the partial wave analysis (Hoshizaki 1968; Matsuda 1993). In this method, the scattering amplitude is expanded in terms of the eigenfunctions of the complete set of conserving operators and the expansion coefficients are the elements of the scattering matrix S.

These elements are expressed in terms of the phase shifts, which contain complete information on the interaction process. There are several reasons to apply this method. First, the number of the phases directly depends on the maximum orbital angular momentum $L_{\max}$ in the interaction. According to nonrelativistic quantum mechanics, $L_{\max}$ is related to the impact parameter b as $(L_{\max} + \frac{1}{2})\hbar \approx bp_i$, where p_i is the incident particle momentum in the center-of-mass frame. According to this relation, $L_{\max}$ increases with energy; thus, the phase analysis becomes impossible when the number of free parameters is equal to or larger than the number of experimental points. The situation is further complicated at energies above the meson production threshold, when the phases become complex. This is the main reason why the phase analysis is not applied for high energies.

Nevertheless, the phase analysis was widely used for low energies in 1950–1960 for several reasons. First, the number of the phases and, correspondingly, the number of free parameters are small for low energies. For example, taking $b = 1.5\frac{\hbar}{m_\pi c}$ for the impact parameter, we can estimate $L_{max} = 1$ and 3 at the kinetic energy $T = 50$ and 300 MeV, respectively. Hence, it is easy to perform the phase analysis. The second important reason to perform this analysis is that the phases, being dependent on the impact parameter, make it possible to scan the internal structure of the nucleon. If the particles pass at a large distance from each other, the interaction between them is weak and the phases are small. On the contrary, for the central collision ($L = 0$), the phases are expected to be large. Deviations from this picture are obviously possible, for example, in the presence of repulsion forces or when resonances are formed. The third reason to perform the phase analysis at low energies is that the angular dependence of any observable can be calculated (predicting its behavior) using a few phases and, then, the calculation can be compared with experimental data. Fourth, any theoretical model should be tested on the phase analysis data if the phase analysis is unambiguous.

Let us express the scattering matrix elements in terms of the phase.

In view of the unitarity of the S-matrix, it can be written in terms of the phase operator δ as follows:

$$S = e^{i\delta}. \tag{2.238}$$

The nucleon–nucleon interaction matrix should conserve the total angular momentum J, total spin S, and parity $\Pi = (-1)^l$. The requirement of the antisymmetry under the permutation of two nucleons leads to the relation

$$(-1)^{S+1+T+1}\Pi = (-1)^{S+T+L} = -1. \tag{2.239}$$

Here, T is the isospin of the system of two nucleons. Relation (2.239) should be applied separately for the system of two nucleons in the initial and final states. Taking into account this relation, the elements of the S-matrix can be characterized by three quantum numbers: the total angular momentum J, the total spin of the system of two nucleons S, and the orbital angular momentum L, because T is unambiguously determined from the above relation; i.e., T can be omitted when specifying the elements of the S-matrix.

Let us consider the matrix M defined by the expression

$$M(\vec{p}_f, \vec{p}_i) = \frac{2\pi}{ik}\langle\theta_f\varphi_f|S - 1|\theta_i\varphi_i\rangle. \tag{2.240}$$

Here, $|\theta_i\varphi_i\rangle$ and $\langle\theta_f\varphi_f|$ are the wave functions of the initial and final system of two nucleons, respectively; and θ_n, φ_n, where $n = i$, f, are the angles of the momenta $\vec{p}_i$, $\vec{p}_f$. The elements of this matrix in spin space are given by the expression

$$\langle Sm_s|M(\vec{p}_f, \vec{p}_i)|S'm'_s\rangle = \frac{2\pi}{ik}\langle\theta_f\varphi_f, Sm_s|e^{2i\delta} - 1|S'm'_s, \theta_i\varphi_i\rangle. \tag{2.241}$$

This expression can be rewritten in terms of the spherical functions of the angles using the properties of the completeness and orthogonality of the wave functions

$$\frac{2\pi}{ik} \sum \langle \theta_f \varphi_f | Lm_L \rangle \langle Lm_L Sm_S | LSJm_J \rangle \langle LSJm_J | e^{2i\delta} - 1 | L'S'J'm'_J \rangle$$
$$\times \langle L'S'J'm'_J | L'm'_L S'm'_S \rangle \langle L'm'_L | \theta_i \varphi_i \rangle. \tag{2.242}$$

Here,

$$\langle \theta \varphi | Lm_L \rangle = Y_L^{m_L}(\theta, \varphi). \tag{2.243}$$

Below, we use the following notation for the Clebsch–Gordan coefficients appearing in expression (2.242):

$$C_{LS}(Jm_J m_L m_S) = \langle Lm_L Sm_S | LSJm_J \rangle. \tag{2.244}$$

The quantization axis is usually taken in the direction of the incident particle momentum; in this case, the angles θ_i and φ_i are zero and the angles θ_f and φ_f are the scattering angles of the final particles. After these simplifications, since S, J, and m_J are conserving quantum numbers, expression (2.242) can be represented in the form

$$\langle Sm_s | M(\vec{p}_f, \vec{p}_i) | S'm'_s \rangle$$
$$= \delta_{SS'} \frac{4\pi}{ik} \sum_L \sum_{J=|L-S|}^{L+S} \sum_{L'=|J-S|}^{L+S} \sqrt{\frac{2L'+1}{4\pi}} Y_L^{m'_s-m_s}(\theta, \varphi)$$
$$\times C_{LS}(J, m'_s, m'_s - m_s, m_s) C_{LS}(J, m'_s, 0, m_s) \langle LSJm'_s | e^{2i\delta} - 1 | L'SJm'_s \rangle. \tag{2.245}$$

The summation should be performed taking into account antisymmetry condition (2.239) and that the scattering matrix elements are independent of the projection of the total angular momentum, m_J, due to the isotropy of space. The nonzero elements of the matrix $S - 1$ are denoted as

$$R_L = \langle L0Lm_J | e^{2i\delta} - 1 | L0Lm_J \rangle;$$
$$R_{LJ} = \langle L1Jm_J | e^{2i\delta} - 1 | L1Jm_J \rangle; \tag{2.246}$$
$$-R_\pm^J = R^J = \langle J \pm 1, 1, J, m_J | e^{2i\delta} - 1 | J \mp 1, 1, J, m_J \rangle.$$

Time reversal invariance leads to the equality $R_+^J = R_-^J = R^J$. For further simplification, we introduce the following notation for the triplet and singlet elements:

$$M_{m_s m'_s} = \langle 1m_s | M | 1m'_s \rangle, \qquad M_{ss} = \langle 00 | M | 00 \rangle. \tag{2.247}$$

As a result, the nonzero matrix elements for the singlet and triplet states can be written in the form

$$M_{ss} = \frac{2\pi}{ik} \sum \sqrt{\frac{2L+1}{4\pi}} R_L Y_L^0(\theta, \varphi) \tag{2.248}$$

for the singlet transitions and

$$M_{m_s m_s'} = \frac{2\pi}{ik} \sum_L \left[\sum_{J=L-1}^{L+1} \sqrt{\frac{2L+1}{4\pi}} C_{L1}(J, m_s', m_s' - m_s, m_s) C_{L1}(J, m_s', 0, m_s) R_{LJ} \right.$$

$$\left. - \sum_{J=L\pm1}^{1} \sqrt{\frac{2L'+1}{4\pi}} C_{L1}(J, m_s', m_s' - m_s, m_s) C_{L1}(J, m_s', 0, m_s) R^J \right]$$

$$\times Y_L^{m_s' - m_s}(\theta, \varphi) \tag{2.249}$$

for the triplet transitions. Here, $L' = 2J - L$ in the second term. When applying these formulas to pp elastic scattering, it is necessary to take into account two circumstances. First, two protons are identical; therefore, measuring instruments cannot determine which proton, from the beam or from the target, is detected. As a result, the number of counts is twice as large as that following from the above consideration. Thus, to compare with the theoretical cross section, the measured cross sections should be halved. The second circumstance is associated with the antisymmetry condition (Pauli exclusion principle). Owing to this circumstance, the partial amplitudes with spins $S = 0$ and 1 include only even and odd orbital angular momenta, respectively. Since the orbital angular momentum is not conserved, for a given total angular momentum J in the triplet state, there are two states differing in the orbital angular momentum. As an example, we point to the states $^3P_2$–$^3F_2$, $^3F_4$–$^3H_4$, etc. of the pp system. For each pair of such states, the mixed transitions $^3P_2 \leftrightarrow ^3F_2$, $^3F_4 \leftrightarrow ^3H_4$ are possible in addition to the direct transitions $^3P_2 \rightarrow ^3P_2$, $^3F_2 \rightarrow ^3F_2$. Correspondingly, the phases describing the direct transitions should be supplemented by additional parameters for describing the mixed transitions. Such parameters are called mixing parameters and are usually denoted as ε_J. To describe the mixed transitions in the absence of inelastic channels, the following two-dimensional symmetric unitary submatrix is introduced:

$$S_J - 1 = \begin{vmatrix} R_{J-1,J} & -R^J \\ -R^J & R_{J+1,J} \end{vmatrix}. \tag{2.250}$$

An unambiguous method for parameterizing this matrix in terms of the phase is absent. One of the methods was proposed by Blatt and Weisskopf (1952) and Blatt and Biedenharn (1952). This method is the diagonalization of the matrix by means of a unitary transformation

$$S_J = G S_J' G^{-1}, \tag{2.251}$$

where

$$S_J' = \begin{vmatrix} e^{2i\delta_{J-1,j}} & 0 \\ 0 & e^{2i\delta_{J+1,j}} \end{vmatrix}, \qquad G = \begin{vmatrix} \cos\varepsilon_J & -\sin\varepsilon_J \\ \sin\varepsilon_J & \cos\varepsilon_J \end{vmatrix}. \tag{2.252}$$

This set of phase shifts is called the proper phase shift and is convenient if the Coulomb interaction can be neglected.

Another parameterization variant was proposed in Stapp et al. (1957) in the form

$$S_J = \tilde{S}_J' \tilde{G} \tilde{S}_J', \tag{2.253}$$

where

$$\tilde{S}'_J = \begin{vmatrix} e^{i\bar{\delta}_{J-1,j}} & 0 \\ 0 & e^{i\bar{\delta}_{J+1,j}} \end{vmatrix}, \qquad \tilde{G} = \begin{vmatrix} \cos 2\bar{\varepsilon}_J & i\sin 2\bar{\varepsilon}_J \\ i\sin 2\bar{\varepsilon}_J & \cos 2\bar{\varepsilon}_J \end{vmatrix}. \tag{2.254}$$

This parameterization is favorable over parameterization (2.252), because it provides the estimation of the mixing parameters for low orbital angular momenta (where the nuclear interaction prevails over the Coulomb interaction) in the pure form (without the Coulomb contribution). An additional advantage of this parameterization is that it allows one to more clearly separate the nuclear and Coulomb contributions. For this reason, this parameterization is more often used in the phase analysis.

The relation between representations (2.251) and (2.253) can be found by equating them to each other, because they are the elements of the same matrix. This relation is expressed as follows:

$$\delta_{J-1,J} + \delta_{J+1,J} = \bar{\delta}_{J-1,J} + \bar{\delta}_{J+1,J},$$
$$\sin(\delta_{J-1,J} - \delta_{J+1,J}) = \sin 2\bar{\varepsilon}_J / \sin 2\varepsilon_J, \tag{2.255}$$
$$\sin(\bar{\delta}_{J-1,J} - \bar{\delta}_{J+1,J}) = \tan 2\bar{\varepsilon}_J / \tan 2\varepsilon_J.$$

Table 2.1 presents the elements of the matrix M in terms of the partial waves h. These matrix elements can also be used in the case of neutron–proton elastic scattering with three changes: (a) the Coulomb amplitudes are neglected, (b) all sums over the even or odd L values are extended over all L values (even and odd), and (c) the resulting sums are multiplied by a factor of $1/2$.

The partial nuclear amplitudes h are expressed in terms of the scattering phases by the formulas

$$2ikh_l = \left(e^{2i\bar{\delta}_l^N} - 1\right)e^{2i\Phi_l} \tag{2.256}$$

for the singlet states and

$$2ikh_{lj} = \left(e^{2i\bar{\delta}_{lj}^N} - 1\right)e^{2i\Phi_l} \tag{2.257}$$

for the triplet states.

For the mixed singlet–triplet states,

$$2ikh_{j\pm1,j} = \left(\cos 2\varepsilon_j^N e^{2i\bar{\delta}_{j\pm1,j}^N} - 1\right)e^{2i\Phi_l} \tag{2.258}$$

$$2kh^j = \sin 2\varepsilon_j^N e^{i(\bar{\delta}_{j-1,j}^N + \bar{\delta}_{j+1,j}^N)}, \tag{2.259}$$

where ε_j^N is the mixing parameter for the total angular momentum j and the superscript N means that this parameter refers to pure nuclear scattering.

The Coulomb amplitudes are defined as follows:

$$f_c(\theta) = \frac{-n}{k(1-\cos\theta)} e^{-in\log[(1-\cos\theta)/2]}, \tag{2.260}$$

where $n = \frac{e^2}{\hbar v}$ and v is the relative velocity in the center-of-mass frame. The symmetrized and antisymmetrized Coulomb amplitudes used in the partial wave analysis are presented in Table 2.1.

Table 2.1 Singlet–triplet matrix elements for pp elastic scattering in terms of the partial amplitudes h. The Coulomb interaction contributions are presented in the explicit form with allowance for the identity of protons

1	$M_{ss} = f_{c,s} + 2\sum_{even\ l}(2l+1)h_l P_l$
2	$M_{11} = f_{c,a} + \sum_{odd\ l}[(l+2)h_{l,l+1} + (2l+1)h_{l,l} + (l-1)h_{l,l-1}$ $- \sqrt{(l+1)(l+2)}h^{l+1} - \sqrt{(l-1)l}h^{l-1}]P_l$
3	$M_{00} = f_{c,a} + 2\sum_{odd\ l}[(l+1)h_{l,l+1} + lh_{l,l-1} + (l-1)h_{l,l-1}$ $+ \sqrt{(l+1)(l+2)}h^{l+1} + \sqrt{(l-1)l}h^{l-1}]P_l$
4	$M_{01} = \sqrt{2}\sum_{odd\ l}[-\frac{l+2}{l+1}h_{l,l+1} + \frac{2l+1}{l(l+1)}h_{l,l} + \frac{l-1}{l}h_{l,l-1}$ $+ \sqrt{\frac{l+2}{l+1}}h^{l+1} - \sqrt{\frac{l-1}{l}}h^{l-1}]P_l^1$
5	$M_{10} = \sqrt{2}\sum_{odd\ l}[h_{l,l+1} - h_{l,l-1} + \sqrt{\frac{l+2}{l+1}}h^{l+1} - \sqrt{\frac{l-1}{l}}h^{l-1}]P_l^1$
6	$M_{1-1} = \sum_{odd\ l}[\frac{1}{l+1}h_{l,l+1} - \frac{2l+1}{l(l+1)}h_{l,l} + \frac{1}{l}h_{l,l-1}$ $- \frac{1}{\sqrt{(l+1)(l+2)}}h^{l+1} - \frac{1}{\sqrt{(l-1)l}}h^{l-1}]P_l^2$
7	$M_{11} - M_{00} - M_{1-1} - \sqrt{2}ctg\theta(M_{10}+M_{01}) = 0$
8	$f_{c,s} = f_c(\theta) + f_c(\pi-\theta),\ f_{c,a} = f_c(\theta) - f_c(\pi-\theta)$, where f_c is the Coulomb amplitude

P_l, P_l^1, and P_l^2 are the associated Legendre polynomials of the zeroth, first, and second orders, respectively

Let us consider the separation of the Coulomb and nuclear contributions in the phase analysis. Since Coulomb forces are long-range and nuclear forces are short-range, i.e., they weakly overlap, it is usually accepted that the pure nuclear, $\bar{\delta}^N$, and Coulomb, ϕ, phases are added. In this case,

$$\bar{\delta}_L^N = \bar{\delta}_L - \phi_L, \qquad \bar{\delta}_{JL}^N = \bar{\delta}_{JL} - \phi_L, \qquad \bar{\varepsilon}_J^N = \bar{\varepsilon}_J. \tag{2.261}$$

These phases denoted by the overlined symbols with the superscript N are called pure nuclear phases. The Coulomb phases are calculated by the formula (Stapp et al. 1957):

$$\phi_L \equiv \eta_L - \eta_0 = \sum_{x=1}^{L} \text{atan}\left(\frac{n}{x}\right). \tag{2.262}$$

Here, $n = \frac{e^2}{\hbar v}$ and v is the relative velocity. Let us introduce the Coulomb scattering matrix by the formula $R_c = S_c - 1$. Then, the general reaction matrix is written in the form

$$R = S - 1 = \varepsilon + R_c, \qquad \alpha = S - R_c, \tag{2.263}$$

where ε is the mixing parameter for a given j value.

The matrix α corresponding to pure nuclear scattering can be expanded in partial waves, whereas R_c is calculated exactly and is given by the expression

Table 2.2 Expressions for experimentally measured quantities in terms of the amplitude of the pp elastic scattering matrix in the nonrelativistic case

1	$\sigma_0 = |a|^2 + |b|^2 + 2|c|^2 + |e|^2 + |f|^2$
2	$\sigma_0 D_{nn} = |a|^2 + |b|^2 + 2|c|^2 - |e|^2 - |f|^2$
3	$\sigma_0 D_{ll} = |a|^2 - |b|^2 - |e|^2 + |f|^2$
4	$\sigma_0 D_{mm} = |a|^2 - |b|^2 + |e|^2 - |f|^2$
5	$\sigma_0 D_{ml} = 2\,\mathrm{Im}\,c^*(a - b)$
6	$\sigma_0 P_0 = 2\,\mathrm{Re}\,c^*(a + b)$
7	$\sigma_0 C_{ml} = 2\,\mathrm{Im}\,c^*(e - f)$
8	$\sigma_0 K_{ml} = 2\,\mathrm{Im}\,c^*(e + f)$
9	$\sigma_0 C_{nn}\backslash 2 = \mathrm{Re}\,ab^* + |c|^2 - \mathrm{Re}\,ef^*$
10	$\sigma_0 K_{nn}\backslash 2 = \mathrm{Re}\,ab^* + |c|^2 + \mathrm{Re}\,ef^*$
11	$\sigma_0 C_{ll}\backslash 2 = \mathrm{Re}\,af^* - \mathrm{Re}\,be^*$
12	$\sigma_0 K_{ll}\backslash 2 = \mathrm{Re}\,af^* + \mathrm{Re}\,be^*$
13	$\sigma_0 C_{mm}\backslash 2 = \mathrm{Re}\,ae^* - \mathrm{Re}\,bf^*$
14	$\sigma_0 K_{mm}\backslash 2 = \mathrm{Re}\,ae^* + \mathrm{Re}\,bf^*$

$$\langle f|R_c|i\rangle = \frac{ik}{2\pi} f_c(\theta), \quad f_c(\theta) = -\frac{n}{k(1 - \cos\theta)}\exp\left[-in\log\left(\frac{1 - \cos\theta}{2}\right)\right].$$

$$(2.264)$$

The partial amplitudes h are calculated using formulas (2.246), (2.253), and (2.261), as well as the relation $\alpha = 2ikh$. These expressions have the form

$$h_L = \frac{1}{2ik}\left[\exp\left(2i\delta_L^N\right) - 1\right]\exp(2i\phi_L) \qquad (2.265)$$

for the singlet state and

$$h_{LJ} = \frac{1}{2ik}\left[\exp\left(2i\bar{\delta}_{LJ}^N\right) - 1\right]\exp(2i\phi_L),$$

$$h_{J\pm1,J} = \frac{1}{2ik}\left[\cos 2\varepsilon_J^N \exp\left(2i\bar{\delta}_{J\pm1,J}^N\right) - 1\right]\exp(2i\phi_{J\pm1}), \qquad (2.266)$$

$$h^J = \frac{1}{2ik}\left[\sin 2\varepsilon_J^N \exp\left(i\bar{\delta}_{J-1,J}^N + i\bar{\delta}_{J+1,J}^N\right)\right]$$

for the triplet state.

These relations make it possible to express the elements of the matrix M in terms of the phase shifts; hence, experimental observables are determined in terms of the phases. Thus, the phase analysis is possible.

Table 2.2 presents the expressions for measured quantities in terms of the amplitudes a, b, c, e, and f expressed through linear relations with the matrix elements presented in Table 2.1. Hence, experimental data can be used either for the phase analysis or for direct reconstruction of the amplitudes.

2.9 Relativistic Pion–Nucleon Scattering Matrix

In the preceding sections, we consider pion–nucleon scattering in the nonrelativistic case, where the kinetic energy of a particle is much lower than its rest energy. This approach was appropriate in the early 1950s, when synchrocyclotrons accelerated protons up to energies 200–300 MeV. However, the kinetic energy of accelerated protons reached the rest energy in the mid-1950s and, then, became much higher than the rest energy. Theoreticians foresaw this situation and developed the covariant formulation for the density and scattering matrices, which made it possible to analyze processes at relativistic energies. The main results of such an analysis confirmed the applicability of the nonrelativistic approach in the center-of-mass frame, and the relativistic corrections were reduced to an additional rotation angle. Such corrections refer to the observables that involve the polarization components in the scattering plane (parameters A and R), whereas the parameters that involved only polarization components perpendicular to the scattering plane (parameters P, D_{NN}, and C_{NN}) remain unchanged.

Below, we apply the relativistic description to the reaction

$$a(0) + b(1/2) = a(0) + b(1/2) \tag{2.267}$$

(the spins of the particles are given in the parentheses) proposed in Stapp (1956). Since the particle b is a Dirac particle with spin $1/2$, it is described by a four-component wave function ψ. The wave function of a free incident (or initial) particle (with positive energy) can be written in the form

$$\psi = \exp(if \cdot x) \sum_{i=1}^{2} A_i U_i, \tag{2.268}$$

whereas the wave function of an antiparticle (with negative energy) has the form

$$\psi = \exp(-if \cdot x) \sum_{i=3}^{4} A_i U_i. \tag{2.269}$$

Here, $f(\vec{f}, f_0)$ is the four-momentum of the particle in the base frame, where it is measured (for example, in the laboratory frame), such that $f_0 > 0$; x are four-dimensional space coordinates. Each spinor U_i has four components U_{si} given by the expressions

$$U_{si}(f) = (\mp if \cdot \gamma_{si} + m) / [2m(f_0 + m)]. \tag{2.270}$$

Hereinafter, the upper sign $(-)$ refers to the subscripts $i = 1, 2$ (positive energy), whereas the lower sign $(+)$, to the subscripts $i = 3, 4$ (negative energy). The subscripts in the four-vector $\gamma(-i\beta\vec{\alpha}, \beta)$ denote its matrix elements and m is the mass of the Dirac particle. The spinors are normalized in the covariant form as follows:

$$U_i^+(f) U_j(f) = U_i^*(f)\beta U_j(f) = \pm\delta_{ij}. \tag{2.271}$$

The sign $(^+)$ in the spinor means complex conjugation and transposition (interchange of the columns and rows of the matrix), i.e., Hermitian conjugation

$U^+ = U^* \beta$. Using formula (2.270), it is easy to verify that the spinors U_i satisfy the Dirac equation

$$(\pm i f \cdot \gamma + m) U_i(f) = 0. \tag{2.272}$$

To make the below expressions shorter, we introduce the notation

$$\gamma(v) = (\gamma \cdot v)/\sqrt{(v \cdot v)}. \tag{2.273}$$

The denominator on the right-hand side can be either a positive real number or a positive imaginary number. With this symbol, the Dirac equation is represented in the shorter form

$$\gamma(f) U_i = \pm U_i. \tag{2.274}$$

The wave function of the initial state of the pion–nucleon system was introduced above (see Eqs. (2.268) and (2.269)). Let Φ be the wave function of the final state of the same system. The relation between these two functions is determined by the reaction matrix $S(f', t, f)$:

$$\Phi(f') = S(f', t, f) \Psi(f). \tag{2.275}$$

The theory of "holes" requires that the Dirac particle described by a plane wave with the momentum f at time $T = -\infty$ and by a plane wave with the momentum f' at time $T = +\infty$ has the same sign of energy. This means that the transition of the particle to the antiparticle and vise versa is forbidden. This does not mean that the scattering matrix cannot describe the production of a particle. However, in this particular case, we analyze elastic processes and imply this exclusion, which is mathematically written in the form

$$S(f', t, f) = \gamma(f') S(f', t, f) \gamma(f). \tag{2.276}$$

Introducing the new symbol

$$\gamma(u, w) = \gamma\left(u/\sqrt{|u \cdot u|} + w/\sqrt{|w \cdot w|}\right) = [\gamma(u) + \gamma(w)] \tag{2.277}$$

and taking into account the relations

$$\gamma(u)\gamma(u) = 1 = \gamma(w)\gamma(w), \tag{2.278}$$

we obtain

$$\gamma(u)\gamma(u, w) = \gamma(u, w)\gamma(w). \tag{2.279}$$

The new scattering matrix $S_q(k', t, k)$ is introduced through the relation

$$S(f', t, f) = \gamma(f', t) S_q(k', t, k) \gamma(t, f). \tag{2.280}$$

This matrix $S_q(k', t, k)$ is the scattering matrix in the center-of-mass frame with the relative momenta $\vec{k}$ and $\vec{k}'$ before and after scattering, respectively, and the total energy t in the center-of-mass frame.

Then, the substitution of Eq. (2.280) into Eq. (2.276) provides the condition of the exclusion of the particle–antiparticle transition in the form

$$S_q(k', t, k) = \gamma(t) S_q(k', t, k) \gamma(t). \tag{2.281}$$

For a deeper insight into the meaning of this matrix, we substitute relation (2.281) into relation (2.280) and obtain

$$S(f',t,f) = \big(\gamma(f',t)\gamma(t)\big)S_q(k',t,k)\big(\gamma(t)\gamma(t,f)\big). \qquad (2.282)$$

The transformation $(\gamma(t)\gamma(t,f))$ relates the rest frame of the initial particle (R_i frame) with the center-of-mass frame (C frame for brevity). The transformation $(\gamma(f',t)\gamma(t))$ relates the $R_{f'}$ frame to the C frame. To verify this, we write the Lorentz transformation

$$L(f) = \exp\left[-\frac{1}{2}\theta(\vec{\alpha}\cdot\vec{f})/|\vec{f}|\right]. \qquad (2.283)$$

This expression is modified as

$$L(f) = \beta(-i\gamma\cdot f + m\beta)/[2m(f_0+m)]. \qquad (2.284)$$

In the C frame, $t(0,t_0)$ and $\gamma(t) = \beta$ and it can be shown that the following equality takes place:

$$\gamma(t_1)\gamma(t_1,f_1) = L(f_1), \qquad \gamma(f_1',t_1)\gamma(t_1) = L^{-1}(f_1'). \qquad (2.285)$$

Subscript 1 means that the quantities are taken in the C frame. Thus, the scattering matrix can be written in the form

$$S(f_1',t_1,f_1) = L^{-1}(f_1')S_q(k_1',t_1,k_1)L(f_1). \qquad (2.286)$$

Now, this expression can be interpreted as follows. The S-matrix in the C frame is the product of two Lorentz transformations and the scattering matrix S_q. The Lorentz transformation $L(f_1)$ transforms the spinor of the initial particle from the C frame to the R_{f_1} frame, i.e., to the rest frame of the initial particle, where spin is physically defined. Then, the unitary operator S_q describes the effect of scattering on this spinor. Finally, the second Lorentz transformation transforms the spinor of the final particles to the C frame.

Since the S-matrix specified by expression (2.275), as well as the γ matrix, has the covariant form, the new matrix $S_q(k',t,k)$ should also be covariant.

It is a 4×4 matrix and can be expanded in 16 Dirac matrices constituting a complete set:

$$S_q(k',t,k) = A + B_\mu\gamma_\mu + \frac{1}{2}C_{\mu\nu}\sigma_{\mu\nu} + D_\mu(i\gamma_5\gamma_\mu) + E\gamma_5. \qquad (2.287)$$

As expected in the general case, we have 16 coefficients, which are functions of the relative momenta $\vec{k}$ and $\vec{k}'$ and the total energy in the center-of-mass frame, t. These 16 coefficients are the scalar parameter A, pseudoscalar E, 4 components of the vector B, 4 components of the pseudovector D, and 6 components of the antisymmetric tensor C. Condition (2.281) implying the conservation of the sign of the total energy before and after reaction (a particle cannot become the antiparticle and vise versa) leads to the following constraints on the coefficients:

$$B_\mu = -i N_B (B t_\mu),$$

$$C_{\mu\nu} = N_C C \left\{ k_\mu k'_\nu - k_\nu k'_\mu \frac{(m^2 - \mu^2)}{|t \cdot t|} \cdot \left[t_\mu (k'_\nu - k_\nu) - t_\nu (k'_\mu - k_\mu) \right] \right\}, \quad (2.288)$$

$$D_\mu = N_D D(-i) k_\lambda k'_\rho t_\sigma \varepsilon_{\mu\lambda\rho\sigma} \equiv D n_\mu, \qquad E = 0.$$

Here, $\varepsilon_{\mu\lambda\rho\sigma}$ is the antisymmetric tensor and n is the four-dimensional unit pseudovector, which is the generalization of the nonrelativistic three-dimensional pseudovector $\vec{n}$ perpendicular to the scattering plane. This relativistic pseudovector n has the properties

$$(k \cdot n) = (k' \cdot n) = (t \cdot n) = (1 - n \cdot n) = 0. \qquad (2.289)$$

The coefficients B, C, and D, as well as A, are scalar functions of the energy and scattering angle and μ is the pion mass. The normalization coefficients N_B, N_C, and N_D are chosen so that the following equalities are satisfied:

$$B_\mu B^\mu = B^2, \qquad C_\mu C^\mu = 2C^2, \qquad D_\mu D^\mu = D^2. \qquad (2.290)$$

In the center-of-mass frame, expressions (22) are strongly simplified:

$$B_\mu \gamma_\mu = B\beta, \qquad \frac{1}{2} C_{\eta\nu} \sigma_{\mu\nu} = C(\vec{\sigma} \cdot \vec{N}), \qquad D_\mu i \gamma_5 \gamma_\mu = D\beta(\vec{\sigma} \cdot \vec{N}), \qquad E = 0. \qquad (2.291)$$

Here, $(\vec{\sigma} \cdot \vec{N})$ is the dot product of the three-dimensional vectors $\vec{\sigma}$ and $\vec{N}$, where $\vec{N}$ is the unit vector perpendicular to the scattering plane in the center-of-mass frame. The 4×4 Dirac matrix σ_i $(i = 1, 2, 3)$ has the form

$$\sigma_i = \left\| \begin{matrix} \sigma_i & 0 \\ 0 & \sigma_i \end{matrix} \right\|. \qquad (2.292)$$

Here, the two-dimensional Pauli matrices appear in the parentheses. Combining the terms in expressions (2.291) with the term A (see Eq. (2.287)), we obtain the following expression for the scattering matrix:

$$S_q(k', t, k) = \left\| \begin{matrix} (F^+ + G^+(\vec{\sigma} \cdot \vec{N})) & 0 \\ 0 & (F^- + G^-(\vec{\sigma} \cdot \vec{N})) \end{matrix} \right\|, \qquad (2.293)$$

where

$$F^\pm = A \pm B, \qquad G^\pm = D \pm C. \qquad (2.294)$$

Thus, pion–nucleon scattering in the relativistic case is also described by two complex amplitudes F and G; the upper $(+)$ and lower $(-)$ superscripts refer to the scattering of particles and antiparticles, respectively. With the use of the projection operator

$$\Lambda^\pm(t) = \frac{1}{2} \left[1 \pm \gamma(t) \right], \qquad (2.295)$$

the expression for pion–nucleon scattering matrix can be easily transformed to the covariant form

$$S_q(k', t, k) = \sum_{\pm} \Lambda^{\pm}(t)\left[F^{\pm} + G^{\pm}i\gamma_5(\gamma \cdot n)\right]. \tag{2.296}$$

After the substitution of this relation into Eq. (2.287), the resulting S-matrix is covariant, and the states with positive and negative energies are clearly separated. This form of the scattering matrix is often used when discussing the problem of spin interactions.

2.9.1 Covariant Density Matrix

As mentioned above when considering nonrelativistic scattering theory, the density matrix is convenient when a partially polarized beam is used. The same is true for relativistic theory. For the physical system state ψ_α, the density matrix is defined by the expression

$$\rho = \sum_{\alpha} |\psi_\alpha\rangle W_\alpha \langle\psi_\alpha|, \tag{2.297}$$

where W_α is the probability of finding the system in this state; therefore, $\sum_{\alpha} W_\alpha = 1$.

The probability of finding the system in the region R is determined by the formula

$$w(R) = Tr(\rho \Pi), \tag{2.298}$$

where Π is the projection operator separating the region R. If the region is an element of the three-dimensional momentum space ($d\vec{f} = df_1 df_2 df_3$), then

$$w(d\vec{f}) = d\vec{f}\, Tr\, \rho_s(f). \tag{2.299}$$

Here, Tr stands for the trace of the matrix $\rho_s(f)$ in spin space and

$$\rho_s(\vec{f}) = \sum \left|a_\alpha(\vec{f})\right|^2 \left[|U_\alpha(\vec{f})\rangle W_\alpha \langle U_\alpha^*(\vec{f})|\right]. \tag{2.300}$$

The amplitude $a_\alpha(\vec{f})$ is related to the wave function in momentum space as

$$\psi_\alpha(\vec{f}) = a_\alpha(\vec{f})|U_\alpha(\vec{f})\rangle. \tag{2.301}$$

Here, the spinor $U_\alpha(\vec{f})$ of the particle moving with the momentum f can be expressed in terms of the spinor of the particle in its R frame by means of the Lorentz transformation:

$$U_\alpha(f) = L^{-1}(f)U_\alpha(0),$$
$$w(df) = (d\vec{f})|a(\vec{f})|^2(\gamma)^f = (d\vec{f})\left[\sum W_\alpha |a_\alpha(\vec{f})|^2\right](\gamma)^f. \tag{2.302}$$

In this expression, the ratio $(d\vec{f})/(\gamma)^f$ is Lorentz invariant (this is the invariant $d\vec{p}/dE$). Hence, the quantity $|a_\alpha(\vec{f})\gamma^f|^2$ should be Lorentz invariant.

Note that both the density matrix and volume element are not invariant separately. However, if the particle is in a definite energy state, it is possible to make the transformation (sum signs are omitted)

$$
\begin{aligned}
\rho_s(\vec{f}) &= |a_\alpha(\vec{f})|^2 |U_\alpha(\vec{f})\rangle W_\alpha \langle U_\alpha^*(\vec{f})| \\
&= |a_\alpha(\vec{f})|^2 |U_\alpha(\vec{f})\rangle W_\alpha \langle U_\alpha^*(\vec{f})| |U_\alpha(\vec{f})\rangle (\pm) \langle U_\alpha^+(\vec{f})| \\
&= |\gamma^f a_\alpha(\vec{f})|^2 |U_\alpha(\vec{f})\rangle (\pm W_\alpha) \langle U_\alpha^+(\vec{f})| / (\gamma)^f \equiv \rho_f/(\gamma)^f .
\end{aligned} \quad (2.303)
$$

The product of the factor $1/(\gamma)^f$ and the volume element in the momentum space $d\vec{f}$ is invariant. The matrix ρ defined by the above expression is covariant, and its matrix elements are calculated by the formula

$$
\rho_{ij}(f) = \langle U_i^+(f) | \rho(f) | U_j(f) \rangle . \quad (2.304)
$$

The square of the absolute value of the amplitude specifies the probability of finding the system in the spin state α and with the momentum $\vec{f}$. The mean value of the operator $\hat{O}$ in the initial state with the momentum f is given by the expression

$$
\langle \hat{O} \rangle_f Tr \rho(f) = Tr \left[\rho(f) \hat{O} \right] . \quad (2.305)
$$

The mean value of the operator in the final state with the momentum f' is written similarly

$$
\langle \hat{O} \rangle_{f'} Tr \rho(f') = Tr \left[\rho(f') \hat{O} \right] . \quad (2.306)
$$

Two density matrices are related by the scattering matrix as

$$
\rho'(f') = S(f', t, f) \rho(f) S^+(f', t, f) . \quad (2.307)
$$

The differential cross section is determined by the expression

$$
I = Tr \rho'(f') / Tr \rho(f) . \quad (2.308)
$$

A Hermitian conjugate operator $\hat{A}^+$ can be defined as: $(AU)^+ = U^+ A^+$, where U is the spinor and $U^+ = U^* \beta$. In application to the S-matrix, we obtain the relation $S^+ = \beta S^* \beta$, where the asterisk in S means complex conjugation.

The application of relativistic formulas (2.307) and (2.308) leads to the differential cross section in the center-of-mass frame, as in the nonrelativistic case. The same conclusion is also valid for polarization. In the case of the spin rotation parameters in the horizontal plane, only an additional kinematic factor appears.

2.10   Relativistic Nucleon–Nucleon Scattering

In the preceding section, we consider the case where both the initial and final states include only one Dirac particle. Now, we discuss the case where both the initial and final states of the reaction include two particles with spin 1/2. Specifically, we analyze nucleon–nucleon elastic scattering.

Each nucleon has its complete set of the spin operators acting in two independent spin spaces. We present these operators (Bjorken and Drell 1964).

In the spin space of the first particle, these are

$$I^{(1)}, \quad \gamma_\mu^{(1)}, \quad \frac{1}{2}\sigma_{\mu\nu}^{(1)}, \quad i\gamma_5^{(1)}\gamma_\mu^{(1)}, \quad \gamma_\mu^{(1)}, \quad \gamma_5^{(1)}, \tag{2.309}$$

in the spin space of the second particle,

$$I^{(2)}, \quad \gamma_\mu^{(2)}, \quad \frac{1}{2}\sigma_{\mu\nu}^{(2)}, \quad i\gamma_5^{(2)}\gamma_\mu^{(2)}, \quad \gamma_\mu^{(2)}, \quad \gamma_5^{(2)}. \tag{2.310}$$

Each of sets (2.309) and (2.310) contains 16 terms. The scattering matrix is composed of the direct product of these terms, i.e., contains $16 \times 16 = 256$ terms, as in the nonrelativistic case. Correspondingly, the number of amplitudes is the same. It is necessary to impose allowable physical conditions in order to reduce this number to the minimum possible number.

By analogy with the case of pion–nucleon scattering (see the preceding section), but taking into account the presence of four nucleons, we introduce the matrix $S_q(k', t, k)$ through the expression (Stapp 1956)

$$S(f', k', t, f, k) = [\gamma^{(1)}(f', t)\gamma^{(1)}(t)][\gamma^{(2)}(k', t)\gamma^{(2)}(t)]S_q(k', t, k)$$
$$\times [\gamma^{(2)}(t)\gamma^{(2)}(k, t)][\gamma^{(1)}(t)\gamma^{(1)}(f, t)]. \tag{2.311}$$

Here, k and k' are the initial and final relative momenta, respectively. We can introduce the condition of the theory of holes and obtain

$$\gamma^{(1)}(t)S_q(k', t, k)\gamma^{(1)}(t) = S_q(k', t, k),$$
$$\gamma^{(2)}(t)S_q(k', t, k)\gamma^{(2)}(t) = S_q(k', t, k). \tag{2.312}$$

The matrix $S_q(k', t, k)$ can be expanded in the products of 16 spin matrices for particle 1 by 16 spin matrices for particle 2 (see Eq. (2.310)). The main aim is to significantly reduce the number of the resulting 256 terms. To this end, we consider a term appearing from the product of tensor operators:

$$C_{\mu\nu\sigma\rho}\left(\frac{1}{2}\sigma_{\mu\nu}^{(1)}\right)\left(\frac{1}{2}\sigma_{\sigma\rho}^{(2)}\right). \tag{2.313}$$

From the first of expressions (2.312) applied to this term, we obtain

$$t_\mu C_{\mu\nu\sigma\rho} = -t_\mu C_{\nu\mu\sigma\rho} = 0. \tag{2.314}$$

Therefore,

$$C_{\mu\nu\sigma\rho}\left(\frac{1}{2}\sigma_{\mu\nu}^{(1)}\right) = \gamma^{(1)}(t)i\gamma_5^{(1)}\gamma_\lambda^{(1)}C_{\lambda;\sigma\rho}, \tag{2.315}$$

where we take into account that $t_\lambda C_{\lambda\nu\sigma\rho} = 0$. The dependence on $\sigma_{\sigma\rho}^{(2)}$ can be modified to the form

$$C_{\mu\nu\sigma\rho}\left(\frac{1}{2}\sigma_{\mu\nu}^{(1)}\right)\left(\frac{1}{2}\sigma_{\sigma\rho}^{(2)}\right) = C_{\lambda\sigma}\gamma^{(1)}(t)\left(i\gamma_5^{(1)}\gamma_\lambda^{(1)}\right)\gamma^{(2)}(t)\left(i\gamma_5^{(2)}\gamma_\sigma^{(2)}\right). \tag{2.316}$$

Here, the equality $C_{\lambda\eta}t_\eta = -t_\lambda C_{\lambda\eta}$ is taken into account. Excluding similarly all terms containing $\sigma_{\mu\nu}$, we obtain

$$S_q(k',t,k) = \sum_{\pm\pm}(\Lambda^{(1)\pm}(t)\Lambda^{(2)\pm}(t))$$
$$\cdot \left[F^{\pm\pm} + G^{(1)\pm\pm}i\gamma_5^{(1)}\gamma^{(1)} \cdot n + G^{(2)\pm\pm}i\gamma_5^{(2)}\right.$$
$$+ (C^{\pm\pm}n_\lambda n_\rho + D^{\pm\pm}s_\lambda s_\rho + E^{\pm\pm}d_\lambda d_\rho)$$
$$\left.\cdot (i\gamma_5^{(1)}\gamma_\lambda^{(1)})(i\gamma_5^{(2)}\gamma_\rho^{(2)})\right]. \tag{2.317}$$

Here four vectors $n, s,$ and d are defined as follows

$$n_\lambda \propto k'_\rho k_\sigma t_\mu \varepsilon_{\rho\sigma\mu\lambda},$$
$$s_\lambda = N_s\left[k_\lambda + k'_\lambda - t_\lambda\{t_\rho(k_\rho + k'_\rho)\}(t \cdot t)^{-1}\right], \tag{2.317a}$$
$$d_\lambda = N_d\left[k_\lambda - k'_\lambda\right].$$

Here N_i $(i = s, d)$ are normalization coefficients. The vectors n, s, d, t form an orthogonal set. The vector d retains its sign under time inversion, whereas s changes sign. Therefore all terms containing the products of these two vectors should reduce to zero.

The scattering matrix is written as follows:

$$S_q(k',t,k) = F + F^{(1)}\gamma^{(1)}(t) + F^{(2)}\gamma^{(2)}(t) + G_\lambda^{(1)}(i\gamma_5^{(1)}\gamma_\lambda^{(1)})$$
$$+ G_\lambda^{(2)}(i\gamma_5^{(2)}\gamma_\lambda^{(2)}) + C_\lambda^{(1)}\gamma^{(1)}(t)(i\gamma_5^{(1)}\gamma_\lambda^{(1)})$$
$$+ C_\lambda^{(2)}\gamma^{(2)}(t)(i\gamma_5^{(2)}\gamma_\lambda^{(2)}) + G_{\lambda\rho}(i\gamma_5^{(1)}\gamma_\lambda^{(1)}) \cdot (i\gamma_5^{(2)}\gamma_\rho^{(2)})$$
$$+ C_{\lambda\rho}\gamma^{(1)}(t)(i\gamma_5^{(1)}\gamma_\lambda^{(1)})\gamma^{(2)}(t)(i\gamma_5^{(2)}\gamma_\rho^{(2)})$$
$$+ F^{(3)}\gamma^{(1)}(t) \cdot \gamma^{(2)}(t) + E_\lambda^{(1)}\gamma^{(2)}(t)(i\gamma_5^{(1)}\gamma_\lambda^{(1)})$$
$$+ E_\lambda^{(2)}\gamma^{(1)}(t)(i\gamma_5^{(2)}\gamma_\lambda^{(2)})$$
$$+ D_\lambda^{(1)}\gamma^{(2)}(t)\gamma^{(1)}(t)(i\gamma_5^{(1)}\gamma_\lambda^{(1)}) + D_\lambda^{(2)}\gamma^{(1)}(t)\gamma^{(2)}(t)(i\gamma_5^{(2)}\gamma_\lambda^{(2)})$$
$$+ H_{\lambda\rho}^{(1)}(i\gamma_5^{(1)}\gamma_\lambda^{(1)})\gamma^{(2)}(t)(i\gamma_5^{(2)}\gamma_\rho^{(2)})$$
$$+ H_{\lambda\rho}^{(2)}(i\gamma_5^{(2)}\gamma_\lambda^{(2)})\gamma^{(1)}(t)(i\gamma_5^{(1)}\gamma_\rho^{(1)}). \tag{2.318}$$

All parameters appearing in the scattering matrix are functions of the momenta $k, k',$ and t. These parameters are orthogonal to t with respect to all subscripts. For example $t_\lambda H_{\lambda\rho} = t_\rho H_{\lambda\rho} = 0$ and similar is for the other terms.

It is easy to group the terms in expression (2.318). For example, the first two terms can be rewritten in the form

$$F + F^{(1)}\gamma^{(1)}(t) = \sum_\pm \frac{1}{2}[1 \pm \gamma^{(1)}(t)]F^\pm, \tag{2.318a}$$

where $F^+ = F + F^{(1)}, F^- = F - F^{(1)}$. The other terms can be pairwise grouped similarly. As a result, we arrive at the expression

$$S_q(k',t,k) = \sum_{\pm} \frac{1}{2}\left[1 \pm \gamma^{(1)}(t)\right] \cdot \left[F^{\pm} + G_{\lambda}^{(1)\pm}\left(i\gamma_5^{(1)}\gamma_{\lambda}^{(1)}\right) + F^{(2)\pm}\gamma^{(2)}(t)\right.$$
$$+ G_{\lambda}^{(2)\pm}\left(i\gamma_5^{(2)}\gamma_{\lambda}^{(2)}\right) + C_{\lambda}^{(2)\pm}\gamma^{(2)}(t)\left(i\gamma_5^{(2)}\gamma_{\lambda}^{(2)}\right)\right]$$
$$+ \left[G_{\lambda\rho}^{\pm}\left(i\gamma_5^{(1)}\gamma_{\lambda}^{(1)}\right)\left(i\gamma_5^{(2)}\gamma_{\rho}^{(2)}\right)\right.$$
$$+ H_{\lambda\rho}^{(1)\pm}\left(i\gamma_5^{(1)}\gamma_{\lambda}^{(1)}\right)\gamma^{(2)}(t)\left(i\gamma_5^{(2)}\gamma_{\rho}^{(2)}\right)$$
$$\left. + E_{\lambda}^{(1)\pm}\gamma^{(2)}(t)\left(i\gamma_5^{(1)}\gamma_{\lambda}^{(1)}\right)\right]. \tag{2.319}$$

Grouping similarly the terms with respect to $\gamma^{(2)}(t)$, we obtain

$$S_q(k',t,k) = \sum_{\pm\pm}\left\{\frac{1}{2}\left[1 \pm \gamma^{(1)}(t)\right]\frac{1}{2}\left[1 \pm \gamma^{(2)}(t)\right]\right\}$$
$$\cdot \left[F^{\pm\pm} + G_{\lambda}^{(1)\pm\pm}\left(i\gamma_5^{(1)}\gamma_{\lambda}^{(1)}\right) + G_{\lambda}^{(2)\pm\pm}\left(i\gamma_5^{(2)}\gamma_{\lambda}^{(2)}\right)\right.$$
$$\left. + G_{\lambda\rho}^{\pm\pm}\left(i\gamma_5^{(1)}\gamma_{\lambda}^{(1)}\right)\left(i\gamma_5^{(2)}\gamma_{\rho}^{(2)}\right)\right]. \tag{2.320}$$

Here, the following orthogonality relations for $G_{\lambda\rho}$ were used:

$$t_{\lambda}G_{\lambda}^{(1)\pm\pm} = t_{\lambda}G_{\lambda}^{(2)\pm\pm} = t_{\lambda}G_{\lambda\rho}^{\pm\pm} = G_{\lambda\rho}^{\pm\pm}t_{\rho} = 0. \tag{2.320a}$$

Since the scattering matrix should be a scalar function, the parameters $G_{\lambda}^{(1)\pm\pm}$ and $G_{\lambda}^{(2)\pm\pm}$ should be pseudovectors. The existing three momenta can provide the single pseudovector

$$n_{\lambda} \propto k'_{\rho}k_{\sigma}t_{\mu}\varepsilon_{\rho\sigma\mu\lambda}. \tag{2.321}$$

Hence, we can write

$$G_{\lambda}^{(1)\pm\pm} = G^{(1)}n_{\lambda} \quad \text{and} \quad G_{\lambda}^{(2)\pm\pm} = G^{(2)}n_{\lambda}. \tag{2.322}$$

To transform the tensor terms $G_{\lambda\rho}^{\pm\pm}$, two normalized vectors s and d are introduced in addition to t and n as follows:

$$s_{\lambda} = N_s\left\{k_{\lambda} + k'_{\lambda} - t_{\lambda}\left[t_{\rho}\left(k_{\rho} + k'_{\rho}\right)\right](t \cdot t)^{-1}\right\}, \quad d_{\lambda} = N_d\left(k_{\lambda} - k'_{\lambda}\right). \tag{2.323}$$

The four vectors $t, n, s,$ and d constitute a set of orthonormalized vectors in which the second-rank tensor $G_{\lambda\rho}^{\pm\pm}$ can be expanded. Relation (2.320a) imposes the constraints on the number of the terms of this tensor. Additional conditions are imposed by the requirement of the invariance of the scattering matrix under space inversion. As a result, we obtain

$$G_{\lambda\rho}^{\pm\pm} = C^{\pm\pm}n_{\lambda}n_{\rho} + D^{\pm\pm}s_{\lambda}s_{\rho} + E^{\pm\pm}d_{\lambda\rho}$$
$$+ G'^{\pm\pm}\left(s_{\lambda}d_{\rho} + d_{\lambda}s_{\rho}\right) + G^{\pm\pm}\left(s_{\lambda}d_{\rho} - d_{\lambda}s_{\rho}\right). \tag{2.324}$$

Since the matrix should be invariant under time reversal, two last terms should be zero, because the vector d does not change sign under this operation, whereas the vector s changes sign.

Finally, the relativistic nucleon–nucleon elastic scattering matrix is written in the form

$$S_q(k', t, k) = \sum_{\pm\pm} \left[\Lambda^{(1)\pm}(t)\right]\left[\Lambda^{(2)\pm}(t)\right]$$

$$\cdot \left[F^{\pm\pm} + G^{(1)\pm\pm}\left(i\gamma_5^{(1)}\gamma^{(1)} \cdot n\right) + G^{(2)\pm\pm}\left(i\gamma_5^{(2)}\gamma^{(2)} \cdot n\right)\right]$$

$$+ \left[\left(C^{\pm\pm}n_\lambda n_\rho + D^{\pm\pm}s_\lambda s_\rho + E^{\pm\pm}d_\lambda d_\rho\right)\left(i\gamma_5^{(1)}\gamma_\lambda^{(1)}\right)\left(i\gamma_5^{(2)}\gamma_\rho^{(2)}\right)\right].$$

$$(2.325)$$

This is the relativistic formula for nucleon–nucleon scattering in the center-of-mass frame. Several conclusions follow from it. First, the number of free parameters is six, as in the nonrelativistic case. Second, the scattering of antiprotons is described by the same formulas as the scattering of protons. Third, passing to the nonrelativistic case and taking the states only with positive energy, we exactly arrive at Wolfenstein nonrelativistic formulas. As will be shown below, the relativistic case gives only kinematic corrections, which are easily taken into account.

In the center-of-mass frame, the relativistic matrix given by formula (2.325) reduces to the Wolfenstein–Ashkin nonrelativistic matrix (where $I^{(1)}$ and $I^{(2)}$ are the identity matrices in the spaces of particles 1 and 2, respectively):

$$M = aI^{(1)}I^{(2)} + c\left(\sigma_n^{(1)} + \sigma_n^{(2)}\right) + m\sigma_n^{(1)}\sigma_n^{(2)}$$

$$+ g\left(\sigma_{\hat{p}}^{(1)}\sigma_{\hat{p}}^{(2)} + \sigma_{\hat{k}}^{(1)}\sigma_{\hat{k}}^{(2)}\right) + h\left(\sigma_{\hat{p}}^{(1)}\sigma_{\hat{p}}^{(2)} - \sigma_{\hat{k}}^{(1)}\sigma_{\hat{k}}^{(2)}\right). \quad (2.326)$$

This matrix differs from the matrix used above in the notation: $a = a$, $c = c$, $b = m$, $l = g - h$, and $f = g + h$.

The ordinary method provides the following expressions for the measured quantities in terms of the elements of scattering matrix (2.326):

$$I_0 R = \frac{1}{2}\,\mathrm{Re}\left\{(M_{00} + \sqrt{2}\cot\theta\, M_{10})(M_{11} + M_{1-1} - M_{ss})^*\right.$$

$$\left. \times \cos(\theta - \theta_L) - \frac{\sqrt{2}}{\sin\theta}\left[(M_{11} + M_{1-1})M_{10}^* - M_{ss}M_{01}^*\right]\cos\theta_l\right\},$$

$$I_0 A = -\frac{1}{2}\,\mathrm{Re}\left\{(M_{00} + \sqrt{2}\cot\theta\, M_{10})(M_{11} + M_{1-1} + M_{ss})^*\right.$$

$$\left. \times \sin(\theta - \theta_L) - \frac{\sqrt{2}}{\sin\theta}\left[(M_{11} + M_{1-1})M_{01}^* - M_{ss}M_{10}^*\right]\sin\theta_l\right\},$$

$$I_0 R' = \frac{1}{2}\,\mathrm{Re}\left\{(M_{00} + \sqrt{2}\cot\theta\, M_{10})(M_{11} + M_{1-1} + M_{ss})^*\right.$$

$$\left. \times \sin(\theta - \theta_L) + \frac{\sqrt{2}}{\sin\theta}\left[(M_{11} + M_{1-1})M_{10}^* - M_{ss}M_{01}^*\right]\sin\theta_l\right\},$$

$$I_0 A' = \frac{1}{2}\,\mathrm{Re}\left\{(M_{00} + \sqrt{2}\cot\theta\, M_{10})(M_{11} + M_{1-1} + M_{ss})^*\right.$$

$$\left. \times \cos(\theta - \theta_L) + \frac{\sqrt{2}}{\sin\theta}\left[(M_{11} + M_{1-1})M_{01}^* - M_{ss}M_{10}^*\right]\cos\theta_l\right\},$$

$$I_0 R_t = \frac{1}{2}\,\mathrm{Re}\Big\{(M_{00} + \sqrt{2}\cot\theta\, M_{10})(M_{11} + M_{1-1} - M_{ss})^*$$

$$\times \cos(\theta' - \theta'_L) + \frac{\sqrt{2}}{\sin\theta}\big[(M_{11} + M_{1-1})M_{10}^* + M_{ss}M_{01}^*\big]\cos\theta_l\Big\},$$

$$I_0 A_t = -\frac{1}{2}\,\mathrm{Re}\Big\{(M_{00} + \sqrt{2}\cot\theta\, M_{10})(M_{11} + M_{1-1} - M_{ss})^*$$

$$\times \sin(\theta' - \theta'_L) - \frac{\sqrt{2}}{\sin\theta}\big[(M_{11} + M_{1-1})M_{01}^* - M_{ss}M_{10}^*\big]\sin\theta_l\Big\},$$

$$I_0 R'_t = \frac{1}{2}\,\mathrm{Re}\Big\{(M_{00} + \sqrt{2}\cot\theta\, M_{10})(M_{11} + M_{1-1} - M_{ss})^*$$

$$\times \sin(\theta' - \theta'_L) - \frac{\sqrt{2}}{\sin\theta}\big[(M_{11} + M_{1-1})M_{10}^* + M_{ss}M_{01}^*\big]\sin\theta_l\Big\},$$

$$I_0 A'_t = \frac{1}{2}\,\mathrm{Re}\Big\{(M_{00} - \sqrt{2}\cot\theta\, M_{10})(M_{11} + M_{1-1} - M_{ss})^*$$

$$\times \cos(\theta' - \theta'_L) - \frac{\sqrt{2}}{\sin\theta}\big[(M_{11} + M_{1-1})M_{01}^* + M_{ss}M_{10}^*\big]\cos\theta_l\Big\},$$

$$I_0 C_{kp} = \frac{1}{2\sin\theta}\big[|M_{01}|^2 - |M_{10}|^2\big]\cos(\alpha - \alpha') - \frac{1}{4}\big[|M_{11} + M_{1-1}|^2 - |M_{ss}|^2\big]$$

$$\times \cos(\alpha + \alpha') - \frac{1}{4\cos\theta}\big[|M_{11} - M_{1-1}|^2 - |M_{00}|^2\big]\sin(\alpha - \alpha'),$$

$$I_0 C_{kk} = -\frac{1}{2\sin\theta}\big[|M_{01}|^2 - |M_{10}|^2\big]\sin(\alpha - \alpha') + \frac{1}{4}\big[|M_{11} + M_{1-1}|^2 - |M_{ss}|^2\big]$$

$$\times \cos(\alpha + \alpha') - \frac{1}{4\cos\theta}\big[|M_{11} - M_{1-1}|^2 - |M_{00}|^2\big]\cos(\alpha - \alpha'),$$

$$I_0 C_{pp} = -\frac{1}{2\sin\theta}\big[|M_{01}|^2 - |M_{10}|^2\big]\sin(\alpha - \alpha') + \frac{1}{4}\big[|M_{11} + M_{1-1}|^2 - |M_{ss}|^2\big]$$

$$\times \cos(\alpha + \alpha') - \frac{1}{4\cos\theta}\big[|M_{11} - M_{1-1}|^2 - |M_{00}|^2\big]\cos(\alpha - \alpha').$$

Here, θ and θ_l are the scattering angles in the center-of-mass and laboratory frames, respectively, and θ' and θ'_L are the respective angles for the recoil particle.

Under relativistic transformations, the parameters that either are scalars or have only components perpendicular to the reaction plane remain unchanged. These are the following quantities: cross section I; polarization P; depolarization tensors of the scattered and recoil particles, D and D_t, respectively; and correlation parameters C_{nn} and A_{nn}.

In the above formalism of the relativistic reaction matrix, the wave functions are represented in the space of angular momenta, where the quantization z axis is fixed.

Jacob and Wick (1959) proposed a relativistic description of reactions using the wave functions quantized along the momentum of the incident and scattered particles in the center-of-mass frame. In this case, the spin projection on the momentum direction has two values, $+1/2$ and $-1/2$, and this projection is called helicity.

Denoting the helicities of the initial and final nucleons as λ_1, λ_2 and λ'_1, λ'_2, respectively, we can write the scattering matrix in the form

$$\langle \lambda'_1 \lambda'_2 | M | \lambda_1 \lambda_2 \rangle. \tag{2.327}$$

These matrix elements are called helicity amplitudes. Let the scattered particle move along the z' axis inclined to the z axis at the angle θ (scattering angle). By analogy with the spinning-top model, matrix element (2.327) can be expanded in terms of the reduced wave functions of the symmetric top, $d^J_{\mu\mu'}$:

$$\langle \lambda'_1 \lambda'_2 | M | \lambda_1 \lambda_2 \rangle = \frac{1}{2ik} \sum_J (2J+1)(\langle \lambda'_1 \lambda'_2 | S(J,E) - 1 | \lambda_1 \lambda_2 \rangle) d^J_{\mu\mu'}(\theta), \tag{2.328}$$

where

$$\mu = \lambda_1 - \lambda_2, \qquad \mu' = \lambda'_1 - \lambda'_2. \tag{2.329}$$

As shown above, owing to the invariance of the scattering matrix under space rotation and inversion and time reversal, as well as to the isotopic invariance, five matrix elements are nonzero in the general case. The same requirements applied to the scattering matrix in the helicity representation impose the following conditions:

parity conservation

$$\langle \lambda'_1 \lambda'_2 | M | \lambda_1 \lambda_2 \rangle = \langle -\lambda'_1 - \lambda'_2 | M | -\lambda_1 - \lambda_2 \rangle, \tag{2.330}$$

time reversal

$$\langle \lambda'_1 \lambda'_2 | M | \lambda_1 \lambda_2 \rangle = (-1)^{\lambda_1 - \lambda_2 - \lambda'_1 + \lambda'_2} \langle \lambda_1 \lambda_2 | M | \lambda'_1 \lambda'_2 \rangle, \tag{2.331}$$

conservation of the total spin

$$\langle \lambda'_1 \lambda'_2 | M | \lambda_1 \lambda_2 \rangle = \langle \lambda'_2 \lambda'_1 | M | \lambda_2 \lambda_1 \rangle. \tag{2.332}$$

Under these conditions, five matrix elements are also nonzero in the helicity representation, and these elements are denoted as follows:

$$\varphi_1 = \left\langle \frac{1}{2}, \frac{1}{2} \middle| M \middle| \frac{1}{2}, \frac{1}{2} \right\rangle, \qquad \varphi_2 = \left\langle \frac{1}{2}, \frac{1}{2} \middle| M \middle| -\frac{1}{2}, -\frac{1}{2} \right\rangle,$$

$$\varphi_3 = \left\langle \frac{1}{2}, -\frac{1}{2} \middle| M \middle| \frac{1}{2}, -\frac{1}{2} \right\rangle, \tag{2.333}$$

$$\varphi_4 = \left\langle \frac{1}{2}, -\frac{1}{2} \middle| M \middle| -\frac{1}{2}, \frac{1}{2} \right\rangle, \qquad \varphi_5 = \left\langle \frac{1}{2}, \frac{1}{2} \middle| M \middle| \frac{1}{2}, -\frac{1}{2} \right\rangle.$$

The amplitudes φ_i, where $i = 1, 2, 3, 4, 5$, are classified according to the physics of the process as the amplitudes without spin flip (φ_1, φ_3), with single spin flip (φ_5), and with double spin flip (φ_2, φ_4).

The relation between the matrix elements in the helicity representation and angular-momentum representation can be obtained as follows (Jacob and Wick 1959). Let the direction of the incident-particle motion be the quantization axis for the spin wave function. Then, the wave functions of the first and second nucleons in the helicity representation can be written in the form

$$\chi_{1/2}^{(1)} = \begin{Vmatrix} 1 \\ 0 \end{Vmatrix}, \qquad \chi_{-1/2}^{(1)} = \begin{Vmatrix} 0 \\ 1 \end{Vmatrix} \tag{2.334}$$

in the initial state and

$$\chi_{1/2}^{(1)} = \chi_{-1/2}^{(2)} = \begin{Vmatrix} \cos(\theta/2) \\ \sin(\theta/2) \end{Vmatrix}, \qquad \chi_{-1/2}^{(1)} = \chi_{1/2}^{(2)} = \begin{Vmatrix} -\sin(\theta/2) \\ \cos(\theta/2) \end{Vmatrix} \tag{2.335}$$

in the final state, where θ is the particle scattering angle in the center-of-mass frame.

Let us rewrite matrix (2.326) as follows:

$$\begin{aligned} M = a &+ c\left(\vec{\sigma}^{(1)} + \vec{\sigma}^2\right) \cdot \vec{n} + m\left(\vec{\sigma}^{(1)} \cdot \vec{n}\right)\left(\vec{\sigma}^{(2)} \cdot \vec{n}\right) \\ &+ g\left[\left(\vec{\sigma}^{(1)} \cdot \vec{P}\right)\left(\vec{\sigma}^{(2)} \cdot \vec{P}\right) + \left(\vec{\sigma}^{(1)} \cdot \vec{K}\right)\left(\vec{\sigma}^{(2)} \cdot \vec{K}\right)\right] \\ &+ h\left[\left(\vec{\sigma}^{(1)} \cdot \vec{P}\right)\left(\vec{\sigma}^{(2)} \cdot \vec{P}\right) - \left(\vec{\sigma}^{(1)} \cdot \vec{K}\right)\left(\vec{\sigma}^{(2)} \cdot \vec{K}\right)\right]. \end{aligned} \tag{2.336}$$

The unit vectors $\vec{n}$, $\vec{P}$, $\vec{K}$ have the following components:

$$\vec{n}(0, 1, 0), \quad \vec{K}\left(\cos\theta/2, 0, -\sin\theta/2\right), \quad \vec{P}\left(\sin\theta/2, 0, \cos(\theta/2)\right). \tag{2.337}$$

Therefore, the vector $\vec{n}$ is perpendicular to the scattering plane, whereas the vectors $\vec{K}$ and $\vec{P}$ lie in the scattering plane. Applying matrix (2.336) to wave functions (2.334) and (2.335) and taking into account relations (2.337), we obtain the relation between the amplitudes in the helicity and angular representations. They are given below along with the relations with the amplitudes in the singlet–triplet representation.

2.10.1 Relation Between the Amplitudes in Different Representations

A. In the helicity and angular representations:

$$\begin{aligned} \varphi_1 - \varphi_2 &= a - m - 2g, \\ \varphi_1 + \varphi_2 &= (a + m)\cos\theta + 2ic\sin\theta + 2h, \\ \varphi_3 + \varphi_4 &= a - m + 2g, \\ \varphi_3 - \varphi_4 &= (a + m)\cos\theta + 2ic\sin\theta - 2h, \\ \varphi_5 &= -\frac{1}{2}(a + m)\sin\theta + ic\cos\theta. \end{aligned}$$

B. In the angular and helicity representations (inverse to case A):

$$a = \frac{1}{4}\left[(\varphi_1 - \varphi_2 + \varphi_3 + \varphi_4) + (\varphi_1 + \varphi_2 + \varphi_3 - \varphi_4)\cos\theta - 4\sin\theta\varphi_5\right],$$

$$ic = \frac{1}{4}\left[(\varphi_1 + \varphi_2 + \varphi_3 - \varphi_4)\sin\theta + 4\cos\theta\varphi_5\right],$$

$$m = \frac{1}{4}\left[(-\varphi_1 + \varphi_2 - \varphi_3 - \varphi_4) + (\varphi_1 + \varphi_2 + \varphi_3 - \varphi_4)\cos\theta - 4\sin\theta\varphi_5\right],$$

$$g = \frac{1}{4}(-\varphi_1 + \varphi_2 + \varphi_3 + \varphi_4),$$

$$h = \frac{1}{4}(-\varphi_1 - \varphi_2 + \varphi_3 - \varphi_4).$$

C. In the helicity and singlet–triplet representations:

$$\varphi_1 - \varphi_2 = M_{ss},$$
$$\varphi_1 + \varphi_2 = \cos\theta M_{00} - \sqrt{2}\sin\theta M_{10},$$
$$\varphi_3 + \varphi_4 = M_{11} + M_{1-1},$$
$$\varphi_1 - \varphi_2 = \cos\theta M_{11} + \sin\theta M_{01} - \cos\theta M_{1-1},$$
$$\varphi_5 = -\frac{1}{2}\sin\theta M_{11} + \frac{1}{\sqrt{2}}\cos\theta M_{01} - \frac{1}{2}\sin\theta M_{1-1}$$
$$= -\frac{1}{2}\sin\theta M_{00} - \frac{1}{\sqrt{2}}\cos\theta M_{10}.$$

The helicity amplitudes can be expressed in terms of the phase shifts. To this end, the matrix elements in helicity space should be transformed to the matrix elements in singlet–triplet space that are already expressed in terms of the phases.

D. The matrix elements in the helicity representation can be expressed in terms of the elements in the singlet–triplet representation as follows:

$$\langle \lambda_1' \lambda_2' | M | \lambda_1 \lambda_2 \rangle = \sum \langle \lambda_1' \lambda_2' | | S m_s \rangle \langle S m_s | M | S m_s' \rangle \langle S m_s' | | \lambda_1 \lambda_2 \rangle. \qquad (2.338)$$

Since the elements of the matrix M, as well as the Clebsch–Gordan coefficients, are known, the helicity amplitudes can be found from this relation (see "Relation between the amplitudes in different representations").

2.11 Isospin T, C and G Parities

2.11.1 Isotopic Invariance

There are many experimental evidences that the properties of the proton and neutron in nuclear interaction are very similar. It can be assumed that they are the components of the same object called the nucleon. First, their masses are very close: the proton and neutron masses are 938.27 and 939.57 MeV, respectively; hence, the mass difference is as small as 1.3 MeV. This small difference ($\approx$0.15 %) is assumingly attributed to the electromagnetic interaction. Second, they strongly interact with each other and are constituents of nuclei. Third, it is known that, excluding the relatively weak electromagnetic interaction in the pp system, this system is almost equivalent to the system of two neutrons nn. This equality of the interactions between two protons and two neutrons is called *charge symmetry*. In particular, additional experimental evidence of charge symmetry follows from nucleon–nucleon scattering experiments. It is shown that the pp and nn elastic scattering processes are identical, excluding the Coulomb interaction. The same data are also obtained in pion–nucleon and kaon–nucleon elastic scattering processes. In addition, the binding energies (B), energy levels, and other properties of the mirror nuclei in which the protons and neutrons are interchanged are very close to each other. The closeness of the binding energies of two mirror nuclei is exemplified as follows:

$$H^3 = (nnp) \rightarrow B = 8.192 \text{ MeV}, \qquad He^3 = (ppn) \rightarrow B = 7.728 \text{ MeV}. \quad (2.339)$$

Here, the nucleon compositions of the tritium and helium-3 nuclei are given in the parentheses. The difference between the binding energies of these nuclei is as small as $\Delta B \approx 0.5$ MeV and can be attributed to the energy of the Coulomb repulsion between two protons in the helium-3 nucleus:

$$V_C(R) = \frac{1}{2}Z(Z-1)\frac{6e^2}{5R}, \qquad R \approx 1.45 \cdot 10^{-13} \text{ cm}. \quad (2.340)$$

The closeness of the energy levels of the mirror nuclei can be illustrated by the example

$$\begin{aligned} B^{11} = (6n5p) &\rightarrow E = (1.98; 2.14; 4.46; 5.03; 6.76) \text{ MeV}, \\ C^{11} = (6p5n) &\rightarrow E = (-; 1.85; 4.23; 4.77; 6.40) \text{ MeV}. \end{aligned} \quad (2.341)$$

Here, the numbers of protons and neutrons in a given nucleus are presented in the first parentheses and the level energies of this nucleus are indicated in the second parentheses. The observed similarity of the energy levels of these mirror nuclei is most simply explained by the hypothesis of the *charge symmetry* between the proton and neutron. The detailed discussion of these problems can be found in Schiff (1968).

It is substantial that both the proton and neutron have a half-integer spin and are governed by the same statistics (Dirac–Fermi statistics). This means that the system of two protons or two neutrons is described by the wave function $\psi(\vec{r}_1, \vec{s}_1; \vec{r}_2, \vec{s}_2)$ (where $\vec{r}$ and $\vec{s}$ are the radius-vector and spin of the particles, respectively) antisymmetric under the simultaneous permutation of the coordinates and spins of the particles. However, the experimentally observed charge symmetry is only one of the manifestations of a deeper similarity of the proton and neutron. This new type of symmetry was introduced by Heisenberg (1932) and was called isotopic invariance. Its main meaning is that the forces between the pp, np, and nn pairs are identical and the proton and neutron are two components of the nucleon. Let us illustrate this hypothesis by an example from low-energy scattering. The interaction of the proton with the proton or the neutron at low energies is characterized by two parameters: the scattering length a and effective scattering radius r_0. They determine the scattering phase δ in the S-state through the relation

$$k \cot \delta = -\frac{1}{a} + \frac{1}{2} r_0 k^2, \qquad (2.342)$$

where k is the wavenumber. The following parameters are experimentally determined (Nishijima 1964):

$$
\begin{aligned}
np: \quad &{}^3S_1, \quad r_{0t} = (1.704 \pm 0.028) \cdot 10^{-13} \text{ cm}, \\
&a_t = (5.39 \pm 0.03) \cdot 10^{-13} \text{ cm}; \\
&{}^1S_1, \quad r_{0S} = (2.670 \pm 0.023) \cdot 10^{-13} \text{ cm}, \\
&a_S = (-23.74 \pm 0.09) \cdot 10^{-13} \text{ cm}; \\
pp: \quad &{}^1S_1, \quad r_{0S} = (2.77) \cdot 10^{-13} \text{ cm}, \quad a_S = (-17.77) \cdot 10^{-13} \text{ cm}.
\end{aligned}
\qquad (2.343)
$$

The parameters a and r_0 of the np- and pp-scattering processes in the same ${}^1S_1$ states are in qualitatively agreement. The quantitative difference can be ignored, because this difference can be eliminated by changing the potential-well depth by 3 %.

These experimental facts led to the notion of the nucleon isospin τ and to the discovery of isotopic invariance in strong interactions. The isospin is a vector in isotopic space and has the same properties as the Pauli matrices. Isospin space is not a real space, but only a mathematical notion. Correspondingly, the nature of isospin is unknown.

The isotopic invariance of strong interactions can be formulated as invariance under "rotations in isospin space." The isospin operator $T(T_1, T_2, T_3)$ and isospinor ψ are written in the explicit form

$$T_1 = \frac{1}{2} \begin{Vmatrix} 0 & 1 \\ 1 & 0 \end{Vmatrix}, \qquad T_2 = \frac{1}{2} \begin{Vmatrix} 0 & -i \\ i & 0 \end{Vmatrix}, \qquad T_3 = \frac{1}{2} \begin{Vmatrix} 1 & 0 \\ 0 & -1 \end{Vmatrix}, \qquad \psi = \begin{Vmatrix} \psi_1 \\ \psi_2 \end{Vmatrix}. \qquad (2.344)$$

Let us consider the rotation by angle π about the x_2 axis in this space. The isospinor is transformed as:

$$\psi \rightarrow e^{i\pi T_2}\psi = e^{i\frac{\pi}{2}\tau_2}\psi = \left(\cos\frac{\pi}{2} + i\tau_2 \sin\frac{\pi}{2} \right)\psi = i\tau_2 \psi. \qquad (2.345)$$

Here, $T_i = 1/2\tau_i$, where $i = 1, 2, 3$; T is the nucleon isospin equal $1/2$ with the eigenvalues $\pm 1/2$; and τ is the analog of the Pauli operator in isotopic space. It is usually accepted that the eigenvalue $+1/2$ corresponds to the proton and is described by the function ψ_1, whereas $-1/2$ corresponds to the neutron and is described by the function ψ_2 in Eq. (2.344). In particular, for the isospinors corresponding to two components of the isodoublet (to the proton and neutron), we have

$$|p\rangle = \begin{Vmatrix} 1 \\ 0 \end{Vmatrix} \rightarrow \begin{Vmatrix} 0 \\ -1 \end{Vmatrix} = |n\rangle, \quad |n\rangle = \begin{Vmatrix} 0 \\ 1 \end{Vmatrix} \rightarrow \begin{Vmatrix} 1 \\ 0 \end{Vmatrix} = |p\rangle, \qquad (2.346)$$

which correspond to the transformations

$$\tau_+ \psi = \frac{1}{2}(\tau_1 + i\tau_2)\psi = \left\| \begin{matrix} 0 & 0 \\ -1 & 0 \end{matrix} \right\| \psi, \qquad n \to p,$$

$$\tau_- \psi = \frac{1}{2}(\tau_1 - i\tau_2)\psi = \left\| \begin{matrix} 0 & 0 \\ -1 & 0 \end{matrix} \right\| \psi, \qquad p \to -n, \tag{2.347}$$

where p and n mean the charge states $|p\rangle$ and $|n\rangle$ of the nucleon. Similar transformations for the antinucleon have the form

$$\bar{p} \to -\bar{n}, \qquad \bar{n} \to \bar{p}. \tag{2.348}$$

As seen, strong interactions are charge independent or, more widely, isotopically invariant.

2.11.2 Charge Conjugation

The charge conjugation operation is defined only in relativistic theory. Let us represent the ψ operator in the form of the expansion

$$\psi = \sum \frac{1}{\sqrt{2\varepsilon}} \left(a_p e^{-i(\omega t - \vec{p}\cdot\vec{r})} + b_p e^{i(\omega t - \vec{p}\cdot\vec{r})} \right), \tag{2.349}$$

where a_p and b_p are the annihilation and creation operators for a particle with the momentum p, respectively. The charge conjugation operation reduces to the change of the particles to the antiparticles and vise versa; i.e.,

$$C: \quad a_p \to b_p, \quad b_p \to a_p. \tag{2.350}$$

The action of operation (2.350) on operator (2.349) provides the charge conjugate operator ψ^C; it is easy to see that

$$\psi^C(t, \vec{r}) = \psi^+(t, \vec{r}). \tag{2.351}$$

This equality expresses the property of the charge symmetry of the particles and antiparticles. According to relation (2.349), the operator C changes the particle to the antiparticle that is not identical to the particle. As a result, this operator has no eigenfunctions and eigenvalues. For this reason, the charge conjugation operation does not generally lead to new physical consequences. However, there is exclusion. In the application to a system where the number of particles coincides with the number of antiparticles, the operator C has eigenfunctions and eigenvalues. The following transformations are illustrative:

$$C|\Lambda\rangle = |\bar{\Lambda}\rangle, \qquad C|n\rangle = |\bar{n}\rangle, \qquad C|p\rangle = -|\bar{p}\rangle. \tag{2.352}$$

The first relation is obvious, because the Λ hyperon is the isotopic singlet. To prove the last two relations, we write the wave function of a pair of nucleons and antinucleons and the charge conjugation matrix in the explicit form (Lifshitz and Pitaevskii 1973)

$$\psi = \left\| \begin{matrix} \psi_1 \\ \psi_2 \end{matrix} \right\| = \left\| \begin{matrix} p \\ n \\ \bar{n} \\ \bar{p} \end{matrix} \right\| \tag{2.353}$$

and

$$C = \left\| \begin{matrix} 0 & -i\tau_2 \\ i\tau_2 & 0 \end{matrix} \right\| = \left\| \begin{matrix} 0 & 0 & 0 & -1 \\ 0 & 0 & +1 & 0 \\ 0 & +1 & 0 & 0 \\ -1 & 0 & 0 & 0 \end{matrix} \right\|. \tag{2.354}$$

The action of operator (2.354) on wave function (2.353) gives

$$\psi^C = \left\| \begin{matrix} \psi_1^C \\ \psi_2^C \end{matrix} \right\| = C\psi = \left\| \begin{matrix} 0 & -i\tau_2 \\ i\tau_2 & 0 \end{matrix} \right\| \left\| \begin{matrix} \psi_1 \\ \psi_2 \end{matrix} \right\| = \left\| \begin{matrix} -i\tau_2\psi_2 \\ i\tau_2\psi_1 \end{matrix} \right\|. \tag{2.355}$$

We obtain two matrix equations of the second rank, which are solved separately:

$$\psi_1^C = \left\| \begin{matrix} C(p) \\ C(n) \end{matrix} \right\| = \left\| \begin{matrix} 0 & -1 \\ 1 & 0 \end{matrix} \right\| \cdot \left\| \begin{matrix} \bar{n} \\ \bar{p} \end{matrix} \right\| = \left\| \begin{matrix} -\bar{p} \\ \bar{n} \end{matrix} \right\|. \tag{2.356}$$

Therefore,

$$C(p) = -\bar{p}, \qquad C(n) = \bar{n}. \tag{2.357}$$

We write and solve the second matrix equation:

$$\psi_2^C = \left\| \begin{matrix} C(\bar{n}) \\ C(\bar{p}) \end{matrix} \right\| = \left\| \begin{matrix} 0 & +1 \\ -1 & 0 \end{matrix} \right\| \cdot \left\| \begin{matrix} p \\ n \end{matrix} \right\| = \left\| \begin{matrix} n \\ -p \end{matrix} \right\|. \tag{2.358}$$

Hence (Pilkuhn 1979),

$$C(\bar{n}) = n, \qquad C(\bar{p}) = -p. \tag{2.359}$$

Since $C^2 = 1$, it is easy to verify that results (2.358) and (2.359) coincide.

2.11.3 G Transformation

The simultaneous application of two conservation laws leads to new selection rules, which do not follow from any individual law (Lee and Yang 1956).

The joint application of the isotopic transformation T and charge conjugation C is described by the product of both operators and is denoted as G:

$$G = Ce^{i\pi T_3}. \tag{2.360}$$

Since $p \leftrightarrow -\bar{p}$ and $n \leftrightarrow \bar{n}$ under charge conjugation, operator (2.360) provides (Lifshitz and Pitaevskii 1973)

$$G: \quad p \to -\bar{n}, \quad n \to \bar{p}, \quad \bar{p} \to -n, \quad \bar{n} \to p. \tag{2.361}$$

The operator G commutes with the operators of all three isospin components T_1, T_2, and T_3. This is directly verified by writing the explicit expressions for the operators in the form of four-row matrices transforming the nucleon and antinucleon states. Let us represent these states in the form of the column

$$\begin{Vmatrix} p \\ n \\ \bar{n} \\ \bar{p} \end{Vmatrix}$$

and generalize isospin to this case:

$$T_1 = \frac{1}{2} \begin{Vmatrix} \tau_1 & 0 \\ 0 & \tau_1 \end{Vmatrix}, \qquad T_2 = \frac{1}{2} \begin{Vmatrix} \tau_2 & 0 \\ 0 & \tau_2 \end{Vmatrix}, \qquad T_3 = \frac{1}{2} \begin{Vmatrix} \tau_3 & 0 \\ 0 & \tau_3 \end{Vmatrix},$$

$$C = \begin{Vmatrix} 0 & -i\tau_2 \\ i\tau_2 & 0 \end{Vmatrix}, \qquad G = \begin{Vmatrix} 0 & I \\ -I & 0 \end{Vmatrix}. \tag{2.362}$$

Here, 0 and I are two-row matrices.

If the operation G transforms a particle (or a system of particles) to itself, the notion of the G parity appears: the state can remain unchanged or change sign. For this, the baryon number and hypercharge Y ($Y = B + S$, where B is the baryon number and S is strangeness) of the particles should be zero. Indeed, charge conjugation (transition from the particles to the antiparticles) changes the signs of the electric charge Z and hypercharge. Rotation in isospace changes Z, but does not change Y and B. Therefore, the joint application of both transformations changes the numbers Y and B if they are nonzero.

An important property of the G parity is that it is the same for all components of the same isomultiplet. This follows from the commutativity of the operator G with all components of T and, therefore, with all rotations in isospace.

At $Y = 0$, we have $Z = T_3$; therefore, T_3 and thereby T are integers. The isomultiplet with an integer T value is described by a symmetric isospinor of the even rank $2T$, which is equivalent to the irreducible isotensor of the rank T. One of the components of such isomultiplet is a neutral particle ($T_3 = 0$). It corresponds to the isotensor $'\psi_{ih}$ with the nonzero component $'\psi_{33}$. The rotation by the angle π about the x_2 axis leads to the multiplication of this isotensor by $(-1)^T$. The G parity of a neutral particle with the charge parity C is given by the expression

$$G = C(-1)^T. \tag{2.363}$$

According to the above consideration, the G parity of all components of the isomultiplet is thus defined.

For example, let us consider the pion isotriplet ($T = 1$). The charge parity of the π^0 meson is $C = +1$. This follows from the fact that the π^0 meson decays into an even number of particles, namely, into two charge-odd particles (photons). Therefore, the G parity of the pions is $G = -1$. In particular, it follows from this that strong interactions can transfer the system of pions to another system of pions only without change of the parity of the number of particles.

Table 2.3 G and C parities of mesons and baryons

	π^+	π^0	π^-	η	K^+	K^0	$\bar{K}^0$	K^-
G	$-\pi^+$	$-\pi^0$	$-\pi^-$	η	$\bar{K}^0$	K^-	$-K^+$	$-K^0$
C	$-\pi^-$	π^0	$-\pi^+$	η	$-K^-$	$\bar{K}^0$	K^0	$-K^+$

	Σ^+	Σ^0	Σ^-	Λ	p	n	$\bar{n}$	$\bar{p}$
G	$-\bar{\Sigma}^+$	$-\bar{\Sigma}^0$	$-\bar{\Sigma}^-$	$\bar{\Lambda}$	$\bar{n}$	$\bar{p}$	$-p$	$-n$
C	$-\bar{\Sigma}^-$	$\bar{\Sigma}^0$	$-\bar{\Sigma}^+$	$\bar{\Lambda}$	$-\bar{p}$	$\bar{n}$	n	$-p$

The η meson is an isosinglet ($T = 0$), and its charge parity is $C = +1$, because the η meson, as well as the π^0 meson, decays into two photons. Therefore, the η meson has the positive G parity ($G = +1$). Hence, strong interactions cannot lead to the $\eta \rightarrow 3\pi$ decay.

It is desirable to extend the notion of the G parity to other single-particle states. The state of the particle m of the charge multiplet with the momentum p, helicity λ, and third isospin component T_3 is described by the function $|m, T_3, p, \lambda\rangle$, and the state of the corresponding antiparticle, by the function $|\bar{m}, T_3, p, \lambda\rangle$. Since the operation G changes the particle to the antiparticle and does not change the other variables, we can write (Pilkuhn 1979):

$$G|m, T_3, p, \lambda\rangle = \eta_G|\bar{m}, T_3, p, \lambda\rangle. \qquad (2.364)$$

Here, η_G is independent of T_3; the function of the state of the antiparticle is transformed in isotopic space similarly to the function of the state of the particle. As a result, for the multiplet of the particles, we have the relation

$$\eta_C = \eta_G(-1)^{T+T_3} \qquad (2.365)$$

or in the other form

$$\eta_G = \eta_C(-1)^{T+T_3}. \qquad (2.366)$$

For the π^0 meson, $C = +1$, $T = 1$, and $T_3 = 0$; therefore, $G = -1$. At the same time, although the parity of the η meson is $C = +1$, but $G = 1$ because $T = 0$ and $T_3 = 0$. This agrees with the above results.

Formula (2.366) can be considered as the expansion of formula (2.363) to the charged components of the multiplet, which should have the same G parity as the truly neutral component of the multiplet. For the pion isotriplet, $C|\pi^\pm\rangle = G|\pi^\pm\rangle = -|\pi^\mp\rangle$. Let us consider the G and C parities for the hyperons and K mesons. It is substantial that a baryon can emit (virtually) one pion and hyperon can emit one K meson with the satisfaction of the selection rules in the isospin, hypercharge, and baryon number. From the $\Sigma \rightarrow \pi \Lambda$ decay, it follows that $\eta_G(\Sigma) = \eta_G(\pi)\eta_G(\Lambda) = \eta_G(\pi)$, because the G parity of the Λ hyperon is positive. From the $N \rightarrow K\Lambda$ decay, it follows that $\eta_G(K) = \eta_G(N)\eta_G(\Lambda) = \eta_G(N)$. From the $\Xi \rightarrow K\Lambda$ decay, it follows that $\eta_G(\Xi) = \eta_G(K) = \eta_G(N)$. The results thus obtained are presented in Table 2.3.

References

Azhgirey, L.S., et al.: Phys. Lett. **6**, 196 (1963)

Baz, L.: Zh. Eksp. Teor. Fiz. **32**, 628 (1957)

Bilenky, S.M., Ryndin, R.M.: Phys. Lett. **6**, 217 (1963)

Bilenky, S.M., Lapidus, L.I., Ryndin, R.M.: Usp. Fiz. Nauk **84**, 243 (1964)

Bilenky, S.M., Lapidus, L.I., Ryndin, R.M.: Zh. Eksp. Teor. Fiz. **49**, 1653 (1965)

Bilenky, S.M., Lapidus, L.I., Ryndin, R.M.: Zh. Eksp. Teor. Fiz. **51**, 891 (1966)

Bjorken, J.D., Drell, S.D.: Relativistic Quantum Mechanics. McGraw-Hill, New York (1964) (Nauka, Moscow, 1978)

Blatt, J.M., Biedenharn, L.C.: Rev. Mod. Phys. **24**, 258 (1952)

Blatt, J.M., Weisskopf, V.F.: Theoretical Nuclear Physics. Wiley, New York (1952)

Dalitz, R.H.: Proc. Phys. Soc. A **65**, 175 (1952)

Golovin, B.M., Dzelepov, V.P., Nadezhdin, V.S., Satarov, V.I.: Zh. Eksp. Teor. Fiz. **36**, 433 (1959)

Heisenberg, W.: Z. Phys. **77**, 1 (1932)

Hoshizaki, N.: Prog. Theor. Phys. Suppl. **42**, 107 (1968)

Jacob, M., Wick, G.C.: Ann. Phys. **7**, 404 (1959)

Kazarinov, Yu.M.: Candidate's Dissertation in Mathematical Physics, Dubna (1956)

Kumekin, Yu.P., et al.: Zh. Eksp. Teor. Fiz. **46**, 51 (1954)

Lee, T.D., Yang, C.N.: Nuovo Cimento **3**, 749 (1956)

Lifshitz, E.M., Pitaevskii, L.P.: Relativistic Quantum Theory, Part 2. Pergamon, Oxford (1973)

Martin, A.D., Spearman, T.D.: Elementary Particle Theory. North-Holland, Amsterdam (1970)

Matsuda, M.: In: Proceedings of Fifth Workshop on High Energy Spin Physics, Protvino, Russia, p. 224 (1993)

Nishijima, K.: Fundamental Particles. Benjamin, New York (1964) (Mir, Moscow, 1965)

Nurushev, S.B.: Zh. Eksp. Teor. Fiz. **37**(7), 301 (1959), vyp. 1

Nurushev, S.B.: Preprint No. 83-192, IFVE, Inst. for High Energy Physics, Serpukhov, Russia (1983)

Philips, R.T.N.: Nucl. Phys. **43**, 413 (1963)

Pilkuhn, H.: Relativistic Particle Physics. Springer, New York (1979) (Mir, Moscow, 1983)

Puzikov, L.D., Ryndin, R.M., Smorodinskii, Ya.A.: Zh. Eksp. Teor. Fiz. **32**, 592 (1957)

Schiff, L.I.: Quantum Mechanics, 3rd edn. McGraw-Hill, New York (1968) (Russ. transl. of 2nd edn., Inostrannaya Literatura, Moscow, 1957)

Schumacher, C.R., Bethe, H.A.: Phys. Rev. **191**, 1534 (1961)

Smorodinskii, Ya.A.: In: Proceedings of Ninth International Conference on High Energy Physics, Kiev, USSR (VINITI, Moscow, 1960)

Stapp, H.P.: Phys. Rev. **103**, 425 (1956)

Stapp, H.P., et al.: Phys. Rev. **105**, 302 (1957)

Vinternitts, P., Legar, F., Yanout, E.: Preprint No. R-2407, OIYaI, Joint Inst. for Nuclear Research, Dubna (1965)

Wolfenstein, L., Ashkin, J.: Phys. Rev. **85**, 947 (1952)

Chapter 3
Theoretical Models

Since the theory of strong interactions is absent, an attempt to describe a certain characteristic of a reaction, for example, polarization requires the development of various models. Many famous theoreticians were involved in this problem at the initial stage of polarization investigations at accelerators. Their ideas were understandable, because they originated from well established facts. They derived simple analytical formulas for calculating observables, for example, the cross sections or polarizations in a particular reaction. These formulas were used as guiding stars when analyzing experimental data and designing new experiments. As an example, we point to the Fermi model proposed in 1954. As energy increases, theoretical models are painfully complicated. Most of them do not provide analytical dependences of observables on arguments, but express these observables in terms of multiple integrals with numerous fitting parameters. For this reason, any physical picture of a process is lost, and these calculations can likely be performed only by their authors themselves. For example, we point to modern calculations of single-spin asymmetry in the perturbative QCD model. Such complicated numerical calculations of observables, which can be represented by very simple functions in experiments, seem very strange: why all are simple in experiments and are very complicated in theory? In view of such a situation, asymptotic predictions are very attractable. These are the hypothesis of γ_5 invariance or asymptotic relations between amplitudes in cross channels, which are derived on the basis of the Phragmén–Lindelöf theorem. They are also presented in this section.

3.1 Fermi Model

Oxley et al. (1953) reported that a polarized proton beam with a polarization of about 20 % was obtained at the Rochester cyclotron with an energy of $\sim$200 MeV by means of the diffractive scattering of a circulating proton beam by an internal proton target; the polarization was almost doubled when a nuclear target was used. A year later, similar beams were obtained in Berkeley at 310 MeV (Chamberlain

S.B. Nurushev et al., *Introduction to Polarization Physics*,
Lecture Notes in Physics 859, DOI 10.1007/978-3-642-32163-4_3,
© Moskovski Inzhenerno-Fisitscheski Institute, Moscow, Russia 2013

et al. 1954) and almost simultaneously in Dubna at the synchrocyclotron of the Institute for Nuclear Problems, Academy of Sciences of the USSR, at an energy of ~600 MeV (Stoletov and Nurushev 1954). The era of polarization physics at accelerators began.

Fermi (1954) was among those who immediately estimated the importance of these events in spin physics. He proposed a model for explaining the appearance of the polarization of protons in nuclear scattering. Fermi considered the simplest reaction

$$a(1/2) + b(0) \rightarrow a(1/2) + b(0) \tag{3.1}$$

(particle spins are indicated in the parentheses). The elastic scattering of unpolarized protons on carbon nuclei is an example of such a reaction. This process is described by two complex potentials: spin-independent central potential V_c and spin–orbit potential V_s

$$V(r, s) = V_c(r) + V_s(r, s). \tag{3.2}$$

The reaction matrix for reaction (3.1) can be written at a given initial energy in the general form

$$M(\theta) = g(\theta) + h(\theta)\vec{\sigma} \cdot \vec{n}, \tag{3.3}$$

where θ is the nucleon scattering angle in the center-of-mass frame and $\vec{n}$ is the unit vector perpendicular to the scattering plane. It can be shown that observables for experiment (3.1) are as follows:

the differential cross section

$$\frac{d\sigma(\theta)}{d\omega} = \left(|g(\theta)|^2 + |h(\theta)|^2 \right), \tag{3.4}$$

the polarization

$$\frac{d\sigma(\theta)}{d\omega} P(\theta) = 2 \operatorname{Re}\left[g^*(\theta) h(\theta) \right], \tag{3.5}$$

the spin rotation parameter

$$\frac{d\sigma(\theta)}{d\omega} R(\theta) = \left[|g(\theta)|^2 - |h(\theta)|^2 \right] \cos\theta - 2 \operatorname{Im}\left[g^*(\theta) h(\theta) \right] \sin\theta, \tag{3.6}$$

and the longitudinal polarization parameter

$$\frac{d\sigma(\theta)}{d\omega} A(\theta) = \left[|g(\theta)|^2 - |h(\theta)|^2 \right] \sin\theta - 2 \operatorname{Im}\left[g^*(\theta) h(\theta) \right] \cos\theta. \tag{3.7}$$

These four observables constitute the complete set for this experiment; i.e., any new measurement reduces to these four observables.

Since formulas (3.2) and (3.3) describe the same process, they should be related to each other. This relation will be obtained below.

The central potential V_c is well studied in the theory of the scattering of particles by nuclei. The spin–orbit potential was studied in the theory of the shell structure of nuclei and was used only for the bound states of a physical system. To describe reaction (3.1) (the system with continuous spectrum), reasonable physical assumptions on the interaction potentials are necessary. Starting with the Meyer–Jensen shell-model equation and taking into account the Thomas spin precession correction (Thomas 1926), Fermi assumed that the spin–orbit potential is proportional to the gradient of the real part of the central potential, as allowed in the model of the shell structure of nuclei. Certainly, it was very risky to extend the idea of the shell structure of nuclei to energies of several hundreds of MeVs. Fermi also took into account that the Thomas correction is approximately 15 times weaker than the effect necessary in the shell model. Then, relation (3.2) can be rewritten in the form

$$V(r, s) = V_c(r) + V_s(r, s) = V_c \rho(r) + \left(\frac{\hbar}{\mu c}\right)^2 V_s \frac{d\rho(r)}{r\,dr} \vec{\sigma} \cdot \vec{L}. \qquad (3.8)$$

Here, $\rho(r)$ is the nuclear matter distribution density, V_c and V_s are the numerical parameters determined by fitting the experimental data, and $(\frac{\hbar}{\mu c})$ is the Compton wavelength of the π meson, which is introduced in order to express the parameter V_s in the same units as V_c. In practice, the function $\rho(r)$ can have different forms for different potentials (Nurushev 1962; Azhgirei et al. 1963).

In the Born approximation, the amplitudes g and h are related to the interaction potentials by the following transformation (Schiff 1968):

$$g(\theta) = -\frac{2\mu}{\hbar^2} \int V_c(\vec{r}') \exp(i\vec{K} \cdot \vec{r}')d\vec{r}', \qquad (3.9)$$

$$h(\theta) = \frac{2\mu}{\hbar^2} \int V_s(\vec{r}') \exp(i\vec{K} \cdot \vec{r}')d\vec{r}'. \qquad (3.10)$$

Fermi took the central potential in the form of a square well and, performing integration, obtained

$$g(\theta) = \frac{k}{2\pi \hbar v} V_C F(q), \qquad h(\theta) = \frac{k}{2\pi \hbar v} \left(\frac{\hbar k}{\mu c}\right) V_S F(q) \sin\theta. \qquad (3.11)$$

Here, $F(q)$ is the form factor of a spinless particle. According to Eq. (3.11), two amplitudes are related as

$$h(\theta) = c_0 \sin\theta \cdot g(\theta), \qquad (3.12)$$

where the coefficient c_0 is an angle-independent complex parameter. To prove this relation in the general case of an arbitrary well, we write the scattering amplitudes in the Born approximation:

$$g(\theta) = \frac{2\mu}{\hbar^2} \int e^{-i\vec{k}' \cdot \vec{r}} \cdot u_0 e^{i\vec{k} \cdot \vec{r}} d\vec{r}, \qquad h(\theta) = \frac{2\mu}{\hbar^2} \int e^{-i\vec{k}' \cdot \vec{r}} \cdot u_s e^{i\vec{k} \cdot \vec{r}} d\vec{r}, \qquad (3.13)$$

where μ is the reduced mass and $\vec{k}$ and $\vec{k}'$ are the center-of-mass momenta before and after scattering, respectively. Let the central, u_0, and spin–orbit, u_s, potentials be given by the expressions

$$u_0 = v\rho(r), \qquad u_s = v_s \frac{1}{r} \frac{d\rho(r)}{dr} \vec{\sigma} \cdot \vec{L} \qquad (3.14)$$

In the nonrelativistic approximation, this form of the potential follows from the Dirac equation. Substituting Eq. (3.14) into Eq. (3.13) and integrating with respect to the angles, we arrive at the expressions

$$g(\theta) = \frac{2\mu}{\hbar^2} v \int \rho(r) \cdot j_0(qr) r^2 dr, \tag{3.15}$$

$$h(\theta) = i \frac{2\mu k \sin\theta}{\hbar^2 q} v_s \int j_1(qr) \left(\frac{1}{r} \frac{d\rho(r)}{dr} \right) r^3 dr. \tag{3.16}$$

Using the relation

$$x^2 j_0(x) = \frac{d}{dx} [x^2 j_1(x)],$$

we obtain

$$g(\theta) = -\frac{2\mu}{\hbar^2 q} v \int j_1(qr) \left[\frac{1}{r} \frac{d}{dr} \rho(r) \right] r^3 dr. \tag{3.17}$$

Comparison with the expression for the amplitude $h(\theta)$ shows that

$$h(\theta) = -i \frac{v_s}{v} \sin\theta k^2 g(\theta). \tag{3.18}$$

Here, k is the momentum in the center-of-mass frame of the reaction and $\vec{q} = \vec{k} - \vec{k}'$. More recently, the validity of relation (3.18) was proved in a weaker approximation than that used by Fermi, more precisely, under the assumption that the spin–orbit potential is much weaker than the central potential, which can be arbitrary. It is substantial that the spin–orbit potential is proportional to the gradient of the central potential. An additional condition is that the scattering angles are small. The semiclassical approximation was used in the proof given in Koeler (1955), Levintov (1956).

From the definition of polarization, using Fermi relation (3.18), we obtain

$$P(\theta) = -\frac{2(\frac{\hbar k}{\mu c})^2 \operatorname{Re}(\frac{iv^* \cdot v_s}{|v|^2}) \cdot \sin\theta}{1 + (\frac{|v_s|}{|v|})^2 \sin^2\theta (\frac{\hbar k}{\mu c})^4}. \tag{3.19}$$

This formula provides several conclusions. If the spin–orbit potential is a real function, polarization can appear only when the central potential contains the imaginary part. The energy dependence of polarization at a given scattering angle is determined only by the ratio of the potential to the square of the particle momentum in the center-of-mass frame. Polarization vanishes if the imaginary part of the central potential increases unboundedly with energy. Polarization reaches the maximum value P_{max}, which is given by the expression

$$P_{max} = \frac{\operatorname{Re}(iv \cdot v_s)}{|v||v_s|}. \tag{3.20}$$

The position of the polarization maximum is determined by the relation

$$\sin\theta_{max} = \left(\frac{\mu c}{\hbar k} \right)^2 \cdot \frac{|v|}{|v_s|}. \tag{3.21}$$

These predictions are in quite good agreement with experimental data in the energy range 100–1000 MeV.

Simple transformations of formula (3.18) with the Gaussian form of the amplitude g give the relation

$$P(\theta) \propto \frac{d \ln \sigma(\theta)}{d\theta}. \tag{3.22}$$

Physicists multiply "rediscovered" this formula, in particular, recently for quark–quark scattering.

For particular calculations, Fermi took the potentials in the following form:

the central potential

$$V = V_1 + i V_2 \tag{3.23}$$

and the spin-orbit potential

$$H_s = -15 \frac{\hbar}{M^2 c^2} \frac{V_1'(r)}{r} \vec{\sigma} \cdot \vec{x} \times \vec{p}. \tag{3.24}$$

The imaginary part of the potential determines the absorption of a proton in nuclear matter (carbon in this particular case). For simplicity, both potentials were taken in the form of a square well:

$$V_1 + i V_2 = \begin{cases} -B - i B_a & \text{for } r < r_0, \\ 0 & \text{for } r > r_0. \end{cases} \tag{3.25}$$

Fermi performed calculations in the Born approximation with potentials (3.24) and (3.25). The matrix elements of the corresponding potentials were calculated in the coordinate system with the x axis along the incident proton momentum. Scattering with a positive angle θ occurs to the left from the beam in the horizontal xy plane, whereas the polarization vector is directed upwards. Plane waves with the same magnitude of the momenta are taken as the initial and final states. The final momentum is directed along the scattering angle. The gradient of the real part of the central potential is written in the form

$$V_1' = B\delta(r - r_0). \tag{3.26}$$

This relation means that the spin–orbit interaction in the Fermi model is a surface effect. Direct calculations give the following expressions for the matrix elements:

$$\langle 2|V|1\rangle = -4\pi r_0^3 (B + i B_s) \left\{ \frac{\sin q}{q^3} - \frac{\cos q}{q^2} \right\} \tag{3.27}$$

and

$$\langle 2|H_s|1\rangle = -i30\pi \left(\frac{p}{Mc} \right)^2 B r_0^3 \sin\theta \left\{ \frac{\sin q}{q^3} - \frac{\cos q}{q^2} \right\}, \tag{3.28}$$

where

$$q = \frac{2 p r_0}{\hbar} \sin \frac{\theta}{2}. \tag{3.29}$$

Matrix element (3.27) does not flip the proton spin, whereas matrix element (3.28) flips it. The differential cross section for the scattering of a proton by a nucleus is proportional to the sum of the squares of the absolute values of the amplitudes, and the proportionality coefficient is $(\frac{M}{2\pi\hbar^2})^2$:

$$\frac{d\sigma}{d\omega} = \frac{4M^2}{\hbar^4} r_0^6 B^2 \left(\frac{\sin q}{q^3} - \frac{\cos q}{q^2} \right)^2 \left\{ 1 + \left[\frac{B_s}{B} + \frac{15}{2} \left(\frac{p}{Mc} \right)^2 \sin\theta \right] \right\}. \quad (3.30)$$

Formula (3.30) shows that the differential cross section depends on the sign of the scattering angle θ. Therefore, asymmetry can be determined from formula (3.30) as

$$\varepsilon = \frac{\frac{d\sigma}{d\omega}(+) - \frac{d\sigma}{d\omega}(-)}{\frac{d\sigma}{d\omega}(+) + \frac{d\sigma}{d\omega}(-)} = \frac{15(\frac{p}{Mc})^2 \frac{B_a}{B} \sin\theta}{1 + (\frac{B_a}{B})^2 + \frac{225}{4}(\frac{p}{Mc})^4 \sin^2\theta}. \quad (3.31)$$

According to this formula, polarization appears due to interference between the real spin–orbit potential and the imaginary part of the central potential. As mentioned above, when the absorption part of the central potential increases unboundedly, polarization vanishes (the model of an absolutely black nuclear sphere).

For numerical calculations, the cross sections for the pp and pn interactions at a proton energy of 340 MeV were taken to be 24 and 32 mb, respectively. Accepting $r = 1.4 \cdot 10^{-13} A^{1/3}$ cm for the nuclear radius, we can estimate the mean free path in the nucleus as $\lambda = 1.1 \cdot 10^{-13}$ cm. Correspondingly,

$$B_a = \frac{\hbar v}{2\lambda} = 16 \text{ MeV}. \quad (3.32)$$

Here, $v = 0.68c$ is the proton velocity at an energy of 340 MeV. The real part of the potential was taken to be 27 MeV. Table 3.1 presents the angular dependences of the differential cross section and polarization of 340-MeV protons scattered from the carbon target. These predictions of the Fermi model are in quite good agreement with experimental data. Fermi pointed out that caution is necessary when comparing with data for large scattering angles. Inelastic interactions prevail at these angles, but the Fermi model is applicable only for elastic scattering.

Fermi emphasized the importance of the determination of the polarization sign. If his hypothesis based on the nuclear shell model is correct, the polarization sign should be positive. Special experiments confirmed his expectation.

3.2 Chirality Conservation Hypothesis

In 1961 and 1962, Logunov et al. (1962) and Nambu and Jona-Lasinio (1961a) considered some consequences of the hypothesis of the approximate γ_5 invariance of strong interactions. According to this hypothesis (so called the chirality conservation hypothesis), the matrix elements of all physical processes at high energies and momentum transfers s, $|t| \gg m^2$ (m is the highest mass of the particles involved in reactions) can be invariant under the γ_5 transformations of the spinor functions:

$$\Psi_i \to \gamma_5 \Psi_i, \qquad \bar{\Psi}_i \to \bar{\Psi}_i \gamma_5, \qquad \gamma_5^2 = -1. \quad (3.33)$$

Table 3.1 Angular dependence of the differential cross section and polarization of 340-MeV protons scattered from the carbon target

Scattering angle (deg)	Asymmetry $A(\theta)$	$\frac{d\sigma}{d\omega} \cdot 10^{24}$ (for carbon)
0	0	2.7
5	0.40	2.2
10	0.51	1.2
15	0.49	0.3
20	0.42	0.02
30	0.33	0.01
40	0.27	0.03
50	0.23	0.01

As an example, the electromagnetic form factor of the nucleon is considered. The most general expression for the electromagnetic vertex of the nucleon that is invariant under Lorentz and gradient transformations has the form

$$F_\mu(q^2) = F_1(q^2)\gamma_\mu + i\sigma_{\mu\nu}q_\nu F_2(q^2).$$ (3.34)

In view of the anticommutation of γ_5 with γ matrices and the equality $\gamma_5^2 = -1$, the application of the γ_5-invariance condition to this relation gives

$$\lim_{q\to\infty} |q| F_2(q^2) = 0.$$ (3.35)

This relation means that the magnetic form factor of the nucleon decreases at sufficiently high momentum transfers.

Let us discuss the consequences of this hypothesis for the reaction

$$0 + \frac{1}{2} \to 0 + \frac{1}{2},$$ (3.36)

where the numbers are the spins of the particles involved in the reaction. In the relativistic case, if the internal parities of the initial and final states are the same, the reaction matrix for reaction (3.36) has the form (analog of the nonrelativistic case)

$$M = \bar{u}_2(p_2)\left[A(s,t) + \frac{\hat{q}_1 + \hat{q}_2}{2} B(s,t)\right] u_1(p_1).$$ (3.37)

Here, q_1 and q_2 are the boson momenta, p_1 and p_2 are the fermion momenta, and $\hat{q} = \gamma q$.

The relativistic amplitudes $A(s,t)$ and $B(s,t)$ are related to the amplitudes a and b for the nonrelativistic case as:

$$4\pi a = \frac{(\sqrt{s} + m_N)^2 - m_\pi^2}{4s}\left[A + (\sqrt{s} - m_N)B\right],$$

$$4\pi b = \frac{(\sqrt{s} - m_N)^2 - m_\pi^2}{4s}\left[-A + (\sqrt{s} + m_N)B\right].$$ (3.38)

Here, the arguments of the amplitudes $A(s,t)$ and $B(s,t)$ are omitted in order to simplify the expressions.

As follows from the requirement of γ_5 invariance,

$$\lim_{s,t \to \infty} A(s,t) = 0. \tag{3.39}$$

Since only one amplitude $B(s,t)$ remains nonzero, polarization in this case is zero in all binary reactions satisfying the γ_5-invariance condition. In particular, the polarization of the hyperons should asymptotically vanish in the reactions

$$\pi^- + p \to \Lambda + K^0, \qquad \pi^- + p \to \Sigma^- + K^+. \tag{3.40}$$

Relation (3.39) makes it possible to formulate the hypothesis of chirality conservation: if the initial fermion is longitudinally polarized, the final fermion is also longitudinally polarized.

A similar consideration was also performed for nucleon–nucleon scattering. For this case, the relativistic matrix is generally written in the form

$$M(s,t) = \bar{u}(p_2)\bar{u}(k_2)\big[G_1 - G_2\big(\gamma^{(1)} P + \gamma^{(2)} K\big) + G_3\big(\gamma^{(1)} P\big)\big(\gamma^{(2)} K\big)$$
$$- G_4\big(\gamma_5^{(1)}\gamma^{(1)} P\big)\big(\gamma_5^{(2)}\gamma^{(2)} K\big) - G_5\gamma_5^{(1)}\gamma_5^{(2)}\big]u(p_1)u(k_1). \tag{3.41}$$

Here,

$$K = \frac{k_1 + k_2}{2}, \qquad P = \frac{p_1 + p_2}{2}, \qquad Q = k_1 - k_2. \tag{3.42}$$

The requirement of the invariance of nucleon–nucleon elastic scattering matrix (3.41) under the γ_5 transformation noticeably simplifies this matrix to the form

$$M(s,t) = \bar{u}(p_2)\bar{u}(k_2)\big[G_3\big(\gamma^{(1)} P\big)\big(\gamma^{(2)} K\big)$$
$$- G_4\big(\gamma_5^{(1)}\gamma^{(1)} P\big)\big(\gamma_5^{(2)}\gamma^{(2)} K\big)\big]u(p_1)u(k_1). \tag{3.43}$$

It is easy to see that, if the nucleons in the initial state are unpolarized, they remain unpolarized in the final state. The appearance of polarization in pp elastic scattering up to an energy of 300 GeV/c (Kline et al. 1980; Fidecaro et al. 1980, 1981) indicates that γ_5 invariance incompletely appears below this energy and momentum transfers $t = 2\text{--}4$ (GeV/c)^2.

The main consequence of γ_5 invariance is chirality conservation in all reactions where this hypothesis is applicable.

The first indications of approximate γ_5 invariance were obtained in weak interaction processes; a lepton is coupled with the almost conserving axial vector current, as well as with the vector current. In view of this circumstance, we point to the Goldberger–Treiman relation (Goldberger and Treiman 1958) between the pion decay constant and the nuclear β-decay constant, which supports the hypothesis that the axial vector current is asymptotically conserved when the momentum transfer is much higher than the pion mass. However, γ_5 invariance or, in other words, the chirality conservation law strongly differs in nature from the convenient conservation laws. Indeed, chirality, being the fourth component of the axial vector, is not a diagonal matrix on real states such as the nucleon and pion. Thus, chirality can be treated as the mean value of the γ_5 operator, which conserves in time. Such a symmetry can be called "hidden symmetry" (Nambu 1962). In addition to the problem of physical interpretation of chirality, it is very important to experimentally test the helicity conservation hypothesis. As mentioned above, polarization is an important test of the chirality conservation law: the presence of polarization in a certain reaction means the absence of the γ_5 symmetry in this reaction. Let us additionally consider a number of processes that are also sensitive to chirality conservation (Nambu 1962).

Nambu and Lurie (1961b) derived a number of interesting relations, which can be most appropriately verified at high energies and, as mentioned in Nambu (1962), in polarization experiments. Namely, if the pion–nucleon system conserves chirality and the pion mass can be neglected as compared to the momentum transfer, the amplitudes of the reactions $a \rightarrow b$ and $a \rightarrow b + \pi$, where the pion is produced at rest, are generally related to each other as

$$i M_{rad}^{\alpha} = f\left[\chi_N^{\alpha, in}, M\right]. \tag{3.44}$$

Here, f is the pion–nucleon coupling constant and $\chi_N^{\alpha, in} = \tau^{\alpha} \vec{\sigma} \cdot \vec{p}/E_p$ is the isotopic operator of nucleon chirality. The above result was obtained under the assumption that $\chi_N^{\alpha, in} = \chi_N^{in}$, where

$$\chi^{\alpha, in} = \chi_N^{\alpha, in} + \chi_\pi^{\alpha, in} = \chi_N^{\alpha, in} + \frac{1}{f} \int \phi^{\alpha, in} d^3 x. \tag{3.45}$$

Here, the time component of the conserving axial vector current is written. The expression for χ^{α} depends on the accepted model of pion–nucleon interaction, but it can be expected that its asymptotic expressions χ^{in} and χ^{out} are independent of the accepted model.

The application of formula (3.44) to the reactions

$$N + \pi \rightarrow N + \pi, \qquad N + \pi \rightarrow N + \pi + \pi, \tag{3.46}$$

at an energy of about 300 MeV provided reasonable results in agreement with experimental data. However, this agreement is not a convincing reason, because this model is applicable at energies much higher than the pion mass.

The authors modified their model in order to describe similar pion emission processes in electromagnetic and weak interactions. We refer readers interested in these problems to original works cited in Nambu (1962).

3.3 Asymptotic Relations Between Polarizations in the Cross Channels of a Reaction

In the early 1960s, theoreticians derived a number of important asymptotic relations between scattering amplitudes in local field theory (Logunov et al. 1964a, 1964b). These results have been obtained with the use of the following basic principles of local relativistic field theory:

1. Invariance with respect to the inhomogeneous Lorentz group.
2. Microcausality.
3. Spectrality condition (the existence of a complete physical system with positive energy).
4. Unitarity of the S-matrix.
5. Scattering matrix elements are generalized, boundedly increasing functions.

If the general principles of the theory are supplemented by the assumptions that oscillations in scattering amplitudes are absent and the amplitudes increases with energy by a certain law (power or logarithmic), certain experimentally testable relations, for example, the Pomeranchuk relation for total cross sections for particles and antiparticles can be obtained.

Theoreticians also widely used the Phragmén–Lindelöf theorem known in the theory of complex functions. According to this theorem, if a function $f(z)$ is analytic in the upper complex energy half-plane and increases at infinity no faster than z^n, it cannot tend to different limits along the positive and negative semiaxes.

This Phragmén–Lindelöf theorem was used in Logunov et al. (1963, 1964b), Bilenky et al. (1964), and Nguyen-Van-Hieu (1964) to study asymptotic relations between polarizations in cross channels of a reaction. Using the crossing symmetry condition, the authors of those works arrived at the following results.

1. Proton polarizations in the $\pi^+ p$ and $\pi^- p$ scattering processes at a given energy and a given angle are equal in magnitude and opposite in sign.
2. Neutron polarization in the $\pi^- + p \to \pi^0 + n$ charge-exchange process vanishes.
3. Hyperon polarizations in the $\pi + p \to K + Y$ and $\bar{K} + p \to \bar{\pi} + Y$ processes are opposite to each other irrespectively of the relative internal parity of the particles; the same is true for the $K^- + p \to K^0 + \Xi^0$ and $\bar{K}^0 + p \to K^+ + \Xi^0$ processes.
4. The polarizations of the final particles in the $\Sigma + \mathrm{He} \to \mathrm{He}_\Lambda + p$ and $\bar{p} + \mathrm{He} \to \mathrm{He}_\Lambda + \bar{\Sigma}$ processes are opposite to each other if the relative parity of the Σ and Λ particles is $+1$ and are the same if the relative parity is -1.
5. The polarizations of the final particles in the $N + N \to N + N$ and $\bar{N} + N \to \bar{N} + N$ elastic scattering processes, as well as in the elastic scattering of strange particles, $Y + N \to Y + N$ and $\bar{Y} + N \to \bar{Y} + N$, are opposite to each other; moreover, the polarizations of the recoil neutrons, for example, in the $\Sigma^- + p \to \Lambda + n$ and $\bar{\Lambda} + p \to \bar{\Sigma}^- + n$ processes are also opposite to each other.
6. The polarization of the Ξ^- hyperon in the $K^- + p \to K^+ + \Xi^-$ process vanishes.
7. The polarization of recoil protons in the elastic scattering of γ-ray photons by protons also vanishes.

All above statements refer to the polarizations appearing in the collisions between unpolarized particles. The proofs of these statements as they were presented by the authors will be given later. Here, we make a brief comment.

First, we emphasize that the discussion concerns only binary reactions. As known, the cross sections for binary reactions without pomeron exchange decrease rapidly with an increase in energy. This concerns most above reactions. At the same time, the proposed model is valid for asymptotic energies. Thus, the question arises:

What is the asymptotic energy in this model? In order to emphasize the importance of this question, we give an example. According to above item 2, the neutron polarization in the $\pi^- + p \to \pi^0 + n$ reaction should vanish. The polarization in this reaction was measured in the mid-1980s by the PROZA collaboration and appeared to be nonzero. What does this mean? This means that the asymptotic energy for this reaction was not reached. However, measurements at higher energies are impossible, because the cross sections are small. Thus, the problem of the experimental test of this item at asymptotic energies arises.

One of the predictions of the model concerns item 1, i.e., the $\pi^+ p$ and $\pi^- p$ elastic scattering processes. The predicted relation between the polarizations in these processes was confirmed by HERA collaboration measurements in the mid-1970s at a momentum of 40 GeV/c. Therefore, this implies that the asymptotic energy for these reactions is reached. However, both proofs and tests of these predictions refer to low momentum transfers for which many models including nonasymptotic models provide almost the same predictions. Further advance in the development of theoretical models is obviously required.

Note that interest in these theoretical results after more than forty years is still great because many predictions have not yet been tested.

The below proofs of items 1–7 follow Bilenky et al. (1964).

I. To obtain asymptotic relations between the cross sections for various processes, the Phragmén–Lindelöf theorem was used (Nevanlinna 1953). In particular, using this theorem, Meiman (1962) derived the relations between the total cross sections for interaction of particles and antiparticles at high energies, which were previously proved by Pomeranchuk using the dispersion relation technique (Pomeranchuk 1958). Using the general principles of local relativistic field theory and the Phragmén–Lindelöf theorem, Logunov et al. (1963, 1964a, 1964b) generalized the Pomeranchuk relations to the case of differential cross sections for nonzero momentum transfers. Using the same technique, the asymptotic relations between polarizations in cross reactions were derived. Note that the creation of polarized hydrogen targets (Abragam et al. 1962; Chamberlain et al. 1963) and polarized colliding beams (RHIC) significantly facilitates the measurement of polarizations at high energies and can ensure a test of these relations already in near future. We consider the simplest reactions with particles of spins 0 and 1/2. This consideration is purely phenomenological, and we do not discuss the mechanism of the appearance of polarizations at high energies. The basic results of the work are listed above. Below, we give their proofs.

II. We begin with detailed consideration of the simplest case of the scattering of $\pi^\pm$ mesons by nucleons. The amplitudes of the processes

$$\pi^+ + p = \pi^+ + p, \tag{3.47a}$$

$$\pi^- + p = \pi^- + p \tag{3.47b}$$

are given by the expression

$$M_\pm(p'q'; pq) = a_\pm + ib_\pm\left(\frac{\hat{q} + \hat{q}'}{2}\right) \cdot \gamma, \tag{3.48}$$

where q and q' are the initial and final four-momenta of a meson, respectively; $\hat{q} = q \cdot \gamma$ is the product of q by the four-dimensional Dirac matrix γ; p and p' are the respective momenta of the proton; $a_\pm$ and $b_\pm$ are the functions of $s = -(p+q)^2$ and $t = -(p - p')^2$; and subscripts $+$ and $-$ refer to the scattering of the positive and negative mesons, respectively.

The polarization of the recoil proton in the scattering of mesons by unpolarized protons is easily found using the formula (Michel and Wightman 1955; Bilenky and Ryndin 1959)

$$\xi_\mu = \frac{Tr[i\gamma_5\gamma_\mu M \Lambda(p)\bar{M}\Lambda(p')]}{Tr[M\Lambda(p)\bar{M}\Lambda(p')]}, \tag{3.49}$$

where ξ_μ is the four-dimensional polarization vector s orthogonal to the momentum p' and $\Lambda(p)$ and $\Lambda(p')$ are the projection operators separating the states with positive energy. The polarization operator ξ for the particle with the momentum p in any reference frame satisfies the relations $\xi \cdot p = 0$ and $\xi^2 = -P_R^2$, where P_R is the polarization in the R frame of the particle.

From Eqs. (3.48) and (3.49), we obtain the following expression for the polarizations:

$$\xi_\mu^\pm = \frac{2\,\mathrm{Im}\,a_\pm b_\pm^*[t(su - (M^2 - \mu^2)^2)]^{1/2}}{|a_\pm|^2(4M^2 - t) + 2\,\mathrm{Re}\,a_\pm b_\pm^* M(u - s) + 1/4|b_\pm|^2[(u - s)^2 - t(t - 4\mu^2)]}$$
$$\times n_\mu, \tag{3.50}$$

where M and μ are the nucleon and meson masses, respectively; $u = -(p - q')^2 = 2(M^2 - \mu^2) - s - t$; and n_μ is the spacelike unit four-vector proportional to $i\varepsilon_{\mu\nu\rho\sigma}p_\nu q_\rho p_\sigma'$. In the center-of-mass frame, $n_4 = 0$ and $\vec{n} = \frac{\vec{p} \times \vec{p}'}{|\vec{p} \times \vec{p}'|}$.

When $s \gg t$ and M^2, from Eq. (3.50) we obtain

$$\xi_\mu^\pm = \frac{2\,\mathrm{Im}\,a_\pm b_\pm^* \sqrt{-t}}{|2Ma_\pm - sb_\pm|^2 - t|a_\pm|^2} n_\mu. \tag{3.51}$$

According to this expression, polarization is nonzero in the limit $s \to \infty$ and for a given t value only when the behaviors of a and $\frac{sb}{M}$ are the same in the indicated region of the variables s and t.

Let us assume that the asymptotic behavior of a_+ and $\frac{sb_+}{M}$ is given by the expression

$$s^{\alpha(t)}\phi(s, t), \tag{3.52}$$

where $a(t)$ and $\phi(s, t)$ are the functions determined in Logunov et al. (1964a, 1964b). Then, as shown in Logunov et al. (1964a), it follows from the Phragmén–Lindelöf theorem that

$$a_+(-s, t) = e^{i\pi\alpha(t)}a_+(s, t),$$
$$b_+(-s, t) = -e^{i\pi\alpha(t)}b_+(s, t). \tag{3.53}$$

Let us use the cross symmetry condition that relates the amplitudes M_+ and M_-:

$$M_-(p'q'; pq) = \gamma_4 M_+(p - q'; p' - q)\gamma_4. \tag{3.54}$$

According to Eq. (3.54), the functions a and b satisfy the relations

$$a_-(s,t) = a_+^*(u,t),$$
$$b_-(s,t) = -b_+^*(u,t). \tag{3.55}$$

Combining Eq. (3.55) with Eq. (3.53), in the limit $s \to \infty$ and for a given t value, we arrive at the relations

$$a_-(s,t) = e^{-i\pi\alpha(t)} a_+^*(s,t),$$
$$b_-(s,t) = e^{-i\pi\alpha(t)} b_+^*(s,t). \tag{3.56}$$

From Eqs. (3.56) and (3.51), we obtain the following asymptotic relation between the polarizations:

$$\xi_\mu^+(s,t) = -\xi_\mu^-(s,t). \tag{3.57}$$

Thus, if the polarization of the recoil proton in the scattering of π^+ mesons by protons at high energies is nonzero, the polarization of the recoil proton in $\pi^- - p$ scattering is also nonzero and differs from the polarization in $\pi^+ - p$ scattering only in sign.

Let us now consider the charge-exchange process

$$\pi^- + p \to \pi^0 + n. \tag{3.58}$$

Applying cross symmetry, we relate the amplitude of reaction (3.58) in the non-physical region to the amplitude of the process

$$\pi^+ + n \to \pi^0 + p. \tag{3.59}$$

However, in view of charge symmetry, the amplitudes of processes (3.58) and (3.59) coincide with each other. Using cross symmetry and the Phragmén–Lindelöf theorem, we obtain the following relations for the charge-exchange process:

$$a_0(s,t) = e^{-i\pi\alpha(t)} a_0^*(s,t),$$
$$b_0(s,t) = e^{-i\pi\alpha(t)} b_0^*(s,t). \tag{3.60}$$

These relations mean that $\mathrm{Im}\, a_0 b_0^*$ is zero and that, as a result, the polarization of the recoil neutron in process (3.58) vanishes at high energies if the asymptotic behaviors of the functions a_0 and $\frac{sb_0}{M}$ are the same. The different asymptotic behaviors of these two terms constitute a possible cause of nonzero neutron polarization measured by the PROZA collaboration. However, there is another possible cause presented at the beginning of this section; this cause is that 40 GeV is not yet the asymptotic energy.

III. Let us consider the reactions

$$\pi + p \to Y + K, \tag{3.61a}$$
$$\bar{K} + p \to Y + \bar{\pi}. \tag{3.61b}$$

If the internal parities I_i and I_f of the initial and final particles coincide with each other, the amplitude of process (3.61a) is represented in the form (scattering matrix should be a scalar function):

$$M(p'q'; pq) = a + ib\frac{\hat{q} + \hat{q}'}{2}, \tag{3.62}$$

where q and q' are the momenta of π and K mesons, respectively; p and p' are the momenta of the nucleon and hyperon, respectively; and $\hat{q} = \gamma q$ and $\hat{q}' = \gamma q'$.

If the internal parities are opposite ($I_i = -I_f$), the amplitude of this process is written in the form (pseudoscalar matrix)

$$M(p'q'; pq) = c\gamma_5 + id\gamma_5 \frac{\hat{q} + \hat{q}'}{2}. \tag{3.63}$$

In the limit $s \to \infty$ and at a given t value, the polarization in the case $I_i = I_f$ is given by the expression

$$\xi_\mu = \frac{2 \operatorname{Im} ab^* s \sqrt{-t}}{|a(m + m') - sb|^2 - t|a|^2} n_\mu, \tag{3.64}$$

where m and m' are the nucleon and hyperon masses, respectively, and n_μ is defined above. Similarly, the asymptotic expression for the polarization in the case $I_i = -I_f$ has the form

$$\xi_\mu = -\frac{2 \operatorname{Im} cd^* s \sqrt{-t}}{|c(m - m') - sd|^2 - t|c|^2} n_\mu. \tag{3.65}$$

The cross symmetry condition of type (3.54) relates the amplitude of reaction (3.61a) in the nonphysical region to the amplitude of the reaction $\bar{\pi} + Y \to K + p$ reverse with respect to reaction (3.61b). The amplitude of this process is easily related to the amplitude of reaction (3.61b) if PT invariance is used. In view of cross symmetry of form (3.54) and PT invariance, we obtain

$$M(p'q'; pq) = \eta\gamma_4 U M^*(p' - q; p - q') U^{-1} \gamma_4. \tag{3.66}$$

Here, M_c is the amplitude of process (3.61b), U is the matrix satisfying the condition

$$U\gamma_\mu^T U^{-1} = \gamma_\mu,$$

η is the phase factor appearing in the PT transformation. In the case $I_i = I_f$, relation (3.66) yields

$$\begin{aligned} a_c(s, t) &= \eta a^*(u, t), \\ b_c(s, t) &= -\eta b^*(u, t). \end{aligned} \tag{3.67}$$

For the case $I_i = -I_f$, we obtain

$$\begin{aligned} c_c(s, t) &= -\eta c^*(u, t), \\ d_c(s, t) &= \eta d^*(u, t). \end{aligned} \tag{3.68}$$

As in item II, assuming that the asymptotic behaviors of the amplitudes a and c coincide with the behaviors of the amplitudes $\frac{sb}{m}$ and $\frac{sd}{m}$, respectively, and applying the Phragmén–Lindelöf theorem in the limit $s \to \infty$ and at a given t value, we obtain

$$\begin{aligned} I_f = I_i, \qquad a_c(s, t) &= \eta e^{-i\pi\alpha_1(t)} a^*(s, t), \\ b_c(s, t) &= \eta e^{-i\pi\alpha_1(t)} b^*(s, t); \end{aligned} \tag{3.69}$$

$$I_f = -I_i, \qquad b_c(s,t) = -\eta e^{-i\pi\alpha_2(t)} c^*(s,t),$$
$$d_c(s,t) = -\eta e^{-i\pi\alpha_2(t)} d^*(s,t). \tag{3.70}$$

The expressions for the polarization ξ_μ^c appearing in reaction (3.61b) can be obtained from Eqs. (3.64) and (3.65) by changing $a \to a_c$ etc. In view of this circumstance and according to Eqs. (3.64), (3.65), (3.69) and (3.70), polarizations in reactions (3.61a) and (3.61b) for any relative parity are equal to each other in magnitude and are opposite to each other in sign:

$$\xi_\mu^c = -\xi_\mu. \tag{3.71}$$

This also naturally refers to the $K^- + p \to K^0 + \Xi^0$ and $K^0 + p \to K^+ + \Xi^0$ reactions if the spin of the Ξ hyperon is $1/2$.

Note also that the application of cross symmetry condition (3.66) and the Phragmén–Lindelöf theorem to the $K^- + p \to K^+ + \Xi^-$ reaction indicates that the polarization of the Ξ^- hyperon in the limit $s \to \infty$ and at a given t value vanishes for any asymptotic behaviors of individual terms of the amplitude.

Let us show that in the reactions

$$Y_1 + A \to Y_2 + B, \tag{3.72a}$$
$$\bar{Y}_2 + A \to \bar{Y}_1 + B, \tag{3.72b}$$

where A and B are the particles with spin 0 and Y_1 and Y_2 are the particles with spin $1/2$, the polarizations ξ_μ and ξ_μ^c of the particles Y_2 and $\bar{Y}_1$ are opposite to each other:

$$\xi_\mu^c = -\xi_\mu, \tag{3.73}$$

if $I_i = I_f$, and are the same:

$$\xi_\mu^c = \xi_\mu \tag{3.74}$$

if $I_i = -I_f$. To this end, we write the matrix elements of processes (3.72a) and (3.72b) in the form

$$\bar{u}(p')N(p'q'; pq)u(p),$$
$$\bar{u}(p')N_c(p'q'; pq)u(p),$$

where $u(p')$ and $u(p)$ are the spinors with positive energy. The amplitudes N and N_c have the form of Eqs. (3.62) and (3.63) in dependence on the relative parity of the particles. In this case, the cross symmetry condition has the form

$$N_c(p'q'; pq) = \eta' \gamma_4 N^+(p - q'; p' - q)\gamma_4, \tag{3.75}$$

where η' is the phase factor appearing in charge conjugation. If the parity remains unchanged, relation (3.75) leads to relations of form (3.67). If the internal parity changes, the resulting relations differ from relations (3.68) only in the sign in the second relation. Since the expressions for polarizations have form (3.64) and (3.65), we arrive at relations (3.73) and (3.74). Examples of reactions (3.72a), (3.72b) are as follows:

$$\Sigma^+ + He \to He_\Lambda + p \quad \text{and} \quad \bar{p} + He \to He_\Lambda + \bar{\Sigma}^+,$$
$$\Xi^- + He \to He_\Lambda + \Sigma^- \quad \text{and} \quad \bar{\Xi}^- + He \to He_\Lambda + \bar{\Xi}^-. \tag{3.76}$$

IV. Let us consider a more complex case of reactions with particles of spin $1/2$. First, we discuss the reactions

$$\Sigma^- + p \to \Lambda + n \quad \text{and} \quad \bar{\Lambda} + p \to \bar{\Sigma}^- + n. \tag{3.77}$$

The amplitude of processes (3.77) can be written in the form

$$M(p_1' p_2'; p_1 p_2) = a + b\gamma_5^{(2)} + c\gamma^{(2)} K_1 + d\gamma_5^{(2)}\gamma^{(2)} K_1, \tag{3.78}$$

where p_1 and p_1' are the proton and neutron momenta, respectively; p_2 and p_2' are the momenta of the Σ and Λ hyperons, respectively; $K_1 = 1/2(p_1 + p_1')$; and $a, b, c,$ and d are the matrices acting on the spin variables of the nucleons.

Let us assume that the internal parities of the Σ and Λ hyperons are the same; in this case,

$$a = a_1 + ia_2\gamma^{(1)} \cdot K_2, \qquad c = c_1 + ic_2\gamma^{(1)} \cdot K_2,$$
$$b = b_1\gamma_5^{(1)} + ib_2\gamma_5^{(1)}\gamma^{(1)} \cdot K_2, \qquad d = d_1\gamma_5^{(1)} + id_2\gamma_5^{(1)}\gamma^{(1)} \cdot K_2, \tag{3.79}$$

where

$$K_2 = 1/2(p_2 + p_2').$$

The expression for the polarization of the final neutron can be found using a formula similar to formula (3.49). The final result for $s \to \infty$ and a given t value has the form

$$\begin{aligned}
\xi_\mu = \frac{2s\sqrt{-t}}{\sigma} n_\mu \Big\{ &\big[(m + m')^2 - t\big] \operatorname{Im} a_1 a_2^* \\
&- \big[(m' - m)^2 - t\big] \operatorname{Im} b_1 b_2^* + s^2 \operatorname{Im} c_1 c_2^* - s^2 \operatorname{Im} d_1 d_2^* \\
&+ (m + m')s \operatorname{Re}(a_2 c_1^* - a_1 c_2^*) + (m - m')s \operatorname{Re}(b_1 d_2^* - b_2 d_1^*) \Big\}, \tag{3.80}
\end{aligned}$$

where

$$\begin{aligned}
\sigma = &\big[(m + m')^2 - t\big]\big[|a_1|^2(4M^2 - t) - 4Ms \operatorname{Re} a_1 a_2^* + s^2|a_2|^2\big] \\
&+ \big[(m - m')^2 - t\big]\big[-t|b_1|^2 + 2(m'^2 - m^2)M \operatorname{Re} b_1 b_2^* + s^2|b_2|^2\big] \\
&+ s^2\big[|c_1|^2(4M^2 - t) - 4Ms \operatorname{Re} c_1 c_2^* + s^2|c_2|^2\big] \\
&+ s^2\big[-t|d_1|^2 + 2(m'^2 - m^2)M \operatorname{Re} d_1 d_2^* + s^2|d_2|^2\big] \\
&+ (m + m')s\big[(4M^2 - t) \operatorname{Im} a_1 c_1^* - 2Ms \operatorname{Im}(a_1 c_2^* + a_2 c_1^*) + s^2 \operatorname{Im} a_2 c_2^*\big] \\
&+ (m - m')s\big[-t \operatorname{Im} b_1 d_1^* + (m'^2 - m^2) \operatorname{Im}(b_2 d_1^* + b_1 d_2^*) \\
&+ s^2 \operatorname{Im} b_2 d_2^*\big], \tag{3.81}
\end{aligned}$$

m and m' are the masses of the Σ and Λ hyperons, respectively, and M is the nucleon mass.

The cross symmetry condition has the form

$$M_c(p'_1 p'_2; p_1 p_2) = \gamma_4^{(2)} \gamma_4^{(1)} C^{(1)} M^{*T^{(2)}} (-p'_1 p_2; -p_1 p'_2) C^{(1)^{-1}} \gamma_4^{(1)} \gamma_4^{(2)}. \quad (3.82)$$

Here, $M_c(p'_1 p'_2; p_1 p_2)$ is the amplitude of reaction (3.77), $T^{(2)}$ means the transposition of the spin indices of the hyperons, and C is the charge conjugation matrix satisfying the conditions $C\gamma_\mu^T C^{-1} = -\gamma_\mu$ and $C^T = -C$. Note that in expression (3.82) we omit a phase factor appearing in charge conjugation that is insignificant for the further presentation. The amplitude $M_c(p'_1 p'_2; p_1 p_2)$ obviously has the same form as amplitude (3.78). The corresponding coefficients are denoted as a_1^c, a_2^c, etc. From Eq. (3.82), we obtain

$$
\begin{aligned}
a_1^c(s,t) &= a_1^*(u,t), & a_2^c(s,t) &= -a_2^*(u,t), \\
b_1^c(s,t) &= b_1^*(u,t), & b_2^c(s,t) &= -b_2^*(u,t), \\
c_1^c(s,t) &= c_1^*(u,t), & c_2^c(s,t) &= -c_2^*(u,t), \\
d_1^c(s,t) &= -d_1^*(u,t), & d_2^c(s,t) &= d_2^*(u,t),
\end{aligned}
\quad (3.83)
$$

where $u = 2M^2 + m^2 + m'^2 - s - t$.

The neutron polarization in reaction (3.77) is obtained from Eqs. (3.80) and (3.81) by changing $a_1 \to a_1^c$ etc. and changing $m \rightleftarrows m'$.

First, we assume that the functions

$$a_1, \quad sa_2, \quad b_1, \quad sb_2, \quad c_1, \quad sc_2, \quad d_1, \quad sd_2 \quad (3.84)$$

have the same behavior in the limit $s \to \infty$ and for a fixed t value. In this case, according to expressions (3.80) and (3.81), polarization is nonzero. From relations (3.83) and the Phragmén–Lindelöf theorem, we obtain in the limit $s \to \infty$:

$$
\begin{aligned}
a_1^c &= e^{-i\pi\alpha(t)} a_1^*, & a_2^c &= e^{-i\pi\alpha(t)} a_2^*, \\
b_1^c &= e^{-i\pi\alpha(t)} b_1^*, & b_2^c &= e^{-i\pi\alpha(t)} b_2^*, \\
c_1^c &= e^{-i\pi\alpha(t)} c_1^*, & c_2^c &= e^{-i\pi\alpha(t)} c_2^*, \\
d_1^c &= e^{-i\pi\alpha(t)} d_1^*, & d_2^c &= e^{-i\pi\alpha(t)} d_2^*.
\end{aligned}
\quad (3.85)
$$

According to relations (3.85) and the expressions for the polarizations of the neutrons in reactions (3.76) and (3.77), the polarizations in this case are obviously opposite to each other:

$$\xi_\mu^c = -\xi_\mu. \quad (3.86)$$

Let us assume now that the behaviors of functions (3.84) in the limit $s \to \infty$ are not necessarily the same. In this case, for the appearance of polarization, the behaviors of at least two most rapidly increasing functions should be the same (this pair of functions should naturally enter in the form of the product into the numerator of the expression for polarization). Relation (3.86) is obviously satisfied in this case.

We consider the case of the same internal parities of the Σ and Λ hyperons. It can be shown that neutron polarizations in processes (3.76) and (3.77) in the case of opposite internal parities are related through Eq. (3.86).

Let us now consider the elastic scattering of hyperons and antihyperons by nucleons:

$$Y + p \to Y + p, \tag{3.87}$$

$$\bar{Y} + p \to \bar{Y} + p. \tag{3.88}$$

The amplitudes of these processes are given by Eqs. (3.78) and (3.79) with $b_2 = d_1 = 0$. The last conditions follow from time reversal invariance. Therefore, all above relations are also valid for elastic scattering processes (3.87) and (3.88), and the polarizations of the recoil protons in these processes are also related through Eq. (3.86).

It can be shown similarly that the polarizations of the final hyperons and antihyperons in (3.87) and (3.88) satisfy relation (3.86). Note that in the case of eight-term amplitudes describing processes (3.76) and (3.77), conclusions on the relation between the polarizations of the hyperons and antihyperons are impossible without the assumptions on the asymptotic behavior of individual terms of the amplitudes in the limit $s \to \infty$ and at a fixed t value.

From the consideration of the elastic scattering of hyperons, it is clear that the polarizations of nucleons and antinucleons in the nucleon–nucleon and antinucleon–nucleon scattering processes are related to the recoil nucleon polarization through Eq. (3.86).

V. To conclude, we briefly discuss the Compton effect on the proton. The amplitude of the process can be written in the form

$$M(p'k'; pk) = \frac{\varepsilon' \cdot P' \cdot \varepsilon \cdot P'}{P'^2}[A_1 + i A_2 \hat{K}]\frac{\varepsilon' \cdot N \cdot \varepsilon \cdot N}{N^2}[A_3 + i A_4 \hat{K}]$$
$$\cdot \frac{\varepsilon' \cdot P' \cdot \varepsilon \cdot N - \varepsilon' \cdot N \cdot \varepsilon \cdot P'}{\sqrt{2P'^2 N^2}} i \gamma_5 A_5$$
$$+ \frac{\varepsilon' \cdot P \cdot \varepsilon \cdot N - \varepsilon' \cdot N \cdot \varepsilon \cdot P'}{\sqrt{2P'^2 N^2}} \gamma_5 \hat{K} A_6, \tag{3.89}$$

where p and p' are the momenta of the initial and final protons, respectively; k and ε (k' and ε') are the momentum and polarization of the initial (final) photon, respectively; $K = 1/2(k + k')$; $P' = P - \frac{PK}{K^2}K$; $P = 1/2(p + p')$; and $N_\alpha = i\varepsilon_{\alpha\beta\gamma\delta}P'_\beta K_\gamma (k - k')_\delta$.

In the limit $s \to \infty$ and at a fixed t value, the recoil proton polarization is given by the expression

$$\xi_\mu = s\sqrt{-t}2 \operatorname{Im}(A_1 A_2^* + A_3 A_4^*)n_\mu \big((|A_1|^2 + |A_3|^2)(4M^2 - t) + (|A_2|^2 + |A_4|^2)s^2$$
$$- 4\operatorname{Re}(A_1 A_2^* + A_3 A_4^*)Ms - t|A_5|^2 + s^2|A_6|^2\big)^{-1}. \tag{3.90}$$

Applying the cross symmetry condition

$$A_{1,3,5,6}(s,t) = A_{1,3,5,6}^*(u,t),$$
$$A_{2,4}(s,t) = -A_{2,4}^*(u,t) \tag{3.91}$$

and the Phragmén–Lindelöf theorem, we can verify that the proton polarization vanishes in the limit $s \to \infty$ and at a fixed t value irrespectively of the assumptions on the asymptotic behaviors of the amplitudes.

3.4 Regge Model

In the early 1960s, physicists discussed with great enthusiasm the idea of Italian theoretician Regge about the analytic continuation of the scattering amplitude in the complex plane of the orbital angular momentum l (de Alfaro and Regge 1965).

As known from works by Mandelstam, the physical amplitudes in the s, t, and u channels constitute a united analytic function satisfying the cross symmetry condition. This means that, knowing the amplitude in one of these channels, one can determine the amplitudes in other channels. Regge's idea was very productive because it provided the explicit prediction of the energy dependence in the s channel in terms of the poles of the amplitude in the t channel. Below, we give brief presentation of the mathematical implementation of Regge's idea.

As we know from Sect. 2.8, any of five physical amplitudes can be expanded in partial waves. For simplicity, we consider the case of the spinless scattering amplitude in the t channel and expand it in the Legendre polynomials:

$$a(s, t) = \sum_l (2l + 1) a_l(t) P_l(\cos \theta_t). \tag{3.92}$$

Here, s is the square of the total energy in the s channel, t is the square of the total energy in the t channel, and θ_t is the scattering angle in the t channel. Extending the region of determining the parameter l to the entire complex plane and using the Sommerfeld–Watson transformation, we can arrive at the formula

$$a(s, t) = \sum_i \beta_i(t) \eta_i(t) s^{\alpha_i(t)-1}; \tag{3.93}$$

where $\alpha_i(t)$ are called Regge pole trajectories. It is assumed that only poles exist in the complex l plane and they lie on one linear trajectory

$$\alpha(t) = \alpha(0) + \alpha'(0)t. \tag{3.94}$$

Experimental evidences allow the assumption that poles exist only at integer l values and $t = m_l^2$ at these points (nonphysical region). Under these assumptions, the partial amplitude $a_l(t)$ is expressed as

$$a_l(t) \approx \frac{\beta(t)}{1 - \alpha(t)} \approx \frac{\beta(t)}{\alpha'(0)(m_l^2 - t)}. \tag{3.95}$$

Here, the function $\beta(t)$ is the residue of the amplitude at the pole. Substituting this expression into expansion (3.92) and passing to the limit $s \to \infty$ at a fixed t value, we arrive at the following form of the amplitude in the s channel:

$$a(s, t) = a(t) s^{\alpha(t)}. \tag{3.96}$$

This expression has several interesting consequences. The first of them refers to the differential cross section for elastic scattering. Let us write this cross section:

$$\frac{d\sigma}{dt} \approx \frac{1}{s^2}|a(s,t)|^2 \approx f(t)s^{2\alpha(t)-2}. \tag{3.97}$$

Using relation (3.94), this expression can be represented in the form

$$\frac{d\sigma}{dt} \approx f(t)s^{2\alpha(t)-2} \approx f(t)\left(\frac{s}{s_0}\right)^{2\alpha(0)-2} e^{\{[2\alpha'(0)\ln(\frac{s}{s_0})]t\}} \approx f(t)\left(\frac{s}{s_0}\right)^{2\alpha(0)-2} e^{Bt}, \tag{3.98}$$

where $B = 2\alpha'(0)\ln\frac{s}{s_0}$ and s_0 is the normalization parameter.

Therefore, the slope parameter of the cross section for elastic scattering should increase logarithmically with energy if the trajectory slope is $\alpha'(0) \neq 0$, and the diffraction scattering cone should be similarly narrowed. This prediction was first confirmed in the experiments on pp elastic scattering at the U-70 accelerator in the energy range 10–70 GeV and in the range of the squares of momentum transfers $t = 0.02$–0.2 (GeV/c)^2. Similar measurements of the slope parameter in $\pi^- p$ elastic scattering were also carried out on the secondary beams of the accelerator U-70 at the Institute for High Energy Physics in the energy range 25–55 GeV and in the range of the squares of momentum transfers 0.05–0.5 (GeV/c)^2. These experiments show that the diffraction cone is not narrowed in pion–nucleon scattering. This means that poles of different types are leading in these two processes, although the largest contribution is expected from the vacuum (pomeron) pole. A contradiction between experimental results and predictions of the Regge model is outlined.

The second consequence was obtained by comparing the cross section for the charge-exchange reaction

$$\pi^- + p = \pi^0 + n \tag{3.99}$$

with the prediction of the Regge model. This reaction should proceed through the exchange by one ρ pole with isospin $I = 1$ and parity $P = G = +$. If this is the case, polarization in this reaction should be zero. This is due to the fact that both amplitudes of reaction (3.99) with and without spin flip in the process of exchange by one pole acquire the same phase and the product of these amplitudes is real. Since the polarization is proportional to the imaginary part of the product of these amplitudes, it is zero. We will return to this problem later.

Let us discuss some consequences of the application of the Regge pole model to the polarization parameters in the πp and pp elastic scattering processes.

The pion–nucleon scattering matrix in the center-of-mass frame is written in the form

$$M = G \pm iH(\vec{\sigma} \cdot \vec{n}), \tag{3.100}$$

where G and H are the spin-independent and spin-dependent amplitudes in the s channel, respectively, which are functions of s and t; $\vec{n} = \vec{k}_i \times \vec{k}/|\vec{k}_i \times \vec{k}|$ is the unit vector perpendicular to the scattering plane; and signs $+$ and $-$ refer to the $\pi^+ p$- and $\pi^- p$-scattering processes, respectively.

As mentioned above, when only one pole exists, polarization in the Regge model is zero. Let us consider the case with two poles: vacuum and nonvacuum. The amplitude in the s channel is represented in the form of the sum of the amplitudes from two poles in the t channel:

$$G = G_1^t \pm G_2^t, \qquad H = H_1^t \pm H_2^t. \tag{3.101}$$

By the definition of the polarization P,

$$I_0 P = \mathrm{Im}\big(GH^*\big) = \mathrm{Im}\big[\big(G_1^t + G_2^t\big)\big(H_1^t + H_2^t\big)^*\big] = \mathrm{Im}\big(G_1^t H_2^{t*} + G_2^t H_1^{t*}\big) \tag{3.102}$$

for $\pi^+ p$ scattering. Here, we take into account that the product of two functions of the same pole is real and, therefore,

$$\mathrm{Im}\big(G_k H_k^*\big) = 0, \tag{3.103}$$

where $k = 1, 2, 3, \ldots, n$, and I_0 is the differential cross section for the scattering of unpolarized protons. Polarization in the case of $\pi^- p$ scattering is given by the formula

$$I_0 P = \mathrm{Im}\big(GH^*\big) = \mathrm{Im}\big[\big(G_1^t - G_2^t\big)\big(H_1^t - H_2^t\big)^*\big] = -\mathrm{Im}\big(G_1^t H_2^{t*} + G_2^t H_1^{t*}\big). \tag{3.104}$$

Comparing expressions (3.102) and (3.104) under the assumption of the equality of the cross sections I_0 for $\pi^+ p$- and $\pi^- p$-scattering processes, we arrive at the relation

$$P^+(s, t) = -P^-(s, t). \tag{3.105}$$

This relation means that the polarizations in the processes of the elastic scattering of π^+ and π^- mesons should be mirror symmetric. Experimental data confirm this prediction to a certain extent (at low t values).

Relation (3.105) can be applied to the polarization in the elastic scattering of protons and antiprotons. The highest energy at which data on the polarizations of particles and antiparticles is 40 GeV for the $\bar{p}p \to \bar{p}p$ reaction (Bruneton et al. 1976b) and 45 GeV for the $pp \to pp$ reaction (Bruneton et al. 1975, 1976a). These data are in sharp disagreement with relation (3.105); the signs of the polarizations for these reactions in the experiment are the same, whereas they should be different according to the Regge pole model.

The next problem is the energy dependence of the polarization at a fixed t value and in the asymptotic s limit. We again consider the case of two poles, vacuum and nonvacuum (e.g., the ρ pole in the case of pion–nucleon scattering or the ω pole in the case of nucleon–nucleon scattering).

We write the pion–nucleon scattering matrix in terms of the contributions from Regge poles in the t channel in the parameterization accepted in Rarita et al. (1968):

$$M(\pi N) = \sum_j (\pm) \frac{s^{1/2}}{8\pi} \xi_j \left(\frac{s}{s_0}\right)^{\alpha_j - 1} \eta_{\pi j}(\eta_{Nj} + i\phi_{Nj}\vec{\sigma} \cdot \vec{N}),$$

$$G = \sum_j (\pm) \frac{s^{1/2}}{8\pi} \xi_j \left(\frac{s}{s_0}\right)^{\alpha_j - 1} \eta_{\pi j}(\eta_{Nj}), \qquad (3.106)$$

$$H = \sum_j (\pm) \frac{s^{1/2}}{8\pi} \xi_j \left(\frac{s}{s_0}\right)^{\alpha_j - 1} \eta_{\pi j}(i\phi_{Nj}\vec{\sigma} \cdot \vec{N}),$$

where the signs $(\pm)$ are determined by the signature of poles ξ_j,

$$\xi_j(t) = \frac{1 + \tau e^{-i\pi\alpha(t)}}{\sin[\pi\alpha(t)]},$$

and by a particular process; j marks poles and the summation is performed over all poles; α_j is the trajectory of the j-th pole; under the assumption that the residue function is factorized, the parameter $\eta_{\pi j}$ is the residue at the upper vertex of the t-channel diagram (pion vertex) and η_{Nj} and ϕ_{Nj} parameterize the residue at the lower vertex of the diagrams (nucleon vertex).

Let us determine the differential cross section I_0, polarization P, and tensor polarization parameters D and R for pion–nucleon elastic scattering. To this end, we recall some general formulas.

The density matrices of the initial, $\hat{\rho}_i$, and final, $\hat{\rho}_f$, states of interacting particles are related by the scattering matrix M as

$$\hat{\rho}_f = M\hat{\rho}_i M^+. \qquad (3.107)$$

Then, the mean value of any spin operator $\hat{\Sigma}$ in the final state is determined by the expression

$$\bar{\Sigma} = Tr(\hat{\rho}_f \hat{\Sigma})/Tr(\hat{\rho}_f). \qquad (3.108)$$

The differential cross section is given by the formula

$$I_0 = Tr(\hat{\rho}_f)/Tr(\hat{\rho}_i) = |G|^2 + |H|^2. \qquad (3.109)$$

Polarization is determined from formula (3.108) where the operator $\hat{\Sigma}$ is replaced by the spin operator $\hat{s} = \frac{1}{2}\hat{\sigma}$:

$$I_0 P = Im(GH^*). \qquad (3.110)$$

The substitution of formula (3.106) into relation (3.110) shows again that polarization is zero in the presence of only one pole.

We return now to the energy dependence of polarization in the presence of two Regge poles. For cross section, we take into account only the vacuum pole. In this case, the energy dependence of cross section has the form

$$I \propto s \left(\frac{s}{s_0}\right)^{2\alpha_P(t) - 2}. \qquad (3.111)$$

The term $\mathrm{Im}(GH^*)$ appearing in polarization has the energy dependence

$$\mathrm{Im}(GH^*) \propto s\left(\frac{s}{s_0}\right)^{\alpha_P + \alpha_R - 2}. \tag{3.112}$$

Here, $\alpha_R(t)$ and $\alpha_P(t)$ are the reggeon and pomeron trajectories, respectively.

The ratio of expressions (3.112) and (3.111) gives the energy behavior of polarization

$$P \propto \left(\frac{s}{s_0}\right)^{\alpha_R(t) - \alpha_P(t)}. \tag{3.113}$$

This theoretical prediction of the Regge pole model agrees as a whole with the experimental data in the energy range below 40 GeV for $-t$ values below $0.5\ (\mathrm{GeV}/c)^2$. For higher values of the invariant momentum transfer, the experimental accuracy is insufficient for a quantitative test of relation (3.113). This statement is illustrated below.

We consider now nucleon–nucleon elastic scattering. In this case, the scattering matrix in the Wolfenstein representation with the inclusion of spins of both interacting particles is written in the general form

$$M(NN) = a + ic(\vec{\sigma}^{(1)} + \vec{\sigma}^{(2)}) \cdot \vec{N} + m(\vec{\sigma}^{(1)} \cdot \vec{N})(\vec{\sigma}^{(2)} \cdot \vec{N})$$
$$+ (g + h)(\vec{\sigma}^{(1)} \cdot \vec{P})(\vec{\sigma}^{(2)} \cdot \vec{P}) + (g - h)(\vec{\sigma}^{(1)} \cdot \vec{K})(\vec{\sigma}^{(2)} \cdot \vec{K}).$$

In the Regge pole model, this matrix has the form (Rarita et al. 1968)

$$M(NN) = \sum_j (\pm)\frac{s^{1/2}}{8\pi}\xi_j\left(\frac{s}{s_0}\right)^{\alpha_j - 1}\left(\eta_{Nj} + i\phi_{Nj}\vec{\sigma}^{(1)} \cdot \vec{N}\right)$$
$$\times \left(\eta_{Nj} + i\phi_{Nj}\vec{\sigma}^{(2)} \cdot \vec{N}\right),$$

$$a = \sum_j (\pm)\frac{s^{1/2}}{8\pi}\xi_j\left(\frac{s}{s_0}\right)^{\alpha_j - 1}\eta_{Nj}^2, \tag{3.114}$$

$$c = \sum_j (\pm)\frac{s^{1/2}}{8\pi}\xi_j\left(\frac{s}{s_0}\right)^{\alpha_j - 1}i\eta_{Nj}\phi_{Nj}(\vec{\sigma}^{(1)} \cdot \vec{N} + \vec{\sigma}^{(2)} \cdot \vec{N}),$$

$$m = \sum_j (\pm)\frac{s^{1/2}}{8\pi}\xi_j\left(\frac{s}{s_0}\right)^{\alpha_j - 1}(-\phi_{Nj}^2)(\vec{\sigma}^{(1)} \cdot \vec{N})(\vec{\sigma}^{(2)} \cdot \vec{N}).$$

In this model, poles with integer spins are considered taking into account the following factors:

- the factorization of the residue function; this allows the use of the same parameters for the pion–nucleon and nucleon–nucleon scattering processes;
- in the asymptotic region, $|g - h| \ll |g + h|$ and $|g + h|$ decreases as s^{-1} as compared to the amplitudes a, c, and m (Sharp and Wagner 1963; Wagner 1963); hence, we can take $g = h = 0$;

- according to relation (3.114), $c_j^2 = a_j m_j$. Thus, the nucleon–nucleon scattering matrix in the Regge pole model includes only two independent amplitudes; for this reason, the calculation of observables is significantly simplified.

The antinucleon–nucleon scattering matrix is given by a similar formula, but the contribution from the pole with the odd signature should appear with the opposite sign as compared to nucleon–nucleon scattering.

The results of Bruneton et al. (1976c) on the comparison of experimental data with Regge-model predictions are presented in Fig. 3.1. The experimental results on measuring polarization were processed by the Regge model formula

$$P(t) = A(t)s^{\alpha_{eff}(t)}, \quad \text{where } \alpha_{eff} = \alpha_R + \alpha_P - 2.$$

All data from 6 to 45 GeV/c were involved in procession for determining $\alpha_{eff}(t)$. For each fixed $-t$ value, the data on polarization as a function of s are fit. In this way, five $\alpha_{eff}(t)$ values for the $K^+ p \rightarrow K^+ p$ reaction (closed points with error bars in Fig. 3.1) and seven points for the $pp \rightarrow pp$ reaction (open points with error bars in Fig. 3.1) were obtained in the range 0.1 $(\text{GeV}/c)^2 \leq |t| \leq 0.5$ $(\text{GeV}/c)^2$. Good agreement with the Regge model is observed for the first reaction with the following dependences for trajectories: $\alpha_R = 0.52 + 0.93t$ for the ρ pole and $\alpha_P = 1 + 0.27t$ for the pomeron. These trajectories are shown by the solid lines in Fig. 3.1. The dependence $\alpha_{eff} = \alpha_R + \alpha_P - 2$, which is compared to the points extracted from the experiments, is also shown by the solid line. As seen, the Regge model predictions are in good agreement with the results on polarization in $K^+ p$ elastic scattering in the energy range 6–45 GeV. The same agreement was observed for the elastic scattering of positive and negative pions by protons in the same energy range in Bruneton et al. (1976c), where another prediction of the Regge pole model, namely, mirror symmetry between polarizations in the $\pi^+ p \rightarrow \pi^+ p$ and $\pi^- p \rightarrow \pi^- p$ reactions, is also confirmed experimentally. Thus, there are some facts in favor of the simple Regge pole model.

However, Fig. 3.1 indicates that the data on polarization in proton–proton elastic scattering are in sharp contradiction with the predictions of the simple Regge pole model. The open points in this figure lie systematically below the theoretical line. As stated in Bruneton et al. (1976c): "A fast decrease in polarization in the range $|t| = 0.5–1.0$ $(\text{GeV}/c)^2$ with an increase in energy can be explained under the assumption that, in contrast to the ordinary picture, interference between the amplitudes with and without change in the helicity of the pomeron exchange makes contribution to polarization in pp elastic scattering at 45 GeV/c." Previously in 1975, the same team (HERA collaboration) arrived at the same conclusion studying the energy dependence of the ratio of the spin-flip to spin-nonflip amplitudes. In addition, taking into account the presence of polarization in the pion charge-exchange reaction, which should be absent according to the Regge pole model, this model should be significantly modified. Some problems in this way will be discussed later.

Pion–pion scattering in the pole model is interesting from the theoretical point of view. Since pions are spinless particles, the formulas are shorter, but the behavior of

Fig. 3.1 Experimental test of the Regge pole model predictions or the energy dependence of polarization

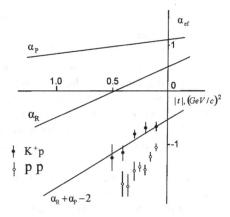

the scattering amplitude is characteristic for more complex cases. In particular, the amplitude for spinless particles has the asymptotic form

$$M = \xi(2s_0)(s/s_0)^\alpha \eta_\pi \eta'_\pi. \tag{3.115}$$

Here, η and η' are the residue functions in the upper and lower vertices of the Feynman diagram in the t channel of the reaction, respectively. These functions are sometimes called the t-dependent couplings between pairs of the initial and final pions. From the optical theorem, the total cross section for $\pi\pi$ interaction is expressed in the form

$$\sigma_{\pi\pi}(s) = \operatorname{Im} M(s, t = 0)/\left[s\left(s - 4m_\pi^2\right)\right]^{1/2} \to \sum_i \tau_i (s/s_{0i})^{\alpha_i - 1} (\eta_{\pi i})^2. \tag{3.116}$$

When deriving this formula, it is assumed that the function η_π is real for energies t below the particle production threshold.

The differential cross section for $\pi\pi$ scattering is written in the form

$$\frac{d\sigma_{\pi\pi}}{dt} = \frac{1}{4\pi} \sum \operatorname{Re}(\xi_i^* \xi_j) \left(\frac{s}{s_{0i}}\right)^{\alpha_i - 1} \left(\frac{s}{s_{0j}}\right)^{\alpha_j - 1} (\eta_{\pi i})^2 (\eta_{\pi j})^2. \tag{3.117}$$

Below, by analogy with the formulas for $\pi\pi$ scattering, in the framework of the Regge pole model (Rarita et al. 1968), (Wagner 1963), we present the determinations of some polarization tensors, relativistic formulas expressing these quantities in terms of the amplitudes, and a scheme for their experimental measurement.

1. For the total cross section:
   πN scattering

$$\sigma_{\pi N}(s) = \operatorname{Im} M(s, t = 0)/\left[s\left(s - (m_N + m_\pi)^2\right)\right]^{1/2}$$
$$\to \sum_i \tau_i (s/s_{0i})^{\alpha_i - 1} \eta_{\pi i} \eta_{Ni}; \tag{3.118}$$

   NN scattering

$$\sigma_{NN}(s) = \operatorname{Im} M(s, t = 0)/\left[s\left(s - 4m_N^2\right)\right]^{1/2} \to \sum_i \tau_i (s/s_{0i})^{\alpha_i - 1} \eta_{Ni}^2. \tag{3.118a}$$

Here, only spin-nonflip amplitudes make contributions and, correspondingly, only the functions η appear.

Formulas (3.118) and (3.119) were widely used (see, e.g., Rarita et al. 1968 discussed here) to process data on total cross sections in the early 1970s. The Regge model including the pomeron with $\alpha_P(0) = 1$ gives constant total cross sections in agreement with the measurements available at that time at energies below 30 GeV. However, an increase in total cross sections at energies above 40 GeV (Serpukhov effect) observed at the U-70 accelerator (IHEP) was the strongest stroke on the Regge pole model: this effect could not be explained in this model.

The differential cross section for the scattering of an unpolarized particle on an unpolarized target:

for pion–nucleon elastic scattering

$$I_0(\pi N) = |G|^2 + |H|^2$$
$$= \frac{1}{4\pi} \sum \mathrm{Re}(\xi_i^* \xi_j) \left(\frac{s}{s_{0i}}\right)^{\alpha_i - 1} \left(\frac{s}{s_{0j}}\right)^{\alpha_j - 1}$$
$$\times \eta_{\pi i} \eta_{\pi j} (\eta_{Ni} \eta_{Nj} + \phi_{Ni} \phi_{Nj}), \tag{3.119}$$

for nucleon–nucleon elastic scattering

$$I_0 = |a|^2 + 2|c|^2 + |m|^2$$
$$= \frac{1}{4\pi} \sum \mathrm{Re}(\xi_i^* \xi_j) \left(\frac{s}{s_{0i}}\right)^{\alpha_i - 1} \left(\frac{s}{s_{0j}}\right)^{\alpha_j - 1}$$
$$\times (\eta_{Ni} \eta_{Nj} + \phi_{Ni} \phi_{Nj})^2. \tag{3.120}$$

2. Polarization for elastic scattering:

   pions by nucleons

$$(I_0 P)_{\pi N} = Tr[MM^+ \sigma_N^{(1)}] = 2\,\mathrm{Im}(GH^*)$$
$$= \frac{1}{2\pi} \sum_{ij} \mathrm{Im}(\xi_i^* \xi_j) \left(\frac{s}{s_{0i}}\right)^{\alpha_i - 1} \left(\frac{s}{s_{0j}}\right)^{\alpha_j - 1}$$
$$\times (\eta_{Ni} \eta_{Nj} \phi_{Ni} \phi_{Nj}), \tag{3.121}$$

   nucleons by nucleons

$$(I_0 P)_{NN} = Tr[MM^+ \sigma_N^{(1)}] = 2\,\mathrm{Im}[(a+m)c^*]$$
$$= \frac{1}{2\pi} \sum_{ij} \mathrm{Im}(\xi_i^* \xi_j) \left(\frac{s}{s_{0i}}\right)^{\alpha_i - 1} \left(\frac{s}{s_{0j}}\right)^{\alpha_j - 1}$$
$$\times (\eta_{Ni} \eta_{Nj} + \phi_{Ni} \phi_{Nj}) \phi_{Ni} \phi_{Nj}. \tag{3.122}$$

When one dominant pole makes contribution, the phases of all scattering amplitudes are the same and polarization vanishes. If contribution from only one dominant pole, for example, pomeron remains at asymptotic energies, polarization should vanish.

3.5 Elements of Quantum Chromodynamics

According to modern representations, strongly interacting particles (hadrons) consist of quarks interacting between each other through gluon exchange. Presently known light and heavy quarks are classified into three generations. Each generation contains two quarks. As known, leptons are also classified among three generations each with two leptons. The central point in this scheme is the introduction of a new degree of freedom: it is assumed that each quark, in addition to the other quantum numbers, is characterized by one of three colors, conventionally red, yellow, and blue (Eidelman et al. 2004). In turn, gluons have eight colors. The concept of colors was based on a number of experimental results, some of which are listed below (Close 1979).

1. The large experimental cross section for the annihilation of electron–positron pairs into hadrons. The theory predicts the ratio

$$R = \frac{\sigma(e^+e^- \to \text{hadrons})}{\sigma(e^+e^- \to \mu^+\mu^-)} = \sum_i e_i^2 = \begin{cases} 2/3 & \text{without color,} \\ 2 & \text{for three colors.} \end{cases} \quad (3.123)$$

The experiments agree with the hypothesis of three colors. The contribution from charmed particles was disregarded in these calculations. Under the assumption that these particles are also three-color, the inclusion of this contribution improves the agreement with the experimental data as compared to the case of one color.

2. The decay

$$\pi^0 \to \gamma + \gamma, \quad (3.124)$$

is also in agreement with the hypothesis of three colors of the quarks.

3. The cross section for the production of lepton pairs by hadrons (Drell–Yan process) under the hypothesis of three colors of the quarks should be one third of the value for the single-color quarks. Experiments confirm such a suppression of the cross sections.

4. Classical evidence in favor of colored quarks is the existence of the Δ^{++} nucleon resonance that consists of three u quarks and has spin $3/2$. As a fermion, it should be governed by Fermi–Dirac statistics; however, owing to the identity of all constituent quarks, it is an explicitly symmetric system. Only by introducing a new degree of freedom, color, the wave function of this isobar can be antisymmetrized in the form

$$\psi(\Delta^{++}) \propto \left(u_R^\uparrow u_B^\uparrow u_Y^\uparrow\right)\varepsilon_{RBY}, \quad (3.125)$$

where $u_k^\uparrow$, $k = R, B, Y$ are the quark wave functions and ε_{RBY} is the antisymmetric tensor of the third rank.

Table 3.2 presents the symbols of the quarks, their flavors (additive quantum numbers I_z, S, C, B, and T) and electric charges Q. We accept the convention that any flavor quantum number of a quark has the same sign as the charge of this quark.

Table 3.2 Quark parameters

Parameter	d	u	S	c	b	T
Electric charge Q	$-1/3$	$+2/3$	$-1/3$	$+2/3$	$-1/3$	$+2/3$
Isospin z-component I_z	$-1/3$	$+1/2$	0	0	0	0
Strangeness S	0	0	-1	0	0	0
Charm C	0	0	0	$+1$	0	0
Beauty B	0	0	0	0	-1	0
Top T	0	0	0	0	0	$+1$
Quark mass M	4–8 MeV	1.5–4 MeV	80–130 MeV	1.15–1.35 MeV	4.1–4.4 MeV	174.3 ± 5.1 GeV

This convention is convenient. Thus, in particular, the strangeness of the K^+ meson is $+1$, whereas both the charm and strangeness of D_s^- are -1. All quarks have spin $1/2$ and baryon number $1/3$. It is accepted that a quark has positive parity. Each quark has three colors, and quarks interact between each other through color-gluon exchange. A gluon has eight colors, its spin is 1, and gluon mass is zero. Since the gluon has no electric charge, it is not subjected to electromagnetic interaction directly, but only indirectly. Thus, the gluon is an analog of the photon as an interaction carrier. In contrast to the photon carrying the electromagnetic interaction, the gluon carries the strong color interaction. The gluon does not interact with usual electric charge, but has eight colors, which are sources of the strong interaction.

The theory of strong interactions, as well as the theory of electroweak interaction, is non-Abelian. It is based on the assumption that a quark with an arbitrary flavor is a three-colored object. A gauge boson (gluon) ensures interaction between quarks and carries color. Three colors constitute the $SU(3)_C$ color symmetry group.

At present, the commonly accepted theory of strong interactions is quantum chromodynamics (QCD). This theory is one of the components of the Standard Model based on the $SU(3) \times SU(2) \times U(1)$ group. In this representation, the quark q is described by a four-component Dirac spinor $\psi_a^i(x)$, where i denotes the quark color and passes three values and x is the spacetime coordinates of the quark. Gluons are represented by the four-dimensional vector-potential $A_\mu^a(x)$ (Yang–Mills field), where a means the gluon color and passes eight values: $a = 1, 2, 3, \ldots, 8$.

The Lagrangian of the interaction between quarks and gluons has the form (up to a gauge factor)

$$L_{QCD} = -\frac{1}{4} F_{\mu\nu}^a F_{\mu\nu}^a + i \sum_q \bar{\psi}_q^i \gamma^\mu (D_\mu)_{ij} \psi_q^j - \sum_q m_q \bar{\psi}_q^i \psi_{qi},$$

$$F_{\mu\nu}^a = \partial_\mu A_\nu^a - \partial_\nu A_\mu^a - g_s f_{abc} A_\mu^b A_\nu^c, \tag{3.126}$$

$$(D_\mu)_{ij} = \delta_{ij} \partial_\mu + i g_s \sum_a \frac{1}{2} \lambda_{ij}^a A_\mu^a.$$

Here, $F_{\mu\nu}^a$ is the generalized gluon field tensor (i.e., the non-Abelian generalization of the electrodynamic field tensor in quantum electrodynamics, m_q is the quark mass, g_s is the coupling constant, λ^a is the Gell-Mann matrix, and f_{abc} are the structure constants of the $SU(3)_C$ color group. The generators of this group satisfy the relations

$$[T_a, T_b] = i f_{abc} T_c. \tag{3.127}$$

Quantum chromodynamics has a number of features. First, the coupling constant $\alpha_s = \frac{g^2}{2\pi}$ is a function of the impact parameter of interacting quarks. This constant decreases as two quarks approach each other; i.e., the constant α_s tends to zero with an increase in the momentum transfer. This means that the so-called asymptotic freedom occurs at very high energies when very high momentum transfers can be reached. In this case, QCD is the theory of free, almost noninteracting particles. Since the coupling constant in this case is much smaller than unity, perturbation theory can be applied. Many interesting effects, including polarization, are theoretically predicted for this region.

Table 3.2 presents the quark masses as they were known in 2004 (Eidelman et al. 2004). The mass directly measured in events is given for the top quark.

Let us define exclusive and inclusive reactions. Exclusive reactions are reactions in which all particles are identified and all their momenta are known. Inclusive reactions, $a + b \rightarrow c + X$, are reactions in which the particle c is identified and its momentum is determined (it is assumed that the initial particles a and b are determined). Such an approach becomes inevitable at high energies, when the probability of the processes with the production of two or more unobservable particles increases.

Measurements of single-spin transverse asymmetry in inclusive processes at high energies often gave very surprising results. This particularly concerns asymmetry in the fragmentation region of polarized particles or, in other words, in the region of soft collisions (low momentum transfers). We discuss the possibility of applying QCD with the inclusion of the generalized factorization scheme to the consistent phenomenological description of single-spin phenomena at high energies.

Polarization experiments ensure a deeper test of the theory than experiments with unpolarized observables, because spin, which is a new degree of freedom and a pure quantum-mechanical object, is involved in interactions. Polarization observables, in contrast to the total and differential cross sections, cannot be described using classical methods. Among polarization parameters, single-spin asymmetry is of particular importance: its measured value is much larger than that expected in parton interactions with the inclusion of their distribution functions and hadronization.

Perturbative quantum chromodynamics (PQCD) and the factorization hypothesis make it possible to represent the differential cross section for a process at high energies as the convolution of two types of interactions, hard at small distances and soft at large distances. The first process is calculated in PQCD with the use of perturbation theory or phenomenologically. At the current status of the theory, the soft process cannot be calculated quantitatively without additional information from experimental data in the form of the parton distribution functions and/or in the form of the fragmentation function. A success of QCD is that the knowledge of the universal distribution functions and fragmentation function at a given Q^2 value from experiments allows one to predict these functions at any other Q^2 values using the evolution equations. This means that measurement of the parton distribution functions and their fragmentation function in one process at a fixed Q^2 value is sufficient for prediction of these functions for other processes and other Q^2 values.

3.6 Single-Spin Asymmetry in the Inclusive Production of Hadrons

At the parton level, helicity is a good natural quantum number for the description of interaction processes. Can this approach be extended at the level of complex systems such as hadrons and ensure the same simple description of experimental observables? For reached energies and existing experimental data, the answer is negative. Let us analyze the current status of experimental data.

- The polarization of hyperons, primarily Λ hyperons, produced in the $p + N \rightarrow \Lambda \uparrow + X$ process with unpolarized initial nucleons (Heller 1997; Panagiotou 1990). The polarization of Λ hyperons reaches 20 %, whereas almost zero polarization is expected at the parton level (Felix 1999). It is convenient to deal with hyperons, particularly with Λ hyperons, because the leading decay mode $\Lambda \rightarrow p + \pi^-$ proceeds with parity violation, and the angular distribution of protons in the rest frame of Λ hyperons has the form

$$W(\theta, \varphi) = \frac{1}{4\pi}\left(1 + \alpha P_\Lambda(\theta) \cdot \cos\varphi\right). \qquad (3.128)$$

  Here, $\alpha = 0.64$, θ is the Λ-hyperon production angle, and φ is the proton emission angle with respect to the polarization of the Λ hyperon. Analyzing this angular distribution, one can determine the Λ-hyperon polarization P_Λ.
- Other important information was acquired in the E704 experiment at Fermilab (Adams et al. 1991d; Bravar et al. 1996). Asymmetry in the inclusive production of pions when bombarding a liquid-hydrogen target by polarized protons and antiprotons was studied. Asymmetry observed in a certain kinematic region, so-called the fragmentation region of the polarized initial particle, was 20–30 %.
- Another interesting fact is the observation of azimuthal asymmetry in the emission of pions in the $l + p \uparrow \rightarrow l + \pi + p$ reaction. This reaction is called "semi-inclusive deep inelastic scattering (SIDIS)." The process is almost exclusive and proceeds with unpolarized leptons and a transversely polarized proton target. Asymmetry observed in the system $(\gamma^* p)$ appeared to be about 10 % (Avakian 1999; Airapetian et al. 2000; Bravar 1999).

The modern generalized version of the factorization scheme in QCD (factorization is the separation of the parton distribution function into parts corresponding to small and large distances between the interacting particles) can possibly ensure a common interpretation of all experimental facts listed above. In this scheme, the transverse momentum distribution of quarks in a hadron, as well as hadrons in the final fragmenting parton, is allowed. Thus, noncollinear kinematics is used in the new scheme and, as a result, spin phenomena, which are absent in the collinear configuration, can be observed.

To introduce the modified formalism for describing single-spin inclusive asymmetry in the general case, it is convenient to begin with the consideration of the inclusive process $p \uparrow + p \to \pi + X$. The reaction is considered in the center-of-mass frame, where the z axis is oriented along the beam direction and the (xz) plane is the reaction plane. The beam polarization vector has signs $+$ and $-$ if it is oriented upwards and downwards along the y axis, respectively. The pion momentum is denoted as $\vec{p}_\pi$ and its longitudinal and transverse components, as $\vec{p}_L$ and $\vec{p}_T$, respectively. The measured asymmetry is given by the formula

$$A_N(x_F, p_T) \equiv \frac{d\sigma \uparrow (x_F, p_T) - d\sigma \downarrow (x_F, p_T)}{d\sigma \uparrow (x_F, p_T) + d\sigma \downarrow (x_F, p_T)}. \tag{3.129}$$

Here, $d\sigma = E_\pi d^3\sigma/d^3 p_\pi$ is the invariant differential cross section and $x_F = \frac{p_L}{p_{L\,\mathrm{max}}} \approx \frac{E + p_L}{(E + p_L)_{\mathrm{max}}}$ $(p_T \ll p_L)$ is the Feynman variable in the center-of-mass frame of the colliding particles. The invariance of the interaction under rotations leads to the equality $d\sigma^{\downarrow}(x_F, p_T) = d\sigma^{\uparrow}(x_F, -p_T)$, which provides another determination of A_N as left–right asymmetry.

Let us introduce the notion "twist." In Leader (2001), twist is defined as the difference between the mass dimension of the operator and spin. This definition can be illustrated as follows. In experiments, it is accepted that all structure functions in the leading approximation are independent of the square of the invariant momentum transfer. However, from the theoretical point of view, the dependence of the structure functions on Q^2 can be represented as the expansion in the inverse powers of Q. Twist is the ordinal number of the term of the expansion beginning with $n = 2$ (twist-2).

In the framework of the leading twist (twist-2) and collinear configuration, formula (3.129) for unpolarized differential cross sections for the $pp \to \pi X$ process at high energies and high momentum transfers can be written in the compact form (sign $\otimes$ means convolution)

$$d\sigma = \sum_{a,b,c,} f_{a/p} \otimes f_{b/p} \otimes d\hat{\sigma}^{ab \to c\cdots} \otimes D_{\pi/c}. \tag{3.130}$$

This formula should obviously be modified in order to apply it to the description of polarization phenomena. For interpretation of the results of the E704 experiment, theoreticians developed various approaches to the problem. They extended the factorization scheme by including the correlation functions of higher order twists (Efremov et al. 1995; Qiu and Sterman 1999; Kanazawa and Koike 2000) or by introducing the internal transverse momentum and spin dependence to the distribution functions (Sivers 1990, 1991; Anselmino et al. 1995; Anselmino and Murgia 1998; Boglione and Mulders 1999; Boer 1999) and to the fragmentation functions (Boglione and Mulders 1999; Collins 1993; Anselmino et al. 1999; Boglione and Leader 2000; Suzuki 2000).

There is also semiclassical model of quarks rotating inside the hadron (Boros et al. 1993; Boros and Zuo-Tang 2000), which is similar to the model of rotating hadron matter.

Following Anselmino (2002), we considered only the approaches based on the generalization of the QCD factorization scheme. The inclusion of the internal motion of quarks is also important in the calculation of unpolarized cross sections (Wang 2000; Apanasevich et al. 1998).

Qiu and Sterman (1999) demonstrated that Eq. (3.130) can be generalized by including higher twists in the distribution and fragmentation functions. The difference between the cross sections for two polarization orientations of the initial proton is represented in the form

$$
d\sigma \uparrow - d\sigma \downarrow = \sum_{a,b,c} \left\{ \Phi_{a/p}^{(3)} \otimes f_{b/p} \otimes \hat{H} \otimes D_{\pi/c} \right\}
$$
$$
+ h_1^{\frac{a}{p}} \otimes \Phi_{b/p}^{(3)} \otimes \hat{H}' \otimes D_{\pi/c}
$$
$$
+ h_1^{\frac{a}{p}} \otimes f_{b/p} \otimes \hat{H}'' \otimes D_{\pi/c}^{(3)}, \tag{3.131}
$$

where $\Phi^{(3)}$ and $D^{(3)}$ are the higher-twist parton correlation functions; $\hat{H}$ is the interaction between partons; and h_1 is the transverse-spin distribution (below, transversity for simplicity). By analogue with distributions (3.131), transversity can be written in the form

$$
h_1^{a/N}(x, Q^2) \equiv f_{a\uparrow/N\uparrow}(x, Q^2) - f_{a\uparrow/N\downarrow}(x, Q^2). \tag{3.132}
$$

Contributions from higher twists are unknown, but they can be estimated under certain simplifications, e.g., as in Kanazawa and Koike (2000):

$$
\Phi_{a/p}^{(3)} \sim \int \frac{dy^-}{4\pi} e^{ixp^+y^-} \left\langle p, s_T \left| \bar{\psi}_a(0)\gamma^+ \right. \right.
$$
$$
\cdot \left[\int dy_2^- \varepsilon_{\rho\sigma\alpha\beta} s_T^\rho p^\alpha p'^\beta F^{\sigma+}(y_2) \right] \psi_a(y^-) \left| p, s_T \right\rangle
$$
$$
= k_a C f_{a/p}. \tag{3.133}
$$

The contribution from higher twists (given in the square brackets) depends on many parameters (final and initial momenta p and p', transverse proton spin s_T, and gluon field $F^{\mu\nu}$), but it can be simplified under certain assumptions as shown in the last term with the constant parameter C. In this expression, coefficient is $k_a = +1$ and -1 for the u and d quarks, respectively. This model provides satisfactory agreement with the results of the E704 experiment (Adams et al. 1991d) and gives an estimate for single-spin asymmetry at RHIC.

Another approach to factorization was analyzed in Sivers (1990), Anselmino et al. (1995, 1999), Boer (1999), and Boglione and Leader (2000) on the basis of Eq. (3.130) corresponding to the leading twist and collinear configuration. Then, the equation was generalized by introducing the internal transverse momentum of the parton in the distribution function and the same was made for the hadron in the fragmentation functions. As a result, modified equation (3.130) is represented in the form ("hats" denote the parameters referring to a subprocess, for example, parton–parton scattering)

$$d\sigma = \sum_{a,b,c,} f_{a/p}(x_1, k_{\perp 1}) \otimes f_{b/p}(x_2, k_{\perp 2})$$

$$\otimes \, d\hat{\sigma}^{ab \to c \cdots}(x_1, x_2, k_{\perp 1}, k_{\perp 2}) \otimes D_{h/c}(z, k_{\perp h}). \tag{3.134}$$

The introduction of $k_\perp$ and spin dependence leads to the appearance of new measurable spin distribution functions, namely,

$$\Delta^N f_{q/p\uparrow} \equiv \hat{f}_{q/p\uparrow}(x, k_\perp) - \hat{f}_{q/p\downarrow}(x, k_\perp)$$

$$= \hat{f}_{q/p\uparrow}(x, k_\perp) - \hat{f}_{q/p\uparrow}(x, -k_\perp), \tag{3.135}$$

$$\Delta^N f_{q\uparrow/p} \equiv \hat{f}_{q\uparrow/p}(x, k_\perp) - \hat{f}_{q\downarrow/p}(x, k_\perp)$$

$$= \hat{f}_{q\uparrow/p}(x, k_\perp) - \hat{f}_{q\uparrow/p}(x, -k_\perp), \tag{3.136}$$

and new fragmentation functions

$$\Delta^N D_{h/q\uparrow} \equiv \hat{D}_{h/q\uparrow}(z, k_\perp) - \hat{D}_{h/q\downarrow}(z, k_\perp)$$

$$= \hat{D}_{h/q\uparrow}(z, k_\perp) - \hat{D}_{h/q\uparrow}(z, -k_\perp), \tag{3.137}$$

$$\Delta^N D_{h\uparrow/q} \equiv \hat{D}_{h\uparrow/q}(z, k_\perp) - \hat{D}_{h\downarrow/q}(z, k_\perp)$$

$$= \hat{D}_{h\uparrow/q}(z, k_\perp) - \hat{D}_{h\uparrow/q}(z, -k_\perp). \tag{3.138}$$

In order to understand the meaning of these functions, it is necessary to note the arrows indicating which particle is polarized (for details, see Anselmino et al. 2000). All these functions vanish at $k_\perp = 0$. They are T-odd. If these functions are represented in the helicity basis, functions in formulas (3.136) and (3.137) mix quarks with different helicities, i.e., are chirally odd, whereas functions (3.135) and (3.138) are chirally even. The chirality operator separates spin states oriented along and against the momentum in the particle wave function. Similar functions in another notation were introduced previously. In particular, there is a direct relation (Boglione and Mulders 1999) between $f_{1T}^\perp$ in Boer and Mulders (1998), $h_1^\perp$ in Boer (1999), and $H_1^\perp$ and $D_{1T}^\perp$ in Boer and Mulders (1998) and Jacob and Mulders (1996). More detailed information can be found in Mulders (2001).

Distribution function (3.135) was introduced by Sivers (1990) and fragmentation function (3.137), by Collins (1993). These functions are named after their authors.

Substituting new functions into relation (3.130) and retaining only leading terms in the expansion in $k_\perp$, we obtain

$$d\sigma^\uparrow - d\sigma^\downarrow = \sum_{a,b,c,} \left\{ \Delta^N f_{a/p\uparrow}(k_\perp) \otimes f_{b/p} \otimes d\hat{\sigma}(k_\perp) \otimes D_{\pi/c} + h_1^{a/P} \otimes f_{b/p} \right.$$

$$\otimes \, \Delta\hat{\sigma}(k_\perp) \otimes \Delta^N D_{\pi/c}(k_\perp) + h_1^{a/p} \otimes \Delta^N f_{b\uparrow/p}(k_\perp)$$

$$\left. \otimes \, \Delta'\hat{\sigma}(k_\perp) \otimes D_{\pi/c}(z) \right\}. \tag{3.139}$$

Here, convolution with respect to $k_\perp$ is implied. The cross sections for elementary processes $\Delta\hat{\sigma}$ are determined as

$$\Delta\hat{\sigma} = d\hat{\sigma}^{a\uparrow b \to c\uparrow d} - d\hat{\sigma}^{a\uparrow b \to c\downarrow d}, \tag{3.140}$$

$$\Delta'\hat{\sigma} = d\hat{\sigma}^{a\uparrow b\uparrow \to cd} - d\hat{\sigma}^{a\uparrow b\downarrow \to cd}. \tag{3.141}$$

These cross sections are calculated in PQCD.

Only an even product of chirally-odd functions appears in physically measured quantity (3.130).

The above relations were successfully applied to the description of the data of the E704 experiment with the use of only Sivers effect (3.130):

$$d\sigma^\uparrow - d\sigma^\downarrow = \sum_{a,b,c} \Delta^N f_{a/p\uparrow}(k_\perp) \otimes f_{b/p} \otimes d\hat\sigma(k_\perp) \otimes D_{\pi/c}, \qquad (3.142)$$

or Collins effect (Anselmino et al. 1999; Boglione and Leader 2000),

$$d\sigma^\uparrow - d\sigma^\downarrow = \sum_{a,b,c} \Delta^N f_{a/p}(k_\perp) \otimes f_{b/p} \otimes d\hat\sigma(k_\perp) \otimes D_{\pi/c\uparrow}. \qquad (3.143)$$

A certain explanation is required for the Sivers function $\Delta^N f_{q/p\uparrow}$. In the helicity basis, this function is proportional to the off-diagonal elements of the expected values of quark operators acting on proton states:

$$\Delta^N f_{a/p\uparrow} \sim \langle p^+ | \bar\psi \gamma^+ \psi | p^- \rangle. \qquad (3.144)$$

Using, as usual, parity and time reversal conservation laws for free states, one can show that the Sivers function is zero (Collins 1993).

A similar remark can be given for function (3.136). However, the Sivers effect can be held either by the inclusion of the interaction between partons in the initial state or by small redefinition of the time reversal rule, as shown in Anselmino et al. (2001). As mentioned in Wang (2000) and Apanasevich et al. (1998), the Sivers effect should be taken into account in the calculation of unpolarized cross sections for their correct normalization.

We consider semi-inclusive deep-inelastic asymmetry in the $lp^\uparrow \to l\pi X$ reaction measured at the HERMES and SMC setups (Avakian 1999; Bravar 1999). Such measurements are directly associated with the Collins function. In particular, Eq. (3.137) can be rewritten in the form

$$\hat D_{h/q\uparrow}(z, k_\perp; P_q) = \hat D_{h/q}(z, k_\perp) + \frac{1}{2} \Delta^N \hat D_{h/q\uparrow}(z, k_\perp) \frac{P_q \cdot (p_q \times k_\perp)}{|p_q \times k_\perp|} \qquad (3.145)$$

for the final quark with the momentum p_q and transverse polarization P_q (four-dimensional dot product $p_q \cdot P_q = 0$). This quark is fragmented into a hadron with the momentum $p_h = zp_q + k_\perp$ ($p_q \cdot k_\perp = 0$). The function $\hat D_{h/q}(z, k)$ is an unpolarized fragmentation function depending on $k_\perp$. The spin-dependent part of the $\hat D$ function appears only from polarization perpendicular to the plane determined by the parent quark and daughter hadron. In the general case, we have

$$P_q \cdot \frac{p_q \times k_\perp}{|p_q \times k_\perp|} = P_q \sin \Phi_C, \qquad (3.146)$$

where $P_q = |\vec P_q|$ and Φ_C is the Collins angle. When $P_q = 1$ and $\vec P_q$ is perpendicular to the (q, h) plane ($P_q = \uparrow, -P_q = \downarrow$), we have $P_q \sin \Phi_C = 1$.

Then, according to Eq. (3.145), the quark has the analyzing power given by the expression

$$A_q^h(z, k_\perp) \equiv \frac{\hat{D}_{h/q\uparrow}(z, k_\perp) - \hat{D}_{h/q\downarrow}(z, k_\perp)}{\hat{D}_{h/q\uparrow}(z, k_\perp) + \hat{D}_{h/q\downarrow}(z, k_\perp)} = \frac{\Delta^N \hat{D}_{h/q\uparrow}(z, k_\perp)}{2\hat{D}_{h/q}(z, k_\perp)}. \tag{3.147}$$

Retaining only leading $k_\perp$ terms in Eq. (3.134), we obtain

$$d\sigma^\uparrow - d\sigma^\downarrow = \sum_q f_{q/p} \otimes d\hat{\sigma} \otimes \Delta^N \hat{D}_{\pi/c}(k_\perp). \tag{3.148}$$

The single-spin asymmetry of the production of the hadron h in the $(\gamma^* - p)$ system is represented in the form (Anselmino et al. 2000):

$$A_N^h(x, y, z, \Phi_C, p_T)$$
$$= \frac{d\sigma^{l+p, P \to l+h+X} - d\sigma^{l+p, -P \to l+h+X}}{d\sigma^{l+p, P \to l+h+X} + d\sigma^{l+p, -P \to l+h+X}}$$
$$= \frac{\Sigma_q e_q^2 h_1^{q/P}(x) \Delta^N \hat{D}_{h/q\uparrow}(z, p_T)}{2\Sigma_q e_q^2 f_{q/p}(x) \hat{D}_{h/q}(z, p_T)} \frac{2(1-y)}{1+(1-y)^2} P \sin \Phi_C. \tag{3.149}$$

Here, P is the proton polarization perpendicular to the orientation of a virtual γ-ray photon. In the $(\gamma^* - p)$ system, the measured pion transverse momentum p_T is equal to the internal pion momentum in the fragmenting parent quark. Expression (3.149) explicitly includes the standard parameters of the deep-inelastic scattering process:

$$x = \frac{Q^2}{2p \cdot q}, \qquad y = \frac{Q^2}{s \cdot x}, \qquad z = \frac{p \cdot p_h}{p \cdot q}, \tag{3.150}$$

where p, q, and p_h are the four-momenta of the proton, virtual photon, and final hadron h, respectively.

Anselmino et al. (2000) carefully considered the analyzing power of the quark $A_q^h(z, p_T)$; they discussed the data on asymmetry reported in Avakian (1999) and Bravar (1999). Below, we briefly present the contents of Anselmino et al. (2000).

Under certain realistic assumptions, isotopic invariance, and charge-conjunction invariance, expression (3.149) provides the relation $(i = +, -, 0)$

$$A_N^{\pi^i}(x, y, z, \Phi_C, p_T) = \frac{h_i(x)}{f_i(x)} A_q^\pi(z, p_T) \frac{2(1-y)}{1+(1-y)^2} P \sin \Phi_C, \tag{3.151}$$

where

$$i = +: \quad h_+ = 4h_1^{u/p} \qquad f_+ = 4f_{u/p} + f_{\bar{d}/p}; \tag{3.152}$$

$$i = -: \quad h_- = h_1^{d/p} \qquad f_- = f_{d/p} + 4f_{\bar{u}/p}; \tag{3.153}$$

$$i = 0: \quad h_0 = 4h_1^{u/p} + h_1^{d/p} \qquad f_0 = 4f_{u/p} + f_{d/p} + 4f_{\bar{u}/p} + f_{\bar{d}/p}. \tag{3.154}$$

Here, f are unpolarized distribution functions and h_1 is the transversity distribution. In the above equations, it is assumed that $A_N^+ \approx A_N^0$ at large x values, as observed in the HERMES experiment (Airapetian et al. 2001).

Measured asymmetry (3.151) depends on two unknown functions: the transversity distribution and analyzing power of quarks or, equivalently, the Collins function. These functions depend on different variables ignoring their smooth evolution with Q^2. In order to determine both functions separately, the HERMES collaboration proposes the program of measurements of semi-inclusive asymmetries in various kinematic regions in z, x, and p_T (Korotkov et al. 2001).

To obtain an estimate of the analyzing power of the quark A_q^π in the absence of experimental data, one can use the Soffer inequality in application to the transversity distribution (Soffer 1995)

$$|h_{1q}| \leq \frac{1}{2}(f_{q/p} + \Delta_q). \tag{3.155}$$

Substituting this inequality into relation (3.151) and comparing with the result of the SMC experiment (Bravar 1999)

$$A_N^{\pi^+} \cong -(0.10 \pm 0.06)\sin\Phi_C, \tag{3.156}$$

we obtain a quite low bound for the analyzing power of the valence quark in the positive pion

$$\left|A_q^\pi(\langle z\rangle, \langle p_T\rangle)\right| \geq (0.24 \pm 0.15) \quad \langle z\rangle \cong 0.45, \ \langle p_T\rangle \cong 0.65 \ \text{GeV}/c. \tag{3.157}$$

Similar results were also obtained in the HERMES experiment (Avakian 1999). However, we should make two remarks concerning these data. The first concerns the very small transverse polarization of the protons as compared to the SMC experiment. Second, owing to a low energy of the longitudinally-polarized electron beam, the HERMES collaboration performed measurements at low Q^2 values; this requires the inclusion of the contributions from higher twists and such contributions were disregarded in Eq. (3.149). Nevertheless, it is interesting that the estimate of the lower bound of the analyzing power of the quark shows that it can be sufficiently large. However, more accurate experimental data are required to test this statement.

The possibility of the fragmentation of the unpolarized quark on the polarized hadron was considered in Mulders and Tangerman (1996, 1997) and Anselmino et al. (2001) with the use of one of four distribution functions appearing when the internal transverse momentum of the quark is introduced.

As a result, the polarization of the Λ hyperon can be described (Heller 1997; Felix 1999) in the framework of the same approach as was applied above for describing asymmetry. By analogy with Eq. (3.145), we can write

$$\hat{D}_{h\uparrow/q}(z, k_\perp; P_h) = \frac{1}{2}\hat{D}_{h/q}(z, k_\perp) + \frac{1}{2}\Delta^N \hat{D}_{h\uparrow/q}(z, k_\perp)\frac{\hat{P}_h \cdot (p_q \times k_\perp)}{|p_q \times k_\perp|}. \tag{3.158}$$

This function describes the fragmentation of the unpolarized quark with the momentum p_q on the hadron h with spin 1/2, momentum $p_h = zp_q + k_\perp$, and polarization vector $P_{h\uparrow} = P_{h/q\uparrow} \cdot \Delta^N \hat{D}_{h\uparrow/q}(z, k_\perp)$ (denoted as $\hat{D}_{1T}^\perp$ in Mulders and Tangerman 1996) and is new polarization fragmentation function.

This approach can explain the polarization of the Λ hyperon

$$P_\Lambda = \frac{d\sigma^{pN \to \Lambda^\uparrow X} - d\sigma^{pN \to \Lambda^\downarrow X}}{d\sigma^{pN \to \Lambda^\uparrow X} + d\sigma^{pN \to \Lambda^\downarrow X}}, \tag{3.159}$$

inclusively appearing in the collision of unpolarized nucleons. In this case, in the region of kinematic variables $x_F > \sim 0.2$ and $p_T > \sim 1$ GeV/c, P_Λ is noticeable in magnitude, negative, and perpendicular to the reaction plane. Taking into account the presence of $k_\perp$ in the hadronization process and assuming factorization, we obtain

$$d\sigma^{pN \to \Lambda X} P_\Lambda = d\sigma^{pN \to \Lambda^\uparrow X} - d\sigma^{pN \to \Lambda^\downarrow X}$$
$$= \sum_{a,b,c,d} f_{a/p}(x_1) \otimes f_{b/N}(x_2) \otimes d\hat{\sigma}^{ab \to cd}(x_a, x_b, k_\perp)$$
$$\otimes \Delta^N \hat{D}_{\Lambda^\uparrow/c}(z, k_\perp). \tag{3.160}$$

Using a simple parameterization for the unknown polarization fragmentation function $\Delta^N \hat{D}_{h\uparrow/q}$ (3.138), Anselmino et al. (2001) obtained good description of the polarization of the Λ hyperon, including its negative sign in the entire measured kinematic region. An increase in polarization with x_F, an increase with p_T until 1 GeV/c, and saturation are described. A weak energy dependence is also predicted. The model also gives the result in agreement with the experiment for $\bar{\Lambda}$, in particular, its small magnitude.

However, it was noted that PQCD is inapplicable for energies considered in this section, and relation (3.160) cannot be used to describe the differential cross section, polarization, and asymmetry of inclusively produced pions. Only the RHIC energy region ($\sqrt{s} \geq 200$ GeV) is appropriate for such an analysis.

The situation is similar for unpolarized cross sections for other reactions (Wang 2000; Wong and Wang 1998; Zhang et al. 2001); in this case, the internal transverse momentum is introduced to approach the model predictions with experimental data. This agreement with cross section leads to the strong decrease in the polarization effects in contradiction with the experimental data. Thus, the problem of describing polarization data cannot be treated as to be solved.

The approach described in this section is sufficiently general and promising (Henneman et al. 2001). According to this approach, we have the convolution of two processes, the hard process theoretically calculated and the soft process. Information on the latter process should be taken from experiment in the form of the distribution and fragmentation functions. In view of the difficulties in mathematical solution of the problem, certain simplifications should be introduced; the applicability of these simplifications is tested experimentally. In this field, RHIC provides unique possibility for polarization investigations.

The presentation in this section is based on Anselmino (2002).

3.7 *U*-Matrix Method (Fixed *t* Values)

Experimental data on the polarization parameters obtained in the early 1970s by the HERA collaboration at IHEP and later, but at high energies, at CERN, as well as the measurements of pp and $\overline{pp}$ elastic differential cross sections at high momentum transfers, put a number of important physical problems. Below, we list some of them.

1. Do spin effects exist at asymptotically high energies?

   This question naturally arose when the HERA collaboration analyzed the energy dependence of the ratio of the spin-flip to spin-nonflip amplitudes and concluded that this ratio is saturated at momenta above 20 GeV/c. Such a conclusion contradicted predictions of many theoretical models. The Regge model is outstanding among them. According to this model, only one pole, pomeron, remains at asymptotically high energies. However, this model itself forbids the appearance of polarization in the presence of only one pole, because the phases of the spin-flip and spin-nonflip amplitudes are the same in this case. Hence, these amplitudes do not interfere and polarization is zero. Even in the presence of two exchange poles, polarization should decrease rapidly with energy, for example, as $1/\sqrt{s}$. In both cases, predictions of the Regge model contradict experiments.

2. The relations between the polarizations of the particles and antiparticles in binary reactions, in particular, elastic scattering processes. The HERA collaboration results on the measurement of the polarizations of the particles and antiparticles are unique to date. No model explains all these results.

3. The behavior of the differential cross sections for pp and $\overline{p}p$ elastic collisions at high momentum transfers.

These problems and a more general problem of the behavior of the polarization parameters at high energies, when exchange proceeds almost through one pomeron pole and at fixed t values, were analyzed in Troshin and Tyurin (1976, 1984, 1988). These works are briefly presented below.

The basic equation relating the amplitude to the generalized reaction matrix is replaced in the spin case by the system of equations for helicity amplitudes. For $1 + 2 \to 3 + 4$ elastic scattering, this system has the form

$$F_{\lambda_1\lambda_2\lambda_3\lambda_4}(\vec{p}_1, \vec{p}_1') = U_{\lambda_1\lambda_2\lambda_3\lambda_4}(\vec{p}_1, \vec{p}_1')$$
$$+ \frac{i}{8\pi^2} \sum_{\nu_1\nu_2} \int \frac{d\vec{q}_1}{2q_1^0 2q_2^0} U_{\lambda_1\lambda_2\nu_1\nu_2}(\vec{p}_1, \vec{q}_1) F_{\nu_1\nu_2\lambda_3\lambda_4}(\vec{q}_1, \vec{p}_1').$$

$$(3.161)$$

In the impact-parameter representation, system (3.161) reduces to the algebraic system

$$f_{\lambda_1\lambda_2\lambda_3\lambda_4}(s, b) = u_{\lambda_1\lambda_2\lambda_3\lambda_4}(s, b) + i\rho(s) \sum_{\nu_1\nu_2} u_{\lambda_1\lambda_2\nu_1\nu_2}(s, b) f_{\nu_1\nu_2\lambda_3\lambda_4}(s, b),$$

$$(3.162)$$

where $\rho(s) \to 1$ at $s \to \infty$.

The implementation of analytic continuation and relation between the s and t channels with the generalized reaction matrix makes it possible to obtain the Regge form for the functions $U_{\{\lambda_i\}}(s, t)$ in the helicity basis:

$$U_{\{\lambda_i\}}(s, t) = \sum_R g_{\{\lambda_i\}}^R(t) \xi_R(t) (s/s_0)^{\beta_R(t)},$$

$$(3.163)$$

Fig. 3.2 Description of the
polarization parameter in πN
scattering in the *U*-matrix
method (Edneral et al. 1979):
the unit in the ordinate axis
corresponds to a polarization
of 10 %

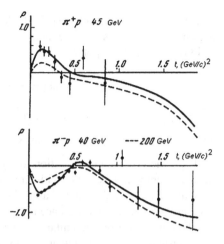

where $\xi_R(t)$ is the signature factor and the summation is performed over all contributions from all Regge trajectories involved in the exchange in this process.

The generalized reaction matrix method provides good agreement with experimental data. As an example, Fig. 3.2 shows the description of the polarization parameter in πN scattering. In addition to the quantitative description, the method leads to important qualitative conclusions. Unitarity leads to nonzero polarization as a result of vacuum exchange disregarding interference with the contributions from other trajectories. We recall that the HERA experimental data indicate the existence of such a contribution to polarization.

Under the assumption that only the vacuum trajectory with even signature makes contribution, the generalized reaction matrix for πN scattering in the impact-parameter representation can be written in the form

$$u_{++}(s,b) = \frac{g_{++}}{a(s)} \left(\frac{s}{s_0}\right)^{\beta(0)-1} \exp\left(-b^2/a(s)\right),$$

$$u_{+-}(s,b) = \frac{2bg_{+-}}{a^2(s)} \left(\frac{s}{s_0}\right)^{\beta(0)-1} \exp\left(-b^2/a(s)\right),$$

$$(3.164)$$

where

$$a(s) = 4\beta'(0)\left[\ln\frac{s}{s_0} - i\frac{\pi}{2}\right], \qquad \beta(t) = \beta(0) + t\beta'(0).$$

The solution of system (3.162) for the case under consideration has the form

$$f_{++}(s,b) = \frac{u_{++}(s,b)[1 - u_{++}(s,b)] - i[u_{+-}(s,b)]^2}{[1 - u_{++}(s,b)]^2[u_{+-}(s,b)]^2},$$

$$f_{+-}(s,b) = \frac{u_{+-}(s,b)}{[1 - u_{++}(s,b)]^2[u_{+-}(s,b)]^2}.$$

$$(3.165)$$

Here, $\rho(s) = 1$ is taken. The calculation of the polarization parameter in the region of low t values with the inclusion of only the pomeron contribution gives the following formula for polarization:

$$p^{\pi^{+-}P}(s,t) \cong -\frac{\sqrt{-t}}{\ln s} \cdot \frac{\varphi_1(s) - t\varphi_2(s)}{\varphi_3(s) - t\varphi_4(s)}, \tag{3.166}$$

where $\varphi_i(s)$ are the functions positive at $s \to \infty$ and

$$\varphi_{1,3} \propto \ln^2 s, \qquad \varphi_{2,4} \propto \ln^6 s.$$

Thus, the pomeron contribution to polarization at high energies is negative and decreases smoothly as $1/\ln s$ with an increase in energy. In this case, the functions F_{++} and F_{+-} have different phases although the phases of the functions U_{++} and U_{+-} are the same. The difference between phases is a consequence of unitarity.

The conclusion on the unitary mechanism of the generation of nonzero polarization from the vacuum exchange is applicable to any process, in particular, to pp- and $p\bar{p}$-scattering processes. This conclusion remains unchanged if the contribution from odderon whose trajectory is degenerate with vacuum exchange is also included in the expression for the U matrix.

If the pomeron and odderon trajectories are nondegenerate, the inclusion of the odderon at low t values provides corrections, which are power function of s, to the above logarithmic behavior of the polarization parameter. Note that analysis of the odderon contribution should be performed taking into account the relation

$$\beta_P(0) \geq \beta_O(0), \tag{3.167}$$

where $\beta_P(0)$ and $\beta_O(0)$ are the intersections of the pomeron and odderon trajectories, respectively. Constraint (3.167) follows from the unitarity condition, which, in the framework of the method under consideration, reduces to the requirement that the imaginary part of the function $u(s,b)$ is nonnegative:

$$\mathrm{Im}\,u(s,b) \geq 0. \tag{3.168}$$

Let us now consider pp elastic scattering. The solution of the system of equations (3.162) for five helicity amplitudes describing pp elastic scattering, has the form

$$f_1(s,b) = \frac{\tilde{u}_1(s,b)[1 - iu_1(s,b)] - i\tilde{u}_2(s,b)u_2(s,b)}{[1 - iu_1(s,b)]^2 - [u_2(s,b)]^2},$$

$$f_2(s,b) = \frac{\tilde{u}_2(s,b)[1 - iu_1(s,b)] - i\tilde{u}_1(s,b)u_2(s,b)}{[1 - iu_1(s,b)]^2 - [u_2(s,b)]^2},$$

$$f_3(s,b) = \frac{\tilde{u}_3(s,b)[1 - iu_3(s,b)] - i\tilde{u}_4(s,b)u_3(s,b)}{[1 - iu_3(s,b)]^2 - [u_4(s,b)]^2}, \tag{3.169}$$

$$f_4(s,b) = \frac{\tilde{u}_4(s,b)[1 - iu_3(s,b)] - i\tilde{u}_4(s,b)u_3(s,b)}{[1 - iu_3(s,b)]^2 - [u_4(s,b)]^2},$$

$$f_5(s,b) = u_5(s,b)\left\{\left[1 - iu_1(s,b) - iu_2(s,b)\right]\left[1 - iu_2(s,b) - iu_4(s,b)\right]\right.$$

$$\left. - 4u_5^2(s,b)\right\}^{-1},$$

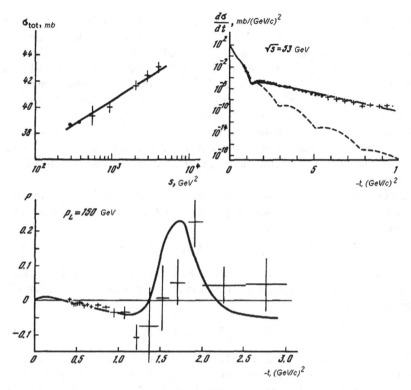

Fig. 3.3 Description of the total and differential cross sections and the polarization parameter for
pp scattering in the single-pole approximation of the U-matrix method

where

$$\tilde{u}_i(s,b) = u_i(s,b) + 2u_5(s,b)f_5(s,b).$$

The use of formulas (3.169) and single-pole Regge parameterization for the
functions $U_{\{\lambda_i\}}$ in the helicity basis provides the description of experimental data
on the differential cross sections, total cross sections, and polarization parameter in pp elastic scattering at high energies (Edneral et al. 1979) (see Fig. 3.3).
The experimental data on $d\sigma/dt$ are described up to momentum transfer squared
$-t \sim 10$ (GeV/c)^2 at ISR energies (CERN).

The experimental data for $d\sigma/dt$ at high t values have a smooth dependence
on the momentum transfer; this dependence is inconsistent with many models that
predict the existence of alternating diffraction minima and maxima in this region.

According to the analysis of the behavior of helicity amplitudes in the approach
under consideration, the smooth behavior of $d\sigma/dt$ at high t values is due to the
increasing role of spin effects in this kinematic region. Such a smooth behavior is
caused by the contribution from the amplitudes with double helicity change. Disregarding these amplitudes, the behavior of $d\sigma/dt$ would exhibit the characteristic
sequence of alternating maxima and minima, because helicity amplitudes F_1, F_3,

and F_5 have maxima and minima in this t region and are close in magnitude in the range 1 (GeV/c)$^2 \leq |t| \leq 10$ (GeV/c)^2. At the same time, the helicity amplitudes with double helicity change, F_2 and F_4, are small as compared to the amplitudes F_1, F_3, and F_5 in the range 0 (GeV/c)$^2 \leq |t| \leq 3$ (GeV/c)^2 and are relatively large in the range 3 (GeV/c)$^2 \leq |t| \leq 10$ (GeV/c)^2, where they have a smooth behavior.

Thus, the angular distribution of pp elastic scattering is smooth. The inclusion of the spin degrees of freedom resulting in the filling of minima at high t values also leads to a significant polarization parameter at energies $\sqrt{s} \sim 50$ GeV and $-t \sim 2$ (GeV/c)^2. Polarization in this region is expected to be 10–20 %. Therefore, the role of spin interaction increases with momentum transfer (Bilenky and Ryndin 1959).

Let us consider some geometrical properties of helicity amplitudes at high fixed t values. The behavior of helicity amplitudes in this region can be represented by the expression

$$F_i(s,t) \cong \varphi_i(s,t) \cos\left[R_i(s)\sqrt{-t} + \tilde{\varphi}_i(s)\right], \qquad (3.170)$$

where the functions $\varphi_i(s,t)$ have no zeros. Oscillation in the amplitudes F_2 and F_4 in the range 3 (GeV/c)$^2 \leq |t| \leq 10$ (GeV/c)^2 is absent because the corresponding radii are

$$R_{2,4} \cong \frac{1}{3} R_{1,3,5} \cong 0.3 \text{ fm.} \qquad (3.171)$$

This relation between the radii can be interpreted as an indication of the existence of the internal region in the proton with sizes of about 0.3 fm, where valence quarks are likely localized. Such representations will be developed below when considering the quark model for the U matrix.

Thus, although the angular distributions are smooth in the region of high momentum transfers, the model predicts the oscillating behavior of polarizations as a function of momentum transfer.

Note that a feature of the method under consideration is the direct inclusion of the unitarity condition in the calculation scheme.

3.8 Other Phenomenological Models

3.8.1 Model of Rotating Hadronic Matter

In the early 1970s, in order to explain the origin of the proton spin, the idea of the rotation of hadronic matter about a separate axis, which could naturally be the proton polarization direction, was proposed in Chou (1973), Yang (1973), and Chou and Yang (1976). This idea attracted particular attention of the HERA collaboration at IHEP, because this model of rotating hadronic matter predicts that the single nonzero polarization parameter at high energies is the spin rotation parameter R; this parameter is the same for all elastic scattering processes for both particles and antiparticles. The HERA collaboration carried out such measurements and published results that agreed with the predictions of the model of rotating hadronic matter. This idea in application to rotating partons in a polarized proton was also used by German physicists in the early 1990s in order to explain the results of the E704 collaboration on single-spin asymmetry in inclusive reactions. The rotating parton model appeared to be efficient in the interpretation of both the single-spin asymmetry of pions and Λ hyperons and polarization transfer from the polarized initial proton to the final Λ hyperon. In some cases, it provided certain predictions, for example, change in the asymmetry sign when the negative pion is changed to the positive pion (or vise versa). The model is richer in predictions. According to its foundations, first, no spin effects should exist in the central region, because the contribution from polarized valence quarks is almost absent in this region. Second, asymmetry cannot exist in the fragmentation region of unpolarized valence quarks. According to this model, symmetry is also absent in the region of small values of the Feynman variable x_F. Polarization symmetry can exist only in the fragmentation region of polarized valence quarks. To date, almost all these predictions of the model have been confirmed. The strongest statement is the model prediction of the absence of polarization effects in the central region. Measurements of single- and double-spin effects in the inclusive production of neutral pions seemingly confirm this conclusion of the model up to the RHIC energies.

Below, we briefly present the basic features of the rotating-quark model in polarized protons (Meng 1991a, 1991b).

The E704 collaboration at Fermilab performed measurements of single-spin asymmetry in the inclusive production of pions by polarized proton and antiproton beams. All measurements revealed noticeable effects (15–39 %) in the fragmentation region of the polarized beam. The results were indeed surprising: theoreticians could not predict such results. All they expected zero effects, as followed from helicity conservation.

The theoreticians of the group led by Prof. Meng (Institute of Theoretical Physics, Free University, Berlin, Germany) compared the predictions of their model (Liang 1991) with experimental data (Adams et al. 1991a, 1991b, 1991c). The model was based on their idea of quarks rotating inside the polarized proton (Liang and Meng 1990). They arrived at the following conclusion.

Left–right asymmetries observed in the E704 experiment with polarized proton and antiproton beams at high energies and high x_F values should be considered as a serious indication of the presence of rotating valence quarks in polarized protons. Omitting details, which can be found in Meng (1991a) by the authors of this model, we point out only its basic features.

1. Valence quarks determine the basic properties of a hadron including its spin. The quarks are considered as relativistic particles in confining field. They lead to the appearance of matter current depending on the color and flavor of partons. This is also valid when quarks are in the ground state, i.e., quarks also rotate in the ground state.
2. The probability of finding such rotating quarks near the hadron surface is much larger than that at its center. The polarization of a valence quark under consideration completely determines the orientation of the mean value of its orbital angular

momentum in the ground state, as well as the direction of its current density at large distances from the hadron center.

3. In the model of rotating hadronic matter, hadron–hadron interactions at high energies are considered as surface interactions of their constituents. This means that constituents on the front surface of the projectile hadron interact with larger probability with constituents of the target hadron on the surface nearest to the projectile.

4. The polarization of valence quarks in the polarized proton is determined by the polarization and wave function of the proton.

The baryon wave function is constructed under the assumption that there is the complete antisymmetry in the color degree of freedom. This means that the wave function of two identical quarks should be symmetric in coordinate–spin space. It is also accepted that the total system of an indefinite number of sea quark–antiquark pairs does not rotate about a certain direction in the polarized proton. It is also assumed that sea quark–antiquark pairs, as well as gluons, fill the entire phase space.

We now consider pp interaction in the center-of-mass frame. The projectile and target protons are denoted as P and T, respectively. Let us introduce the Cartesian coordinate system whose z axis is oriented along the beam direction, the x axis is perpendicular to the reaction plane, and the y axis is perpendicular to both the x and z axes and forms with them a right-handed system. The E704 experiment was devoted to studying symmetry in the $p(\uparrow) + p \to \pi^{\pm,0} + X$ and $\bar{p}(\uparrow) + p \to \pi^{\pm,0} + X$ reactions, where arrows mean that the initial beam is polarized and the polarization vector is oriented along the positive direction of the x axis. The target is unpolarized. The transverse-momentum distributions of partons in the polarized proton P in different coordinate systems are the same if these systems are obtained from each other by a Lorentz boost (translation) along the z axis. Therefore, postulates 2 and 4 are applicable. This means that an observer in the rest system of the T particle observes that surface valence quarks of the P particle undergo sequential motions along the y axis and the direction of this motion depends on quark polarization. To form a meson, this valence quark should find its partner among sea quarks. In this case, the formed pion acquires an additional transverse momentum of a quark appearing due to orbital motion. Assumption 4 also determines the sign of such transverse momentum along the y axis. These results are presented in Tables 3.3 and 3.4 for the proton and antiproton beams, respectively. Both beams are polarized. The valence quark and a pair of sea quarks are denoted as val and sea in the parentheses. The arrows $\leftarrow$ and $\to$ mean the $-y$ and $+y$ orientations of the transverse momentum p_Y of the rotating quark, respectively. These signs in the model correspond to the more probable emission of pions to the left and right in the experiment, respectively. The results in the tables were obtained with the use of the proton wave functions in coordinate, spin, and flavor spaces. A number of conclusions follow from these tables. In the fragmentation region of the polarized proton (see the right part of the tables), asymmetries for the $\pi^+(u\bar{d})$, $\pi^0[\frac{1}{\sqrt{2}}(u\bar{u} - d\bar{d})]$, and $\eta[\frac{1}{\sqrt{2}}(u\bar{u} + d\bar{d} - 2s\bar{s})]$ mesons are positive, whereas asymmetry for $\pi^-(d\bar{u})$ is

Table 3.3 $p(\uparrow) + p(0) \rightarrow \pi^{\pm,0}(\eta) + X$

P (sea)–T (val)					P (val)–T (sea)				
P (sea)	u	$\bar{u}$	d	$\bar{d}$	P (val)	u	u	d	d
p_y	0	0	0	0	p_y	$\leftarrow$	$\rightarrow$	$\leftarrow$	$\rightarrow$
weight	1	1	1	1	weight	5/3	1/3	1/3	2/3
T (val)	d	u			T (sea)	$\bar{d}$	d	$\bar{u}$	u
p_y	0	0			p_y	0	0	0	0
weight	1	2			weight	1	1	1	1
product	$d\bar{u}$	$u\bar{d}$			product	$u\bar{d}$	$u\bar{d}$	$d\bar{u}$	$d\bar{u}$
p_y	0	0			p_y	$\leftarrow$	$\rightarrow$	$\leftarrow$	$\rightarrow$
weight	1	2			weight	5/3	1/3	1/3	2/3
T (val)	u	d			T (sea)	$\bar{u}$	u	$\bar{d}$	d
p_y	0	0			p_y	0	0	0	0
weight	2	1			weight	1	1	1	1
product	$u\bar{u}$	$d\bar{d}$			product	$u\bar{u}$	$u\bar{u}$	$d\bar{d}$	$d\bar{d}$
p_y	0	0			p_y	$\leftarrow$	$\rightarrow$	$\leftarrow$	$\rightarrow$
weight	1	2			weight	5/3	1/3	1/3	2/3

negative. In addition to the quark composition of the mesons, weight factors should be taken into account. According to the model of rotating hadronic matter, polarization effects are determined only by valence quarks whose fraction in the proton increases with the Feynman parameter x_F. Therefore, asymmetry should increase with this parameter; this behavior is observed in the experiments. In the fragmentation region of the unpolarized target (see the left part of the tables), asymmetry should be zero, because valence quarks in unpolarized protons have zero transverse momentum p_y. This prediction is well confirmed at RHIC.

Similar tables can also be obtained for other mesons. Another prediction of this model following from Assumption 4 is that asymmetry at small $x_F (< 0.3)$ and p_T values (<0.2 GeV/c) is not expected. The cause is that these mesons are formed from sea quarks, and they do not provide asymmetry in this model.

Using the parton structure functions from Barger et al. (1974), the authors of this model calculated the analyzing power of pions from the E704 experiment. The results are shown in Fig. 3.4 for reactions with polarized proton beams. As seen in the figure, the predictions of the model of rotating hadronic matter are in qualitative agreement with the experimental data.

More recently, the model of rotating hadronic matter was successfully applied for describing the results of the E704 experiment on the measurement of asymmetry in the production of Λ hyperons in the $p(\uparrow) + p \rightarrow \Lambda + X$ reaction (Boros 1995; Boros and Zuo-Tang 1996), as well as the depolarization parameter D_{NN} in this reaction (Boros et al. 1996).

The presentation in this section is based on Boros (1995), Boros et al. (1996).

Table 3.4 $\bar{p}(\uparrow) + p(0) \to \pi^{\pm,0}(\eta) + X$

P (sea)–T (val)					P (val)–T (sea)				
P (sea)	u	$\bar{u}$	d	$\bar{d}$	P (val)		$\bar{u}$		$\bar{d}$
p_y	0	0	0	0	p_y	←	→	←	→
weight	1	1	1	1	weight	5/3	1/3	1/3	2/3
T (val)	d	u			T (sea)	d	$\bar{d}$	u	$\bar{u}$
p_y	0	0			p_y	0	0	0	0
weight	1	2			weight	1	1	1	1
product	$d\bar{u}$	$u\bar{d}$			product		$d\bar{u}$		$u\bar{d}$
p_y	0	0			p_y	←	→	←	→
weight	1	2			weight	5/3	1/3	1/3	2/3
T (val)	u	d			T (sea)	u	$\bar{u}$	d	$\bar{d}$
p_y	0	0			p_y	0	0	0	0
weight	2	1			weight	1	1	1	1
product	$u\bar{u}$	$d\bar{d}$			product		$u\bar{u}$		$d\bar{d}$
p_y	0	0			p_y	←	→	←	→
weight	1	2			weight	5/3	1/3	1/3	2/3

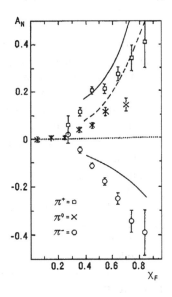

Fig. 3.4 Analyzing power A_N versus x_F in the $p(\uparrow) + p \to \pi^{\pm,0} + X$ reaction at the initial momentum 200 GeV/c (the E704 experiment). The *lines* are the predictions of the model of rotating hadronic matter

3.8.2 DeGrand–Miettinen Model for Polarization Asymmetry in the Inclusive Production of Hadrons

High polarizations surprisingly observed in the inclusive production of hyperons at Fermilab in the mid-1970s (Bunce et al. 1976) had to be theoretically explained. One of the successful models was proposed in DeGrand and Miettinen (1981) on the basis of the parton recombination model and $SU(6)$ symmetry that make it possible to attribute the polarization of inclusive baryons to subprocesses at the level of constituents. Taking into account the Thomas precession of the quark spin in the recombination process, the authors qualitatively interpreted the features of data on the polarizations of baryons and antibaryons in inclusive reactions at high energies and moderate momentum transfers.

The following features of experimental data were considered in these models (Heller 1997; Heller et al. 1978; Erhan et al. 1979; Bunce et al. 1979; Lomanno et al. 1979; Rayachauduri et al. 1980).

1. The polarization of Λ hyperons in pp and p-nucleus collisions at low momentum transfers has negative sign, i.e., is directed along the $\vec{p}_\Lambda \times \vec{p}$ vector.
2. The polarization is independent of the initial energy and depends only slightly on x_F.
3. The polarization increases almost linearly with p_T.
4. $\bar{\Lambda}$ produced in pp and pA interactions are unpolarized.
5. Measurements of the polarizations of Ξ^0, Ξ^-, and Σ^+ hyperons on a proton beam show that the polarizations of Ξ^0 and Ξ^- are the same as the polarization of Λ, whereas the polarization of Σ^+ is opposite in sign to the polarization of Λ, although its magnitude is the same as that for Λ.

Item 2 indicates that polarization data can be appropriately analyzed in the quark recombination model, which was successfully applied to the description of fragmentation processes with low momentum transfers. In this model, the proton in the infinite momentum frame is represented as consisting of three valence quarks and a large number of sea partons. When slow partons interact with the target, the coherence of the wave function is violated and the wave function is decomposed into numerous final hadron states. The pseudorapidity-semilocal transition of a parton into a hadron occurs. This process proceeds as the recombination of quarks; namely, pairs $q\bar{q}$ form mesons and triplets qqq form baryons. Fast hadrons are produced through the recombination of the valence quarks of the beam with the valence quarks of the target or with sea partons. For example, for the production of fast Λ (Σ^+) hyperons, a ud- (uu-) valence pair from the proton beam should be coupled with an s quark from the sea of the proton target. Such a recombination process is denoted as VVS. Note that exchange by one new quark, which does not exist at the beginning, namely, the s quark, occurs. In processes with the exchange by two new quarks such as $p \to \Xi^0$ and $p \to \Xi^-$, the VVS mechanism is impossible. In this case, the production of hadrons occurs through the VSS recombination mechanism. Finally, if the initial proton and final hadron have no common quarks, SSS is the only possible mechanism. In these cases, the cross sections for the processes are small and their slopes are large. As a whole, this prediction is in agreement with experiments.

In the recombination model, it is convenient to analyze the polarization of Λ hyperons in the rest frame of the fragmenting proton of the beam. In this case, the wave function of the proton in this frame has a simple form, as containing three quarks, uud. Then, the $p \to \Lambda$ process proceeds so that a fast particle of the target is scattered by a proton and the slow s-quark of the target recombines with a ud pair in the proton of the beam and forms a Λ hyperon. Since the ud pair in the Λ hyperon is in the singlet state, the spin and, correspondingly, the polarization of the Λ hyperon is completely determined by the s quark. According to the (approximate) hypothesis (Feynman 1972) on short-range forces between partons, the distribution of slow quarks in the particle of the target and the low-energy subprocess $s + p \rightarrow u + \Lambda$ with the appearance of the polarization of the Λ hyperon should depend slightly on the energy of the particle of the target. Therefore, the polarization of the Λ hyperon should be independent of the total energy of the colliding particles in agreement with experimental data.

Let us consider the relation between polarizations in different baryon–baryon transitions (Heller et al. 1977) under the following four simplifying assumptions.

1. The transverse momentum of each quark is parallel to the transverse momentum of the final baryon.
2. The polarization of the quark correlates with its transverse momentum and is independent of the quark flavor.
3. The wave function of the quarks common for the fragmenting and secondary baryons, for example, for the ud pair in the $p \rightarrow \Lambda$ transition is identical.
4. Valence quarks are not depolarized in the recombination process.

Let us first consider the VVS recombination process. The polarization of the final baryon with the transverse momentum p_T can be expressed in terms of two amplitudes A_+ and A_-. The former amplitude corresponds to the recombination of, e.g., ud quarks with a sea quark whose spin is upward with respect to the reaction plane, whereas the latter amplitude, to the recombination of a sea quark whose spin is downward. The calculations with $SU(6)$ symmetry give the expression

$$P(B \rightarrow B') = C \frac{|A_\uparrow|^2 - |A_\downarrow|^2}{|A_\uparrow|^2 + |A_\downarrow|^2}, \qquad (3.172)$$

where C is presented in Table 3.5. Interference between the amplitudes is forbidden due to the conservation of parity and total angular momentum. Taking into account that expected spin effects are small, we can use the expansions $|A_\downarrow|^2 = A(1 + \varepsilon)$ and $|A_\uparrow|^2 \sim A(1 - \varepsilon)$, where ε is on the order of the expected polarization.

However, this model in such a form gives an incorrect result for Σ^+: it predicts $C = -1/3$, whereas $C = -1$ follows from the experiment. Such discrepancy possibly appears because Σ^+ includes the uu diquark, which has spin $j = 1$, in contrast to the ud diquark that enters into Λ and has zero spin. Since correlation between spin and p_T for s quarks in the Λ hyperon is postulated from the very beginning, it is reasonable to assume the same correlation for the leading uu diquark in the Σ^+ hyperon. This means that the probability of recombination into the diquark in the (j, m) state depends on m (see Fig. 3.5). Taking $|A_{1,1}|^2 = B(1 + \delta)$, $|A_{1,-1}|^2 = B(1 - \delta)$, and $|A_{1,0}|^2 = |A_{00}|^2 = B$, we obtain

$$P(p \rightarrow \Sigma^+) = \frac{1}{3}\varepsilon + \frac{2}{3}\delta. \qquad (3.173)$$

Table 3.5 Predictions of polarizations for various transitions in the model of the leading and nonleading quarks; the constant C in Eq. (3.172) for VVS recombination is the coefficient of the term $-\varepsilon$

$B \to B$ transition	Polarization
$p \leftrightarrow n,\ \Sigma^- \leftrightarrow \Xi^-,\ \Sigma^+ \leftrightarrow \Xi^0$	$-(20/21) \cdot \varepsilon + (1/42) \cdot \delta$
$p \leftrightarrow \Sigma^+,\ n \leftrightarrow \Sigma^-,\ \Xi^- \leftrightarrow \Xi^0$	$(1/3) \cdot \varepsilon + (2/3) \cdot \delta$
$p, n \leftrightarrow \Lambda^0$	$-\varepsilon$
$\Sigma^+, \Sigma^-, \Xi^-, \Xi^0 \leftrightarrow \Lambda^0$	$-(2/3) \cdot \varepsilon + (1/6) \cdot \delta$
$p, n \leftrightarrow \Sigma^0$	$(1/3) \cdot \varepsilon + (2/3) \cdot \delta$
$\Sigma^+, \Sigma^-, \Xi^-, \Xi^0 \leftrightarrow \Sigma^0$	$-(20/21) \cdot \varepsilon + (1/42) \cdot \delta$
$p \leftrightarrow \Xi^0, \Xi^-, \Sigma^-$	$-(1/3) \cdot \varepsilon - (2/3) \cdot \delta$
$n \leftrightarrow \Xi^0, \Xi^-, \Sigma^+$	$-(1/3) \cdot \varepsilon - (2/3) \cdot \delta$
$\pi, K^+ \to \Lambda$	$-(1/2) \cdot \delta$
$K^- \to \Lambda$	ε

Fig. 3.5 Diagrams of the $p \to \Sigma^+$ transitions for the amplitude (**a**) $A_\uparrow A_{10}$ and (**b**) $A_\downarrow A_{11}$. Parton $\to$ hadrons transitions occur in ovals

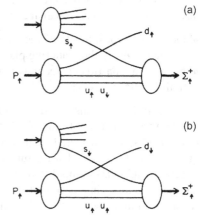

It follows from the experimental data that $\varepsilon = \delta$. Recombination occurs predominantly when the spin of the s quark is downward, and the leading diquark is in the state with the projection $m_j = +1$ in the scattering plane.

This idea can be applied to processes proceeding through VSS recombination. In this case, the V quark is leading and the SS diquark is nonleading. Under the same assumptions for the recombination probability as those used above, but taking into account change in the signs of ε and δ (the leading parton prefers upward spin), we obtain

$$P\left(p \to \Xi^0, \Xi^-, \Sigma^-\right) = -\left(\frac{1}{3}\varepsilon + \frac{2}{3}\delta\right). \tag{3.174}$$

The experimental results reported in Heller et al. (1978) satisfy the relations

$$P\left(p \to \Xi^0\right) = P\left(p \to \Xi^-\right) = P(p \to \Lambda),$$

which follow from formula (3.174) under the assumption that $\varepsilon = \delta$.

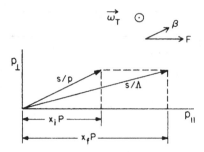

Fig. 3.6 Transverse and longitudinal momenta of the s quark in the proton (s/p) and in the Λ hyperon (s/Λ): the s quark has the longitudinal momentum $x_i P$ in the proton and $x_F P$ in the Λ hyperon; the vectors $\vec{\omega}_T$ and $\vec{F} \times \vec{\beta}$ coincide in orientation, both vectors are perpendicular to the figure plane and are directed toward a reader

According to the above presentation, all discussed experimental data on the polarizations of hyperons are satisfactorily described under the following assumptions:

- the leading constituents of the beam recombine primarily with a positive spin projection m_j in the scattering plane,
- nonleading constituents recombine primarily with a negative spin projection m_j in the scattering plane.

Predictions of the discussed model for other reactions are presented in Table 3.5.

The recombination of diquarks with a quark, rather than the recombination of three quarks, is considered in this model, because the interaction between two quarks with the identical wave functions, for example, VV, differs from their interaction with quarks with different wave functions. The interaction between similar quarks is taken into account by joining them into one object, diquark. All wave functions in this case are determined through the exact $SU(6)$ symmetry.

The above predictions followed from the representation on the recombination of quarks and $SU(6)$ symmetry. In this case, a dynamic mechanism leading to the polarization of sea quarks is disregarded. One of such mechanisms is considered below (Anderson et al. 1979).

Let us consider the $p \rightarrow \Lambda$ transition in the infinite-momentum frame. The longitudinal and transverse momenta of the strange quark before recombination are denoted as $x_l p$ and $k_{\perp s}$, respectively. Since the distribution of sea quarks is very steep such as $|1 - x|^n$, where $n = 7$–9, the Λ hyperon acquires its momentum from the valence ud diquark. As an illustration, we note that the relative momentum of Λ, $x_\Lambda = 0.6$, consists of $x_l \leq 0.1$ and $x_{ud} \geq 0.5$. After recombination and formation of the Λ hyperon, all three quarks have almost equal momenta, including the s quark, $x_l \approx 0.2$ (Fig. 3.6). This means that the sea quark is subjected to a force driving it in parallel to the beam axis.

Since this force is not parallel to the velocity $\vec{\beta}$ of the s quark, the spin of the s quark undergoes Thomas precession. The Thomas interaction potential is $U = \vec{s} \cdot \vec{\omega}_T$, where $\vec{\omega}_T \propto [\gamma/(\gamma + 1)] \cdot (\vec{F} \times \vec{\beta})$ (Thomas 1927). The amplitude of the formation of the Λ hyperon with spin s is proportional to the energy difference between the initial and final states $(\Delta E_0 + U)^{-l}$, where $\Delta E_0(> 0)$ is the difference between the quark energies in the intermediate and final states in the absence of spin-dependent interaction. Thus, the cross section for the formation of the Λ hyperon is larger when the quantity $\vec{s} \cdot \vec{\omega}_T$ is negative. In the infinite-momentum frame, the vector $\vec{\omega}_T$ is directed along the $\vec{p}_p \times \vec{p}_\Lambda$ vector, i.e., is perpendicular to the reaction plane. As a result, the cross section is larger when the spin of the s quark and, hence, Λ hyperon, is opposite to this direction; this conclusion is in agreement with experimental results. Particular calculations of asymmetry show that polarization depends slightly on x_F and is proportional to the transverse momentum of the Λ hyperon. This fact is a consequence of the large difference between the $p_\parallel$ and p_T values in the desired problem. Thus, Thomas precession predicts the basic qualitative characteristics of polarization in the $p \to \Lambda$ transition.

It is now easier to understand why the polarization of the $\bar{\Lambda}$ hyperon is zero, i.e., $P(p \to \bar{\Lambda}) = 0$. The production of $\bar{\Lambda}$ by the proton beam entirely occurs through the recombination of sea quarks, i.e., through the integration of the fast $\bar{s}$ sea antiquark with the $\bar{u}\bar{d}$ sea antidiquark. The number of such combinations is large and they provide zero polarization in average in agreement with experiment.

Thus, the DeGrand–Miettinen model is a model for explaining the polarization of quarks. The basic mechanism of the polarization of quarks is Thomas spin precession appearing in the quark recombination process with the formation of final hadrons. The model provides relations between polarizations in various inclusive baryon–baryon transitions. These relations allow one to check the quark recombination picture irrespectively of the quark polarization mechanism. This model can also be applied to other inclusive processes, for example, the transitions of octet baryons to a baryon from a decuplet, baryon to vector mesons, mesons to baryons, and mesons (scalar and pseudoscalar) to vector mesons. Finally, in complete analogy with the above argumentation, this model can be applied to other processes, for example, to electron–positron annihilation or deep-inelastic scattering. It can be shown that the leading baryon or vector meson from a quark jet, which has a transverse momentum (perpendicular to the jet axis), is polarized. Thus, according to this model, polarization is a very widespread phenomenon, although its magnitude is small.

3.8.3 Lund Model

This model is based on the notion of string taken from QCD and on some commonly accepted assumptions. At the instant of interaction between hadrons, a color string is stretched between their partons. Then, the string breaks with the emission of a quark–antiquark pair, which is separated by the chromomagnetic field. In the case of zero quark mass, this pair can be produced at a point. However, if quarks have mass or transverse momentum $k_\perp$, quarks can classically be produced only at a certain distance l from each other. The energy of the field between these quarks transforms to the transverse quark mass (Fig. 3.7):

$$kl = 2\mu_\perp, \tag{3.175}$$

Fig. 3.7 Diagram of the
Lund model for the
polarization of the sea quark

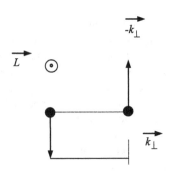

where $k \sim 1$ GeV/fm ~ 0.2 GeV^2 is the string tension energy per unit length, μ is the quark mass, and $\mu_\perp = \sqrt{\mu^2 + k_\perp^2}$ is the transverse quark mass (equal to the antiquark mass). Such a formation of $q\bar{q}$ pair is described in quantum mechanics as the tunnel effect.

For the conservation of the transverse momentum, the quark and antiquark should be produced with transverse momenta equal in magnitude but opposite in sign. According to Fig. 3.7, the appearing pair has orbital angular momentum $\vec{L}$ given by the expression

$$\vec{L} = \vec{k}_\perp \times \vec{l} = 2\frac{k_\perp}{k}\mu_\perp \vec{n}. \tag{3.176}$$

Here, $\vec{n}$ is the unit vector perpendicular to the plane containing the string and transverse momentum $\vec{k}_\perp$. The conservation law for total momentum $\vec{J}$ should also be taken into account. In the initial state, before the appearance of the pair, $\vec{J} = 0$. After the appearance of the $q\bar{q}$ pair (or diquark) with orbital angular momentum $\vec{L}$, the spins of both quarks should be parallel and opposite to $\vec{L}$ in order to compensate orbital angular momentum,

$$\overleftarrow{J} = \vec{L} + \vec{S}, \tag{3.177}$$

where $\vec{S}$ is the total spin of the $q\bar{q}$ pair. Therefore, both the quark and antiquark should be polarized identically in the direction opposite to the orbital angular momentum (see Fig. 3.7). The directions of the orbital angular momentum and spin are indicated in the circles near their symbols. Since the total spin is unity, the orbital angular momentum should also be unity. The theory of tunneling processes provides the condition $L \sim 2/\pi \approx 1$.

The above picture (Lund model) (Anderson et al. 1981, 1983) gives a simple description of the polarizations of Λ hyperons produced in the proton fragmentation process (Anderson et al. 1979). For the production of Λ hyperons with large x_F values, a string should appear; the ud quark pair is usually presented at one of the ends of the string. The string breaks with the formation of the $s\bar{s}$ quark pair (Fig. 3.8). For the uds quark triplet to form the Λ hyperon, the ud pair should be in the singlet isospin ($I = 0$) and spin ($s = 0$) states. In this case, the spin of the Λ hyperon is equal to the spin of the s quark. Then, if the transverse momentum of

Fig. 3.8 Diagram of the formation of the polarized Λ hyperon in the Lund model

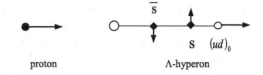

proton Λ-hyperon

the Λ hyperon is nonzero and directed upwards, its polarization coincides with the polarization of the s quark and is directed along the $\vec{p}_\Lambda \times \vec{p}$ vector (Fig. 3.8).

In this picture, confinement of quarks is very important. When the quark–antiquark pair is formed in the string-like field, the string moves as shown in Fig. 3.9, the external forces F and $-F$ (string tension forces) do not generate torque and, therefore, the total angular momentum is conserved.

For comparison, Fig. 3.10 shows the case of the formation of an e^+e^- pair in an external uniform electric field. Since confinement is absent, the particles in the pair are not connected by a string and the external force adds angular momentum. Thus, the conservation law for the total momentum is violated. As a result, the positron and electron are polarized in the directions opposite to each other and are independent of each other.

Let us consider several additional physical processes. First, we analyze the polarization of the $\bar{\Lambda}$ hyperon in the reaction

$$K^+ + p \to \bar{\Lambda} + X. \tag{3.178}$$

The polarization of the $\bar{\Lambda}$ hyperon in this reaction was observed in the experiments reported in Faccini-Turluer et al. (1979) and Barth et al. (1981). If the $\bar{\Lambda}$ hyperon is measured with a large Feynman parameter in the fragmentation region of the initial K^+ meson, it is reasonable to assume that the $\bar{s}$ quark and $(\bar{u}\bar{d})_0$ diquark for the formation of the $\bar{\Lambda}$ hyperon are taken from the K^+ meson and broken string, respectively. In the collision of the proton (its constituents) with the u quark

Fig. 3.9 Action of confining force F on (*closed circles*) sea quarks: the *large arrows* indicate the direction of the color force and the *small arrows* indicate the direction of quark motion

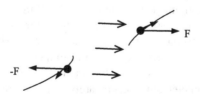

Fig. 3.10 Production of an e^+e^- pair in vacuum: the *closed circles* denote the electron and positron, the *long thick arrows* show the vacuum breaking force, the *short thick arrows* indicate the lepton motion direction, and the *thin arrows* denote the external electric field

from the K^+ meson, a color string appears and rotation motion is induced and leads to two consequences. First, the $\bar{s}$ quark and string connected with it acquire angular momentum and the $\bar{s}$ quark is polarized. Second, the $\bar{s}$ quark acquires transverse momentum in the string motion direction. When the $\bar{\Lambda}$ hyperon is produced due to string breaking and joining the appearing singlet antidiquark $(\bar{u}\bar{d})_0$ to the $\bar{s}$ quark, the $\bar{\Lambda}$ hyperon acquires both these parameters. As a result, it is polarized in the necessary direction (in the $\vec{p}_K \times \vec{p}_\Lambda$ direction). The situation is similar for the reaction

$$K^- + p \rightarrow \Lambda + X. \tag{3.179}$$

Here, the background from proton fragmentation provides some difficulty.

The second process of interest involves Σ^+:

$$p + N \rightarrow \Sigma^+ + X. \tag{3.180}$$

A fast Σ^+ hyperon is composed by a uu pair from the proton in the triplet spin state $(S = 1)$ and an s quark from the chromomagnetic field. Since the spin of the Σ^+ hyperon is parallel to the spin of the uu diquark, the polarization in this case is larger than the polarization of the $\bar{\Lambda}$ hyperon in the preceding reaction. This is in agreement with the experimental relation $P(\Sigma^+) \approx -P(\Lambda)$ (Lundberg et al. 1982; Cooper et al. 1982). If only the s quark were polarized, the polarization of the Σ^+ hyperon would be much smaller, namely, $P(\Sigma^+) \approx -\frac{1}{3}P(\Lambda)$ disregarding depolarizing processes such as the Σ^0, $Y^* \rightarrow \Lambda$ decays.

To conclude, we note that the Lund model predicts significant polarization effects in the production of hadrons.

The presentation in this section was based on report of Gustafson (1984).

3.8.4 Chromomagnetic String Model

The observation of high single-spin effects in inclusive hadron production at high energies $\sqrt{s} \sim 10$ GeV was one of the strong driven factors for polarization physics in the late 1970s–early 1980s. Recent measurements showed that these spin effects hold up to the energy $\sqrt{s} = 200$ GeV (Adams et al. 2003). Since these effects were primarily observed in the soft collision region, the PQCD technique could not be directly applied. For this reason, phenomenological approaches began to be rapidly developed; one of these approaches is the model described below, which was named by the authors the "chromomagnetic string model" (CMSM) (Nurushev and Ryskin 2006). This model was proposed by Ryskin (1988) for explaining inclusive pion asymmetries observed primarily at the U-70 accelerator (IHEP). This model provides very simple analytical dependences for asymmetry in almost the entire kinematic region of inclusive processes. In this model, single-spin asymmetry is predicted without free parameters; this occurs rarely. The essence of the model is briefly as follows. A color tube (string) is stretched between hadrons after their collision. In the simplest case, such a tube contains a color electric field. However, this system is unstable. For the system to be stable, a color magnetic field should circulate around the tube. The interaction of this magnetic field with the color magnetic moment of the quark shifts the quark in the direction perpendicular to the string axis. This shift depends on the quark spin orientation with respect to the magnetic field direction. The estimate of this shift gives the value $\delta p_T \cong 0.1$ GeV/c for the increment of the transverse momentum ("kick") (Ryskin 1988). In terms of the invariant cross section $\rho = E\frac{d\sigma}{d^3 p}$, the asymmetry of the polarized quark can be expressed as

$$A_q = \frac{\rho(+) - \rho(-)}{\rho(+) + \rho(-)},$$
(3.181)

where $(+)$ and $(-)$ in the argument of the function ρ mean the directions of the quark polarization vector. Since $\delta p_T < p_T$, where p_T is the transverse momentum of the final pion into which the polarized quark fragments, expanding in powers of δp_T, we arrive at the following expression for the analyzing power:

$$A_q(x) = \frac{d\sigma(p_T + \delta p_T) - d\sigma(p_T - \delta p_T)}{d\sigma(p_T + \delta p_T) + d\sigma(p_T - \delta p_T)} = \delta p_T \cdot \frac{\delta}{\delta p_T}\left(\frac{d\sigma}{d^3 p}\right) \bigg/ \frac{d\sigma}{d^3 p} = \delta p_T \cdot B.$$
(3.182)

Here, the slope parameter B of the function ρ is determined by the standard expression

$$B = \frac{\delta}{\delta p_T}\left(\frac{d\sigma}{d^3 p}\right) \bigg/ \frac{d\sigma}{d^3 p} = \frac{\delta}{\delta p_T}(\ln \rho).$$
(3.183)

The formula for the analyzing power of the final hadron can now be represented in the general form (Nurushev and Ryskin 2006):

$$A_N(x) = P_q(x) \cdot A_q(x) \cdot w(x).$$
(3.184)

Here, $P_q(x)$ is the polarization of the initial quark carrying the fraction x of the momentum of the initial polarized proton, $A_q(x)$ is the analyzing power of the quark defined by Eq. (3.182), and $w(x)$ is the fraction of the contribution from a given reaction channel to the parton–parton interaction. Transverse quark polarization (transversity) $P_q(x)$ has been already discussed above. However, it has not yet been determined from experiments. It is expected that transversity can be associated with the structure function $g_2(x)$, which is close to zero according to the experimental data (Mcnully 2002) and has features at small x values.

In this situation, to determine $P_q(x)$, it is necessary to use models. The nonrelativistic quark model proposes the dependences

$$P_q(x) = \frac{2}{3} \cdot x \quad \text{for } \pi^+,$$

$$P_q(x) = \frac{1}{3} \cdot x \quad \text{for } \pi^0, \quad \text{and} \quad P_q(x) = -\frac{1}{3} \cdot x \quad \text{for } \pi^-.$$
(3.185)

The nonrelativistic quark model predicts that the behaviors of the polarization of the quark are substantially different for different charges of the final pion. First, the sign of the polarization is positive for π^+ and π^0 and negative for π^-. Second, asymmetry is expected to be maximal for π^+ as compared to π^- and π^0.

These predictions are generally in agreement with experiment, although there are some disagreements. Another source of information on quark polarization is Anselm and Ryskin (1995, 1996), where an attempt to explain "spin crisis" was made. According to this work,

$$P_q(x) = 0.7x \quad \text{for } \pi^+,$$
$$P_q(x) = 0.28x \quad \text{for } \pi^0, \quad \text{and} \quad P_q(x) = -0.55x \quad \text{for } \pi^-. \tag{3.186}$$

The comparison of Eqs. (3.186) and (3.185) shows that a significant difference between these formulas is observed only in the expression for the polarization of π^-. This fact will be taken into account when comparing the analyzing power $A_N(x)$ with experimental data.

The general expression for the weight factor depends on the parton distribution function $V_a(x)$ and parton fragmentation function $D_a(x)$ (a is the parton flavor). Assuming that polarization is primarily transferred by quarks and, in a much lower degree, by gluons, we arrive at the following expression for the weight factor (or the dilution factor):

$$w(q) = \frac{\sigma(q)}{\sigma(q) + \sigma(g)}, \tag{3.187}$$

under the assumption that the leading contribution comes from the interaction qg' (to $\sigma(q)$) and gg' (to $\sigma(g)$) in the t channel with gluon exchange. In this case, it is accepted that an unpolarized nucleon emits the gluon g, whereas a polarized nucleon emits the quark q and gluon g. Then, the contributions from the quark and gluon can be represented in the form

$$\sigma(q) \propto C_F \int_x^1 V^q\left(\frac{x}{z}\right) D^q(z) \frac{dz}{z},$$
$$\sigma(g) \propto C_A \int_x^1 V^g\left(\frac{x}{z}\right) D^g(z) \frac{dz}{z}, \tag{3.188}$$

where $C_F = 4/3$ and $C_A = 3$ are the color factors and z is the fraction of the momentum of the polarized quark that is carried by the final hadron.

The quark and gluon distribution functions were taken in the form

$$V^q(x) = x \cdot v(x) = 2.8 \cdot \sqrt{x} \cdot (1-x)^2,$$
$$V^g(x) = x \cdot g(x) = 3.0 \cdot \sqrt{x} \cdot (1-x)^5. \tag{3.189}$$

The fragmentation functions were taken in the form

$$D^u_{\pi^+}(z) = \frac{4}{3}(1-z), \qquad D^u_{\pi^0}(z) = \frac{2}{3}(1-z), \qquad D^u_{\pi^-}(z) \approx 0;$$
$$D^g_{\pi^+}(z) = D^g_{\pi^0}(z) = D^g_{\pi^-}(z) = (1-z)^2. \tag{3.190}$$

Isotopic invariance and charge conjugation lead to the relations

$$D^u_{\pi^+}(z) = D^{\bar{u}}_{\pi^-}(z) = D^d_{\pi^-}(z) = D^d_{\pi^+}(z),$$
$$D^u_{\pi^-}(z) = D^{\bar{u}}_{\pi^+}(z) = D^d_{\pi^+}(z) = D^d_{\pi^-}(z). \tag{3.191}$$

Weight factor (3.187) was calculated with the use of relations (3.188)–(3.190). The final expression for the weight factor for pions with different charges can be represented in the form

$$w(x) = \frac{\sqrt{x}}{\sqrt{x} + c(1-x)^{4.5}}. \tag{3.192}$$

Here, the constant c is: $c_+ = 0.48$ for π^+, $c_0 = 0.64$ for π^0, and $c_- = 0.96$ for π^-.

The analyzing power of inclusive pions is calculated from Eqs. (3.184) and (3.192) and has no free parameters. The explicit and simple analytical form of the formulas for asymmetry and the absence of fitting parameters are advantages of the chromomagnetic string model. The model is applicable at high energies, $x_F > 0.2$, and $p_T > 0.3$ GeV/c. At low energies and small x_F values, processes with the formation of resonances play a noticeable role (Musulmanbekov and Tokarev 1995). To temporarily avoid this problem, we analyze only the experimental data at energies $\sqrt{s} > 10$ GeV.

Inclusive reactions can be conventionally separated into three kinematic regions: the central region with the Feynman parameter in the range $-0.15 \leq x_F \leq 0.15$, beam fragmentation region $0.3 \leq x_F < 1$, and intermediate region. Polarization results are primarily concentrated in the first two regions. We consider them in the CMSM.

A. Central region

To date, we know three experimental works on the measurement of single-spin asymmetry in the inclusive production of π^0 mesons (we do not know data on charged particles). They are as follows.

1. Inclusive asymmetry in the reaction

$$p(\uparrow) + p \rightarrow \pi^0 + X. \tag{3.193}$$

In the E704 experiment, this reaction was measured at a momentum of 200 GeV/c in the kinematic region $-0.15 \leq x_F \leq 0.15$ and $1.48 \leq p_T$ (GeV/c) ≤ 4.31 (Adams et al. 1996).

2. Inclusive asymmetry in the reaction

$$\bar{p}(\uparrow) + p \rightarrow \pi^0 + X. \tag{3.194}$$

In the E704 experiment, this reaction was measured at a momentum of 200 GeV/c in the kinematic region $-0.15 \leq x_F \leq 0.15$ and $1.48 \leq p_T$ (GeV/c) ≤ 3.35 (Adams et al. 1996).

3. Inclusive asymmetry in the reaction

$$p + p(\uparrow) \rightarrow \pi^0 + X. \tag{3.195}$$

In the PROZA-M experiment, this reaction was measured at a momentum of 70 GeV/c in the kinematic region $-0.15 \leq x_F \leq 0.15$ and $1.05 \leq p_T$ (GeV/c) ≤ 2.74 (Vasiliev et al. 2003).

Below, we discuss these data and compare them with the model predictions.

1. As seen in the above discussions, the CMSM can predict asymmetry if the slope parameter B is known. In the E704 experiment, this parameter in the central region was determined as $B = (4.19 \pm 0.08)$ $(GeV/c)^{-1}$ (Adams et al. 1996) and appeared to be independent of $-t$ in the measured range. Then, the calculation of the analyzing power of the quarks gives $A_q = 0.1$, $B = 0.419$.

According to relations (3.185) and (3.186), taking transversity $P_q(x) = x/3$, and constant $c_0 = 0.64$ (3.192), and substituting all presented relations into Eq. (3.184), we obtain the final expression for the asymmetry of the π^0 meson in the central region for the E704 experiment conditions:

$$A_N(x) = \frac{0.14 \cdot x^{1.5}}{\sqrt{x} + 0.64(1-x)^{4.5}}. \tag{3.196}$$

Several features follow from this formula. First, asymmetry in the central region depends only on $X = x_T = \frac{2p_T}{\sqrt{s}}$ if the slope parameter B is independent of the momentum transfer. Second, asymmetry vanishes at $x = 0$ due to the polarization of the quark and, to a certain degree, to the contribution from the gluon (the second term in the denominator). When x increases and approaches unity, asymmetry increases linearly with x. All listed properties of asymmetry are observed in experiments. The calculation results by formula (3.196) are shown by the solid line in Fig. 3.11. As seen, they are in agreement with the data of the E704 experiment at 200 GeV/c.

2. The second group of data refers to asymmetry in the inclusive production of π^0 mesons on the 200-GeV/c antiproton beam in reaction (3.194) (Adams et al. 1996). This reaction differs from reaction (3.193) by changing all quarks in the proton to antiquarks; thus, the proton is replaced by the antiproton. Therefore, charge conjugation invariance leads to the relation

$$A_N(\bar{p}(\uparrow) + p \to \pi^0 + X) = A_N(p(\uparrow) + p \to \pi^0 + X). \tag{3.197}$$

As seen in Fig. 3.11, this condition corresponds to the experimental data within the measurement errors.

3. New data obtained at the PROZA-M setup (Vasiliev et al. 2003) at 70 GeV refer to reaction (3.195). The data were obtained on the unpolarized proton beam with the polarized proton target. They are presented in Fig. 3.11 with the inverse sign for comparison with the results of the E704 experiment. The expression for asymmetry obtained in the CMSM at 70 GeV has the form

$$A_N(x_T) = 0.2 \cdot \frac{x_T^{1.5}}{\sqrt{x_T} + 0.64(1-x_T)^{4.5}}. \tag{3.198}$$

Here, two corrections are taken into account as compared to formula (3.196): to the energy difference and to the difference in the slope parameter.

At 70 GeV/c, the slope parameter is $B = (5.89 \pm 0.08)$ GeV/c according to (Vasiliev et al. 2003). The calculation results are given in Fig. 3.11 by the dashed line and are in agreement with the experimental data.

However, the qualitative test of the model requires more accurate data on asymmetry in the central region.

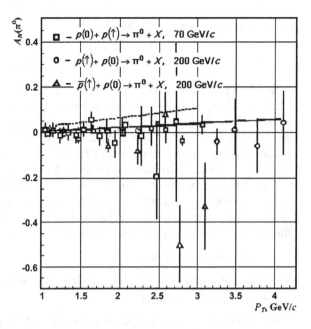

Fig. 3.11 Asymmetry versus P_T for the $p + p \to \pi^0 + X$ reaction in the central region; the target is polarized at 70 GeV/c and the beam is polarized at 200 GeV/c; asymmetry for the $\bar{p} + p \to \pi^0 + X$ reaction at 200 GeV/c is also shown; and the CMSM predictions are shown by the *solid* and *dashed lines* for momenta 200 and 70 GeV/c, respectively

Asymmetry in the $p(\uparrow) + p \to \pi^0 + X$ reaction in the central region at energy $\sqrt{s} = 200$ GeV was measured in the PHENIX experiment at RHIC. The PHENIX collaboration has not yet published its results on asymmetry in this reaction. However, it published the data of the precision measurement of the invariant differential cross section for this reaction in the range $1 \le p_T \le 13$ GeV/c at pseudorapidities $|\eta| \le 0.39$ (Adler et al. 2003). This cross section was parameterized in the form

$$\rho = E \frac{d^3\sigma}{dp^3} = A \cdot \left(1 + \frac{p_T}{p_0}\right)^{-n}. \tag{3.199}$$

Here, $A = 386$ mb GeV^{-2}, $p_0 = 1.219$ GeV/c, and $n = 9.99$. Then, we can calculate the slope parameter

$$B(p_T) = \frac{d\ln\rho}{dp_T} = \frac{n}{p_0 + p_T}. \tag{3.200}$$

According to formula (3.200), the slope parameter decreases as p_T^{-1} with an increase in the momentum transfer; as a result, asymmetry also decreases with an increase in the momentum transfer. This behavior is similar to that observed at lower momenta (≈ 200 GeV/c) (Donaldson et al. 1976). The invariant cross section at 100 and 200 GeV/c was parameterized as follows:

$$\rho = E \frac{d^3\sigma}{d^3 p} \propto \left(p_T^2 + M^2\right)^N \cdot (1 - x_T)^F. \tag{3.201}$$

Fitting the experimental results gives

$$N = -5.4 \pm 0.2, \qquad M^2 = (2.3 \pm 0.3)\,\text{GeV}^2, \quad \text{and}$$
$$F = 7.1 \pm 0.4; \quad \text{where } x_T = 2p_T\sqrt{s}.$$

Then, we can determine the slope parameter

$$B(p_T) = \frac{2Np_T}{p_T^2 + M^2} - \frac{2F}{\sqrt{s}(1 - x_T)}. \tag{3.202}$$

The slope parameter from PHENIX data (3.200) has the same p_T dependence as in the E704 experiment at lower energies $\sqrt{s} \approx 20$ GeV (3.192). Thus, the slope parameter in the energy range $\sqrt{s} = 20$–200 GeV is almost independent of energy. Since two other parameters appearing in the CMSM formula are also slightly dependent on energy, asymmetry at the PHENIX energy is expected to be small. The calculations show that asymmetry is no more than 1 %. Such an asymmetry can be observed if the measurement accuracy is no worse than 0.1 %.

Thus, it can be preliminarily concluded that asymmetry in the inclusive production of π^0 mesons in the central region is almost absent (or very small), beginning with the center-of-mass energy $\sqrt{s} \approx 10$ GeV.

B. Beam fragmentation region

For this region and energies $\sqrt{s} \geq 10$ GeV, the following results on single-spin asymmetry have been reported.

1. The results of the STAR experiment on the $p(\uparrow) + p \to \pi^0 + X$ reaction at $\sqrt{s} = 200$ GeV for the kinematic measurement region $0.18 \leq x_F \leq 0.59$, $1.5 \leq p_T$ (GeV/c) ≤ 2.3, $|\eta| \approx 3.8$ (Adams et al. 2003).
2. The results of the E704 experiment on the $p(\uparrow) + p \to \pi^0 + X$ reaction at $\sqrt{s} \approx 20$ GeV for the kinematic measurement region $0.03 \leq x_F \leq 0.9$, $0.5 \leq p_T$ (GeV/c) ≤ 2.0 (Adams et al. 1992).
3. The results of the E704 experiment on the $\bar{p}(\uparrow) + p \to \pi^0 + X$ reaction at $\sqrt{s} \approx 20$ GeV for the kinematic measurement region $0.03 \leq x_F \leq 0.67$, $0.5 \leq p_T$ (GeV/c) ≤ 2.0 (Adams et al. 1991b).
4. The results of the E704 experiment on the $p(\uparrow) + p \to \pi^\pm + X$ reaction at $\sqrt{s} \approx 20$ GeV for the kinematic measurement region $0.2 \leq x_F \leq 0.9$, $0.2 \leq p_T$ (GeV/c) ≤ 1.5 (Adams et al. 1991d).
5. The results of the E704 experiment on the $\bar{p}(\uparrow) + p \to \pi^\pm + X$ reaction at $\sqrt{s} \approx 20$ GeV for the kinematic measurement region $0.2 \leq x_F \leq 0.9$, $0.2 \leq p_T$ (GeV/c) ≤ 1.5 (Bravar et al. 1996).

All these experimental data are shown in Fig. 3.12. Formulas (3.182) and (3.187) for asymmetry and weight factor $w(x)$, respectively, are also applicable for the beam fragmentation region. The main change is that the argument x is replaced by x_F, which now means the momentum fraction carried by the final hadron after quark decay. Then, the formula for asymmetry is written in the form (see Eq. (3.184))

$$A_N(x_F, p_T, g) = P_q(x_F) \cdot A_q(x_F, p_T, g) \cdot w(x_F, c). \tag{3.203}$$

The parameter c in the dilution factor was defined above (see Eq. (3.192)).

To calculate asymmetry, it is necessary to calculate the slope parameter given by expression (3.202) for each reaction taking into account the kinematic measurement region.

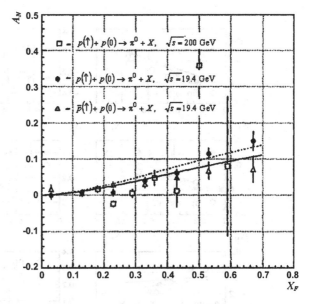

Fig. 3.12 Asymmetry versus x_F in the polarized-beam fragmentation region for the $p(\uparrow) + p \rightarrow \pi^0 + X$ reaction at $\sqrt{s} = 19.4$ and 200 GeV and for the $\bar{p}(\uparrow) + p \rightarrow \pi^0 + X$ reaction at $\sqrt{s} = 19.4$ GeV. The *solid* and *dotted lines* are the calculation results for $\sqrt{s} = 19.4$ and 200 GeV, respectively

The derivation of the formula for asymmetry was discussed above. For this reason, we immediately write the following final expression for asymmetry expected in the STAR experiment:

$$A_N(x_F) = 0.2 \cdot \frac{x_F^{1.5}}{\sqrt{x_F} + 0.64(1 - x_F)^{4.5}}. \tag{3.204}$$

The results of the calculation by this formula are given by the dotted line in Fig. 3.12. As seen, the agreement with the first results of the STAR experiment is satisfactory.

Since the cross sections were not measured in the E704 experiments, we take them from work (Carey et al. 1976) near the kinematic region appropriate for our aim. The value $B = (4.75 \pm 0.07)$ $(\text{GeV}/c)^{-1}$ was obtained by fitting.

Taking $P_q(x_F) = (1/3)x_F$ for quark polarization, we obtain the following final expression for asymmetry at a momentum of 200 GeV/c (the E704 experiment) for π^0 mesons:

$$A_N(x_F) = 0.16 \cdot \frac{x_F^{1.5}}{\sqrt{x_F} + 0.64(1 - x_F)^{4.5}}. \tag{3.205}$$

The results of the calculation by this formula are shown in Fig. 3.12 by the solid line. As seen, qualitative agreement with the E704 data is achieved.

Figure 3.12 also shows the results on A_N for the $\bar{p}(\uparrow) + p \rightarrow \pi^0 + X$ reaction at a momentum of 200 GeV/c. The model prediction is simplified due to relation (3.206), which is a consequence of the change of the quark to the antiquark. As seen in Fig. 3.12, agreement between the model prediction (solid line) and experimental data is good.

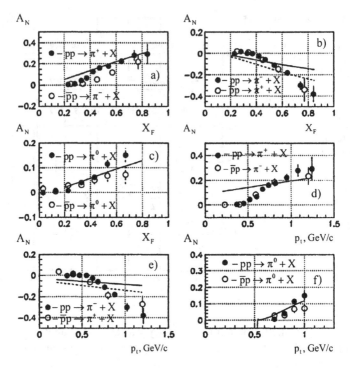

Fig. 3.13 Test of charge conjugation rules (3.197) and (3.209) for single-spin asymmetry A_N in the inclusive production of pions in the (*closed circles*) $p(\uparrow) + p \to \pi^{\pm,0} + X$ and (*open circles*) $\bar{p}(\uparrow) + p \to \pi^{\pm,0} + X$ reactions at 200 GeV/c. Asymmetry A_N in (**a**)–(**c**) and (**d**)–(**f**) is given versus x_F and p_T, respectively. The *solid* and *dashed lines* are the CMSM predictions for the quark distribution by formulas (3.185) and (3.186), respectively

Figure 3.13 shows the data of the E704 experiment on single-spin asymmetry in the inclusive production of pions at an initial momentum of 200 GeV/c in the reactions

$$p(\uparrow) + p \to \pi^{\pm,0} + X, \qquad \bar{p}(\uparrow) + p \to \pi^{\pm,0} + X. \qquad (3.206)$$

Since the cross sections were not measured in the E704 experiments, the slope parameter was estimated from the data reported in Breakstone et al. (1995). As a result, the slope parameters $B_+ = 5.55$ GeV^{-1} and $B_- = 5.33$ GeV^{-1} were obtained. Thus, taking the expression $P_q(x) = \frac{2}{3}x$ for quark polarization, we arrive at the formula

$$A_N^{\pi^+}(x_F) = 0.37 \cdot \frac{x_F^{1.5}}{\sqrt{x_F} + 0.48(1 - x_F)^{4.5}}. \qquad (3.207)$$

The model prediction for positive pions is shown by the solid line in Fig. 3.13a and is in satisfactory agreement with the experimental data (presented by the closed circles) except for small values $x_F < 0.4$.

The prediction for π^- mesons was obtained with the quark polarization $P_q(x) = -\frac{1}{3}x$ in the nonrelativistic quark model by the formula

$$A_N^{\pi^-}(x_F) = -0.18 \cdot \frac{x_F^{1.5}}{\sqrt{x_F} + 0.96(1 - x_F)^{4.5}}. \qquad (3.208)$$

The results are shown in Fig. 3.13b by the solid line. It is seen that agreement is not rather good. In view of this circumstance, the dashed line in the same figure is obtained when the expression for quark polarization was taken in the form $P_q(x) = -0.55x$ from Carey et al. (1976). This variant provides better description at high x values. However, good description at small x_F values is not achieved in both cases.

Figure 3.13c shows the asymmetry of the π^0 mesons produced by the (closed circles) proton and (open circles) antiproton polarized beams. As seen, charge conjugation rule (3.197) is satisfactorily fulfilled.

Asymmetry in the inclusive production of charged pions should satisfy the following charge conjugation rule:

$$A_N\big(\bar{p}(\uparrow) + p \to \pi^\pm + X\big) \approx A_N\big(p(\uparrow) + p \to \pi^\mp + X\big). \qquad (3.209)$$

The validity of this relation can be concluded from Figs. 3.13a and 3.13c, where the proton and antiproton data are given by the closed and open circles, respectively. As a whole, agreement is achieved, although deviations from this rule are observed, particularly at small x_F values.

Figures 3.13d–3.13f show the p_T dependence of asymmetry. This dependence was reconstructed from the tables in Adams et al. (1991b, 1991d, 1992), where the tables of the correlation of the arguments p_T and x_F were given. The solid and dashed lines in these figures show the CMSM predictions for quark distributions (3.185) and (3.186), respectively. As a whole, qualitative agreement is seen, particularly at high arguments. Discrepancy between the predictions and experimental data at small arguments was attributed by the authors to either resonances or threshold phenomena.

3.9 Polarization as a Tool for Studying Hadron Matter Under the Extremum Conditions

In modern nuclear physics, particularly in the field of the interaction of heavy relativistic ions, the search for the so-called quark–gluon plasma (QGP) is a very popular item. A number of methods have been proposed in the literature for applying polarization phenomena to separate the QGP signal from the background processes. These are the following processes.

The appearance of the quark–gluon plasma should be accompanied by the intense production of cascade $\varXi$ ($\bar{\varXi}$) hyperons. The cascade hyperons (anticascade hyperons) decay into $\varLambda$ ($\bar{\varLambda}$) hyperons, which appear to be longitudinally polarized. No other sources of longitudinally-polarized $\varLambda$ ($\bar{\varLambda}$) hyperons are expected. Therefore, in addition to the measurement of the yields of cascade hyperons, the measurement of the polarization of their decay products was proposed with the focus on the longitudinal polarization of Λ hyperons.

The next proposal is similar to a review experiment. Let the polarization of Λ hyperons produced in pp collisions at the ISR accelerator be measured. In this case, the presence of the polarization of Λ hyperons has been experimentally established. What should we expect for the polarization of Λ hyperons from heavy-ion collisions at the same accelerator? We should expect zero polarization, because the presence of the quark–gluon plasma leads to the complete mixing of spin states.

Almost half a century ago, it was pointed out that polarization effects can be used as signatures of quark–gluon plasma formation.

A review of the use of the spin effects as probes for the investigation of quark–gluon plasma formation was given several years ago (Nurushev 1993). Now, when polarized proton beams are available at RHIC, we can make some proposals oriented to new possibilities of the RHIC.

The quark–gluon plasma is a hot substance appearing in heavy-ion collisions (RHIC 1989). As emphasized in STAR (1992), evidence of quark–gluon plasma formation should be based on the combination of several signatures. Several measurements of polarizations as possible signatures of the appearance of the quark–gluon plasma in heavy-ion collisions were proposed (Hoyer 1987; Jacob and Rafelski 1987; Jacob 1988).

Heavy-ion collisions can be classified into the following groups according to the spins of the involved particles: unpolarized initial states, one of the initial nuclei is polarized, and both initial nuclei are polarized.

1. Angular distribution (Hoyer 1987).

The main feature of quark–gluon plasma formation in heavy-ion collisions is the thermalization of the kinetic energy or its part. Thus, in the plasma rest frame (local), there is no separate direction; in particular, the direction of the spin of the initial nuclei "is forgotten."

These reasons are based exclusively on symmetry and, thus, are model independent. The application of these ideas to data on heavy-ion collisions includes uncertainties owing to the determination of the plasma rest frame and possibilities of generating particles by nonthermalized quarks and gluons. In this case, the existence of the perfect plasma at rest is implied.

Hoyer (1987) pointed to three cases in which the polarization of the produced particles can be measured.

(a) Virtual photons, i.e., the generation of lepton pairs. The $q\bar{q} \rightarrow \gamma^* \rightarrow l\bar{l}$ process is of most interest in experiments with heavy ions, because it is direct evidence of the motion of quarks. The use of the angular distribution of the decay of a virtual photon was proposed in Hoyer (1987) as an additional test that the increased rate of the generation of lepton pairs is caused by plasma formation rather than by, i.e., generation in the pre-equilibrium state. A virtual photon is always transversely polarized with respect to the momenta of annihilating quarks. In the actual plasma, quarks move in random directions; hence, the photon should be unpolarized. Therefore, the angular distribution of leptons in the rest frame of the pair is isotropic. This is strictly valid only for lepton pairs whose rest frame coincides with the plasma rest frame. Otherwise, the distribution of leptons with respect to the momentum of the pair is expected to have the form $(1 + \alpha \cos 2\theta)$, where the coefficient α depends on the plasma temperature and on the mass and momentum of the pair, because annihilating quarks tend to be oriented along the direction of the motion of the virtual photon.

(b) Hyperons. Hyperons (Λ, Σ, Ξ) can be produced with polarization that is perpendicular to the scattering plane and is measurable due to their decay with parity violation. This polarization was observed experimentally in collisions of hadrons with protons and nuclear targets (Pondrom 1985).

Significant production of hyperons is expected in the quark–gluon plasma due to the abundance of baryons and the presence of strange quarks. Similar to the case of virtual photons, the only appropriate direction is the hyperon momentum direction with respect to the plasma. However, parity conservation in the production process forbids polarization along the hyperon momentum. Therefore, hyperons produced from the plasma should be unpolarized.

(c) Resonances. The polarization of hadron resonances such as $\rho, \omega, K^*, \phi, \dots$, $\Delta, Y^*, \dots$ can be reconstructed from the angular distributions of their decays. When they are produced at rest in the (local) plasma reference frame, they should be completely unpolarized. Otherwise, the angular distributions in decay can depend on the polar angle with respect to the resonance motion direction. The dependence on the azimuth angle should always disappear.

The reliability of the described measurements of polarization can be improved by comparing the decay distribution for events in which the plasma is hardly formed (e.g., peripheral collisions) with the decay distributions for events that can produce the plasma (central collisions) in the same experiment.

2. The abundance of Ξ ($\bar{\Xi}$) hyperons (Jacob and Rafelski 1987; Jacob 1988).

It is known that the relative abundance of antibaryons with multiple strangeness would ensure key information on quark–gluon plasma formation. Detailed calculations show that the abundance of anticascade $\bar{\Xi}$ ($\bar{s}\bar{s}\bar{q}$) is enriched to half the abundance of various antihyperons $\bar{Y}$ ($\bar{s}\bar{q}\bar{q}$). This prediction can be compared to the ratio $\bar{\Xi}/\bar{Y}$ observed in the standard hadron reactions, which is as small as 0.06 ± 0.02 in the central rapidity region at $\sqrt{s} = 63$ GeV (Akesson et al. 1984). Thus, the state of the quark–gluon plasma would provide the ratio larger by a factor of several tens. Jacob and Rafelski (1987) showed how the measurement of the longitudinal polarization of $\bar{\Lambda}$ should allow one to easily determine the $\bar{\Xi}/\bar{Y}$ abundance ratio.

The central item of this work is a deep difference in the polarization of $\bar{\Lambda}$ hyperons produced in weak decays of $\bar{\Xi}$ hyperons. Weak decay longitudinally polarizes the spin of $\bar{\Lambda}$. In the subsequent weak decay of $\bar{\Lambda}$, this polarization is analyzed giving observable effects.

The longitudinal polarization of $\bar{\Lambda}$ is the unique and indestructible signature of $\bar{\Xi}$ abundance.

We have showed that the measurement of the longitudinal polarization of $\bar{\Lambda}$, which is usually absent in hadron reactions, will make it possible to determine the

$\bar{\Xi}/\bar{Y}$ abundance ratio, which is expected to be a characteristic signature of the formation of a new state of hadronic matter, quark–gluon plasma.

A large asymmetry value would indicate a relatively large abundance of $\bar{\Xi}$ as compared to $\bar{\Lambda}$. This would be unambiguous test on quark–gluon plasma formation.

As known, in the reactions

$$p + p \to \Lambda + X, \tag{3.210}$$

$$p + A \to \Lambda + X, \tag{3.211}$$

$$A + A \to \Lambda + X, \tag{3.212}$$

where p is the proton, Λ is the Λ hyperon, A is a nucleus, and X are other particles, the final Λ hyperons are polarized. The polarization of the Λ hyperon can be determined from the angular distribution of its weak decay

$$\Lambda \to p + \pi^-. \tag{3.213}$$

The angular distribution of this decay can be written in the form

$$W_\Lambda = 1/2(1 + \alpha_\Lambda P_\Lambda \cos\theta), \tag{3.214}$$

where $\alpha_\Lambda = 0.64$ is the decay parameter, P_Λ is the polarization of the Λ hyperon, and θ is the proton emission angle with respect to the proton motion direction in the rest frame of the Λ hyperon.

The QGP state does not "remember" any features of the parent particles; for this reason, any spin transfer processes do not remain. Utility of such reactions as the signatures of plasma formation depends on the spin transfer degree observed in standard hadron collisions without plasma formation. Scarce data on the spin transfer mechanism predict small effects in the reactions with the (nucleon–strange particle) transition, but possibly huge effect in the s-quark transfer reactions.

Comparative investigations of specific reactions with unpolarized and polarized heavy ions can be useful for distinguishing the QGP formation signal.

Further advances in spin physics will undoubtedly lead to a new approach to the QGP problem.

References

Abragam, A., et al.: Phys. Lett. **2**, 310 (1962)
Adams, D.L., et al.: Preprint FERMILAB-Pub-91/13-E, 14-E and ANL-HEP-PR-91-46 (1991a)
Adams, D.L., et al.: Phys. Lett. B **261**, 201 (1991b)
Adams, D.L., et al., E581/704 Collaboration: Preprint ANL-HEP-TR-91-16 (1991c)
Adams, D.L., et al.: Phys. Lett. B **264**, 462 (1991d)
Adams, D.L., et al.: Z. Phys. C **56**, 181 (1992)
Adams, D.L., et al.: Phys. Rev. D **53**, 4747 (1996)
Adams, J., et al.: (2003). arXiv:hep-ex/0310058v1
Adler, S.S., et al.: Phys. Rev. Lett. **91**, 241803 (2003)
Airapetian, A., et al., HERMES Collaboration: Phys. Rev. Lett. **84**, 4047 (2000)
Airapetian, A., et al.: Phys. Rev. D **64**, 097101 (2001)

Akesson, T., et al.: Nucl. Phys. B **246**, 1 (1984)

Anderson, B., Gustafson, G., Ingelman, G.: Phys. Lett. B **85**, 417 (1979)

Anderson, B., et al.: Nucl. Phys. B **178**, 242 (1981)

Anderson, B., et al.: Phys. Rep. **97**, 349 (1983)

Anselm, A.A., Ryskin, M.G.: Z. Phys. C **68**, 297 (1995)

Anselm, A.A., Ryskin, M.G.: Yad. Fiz. **59**, 708 (1996)

Anselmino, M.: 16 January 2002. arXiv:hep-ph/0201150v1

Anselmino, M., Murgia, F.: Phys. Lett. B **442**, 470 (1998)

Anselmino, M., Boglione, M., Murgia, F.: Phys. Lett. B **362**, 164 (1995)

Anselmino, M., Boglione, M., Murgia, F.: Phys. Rev. D **60**, 054027 (1999)

Anselmino, M., Boglione, M., Murgia, F.: (2000). arXiv:hep-ph/0005081

Anselmino, M., et al.: (2001). arXiv:hep-ph/0111044

Apanasevich, L., et al., E706 Collaboration: Phys. Rev. Lett. **81**, 2642 (1998)

Avakian, H., HERMES collaboration: Nucl. Phys. B **79**, 523 (1999)

Azhgirei, L.S., et al.: Zh. Eksp. Teor. Fiz. **44**, 177 (1963)

Barger, E., et al.: Nucl. Phys. B **13**, 269 (1974)

Barth, M., et al.: Z. Phys. C **10**, 205 (1981)

Bilenky, S.M., Ryndin, R.M.: Zh. Eksp. Teor. Fiz. **36**, 1609 (1959)

Bilenky, S.M., Nguyen-Van-Hieu, Ryndin, R.M.: Zh. Eksp. Teor. Fiz. **46**, 1098 (1964)

Boer, D.: Phys. Rev. D **60**, 014012 (1999)

Boer, D., Mulders, P.: Phys. Rev. D **57**, 5780 (1998)

Boglione, M., Leader, E.: Phys. Rev. D **61**, 11400 (2000)

Boglione, M., Mulders, P.: Phys. Rev. D **60**, 054007 (1999)

Boros, C.: In: Proceedings of the VI Workshop on High Energy Spin Physics, p. 122. IHEP, Protvino (1995)

Boros, C., Zuo-Tang, L.: Phys. Rev. D **53**, R2279 (1996)

Boros, C., Zuo-Tang, L.: Int. J. Mod. Phys. A **15**, 927 (2000)

Boros, C., Zuo-Tang, L., Ta-chung, M.: Phys. Rev. Lett. **67**, 1751 (1993)

Boros, C., Zuo-Tang, L., Tao-Chung, M.: Preprint No. FUB-HEP 96-9, Free University Berlin (1996)

Bourrely, C., Leader, E., Soffer, J.: Phys. Rep. **59**, 95 (1980)

Bravar, A., SMC collaboration: Nucl. Phys. B, Proc. Suppl. **79**, 520 (1999)

Bravar, A., et al.: Phys. Rev. Lett. **77**, 2626 (1996)

Breakstone, A., et al.: Z. Phys. C **69**, 17 (1995)

Bruneton, C., et al.: Phys. Lett. B **57**, 389 (1975)

Bruneton, C., et al.: Yad. Fiz. **23**, 769 (1976a)

Bruneton, C., et al.: Phys. Lett. B **61**, 103 (1976b)

Bruneton, C., et al.: Czechoslov. J. Phys., Sect. B **26**, 25 (1976c)

Bunce, G., et al.: Phys. Rev. Lett. **36**, 1113 (1976)

Bunce, G., et al.: Phys. Lett. B **86**, 386 (1979)

Carey, D.C., et al.: Phys. Rev. D **14**, 1196 (1976)

Chamberlain, O., et al.: Phys. Rev. **93**, 1430 (1954)

Chamberlain, O., et al.: Bull. Am. Phys. Soc. **8**, 38 (1963)

Chou, T.T.: In: AIP Conf. Proc. N15, New York, p. 118 (1973)

Chou, T.T., Yang, C.N.: Nucl. Phys. B **107**, 1 (1976)

Close, F.E.: An Introduction to Quarks and Partons. Academic, London (1979) (Mir, Moscow, 1982)

Collins, J.C.: Nucl. Phys. B **396**, 161 (1993)

Cooper, P.S., et al.: In: Proceedings of the Fifth International Symposium on High Energy Spin Physics. BNL, Brookhaven (1982)

de Alfaro, V., Regge, T.: Potential Scattering. North-Holland, Amsterdam (1965) (Mir, Moscow, 1966)

DeGrand, T.A., Miettinen, H.I.: Phys. Rev. D **23**, 1227 (1981)

Donaldson, G., et al.: Phys. Rev. Lett. **36**, 1110 (1976)

Edneral, V.F., Troshin, S.M., Tyurin, N.E.: Pis'ma Zh. Eksp. Teor. Fiz. **30**, 356 (1979)

Efremov, A.V., Korotkiyan, V.M., Teryaev, O.V.: Phys. Lett. B **348**, 577 (1995)

Eidelman, S., et al.: Phys. Lett. B **592**, 37 (2004)

Erhan, S., et al.: Phys. Lett. B **82**, 301 (1979)

Faccini-Turluer, M.L., et al.: Z. Phys. C **1**, 19 (1979)

Felix, J.: Mod. Phys. Lett. A **14**, 827 (1999)

Fermi, E.: Nuovo Cimento **10**, 407 (1954)

Feynman, R.P.: Photon-Hadron Interactions. Benjamin, New York (1972)

Fidecaro, G., et al.: Nucl. Phys. B **173**, 513 (1980)

Fidecaro, G., et al.: Phys. Lett. B **105**, 309 (1981)

Goldberger, M., Treiman, S.B.: Phys. Rev. **111**, 354 (1958)

Gustafson, G.: In: Proceedings of the Second International Seminar on High Energy Spin Physics, p. 212. IHEP, Protvino (1984)

Heller, K.: In: de Jager, C.W., Ketel, T.J., Mulders, P. (eds.) Proceedings of Spin 96. World Sci., Singapore (1997)

Heller, K., et al.: Phys. Lett. B **68**, 480 (1977)

Heller, K., et al.: Phys. Rev. Lett. **41**, 607 (1978)

Henneman, A.A., Boer, D., Mulders, P.J.: (2001). arXiv:hep-ph/0104271

Hoyer, P.: Phys. Lett. B **187**, 162 (1987)

Jacob, M.: Z. Phys. C **38**, 273 (1988)

Jacob, R., Mulders, P.J.: (1996). arXiv:hep-ph/9610295

Jacob, M., Rafelski, J.: Phys. Lett. B **190**, 173 (1987)

Kanazawa, Y., Koike, Y.: Phys. Lett. B **478**, 121 (2000)

Kline, R.V., et al.: Phys. Rev. D **22**, 553 (1980)

Koeler, S.: Nuovo Cimento **2**, 911 (1955)

Korotkov, V.A., Nowak, W.D., Oganesian, K.A.: Eur. Phys. J. C **18**, 639 (2001)

Leader, E.: Spin in Particle Physics. Cambridge Univ. Press, Cambridge (2001)

Levintov, I.I.: Dokl. Akad. Nauk SSSR **197**, 240 (1956)

Liang, Z.-T., Meng, T.-C.: Phys. Rev. D **42**, 2380 (1990)

Liang, Z.-T., Meng, T.-C.: FU-Berlin Preprint No, FUB-HEP/91-8, Free University Berlin (1991)

Logunov, A.A., Meshcheryakov, V.A., Tavkhelidze, A.N.: Dokl. Akad. Nauk SSSR **142**, 317 (1962)

Logunov, A.A., et al.: Phys. Lett. **7**, 69 (1963)

Logunov, A.A., et al.: Zh. Eksp. Teor. Fiz. **46**, 1079 (1964a)

Logunov, A.A., Nguyen-Van-Hieu, Todorov, I.T.: Phys. Lett. **12**, 139 (1964b)

Lomanno, F., et al.: Phys. Rev. Lett. **43**, 1905 (1979)

Lundberg, B., et al.: In: Proceedings of the Fifth International Symposium on High Energy Spin Physics. BNL, Brookhaven (1982)

Mcnully, D.E.: SLAC-R-674. Stanford, California (2002)

Meiman, N.N.: In: Problems of Elementary Particle Physics. Akad. Nauk Arm. SSR, Yerevan (1962)

Meng, T.-C.: In: Proceedings of the IV Workshop on High Energy Spin Physics, p. 112. IHEP, Protvino (1991a)

Meng, T.-C.: In: Proceedings of the IV Workshop on High Energy Spin Physics, p. 121. IHEP, Protvino (1991b)

Michel, L., Wightman, A.: Phys. Rev. **98**, 1190 (1955)

Mulders, P.: (2001). arXiv:hep-ph/0112225

Mulders, P.J., Tangerman, R.D.: Nucl. Phys. B **461**, 197 (1996)

Mulders, P.J., Tangerman, R.D.: Nucl. Phys. B **484**, 538 (1997)

Musulmanbekov, G.J., Tokarev, M.V.: In: Proceedings of the VI Workshop on High Energy Spin Physics, p. 132. IHEP, Protvino (1995)

Nambu, Y.: In: Proceedings of International Conference on High Energy Physics. Geneva, p. 153 (1962)

Nambu, Y., Jona-Lasinio, G.: Phys. Rev. **122**, 345 (1961a)

Nambu, Y., Lurie, D.: Phys. Rev. **125**, 1469 (1961b)

Nevanlinna, R.: Eindeutige Analytische Funktionen, 2nd edn. Springer, Berlin (1953)

Nguyen-Van-Hieu: Phys. Lett. **9**, 81 (1964)

Nurushev, S.B.: Spin interaction of protons with complex nuclei at energies 565–660 MeV. Candidate's Dissertation in Mathematical Physics, Dubna (1962)

Nurushev, S.B.: In: Workshop on Spin Phenomena in High Energy Physics. Protvino, p. 5 (1993)

Nurushev, S.B., Ryskin, M.G.: Phys. Part. Nucl. **69**, 1 (2006)

Oxley, C.L.C., et al.: Phys. Rev. **91**, 419 (1953)

Panagiotou, A.D.: Int. J. Mod. Phys. A **5**, 1197 (1990)

Pomeranchuk, I.Ya.: Zh. Eksp. Teor. Fiz. **34**, 725 (1958)

Pondrom, L.G.: Phys. Rep. **122**, 57 (1985)

Qiu, J., Sterman, G.: Phys. Rev. D **59**, 014004 (1999)

Rarita, W., et al.: Phys. Rev. **165**, 1615 (1968)

Rayachauduri, K., et al.: Phys. Lett. B **90**, 319 (1980)

RHIC Collaboration: Conceptual Design of the Relativistic Heavy Ion Collider RHIC, BNL 52195, UPTON, Long Island, New York 11973 (1989)

Ryskin, M.G.: Yad. Fiz. **48**, 1114 (1988) [Sov. J. Nucl. Phys. **48**, 708 (1988)]

Schiff, L.I.: Quantum Mechanics, 3rd edn. McGraw-Hill, New York (1968) (Russ. transl. of 2nd ed., Inostrannaya Literatura, Moscow, 1957)

Sharp, D.H., Wagner, W.G.: Phys. Rev. **131**, 2226 (1963)

Sivers, D.: Phys. Rev. D **41**, 83 (1990)

Sivers, D.: Phys. Rev. D **43**, 261 (1991)

Soffer, J.: Phys. Rev. Lett. **74**, 1292 (1995)

STAR Collaboration: Conceptual Design Report, LBL, Pub-5347 (1992)

Stoletov, G.D., Nurushev, S.B.: Report of the Institute for Nuclear Problems. Academy of Sciences of the USSR (1954)

Suzuki, K.: (2000). arXiv:hep-ph/0002218

Thomas, L.H.: Nature **117**, 514 (1926)

Thomas, L.H.: Philos. Mag. **3**, 1 (1927)

Troshin, S.M., Tyurin, N.E.: Pis'ma Zh. Eksp. Teor. Fiz. **23**, 716 (1976)

Troshin, S.M., Tyurin, N.E.: Fiz. Elem. Chastits At. Yadra **15**, 53 (1984)

Troshin, S.M., Tyurin, N.E.: Fiz. Elem. Chastits At. Yadra **19**, 997 (1988)

Vasiliev, A.N., et al.: Preprint No. 2003-22, IHEP, Protvino (2003)

Wagner, W.G.: Phys. Rev. Lett. **10**, 202 (1963)

Wang, X.-N.: Phys. Rev. C **61**, 064910 (2000)

Wong, C.-Y., Wang, H.: Phys. Rev. C **58**, 376 (1998)

Yang, C.N.: In: Proceedings of the International Symposium on High Energy Physics, Tokyo, Japan, p. 629 (1973)

Zhang, Y., et al.: (2001). arXiv:hep-ph/0109233

Chapter 4
Deep Inelastic Scattering of Leptons

Deep inelastic scattering (DIS) of leptons by nucleons became of particular interest after the discovery of the internal structure of the nucleon, which was represented by point particles called partons, in the late 1960s.

In the International Symposium on High Energy Physics in Vienna in 1968, Prof. Panofsky (USA) reported the results of the measurements of the cross sections for the inelastic scattering of electrons by nucleons (Panofsky 1968). He pointed out that the probability of electron scattering at large angles is noticeable. Such events were interpreted as scattering by point charges locating inside a nucleon. This experiment in results was similar to the famous Rutherford experiment discovering the existence of nuclei in atoms. The angular and energy distributions of scattered electrons indicate that electrons are scattered by structureless objects with spin $1/2$. These results along with data on neutrino scattering led to the observation of quarks. The real sizes and masses of quarks are open questions. It is not excluded that quarks, as well as nucleons, are composite objects. Moreover, there are many quarks and they are classified in new conserving numbers such as flavor, color, fractional charge, and magnetic moment.

Subsequent experimental and theoretical investigations led to the development of the nonrelativistic quark model. In this model, baryons (nucleons and hyperons) are considered as consisting of three constituent quarks u, d and s. The masses of these quarks are approximately one third of the nucleon mass. When baryons are in the ground state, quarks have zero orbital angular momentum (s state), spin $1/2$, and the corresponding magnetic moments. The magnetic moment of the baryon is the sum of the magnetic moments of constituent quarks. The nonrelativistic Schrödinger equation can provide the description of almost all statistical parameters of baryons at low energies. In this scheme, the gluon degree of freedom does not play almost any role.

According to this model, the baryon spin, more precisely, its projection on a certain direction is the sum of the projections of the spins of constituent quarks on the same direction. For example, let a proton with spin $\vec{s}$ moves along the z axis and its spin projection on this axis be s_z. Then, the following equality should be

S.B. Nurushev et al., *Introduction to Polarization Physics*,
Lecture Notes in Physics 859, DOI 10.1007/978-3-642-32163-4_4,
© Moskovski Inzhenerno-Fisitscheski Institute, Moscow, Russia 2013

satisfied:

$$\langle s_z \rangle = \left\langle \sum_{i=1}^{3} s_{qz}^i \right\rangle = \langle s_{qz} \rangle. \tag{4.1}$$

Almost all parameters of the hadrons, as well as resonance states, are satisfactorily described in such a nonrelativistic model of quarks.

If the baryon structure is analyzed at higher energies and higher momentum transfers, the nucleon should be considered as consisting of point objects, partons (quarks and gluons). Quarks in this parton model are called current quarks, have the same quantum numbers as constituent quarks, and have very small masses in contrast to the latter quarks. The relation between constituent and current quarks remains undetermined. This problem becomes of particular importance in view of the observation of the strong violation of the above relation (4.1) between the spin components of the baryon and its partons. This was found by the EMC collaboration in 1988 (Ashman et al. 1988, 1989). The subsequent series of the experimental measurements of this effect confirmed it with a higher accuracy and provided the following important conclusions.

1. Valence quarks and antiquarks carry only a quarter of the proton spin rather than the entire this spin, as expected;
2. The strange quark (sea) is surprisingly also polarized, though slightly;
3. Gluon spins and orbital angular momenta of quarks and gluons possibly also contribute to the nucleon spin (this has not yet been confirmed experimentally). In this case, relation (4.1) has the form

$$\left\langle \frac{1}{2} \right\rangle = \langle \vec{s}_q \rangle + \langle \vec{s}_G \rangle + \langle \vec{L}_q \rangle + \langle \vec{L}_G \rangle. \tag{4.2}$$

Although numerous theoretical works have been performed, the problem of the nucleon spin has not yet been solved. This problem is called "spin crisis."

In this chapter, we discuss the deep inelastic scattering (DIS) of leptons by nucleons in theoretical models. The kinematics of the process is considered in Sect. 4.1, Sect. 4.2 is devoted to the cross sections for the DIS of leptons by nucleons, the structure functions of nucleons are analyzed in Sect. 4.3, the structure functions and quark–parton model are discussed in Sect. 4.4, the structure function and QCD are considered in Sect. 4.5, Sect. 4.6 concerns the determination of parton distributions, and transversity is analyzed in Sect. 4.7.

4.1 Kinematics of the DIS Process

The diagram of the deep inelastic scattering (DIS) of leptons by nucleons for the process

$$l(k) + N(p) \rightarrow l'(k') + X \tag{4.3}$$

is shown in the general form in Fig. 4.1.

Fig. 4.1 Diagram of the DIS
of a lepton in the parton
model

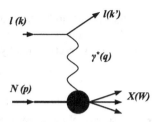

Here, k and k' are the initial and final four-momenta of the lepton l, respectively;
p is the momentum of the initial nucleon N; $q = k - k'$ is the four-momentum
transferred from the lepton to the nucleon; and W is the square of the masses of the
system of recoil particles X. The momentum q in this diagram is transferred by the
virtual photon $\gamma^*(q)$. However, it can be transferred by the Z boson in the case of
neutral current or by $W^{\pm}$ bosons in the case of interaction with charged current.

The subsequent description of deep inelastic scattering requires a number of kinematic parameters. They are denoted and defined below.

- Energy transfer in lepton scattering,

$$\nu = \frac{q \cdot p}{M} = E - E', \tag{4.4}$$

is the energy lost by the lepton in its collision with the nucleon at rest. Here, E
and E' are the initial and final energies of the lepton, respectively, and M is the
nucleon mass.
- The square of the four-momentum of the virtual photon

$$Q^2 = -q^2 = -\left(k - k'\right)^2 = \left(E - E'\right)^2 - \left(\vec{k} - \vec{k}'\right)^2 = 2\left(EE' - \vec{k} \cdot \vec{k}'\right) - m_l^2 - m_{l'}^2.$$

Here, m_l and $m_{l'}$ are the masses of the initial and final leptons, respectively. Neglecting the masses of the leptons compared to their energies, we can represent
the formula in the form

$$Q^2 \approx 4EE' \sin^2(\theta/2), \tag{4.5}$$

where θ is the lepton scattering angle in the rest frame of the initial nucleon.
This angle is measured from the directions of the incident and scattered lepton
momenta.

The parameter Q^2 is a criterion for determining DIS processes: the parameter
Q^2 for DIS should be much larger than the nucleon mass M^2. At the same time,
this parameter specifies the photon virtuality degree: $Q^2 = m_\gamma^2$, where m_γ is the
mass of the virtual photon. The higher the photon virtuality, the farther the photon
is from the mass (physical) shell.
- The Bjorken parameter is defined as

$$x = \frac{Q^2}{2M\nu}. \tag{4.6}$$

In the parton model, this parameter specifies the nucleon-momentum fraction carried by the quark interacting with the virtual photon.

- The energy fraction lost by the lepton colliding with the nucleon is

$$y = \frac{q \cdot p}{k \cdot p} = \frac{v}{E}. \tag{4.7}$$

- The square of the masses of the system of recoil particles X

$$W^2 = (p+q)^2 = M^2 + 2Mv - Q^2. \tag{4.8}$$

- The square of the total energy of the initial lepton–nucleon system is

$$s = (k+p)^2 = \frac{Q^2}{xy} + m_l^2 + M^2. \tag{4.9}$$

The DIS process is a part of an inclusive process with the imposed conditions

$$Q^2 \gg M^2, \qquad W^2 \gg M^2. \tag{4.10}$$

To describe DIS as an inclusive process, it is sufficient to use three parameters at option. The set s, Q^2, and x is used most often although other sets can be used (s is usually fixed in experiment and is rarely changed).

4.2 Cross Sections for the DIS of Leptons by Nucleons

At present, differential cross sections for DIS processes with unpolarized nuclei are sufficiently well studied. The corresponding experiments can be classified into two classes. The first class unites the experiments with fixed targets. They were the first experiments on DIS and led to the discovery of partons. The measurements in these experiments have been carried out in the kinematical regions $6 \cdot 10^{-3} < x < 1$ and $3 \cdot 10^{-1} < Q^2$ (GeV^2) $< 3 \cdot 10^2$. The second class unites all measurements at colliders. In this case, the measurement range is much wider than that in the preceding case. In particular, the measurement ranges for the HERA setup (the $e + p$ collider with an electron energy of 30 GeV and a proton energy of 820 GeV) are $0.7 \cdot 10^{-6} < x < 1$ and $7 \cdot 10^{-2} < Q^2$ (GeV^2) $< 2 \cdot 10^4$. The measurements in the collider energy range are of most theoretical interest for studying the spin structure functions.

Let us write the differential cross section for the DIS process in the invariant form

$$\frac{d^2\sigma}{dxdy} = x(s - M^2)\frac{d^2\sigma}{dxdQ^2} = \frac{2\pi Mv}{E'}\frac{d^2\sigma}{d\Omega_N}. \tag{4.11}$$

In the lowest order of perturbation theory, this cross section is expressed in terms of the product of the lepton and hadron tensors corresponding to the upper and lower vertices of the diagram (see Fig. 4.1):

$$\frac{d^2\sigma}{dxdy} = \frac{2\pi y\alpha^2}{Q^4}\sum_k \eta_k L_k^{\mu\nu} W_{\mu\nu}^k. \tag{4.12}$$

For the interaction of the neutral current, the lepton tensor $L_k^{\mu\nu}$ consists of three terms (with γ and Z exchanges and interference term γZ) over which the summation is performed. This tensor is composed of the available parameters of DIS. In the case of photon exchange, these parameters are the momenta k and k' of the leptons before and after scattering, respectively; the lepton charge $e = \pm 1; 0$; and the helicity of the initial electron $k\lambda = \pm 1$. In the case of interaction through the Z boson (weak current), there are additional two weak-interaction constants: the vector coupling constant g_V^e and pseudovector coupling constant g_A^e:

$$g_V^e = -\frac{1}{2} - 2e\sin^2\theta_W \quad \text{and} \quad g_A^e = -\frac{1}{2}.$$

The interaction of the charged current occurs through $W^\pm$ boson exchange; in this case, the lepton current is expressed in terms of the same parameters as earlier. Finally, the lepton tensor for neutral currents are given by the expressions

$$L_{\mu\nu}^\gamma = 2\left(k_\mu k_\nu' + k_\mu' k_\nu - k \cdot k' g_{\mu\nu} - i\lambda\varepsilon_{\mu\nu\alpha\beta}k^\alpha k'^\beta\right), \tag{4.13}$$

$$L_{\mu\nu}^{\gamma Z} = \left(g_V^e + e\lambda g_A^e\right)L_{\mu\nu}^\gamma, \tag{4.14}$$

$$L_{\mu\nu}^Z = \left(g_V^e + e\lambda g_A^e\right)^2 L_{\mu\nu}^\gamma. \tag{4.15}$$

The charged current tensor (the $eN \to \nu_e X$ reaction) has the form

$$L_{\mu\nu}^W = (1 + e\lambda)^2 L_{\mu\nu}^\gamma. \tag{4.16}$$

This expression is very remarkable. The expression in the parentheses follows from the hypothesis of the left-handed orientation of the lepton spin ($\lambda = -1$). Since the charge of a lepton (electron or muon) is negative, the coefficient in the parentheses is nonzero. For an antilepton (positron or positive muon) or an antiparticle, both the helicity and charge sign change, so that this coefficient does not change. Experiments do not refute the hypothesis of the absence of leptons with right-handed helicity in nature.

Owing to the point structure of the lepton, the lepton tensor $L_{\mu\nu}$ has the simple explicit analytical form as follows from formulas (4.13)–(4.15). However, since the nucleon structure is not point, the hadron tensor $W_{\mu\nu}$ is a complex function, which can be written in the form

$$W_{\mu\nu} = \frac{1}{4\pi}\int d^4z e^{iqz}\langle p, s|[J_\mu^+(z), J_\nu(0)]|p, s\rangle. \tag{4.17}$$

Here, s is the four-dimensional spin of the proton, which satisfies the relations $(ps) = 0$ and $s^2 = -1$.

The coefficients η_k in Eq. (4.12) are the ratios of the propagators and constants of the corresponding intermediate bosons to the propagator and coupling constant of the photon, respectively. They are given by the expressions

$$\eta_\gamma = 1, \qquad \eta_{\gamma Z} = \left(\frac{G_F M_Z^2}{2\sqrt{2}\pi\alpha}\right)\left(\frac{Q^2}{Q^2 + M_Z^2}\right),$$

$$\eta_Z = \eta_{\gamma Z}^2, \qquad \eta_W = \frac{1}{2}\left(\frac{G_F M_W^2}{4\pi\alpha}\frac{Q^2}{Q^2 + M_W^2}\right)^2. \tag{4.18}$$

4.3 Nucleon Structure Functions

In Blümlein and Kochelev (1997), Forte et al. (2001), and Anselmino et al. (1994), the hadron tensor $W_{\mu\nu}$ was represented in terms of the structure functions of the polarized and unpolarized nucleons in the form

$$
\begin{aligned}
W_{\mu\nu} = {}& \left(-g_{\mu\nu} + \frac{q_\mu q_\nu}{q^2}\right)F_1(x, Q^2) + \frac{\hat{p}_\mu \hat{p}_\nu}{p\cdot q}F_2(x, Q^2) \\
& - i\varepsilon_{\mu\nu\alpha\beta}\frac{q^\alpha p^\beta}{2q\cdot p}F_3(x, Q^2) \\
& + i\varepsilon_{\mu\nu\alpha\beta}\frac{q^\alpha}{q\cdot p}\left[s^\beta g_1(x, Q^2) + \left(s^\beta - \frac{s\cdot q}{p\cdot q}p^\beta\right)g_2(x, Q^2)\right] \\
& + \frac{1}{q\cdot p}\left[\frac{1}{2}(\hat{p}_\mu \hat{s}_\nu + \hat{s}_\mu \hat{p}_\nu) - \frac{s\cdot q}{p\cdot q}\hat{p}_\mu \hat{p}_\nu\right]g_3(x, Q^2) \\
& + \frac{s\cdot q}{q\cdot p}\left[\frac{\hat{p}_\mu \cdot \hat{p}_\nu}{p\cdot q}g_4(x, Q^2) + \left(-g_{\mu\nu} + \frac{q_\mu q_\nu}{q^2}\right)g_5(x, Q^2)\right]. \tag{4.19}
\end{aligned}
$$

Here,

$$\hat{p}_\mu = p_\mu - \frac{p\cdot q}{q^2}q_\mu, \qquad \hat{s}_\mu = s_\mu - \frac{s\cdot q}{q^2}q_\mu. \tag{4.20}$$

As seen, the hadron tensor consists of two parts. The first part, so-called symmetric part (with respect to the subscripts μ and ν) of the tensor is independent of the nucleon spin. The convolution of this part with the lepton tensor $L_{\mu\nu}$ gives the unpolarized differential cross section $\frac{d^2\sigma}{dxdy}$. The measurement of this cross section makes it possible to experimentally reconstruct the spinless structure functions F_1, F_2, and F_3. The second part of the tensor $W_{\mu\nu}$ depends on the nucleon spin and is antisymmetric. Its convolution with the lepton tensor provides the polarized differential cross sections whose difference $\Delta\sigma$ is proportional to asymmetry and this asymmetry is measured in experiments. Such measurements should generally allow the determination of five spin structure functions g_1, g_2, g_3, g_4, and g_5.

Thus, the unpolarized cross section for DIS is given by the formula

$$\frac{d^2\sigma^i}{dxdy} = \frac{4\pi\alpha^2}{xyQ^2}\eta^i\left[\left(1-y-\frac{x^2y^2M^2}{Q^2}\right)F_2^i + x/y^2F_1^i \mp \left(1-\frac{y}{2}\right)xyF_3^i\right]. \quad (4.21)$$

Here, the minus sign of the last term refers to the positron e^+ or antineutrino $\bar{\nu}$, and the plus sign corresponds to the case of the electron e or neutrino ν. The superscript $i = NC$ means the DIS process through the neutral current such as the $eN \to eX$ reaction, and $i = CC$ means the DIS process through the charged current such as the $eN \to \nu X$ reaction or the reverse reaction $\nu N \to eX$. The coefficient η^{NC} is 1 for an unpolarized process, and

$$\eta^{CC} = (1 \pm \lambda)^2 \eta_W. \quad (4.22)$$

Here, the sign $\pm$ refers to the lepton charge and λ is the helicity of the incident lepton. The structure functions of the charged current, which are determined only through the exchange by W bosons, are written in the form $F_i^{CC} = F_i^W, i = 1, 2, 3$. The structure functions for the neutral current are defined in Klein and Reimann (1984) by the expression

$$F_2^{NC} = F_2^\gamma - \left(g_V^e \pm \lambda g_A^e\right)\eta_{\gamma Z}F_2^{\gamma Z} + \left(g_V^{e\,2} + g_A^{e\,2} \pm 2\lambda g_V^e g_A^e\right)\eta_Z F_2^Z. \quad (4.23)$$

A similar expression is valid for F_1^{NC}. The expression for F_3^{NC} has the form

$$xF_3^{NC} = -\left(g_A^e \pm \lambda g_V^e\right)\eta_{\gamma Z}xF_3^{\gamma Z} + \left[2g_V^e g_A^e \pm \lambda\left(g_V^{e\,2} + g_A^{e\,2}\right)\right]\eta_Z xF_3^Z. \quad (4.24)$$

Experiments with neutrino beams are very difficult because of low fluxes and small cross sections. Only measurements of the unpolarized structure functions have been performed. Measurements of the spin structure functions on neutrino beams have not yet been carried out primarily because it is unreal to create a polarized target with the necessary mass. However, in view of the importance of studying the spin effects on neutrino beams, physicists are actively working on this problem. We now discuss the scattering of polarized leptons by polarized nucleons. In this case, the difference between the cross sections for protons polarized in the opposite directions is usually measured:

$$\Delta\sigma = \sigma(\lambda_n = -1, \lambda_l) - \sigma(\lambda_n = 1, \lambda_l), \quad (4.25)$$

where λ_l and $\lambda_n(= \pm 1)$ are the helicities of the initial lepton and nucleon, respectively. Measurements involve the collection of statistics for different directions of the nucleon helicity. Note that it is difficult to change the lepton helicity due to physical reasons although the experiment is symmetric with respect to the helicities of the nucleon and lepton. This difference between differential cross sections can be expressed in terms of five spin-dependent structure functions $g_i(x, Q^2)$ $(i = 1 \dots 5)$ by the formula

$$\frac{d^2\Delta\sigma^i}{dxdy} = \frac{8\pi\alpha^2}{xyQ^2}\eta^i\left[-\lambda_l y\left(2-y-\frac{2x^2y^2M^2}{Q^2}\right)xg_1^i + 4\lambda_l x^3 y^2\frac{M^2}{Q^2}g_2^i\right.$$

$$+ 2x^2 y \frac{M^2}{Q^2} \left(1 - y - \frac{x^2 y^2 M^2}{Q^2}\right) g_3^i$$

$$- \left(1 + \frac{2x^2 y M^2}{Q^2}\right) \left[\left(1 - y - \frac{x^2 y^2 M^2}{Q^2}\right) g_4^i + xy^2 g_5^i\right]. \quad (4.26)$$

The superscript $i = NC$ and CC has the same meaning as above. Above expression (4.26) determines the difference between the cross sections for parallel and antiparallel spins of the initial particles in the scattering of the positron and antineutrino. For the scattering of the electron and neutrino, the difference should be taken between the antiparallel and parallel spins. As seen in expression (4.26), the contributions from the structure functions g_2 and g_3 in the case of the longitudinal polarization of the nucleon are suppressed by the coefficient M^2/Q^2. In the case of the transverse polarization of the nucleon, this suppression factor is absent Blümlein and Kochelev (1997), but the difference between the differential cross sections decreases as M/Q; for this reason, it is difficult to measure these structure functions.

The limit of the expressions at $M^2/Q^2 \to 0$ is of great interest for theory. In this limit, scaling should occur according to the Bjorken hypothesis; i.e., the structure functions should depend only on one variable, which is the parameter $x \equiv x_B$, and should be independent of Q^2. This hypothesis is well satisfied in experiments in the region $x_B \geq 0.1$ (Eidelman et al. 2004). From this point of view, it is interesting to determine the form of the unpolarized and polarized differential cross sections given by Eqs. (4.21) and (4.26), respectively. Both cross sections include the same structure functions (spin-dependent and spin-independent) determining the hadron tensor $W_{\mu\nu}$ given by Eq. (4.19). These formulas have the form

$$\frac{d^2\sigma^i}{dx dy} = \frac{2\pi\alpha^2}{xy Q^2} \eta^i \left[Y_+ F_2^i \mp Y_- x F_3^i - y^2 F_L^i\right]. \quad (4.27)$$

Here, $i = NC, CC, Y_\pm = 1 \pm (1 - y)^2$ and

$$F_L^i = F_2^i - 2x F_1^i. \quad (4.28)$$

According to Callan and Gross (1969), in the naive quark model, the function is $F_L^i = 0$. Hence, the unpolarized differential cross section includes only two structure functions F_2 and F_3 and is strongly simplified.

A similar formula can be obtained for polarized cross section by making the changes

$$F_1 \to -g_5, \qquad F_2 \to -g_4, \qquad F_3 \to 2g_1,$$

in expression (4.27) and multiplying the resulting expression by a factor of 2 for averaging over the initial polarizations of the nucleon. Thus, we obtain

$$\frac{d^2\Delta\sigma^i}{dx dy} = \frac{4\pi\alpha^2}{xy Q^2} \eta^i \left[-Y_+ g_4^i \mp Y_- x g_1^i + y^2 g_L^i\right]. \quad (4.29)$$

Fig. 4.2 Diagram of the DIS
of the lepton in the parton
model

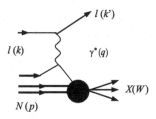

Here, $g_L^i = g_4^i - 2xg_5^i$ and $Y_\pm$ were defined above and it was shown by Dicus
(1972) that $g_L^i = 0$ in the same naive quark model. In this case, the cross section of
interest is simplified and has the final form

$$\frac{d^2\Delta\sigma^i}{dxdy} = \frac{4\pi\alpha^2}{xyQ^2}\eta^i\left[-Y_+g_4^i \mp Y_-xg_1^i\right]. \tag{4.30}$$

4.4 Structure Functions and Quark–Parton Model

The quark–parton model introduces the quark distribution function $q(x, Q^2)$, which
determines the probability of finding the quark with the flavor q ($q = u, d, s, \bar{u}, \bar{d}, \bar{s}$,
etc.) and with the momentum fraction x of the proton momentum in the infinite-
momentum frame. In this model, the DIS of the lepton by the nucleon can be represented by the diagram (see Fig. 4.2)

In this diagram, the point lepton is scattered by one of three quarks of the nucleon.
The quarks are also assumed to be point (so-called current quarks). The scattering of
the lepton by one quark is independent of other quarks. In the quark–parton model,
the following relations between the parton distributions and structure functions were
derived (Bjorken and Paschos 1969; Feynman 1972).

For the processes with neutral currents such as $e + p \rightarrow e + X$:

$$\left[F_2^\gamma, F_2^{\gamma Z}, F_2^Z\right] = x\sum_q\left[e_q^2, 2e_qg_V^q, g_V^{q2} + g_A^{q2}\right](q + \bar{q}), \tag{4.31}$$

$$\left[F_3^\gamma, F_3^{\gamma Z}, F_3^Z\right] = \sum_q\left[0, 2e_qg_A^q, 2g_V^qg_A^q\right](q - \bar{q}), \tag{4.32}$$

$$\left[g_1^\gamma, g_1^{\gamma Z}, g_1^Z\right] = \frac{1}{2}\sum_q\left[e_q^2, 2e_qg_V^q, g_V^{q2} + g_A^{q2}\right](\Delta q + \Delta\bar{q}), \tag{4.33}$$

$$\left[g_5^\gamma, g_5^{\gamma Z}, g_5^Z\right] = \sum_q\left[0, e_qg_A^q, g_V^qg_A^q\right](\Delta q - \Delta\bar{q}). \tag{4.34}$$

Here, $g_V^q = \pm\frac{1}{2} - 2e_q\sin^2\theta_W$ and $g_A^q = \pm\frac{1}{2}$, where the plus and minus signs correspond to the u and d quarks, respectively; and $\Delta q = q\uparrow -q\downarrow$, where $q\uparrow$ and $q\downarrow$
mean the distribution functions of quarks with the spins parallel and antiparallel to
the nucleon spin, respectively.

For processes with charged currents such as $e^- + p \rightarrow \nu + X$ or $\bar{\nu} + p \rightarrow e^+ + X$, the formulas have the form

$$F_2^{W-} = 2x(u + \bar{d} + \bar{s} + c + \cdots), \tag{4.35}$$

$$F_3^{W-} = 2(u - \bar{d} - \bar{s} + c + \cdots), \tag{4.36}$$

$$g_1^{W-} = (\Delta u + \Delta \bar{d} + \Delta \bar{s} + \Delta c + \cdots), \tag{4.37}$$

$$g_5^{W-} = (-\Delta u + \Delta \bar{d} + \Delta \bar{s} - \Delta c + \cdots). \tag{4.38}$$

Only active quarks should be taken into account in these formulas. The Cabibbo–Kobayashi–Maskawa mixing is disregarded. The formulas for the structure functions F^{W+} and g^{W+} (the $e^- + p \rightarrow \bar{\nu} + X$ and $\nu + p \rightarrow e^+ + X$ reactions) are obtained by changing the quark flavors: $d \leftrightarrow u$ and $s \leftrightarrow c$. The structure functions for scattering by neutrons are also obtained from the above formulas by changing $u \leftrightarrow d$. For processes with both neutral and charged currents, the quark-parton model predicts the relations

$$2x F_1^i = F_2^i \quad \text{and} \quad g_4^i = 2x g_5^i. \tag{4.39}$$

When the mass terms are neglected, the structure functions g_2 and g_3 make contributions only to the processes involving transversely polarized nucleons. In the quark–parton model, the usual probability interpretation is inapplicable to these functions. They appear from the off-diagonal elements of the matrix $\langle P, \lambda' || J_\mu^+(z) J_\nu(0) || P, \lambda \rangle$; in this case, the helicities of the protons in the initial and final states of the process are different: $\lambda' \neq \lambda$. These functions contain the contributions from the twist-2 and twist-3 terms of the same order in Q^2. The Wandzura–Wilczek relation (Wandzura and Wilczek 1977) makes it possible to relate the part of the structure function g_2 appearing from twist-2 to the other structure function g_1 as

$$g_2^i(x) = -g_1^i(x) \int_x^1 \frac{dy}{y} g_1^i(y). \tag{4.40}$$

An additional difficulty in analysis of g_2 is that the contribution from twist-3 terms to this amplitude is yet unknown. The situation with the structure function g_3 is similar. Its part determined by the contribution from twist-3 can be expressed in terms of g_4. Detailed analysis of these problems can be found in Blumlein and Tkabladze (1999).

4.5 Structure Function and QCD

In 1969 in the quark–parton model, Bjorken (1960) proposed a strong statement that the structure functions in the limits $Q^2 \rightarrow \infty$ and $\nu \rightarrow \infty$ at a fixed x value tend to the limits $F^i(x, Q^2) \rightarrow F^i(x)$ and $g^i(x, Q^2) \rightarrow g^i(x)$ (so-called Bjorken scaling). This hypothesis is based on the assumption that the transverse momenta of the

Fig. 4.3 Proton structure
function $F_2(x, Q^2)$ versus x
for $Q^2 = 3.5$ and 90 GeV^2

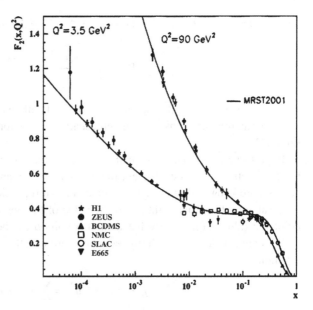

quarks in the infinite-momentum frame for the proton are low and can be neglected.
However, according to QCD, quarks can emit hard gluons and, thus, violate scaling. An experimental test of Bjorken scaling is presented in Fig. 4.3. As seen in this
figure, the violation of scaling is particularly strong at small x values and large Q^2
values.

Owing to this process, scaling is violated and the structure functions and distribution functions of quarks depend on Q^2. As Q^2 increases, the emission of gluons
increases and they produce quark–antiquark pairs. As a result, the initial momentum
distributions of quarks are softened and gluon and sea-quark distribution densities
increase with a decrease in x.

The evolution of the structure functions $F_2(x, Q^2)$ with Q^2 is shown in Fig. 4.3
for $Q^2 = 3.5$ and 90 GeV^2.

Three important conclusions follow from this figure. First, Bjorken scaling occurs only for $x \geq 0.14$ and is violated for lower values. Second, QCD predictions
with allowance for the Q^2 evolution of the structure functions are in rather good
agreement with experimental data (calculations were performed in Martin et al.
2002). Third, it is very interesting to measure the structure functions for very small x
values.

The evolution of the structure functions in QCD, which is shown in the figure, is
described by the parton distribution functions $f(x, \mu^2)$ ($f = q$ or $f = g$), where μ
is the scaling factor of about the four-momentum of the virtual photon. These functions for a fixed x value are obtained by integrating with respect to the transverse
momentum of the parton from zero to μ. The evolution of these functions with μ
is described in QCD by the DGLAP equations (Gribov and Lipatov 1972; Lipatov

1975; Altarelli and Parisi 1977; Dokshitzer 1975) of the form

$$\frac{\partial f}{\partial \ln \mu^2} \approx \frac{\alpha_s(\mu^2)}{2\pi}(P \otimes f) = \frac{\alpha_s(\mu^2)}{2\pi} \int_x^1 \frac{dy}{y} P(y) f\left(\frac{x}{y}\right). \tag{4.41}$$

In the general form, all observables associated with hard hadron interaction such as the structure functions can be represented in the form of convolution, as in Eq. (4.41).

In perturbative QCD, this equation can be solved if the universal parton distribution functions f and coefficient functions P, which are specific for a particular process, are known. However, theory cannot a priori predict a particular value of an observable at the initial point μ_0. If the value of the observable is specified at this point, the value of this observable at another point μ can be theoretically calculated.

For convenience, the evolution equations are usually written separately for two functions

$$q^{NS} = q_i - \bar{q}_i, \qquad q^S = \sum (q_i + \bar{q}_i). \tag{4.42}$$

Here, q^{NS} and q^S are the nonsinglet gluon and singlet quark distributions, respectively. The distribution q^{NS} corresponds to nonzero values of such quantum numbers as flavor, isospin, and baryon number. In this case, the DGLAP equations are written in the form

$$\frac{\partial q^{NS}}{\partial \ln \mu^2} = \frac{\alpha_s(\mu^2)}{2\pi}(P_{qq} \otimes q^{NS}), \tag{4.43}$$

$$\frac{\partial}{\partial \ln \mu^2}\begin{pmatrix} q^S \\ g \end{pmatrix} = \frac{\alpha_s(\mu^2)}{2\pi}\begin{pmatrix} P_{qq} & 2n_f P_{qg} \\ P_{gq} & P_{gg} \end{pmatrix} \otimes \begin{pmatrix} q^S \\ g \end{pmatrix}. \tag{4.44}$$

Here, P is the splitting function and describes the probability of the decay of a given parton into two others and n_f determines the number of active quark flavors. In the leading approximation, these functions were given in Altarelli and Parisi (1977) and have the form

$$P_{qq} = \frac{4}{3}\left[\frac{1+x^2}{(1-x)}\right]_+ = \frac{4}{3}\left[\frac{1+x^2}{(1-x)_+}\right] + 2\delta(1-x), \tag{4.45}$$

$$P_{qg} = \frac{1}{2}[x^2 + (1-x)^2], \tag{4.46}$$

$$P_{gq} = \frac{4}{3}\left[\frac{1+(1-x)^2}{x}\right], \tag{4.47}$$

$$P_{gg} = 6\left[\frac{1-x}{x} + x(1-x) + \frac{x}{(1-x)_+}\right] + \left(\frac{11}{2} - \frac{n_f}{3}\right)\delta(1-x). \tag{4.48}$$

Here, the notation $[F(x)]_+$ means that any sufficiently regular test distribution function $f(x)$ satisfies the relation

$$\int_0^1 dx f(x) [F(x)]_+ = \int_0^1 dx [f(x) - f(1)] F(x). \tag{4.49}$$

The DGLAP approximation is sufficiently well applicable for Q^2 larger than several GeV^2 and it is not yet necessary to use more complex variants of the theory.

Perturbative QCD well predicts the contribution from leading twist-2 to the structure functions. The contribution from higher-order twists should decrease to zero with an increase in Q. For example, the contribution from the n-th order twist should decrease as $1/Q^{n-2}$. The problems of experimental and theoretical investigations of two important cases: (a) small x values and (b) the inclusion of quark masses (e.g., b and c quarks) remain. One of the urgent unsolved problems is the so-called spin crisis. We will return to this problem.

4.6 Determination of Parton Distributions

The processes of the DIS of polarized leptons by nucleons are studied in experiments whose main aim is to reconstruct the parton distributions. To this end, the processes of hard collisions of hadrons can be used. A number of processes that have been already studied experimentally or planned for future experiments are listed in Table 4.1.

All experiments on the determination of the parton distributions are classified into two groups: (a) experiments with fixed targets and (b) experiments at colliders. The kinematic regions of these two groups supplement each other and cover a very wide region of the parameters x and Q^2, as seen in Fig. 4.4.

Table 4.1 Reactions of the DIS of leptons by nucleons and hard hadron processes, as well as the parton distribution functions tested in these processes

No.	Reaction	Subprocess	Tested parameters
1	$l^\pm N \to l^\pm X$	$\gamma * q \to q$	$g(x \leq 0.01), q, \bar{q}$
2	$l^+(l^-)N \to \bar{\nu}(\nu)X$	$W^* q \to q'$	
3	$\nu(\bar{\nu})N \to l^-(l^+)X$	$W^* q \to q'$	
4	$\nu T \to \mu^+ \mu^- X$	$W^* s \to c \to \mu^+$	S
5	$pp \to \gamma$	$qq \to \gamma q$	$g(x \approx 0.4)$
6	$pN \to \mu^+ \mu^- X$	$q\bar{q} \to \gamma^*$	$\bar{q}$
7	$pp, pn \to \mu^+ \mu^- X$	$u\bar{u}, d\bar{d} \to \gamma^*$	$\bar{u} - \bar{d}$
8	$pp, pn \to \mu^+ \mu^- X$	$d\bar{u}, u\bar{d} \to \gamma^*$	
9	$ep, en \to e\pi X$	$\gamma * q \to q$	
10	$p\bar{p} \to W \to l^\pm X$	$u\,\vec{d} \to W$	$gg, gq, qq \to u, d, u/d$
11	$p\bar{p} \to jet + X$	$gg, gq, qq \to 2jet$	$q, g \,(0.01 \leq x \leq 0.5)$

Fig. 4.4 Kinematic (x, Q^2) region studied in two groups of experiments: with fixed targets and at colliders; parton distributions that can be determined in these experiments are also indicated

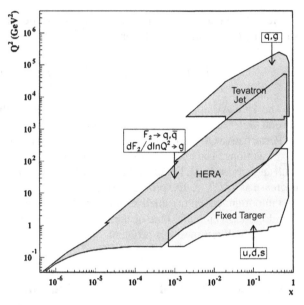

Fig. 4.5 Distributions of unpolarized quarks determined from experimental data with the use of the MRST2001 parameterization scheme (Martin et al. 2002)

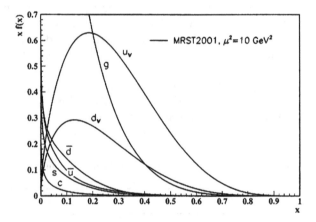

To date, the distributions of unpolarized partons are determined from experiments much better than those for polarized partons. This is due to three circumstances. First, experiments with unpolarized beams and targets are easier. Second, investigation of DIS with unpolarized particles began much earlier. Third, experiments on DIS with unpolarized particles are also carried out at colliders, whereas any experiment on DIS with polarized particles at colliders has not yet been performed. We can hope that first such experiments will be conducted at RHIC.

Some such distributions for light quarks and antiquarks, gluon, and c quark are shown in Fig. 4.5.

As seen in this figure, parton distributions are soft. For example, the distribution of the valence u quark has a maximum for x values near 0.2, and the spectra of other

Fig. 4.6 Distributions of
polarized partons

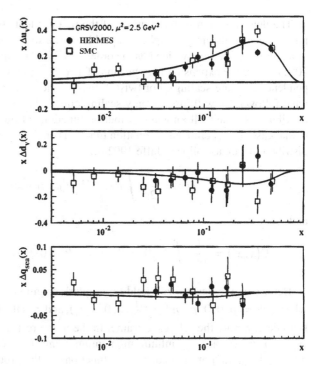

partons are even softer. A fast increase in the gluon density with a decrease in x is particularly remarkable. This effect is one of the factors stimulating investigation of parton distributions for very small x values. Such investigations can be performed only with colliding beams. This was one of the reasons for building of the eRHIC collider in Brookhaven.

Experimental results on the distributions of polarized partons are given in Fig. 4.6.

4.7 Transversity

In the leading PQCD approximation, the nucleon structure in the DIS processes is described by three independent parton distribution functions. Two parton distribution functions were considered above: the distribution function of unpolarized partons, $f_1(x)$, and the helicity distribution $g_1(x)$. These two functions are leading (twist-2 functions). The distribution function of the transverse polarization of the partons, $h_1(x)$, is also a twist-2 function (Jaffe 1992a). This function was first introduced in Ralston and Sopper (1979), but it was forgotten for a long time likely because this function, in contrast to the first two functions, cannot be measured in inclusive DIS processes. The cause of the appearance of this function, its parton interpretation, and possible experimental methods for measuring it are considered below.

The determination of the number of the independent structure functions of the nucleon is a sufficiently complicated problem. The so-called scaling approximation of the twist-2 function, which has the order $n = 2$ in the expansion in Q^{2-n} (Leader 2001). Since the expansion is in powers of $1/Q^{n-2}$, the twist-2 functions are independent of Q (are scaling). For twist-3, twist-4, etc., $n = 3, 4$, etc., respectively. In what follows, we will discuss only scaling effects, i.e., only twist-2 functions and distributions. The well known distribution functions of quarks and gluons are particular cases of generalized correlation functions on the light cone. Two such typical distributions are as follows (Jaffe 1992a):

$$f(x) = \frac{1}{2\pi} \int d\lambda e^{-i\lambda x} \langle P | \phi(0)\phi(\lambda n) | P \rangle \qquad (4.50)$$

and

$$E(x, x') = \frac{1}{4\pi^2} \int d\lambda d\lambda' e^{-i(\lambda x + \lambda' x')} \langle P | \phi(0)\phi(\lambda n)\phi(\lambda' n) | P \rangle. \qquad (4.51)$$

Here, ϕ is the generalized field (e.g., quark or gluon) and $P^\mu = p^\mu + \frac{1}{2}Mn^\mu$, where $p^\mu = (p, 0, 0, p)$, $n^\mu = (\frac{1}{2p}, 0, 0, -\frac{1}{2p})$, and p is the free parameter (momentum) determining the reference frame. In the target rest frame, $p = M/2$, whereas $p \to \infty$ determines the infinite-momentum frame (Close 1979). Both functions $f(x)$ and $E(x, x')$ are the correlation functions of the ground state for the case of two and three interacting partons, respectively, and correlations are calculated along a tangent to the light cone (Leader 2001). The function $f(x)$ allows the probability interpretation after summation over the total system of the intermediate states:

$$f(x) = \sum |\langle P | \phi(0) | X \rangle|^2 \delta(P_X^+ - (1 - x)P^+). \qquad (4.52)$$

The function $f(x)$ can be treated as the probability of finding a quantum of the field ϕ with the momentum $k^+ = xP^+$ in the target. The three-particle correlation function $E(x, x')$ does not allow such a probability interpretation.

To determine the number of the independent twist-2 amplitudes, two operations should be performed. The first operation is the determination of the independent components of quark and gluon fields using the projection operator on the light cone. This decomposition operation for the quark field has the form

$$\psi_\pm = P_\pm \psi, \qquad (4.53)$$

where the projection operator of the light cone is determined by the expression

$$P_\pm = \frac{1}{2}\gamma^\pm \gamma^\mp, \quad \gamma^\pm = \frac{1}{\sqrt{2}}(\gamma^0 \pm \gamma^3). \qquad (4.54)$$

The action of the projection operator on the field separates the independent quark field ψ_+ having two helicity components $\pm 1/2$. A similar decomposition of the gluon field in the directions of the light coordinates leads to an independent gluon field with two helicity components $\vec{A}_\perp$ (with two helicity components ± 1 according to Jaffe 1992a). In view of the above conditions on the light cone, the scattering of the parton q with the momentum k and the helicity h by the target T with the momentum P and the helicity H occurs with zero momentum transfer (i.e., scattering occurs at zero angle):

$$q(k, h) + T(P, H) = q(k, h') + T(P, H'). \tag{4.55}$$

Such a forward scattering along the direction of motion of the initial quark, $\vec{e}_3$, is called collinear scattering. Therefore, helicity, which is the angular momentum along the $\vec{e}_3$ axis, is conserved:

$$h + H = h' + H'. \tag{4.56}$$

Let us denote the amplitudes of reaction (4.55) as $A(h, H \to h', H')$. Then, in view of the parity conservation in strong interactions and time reversibility of the processes, the amplitudes satisfy the relations

$$A(h, H \to h', H') = A(-h, -H \to -h', -H') \tag{4.57}$$

and

$$A(h, H \to h', H') = A(h', H' \to h, H). \tag{4.58}$$

As a result, there are the following three amplitudes of the scattering of quarks by nucleons:

$$A\left(\frac{1}{2}, \frac{1}{2} \to \frac{1}{2}, \frac{1}{2}\right), \tag{4.59a}$$

$$A\left(\frac{1}{2}, -\frac{1}{2} \to \frac{1}{2}, -\frac{1}{2}\right), \tag{4.59b}$$

$$A\left(-\frac{1}{2}, \frac{1}{2} \to \frac{1}{2}, -\frac{1}{2}\right). \tag{4.59c}$$

These three scattering amplitudes correspond to three quark distribution functions (for the case of the spin-1/2 target):

$$f_1(x, Q^2) \propto A\left(\frac{1}{2}, \frac{1}{2} \to \frac{1}{2}, \frac{1}{2}\right) + A\left(\frac{1}{2}, -\frac{1}{2} \to \frac{1}{2}, -\frac{1}{2}\right). \tag{4.60}$$

This function corresponds to the scattering of unpolarized partons by unpolarized nucleons. The distribution function of unpolarized quarks, $f_1(x, \ln Q^2)$, is rather well studied in DIS processes. The other structure function

$$g_1(x, Q^2) \propto A\left(\frac{1}{2}, \frac{1}{2} \to \frac{1}{2}, \frac{1}{2}\right) - A\left(\frac{1}{2}, -\frac{1}{2} \to \frac{1}{2}, -\frac{1}{2}\right) \tag{4.61}$$

corresponds to the scattering of a longitudinally polarized parton by longitudinally polarized nucleon. Its investigation began in the mid-1970s at the SLAC accelerator (Alguard et al. 1976) and continues at present in many laboratories.

The third structure function from the complete set is called function transversity and is specified by the formula

$$h_1(x, Q^2) \propto A\left(-\frac{1}{2}, \frac{1}{2} \rightarrow \frac{1}{2}, -\frac{1}{2}\right). \tag{4.62}$$

It is the only function from the twist-2 group that has not yet been measured. To measure it, a transversely polarized parton should be scattered by a transversely polarized hadron. We will discuss this function in more detail later. Now, we consider the gluon structure function.

In the case of the scattering of a gluon (a massless particle conserving helicity), the following two independent amplitudes remain:

$$A\left(1, \frac{1}{2} \rightarrow 1, \frac{1}{2}\right), \qquad A\left(1, -\frac{1}{2} \rightarrow 1, -\frac{1}{2}\right). \tag{4.63}$$

These two gluon-scattering amplitudes correspond to the following two distribution functions of gluons in the nucleon:

$$G(x, Q^2) \propto A\left(1, \frac{1}{2} \rightarrow 1, \frac{1}{2}\right), \qquad +A\left(1, -\frac{1}{2} \rightarrow 1, -\frac{1}{2}\right) \tag{4.64}$$

and

$$\Delta G(x, Q^2) \propto A\left(1, \frac{1}{2} \rightarrow 1, \frac{1}{2}\right), \qquad -A\left(1, -\frac{1}{2} \rightarrow 1, -\frac{1}{2}\right). \tag{4.65}$$

As follows from the form of the gluon amplitudes when the leading terms (twist-2) are retained in the expansion, gluon scattering occurs without spin flip; i.e., helicity is conserved.

The function $G(x, Q^2)$ is measured in the scattering of unpolarized partons by unpolarized hadrons. It is well studied in DIS processes. The measurement of the spin-dependent gluon distribution function $\Delta G(x, Q^2)$ is of particular interest. In the case of scaling, this function cannot be measured in DIS processes because a gluon does not directly interact with a photon. This interaction occurs through a quark (antiquark) that is produced in the decay of the gluon into a quark–antiquark pair. Such processes are nonscaling. For a long time, attempts have been made to acquire information on $\Delta G(x, Q^2)$ from data on DIS with the inclusion of nonscaling terms, but the errors are very large. At present, the COMPASS and HERMES collaborations are developing a program for determining this function from the experiments with the production of mesons with open and hidden charms.

The structure distribution function of quarks in the transverse polarization, $h_1(x, q^2)$ (for brevity, it will be called transversity in what follows) was discovered in Ralston and Sopper (1979), but it was rediscovered and began to study in detail theoretically only in the early 1990s (Artru and Mekhfi 1990; Artru 1993; Jaffe and Ji 1991, 1992b).

Transversity $h_1(x)$ appears in the expansion of the quark correlation function, which contains the Dirac matrix $\sigma_{\mu\nu}\gamma_5$, on the light cone:

$$\frac{1}{4\pi}\int d\lambda e^{i\lambda x}\langle Ps|\bar{\psi}(0)\sigma_{\mu\nu}i\gamma_5\psi(\lambda n)|Ps\rangle$$

$$= h_1(x)(s_{\perp\mu}p_\nu - s_{\perp\nu}p_\mu)/M$$

$$+ h_L(x)M(p_\mu n_\nu - p_\nu n_\mu)s\cdot n + h_3(x)M(s_{\perp\mu}n_\nu - s_{\perp\nu}n_\mu). \quad (4.66)$$

Here, P and S are the four-vectors of the momentum and spin of the hadron, respectively; p and s are the four-vectors of the momentum and spin of the parton, respectively; $p^\mu = (p,0,0,p)$ is the parton four-momentum; $n^\mu = (\frac{1}{2p})(1,0,0,-1)$; and $s^\mu \equiv (sn)p^\mu + (sp)n^\mu + s_\perp^\mu$.

The function $h_1(x)$ is transversity, i.e., the distribution function of quarks over the transverse spin in the nucleon with transverse polarization. Transversity is a twist-2 function. The distribution functions $h_L(x)$ and $h_3(x)$ appearing in Eq. (4.66) are twist-3 and twist-4 functions, respectively.

Now, we are interested only in the function $h_1(x)$. To give the parton interpretation of this function, we should make several operations. First, it is necessary to decompose the quark field ψ entering into relation (4.66) using the light-cone projection operator $P_\pm$. This operation was discussed above and led to the expression of $h_1(x)$ in terms of one independent amplitude (4.62). The next step is the application of the chirality operator

$$P_{L,R} = \frac{1}{2}(1 \mp \gamma_5) \quad (4.67)$$

to the same quark field ψ. As a result, two quark states appear, namely: the left state L when the quark spin is oriented, e.g., against the quark momentum (negative chirality), and the right state R when the spin is directed along the momentum (positive chirality). The chirality operator commutes with the projection operator on the light cone, i.e.,

$$[P_{L,R}, P_\pm] = 0; \quad (4.68)$$

hence, the eigenvalues of these operators can be determined simultaneously. The chirality operator coincides with the helicity operator in the limit of zero quark mass, which is implicitly assumed in this consideration.

There is an additional spin projection operator $Q_\pm$ that is called the transversity projection operator and is given by the formula (Goldstein and Moravcsik 1976, 1982, 1989)

$$Q_\pm = \frac{1}{2}(1 \mp \gamma_5\gamma^\perp), \quad (4.69)$$

where $\gamma^{\perp}$ is the γ^1 or γ^2 Dirac matrix. The operator $Q_{\pm}$ also commutes with the projection operator on the light cone $P_{\pm}$:

$$[Q_{\pm}, P_{\pm}] = 0. \tag{4.70}$$

Such theoretical analysis shows that "transverse spin effects and longitudinal spin effects are on a completely equivalent footing in perturbative QCD" (Jaffe 1992a). Let us list some well-established properties of the transversity function $h_1(x)$ and compare them with the properties of the function $g_1(x)$ (called the chirality function for brevity). These properties are as follows:

• Inequalities

$$\left|g_1\left(x, Q^2\right)\right| < f_1\left(x, Q^2\right), \qquad \left|h_1\left(x, Q^2\right)\right| < f_1\left(x, Q^2\right). \tag{4.71}$$

These inequalities are valid for each flavor of the quark and antiquark.
• Physical interpretation: $h_1(x, Q^2)$ determines the probability of finding the transversely-polarized quark with the kinematic parameters x and Q^2 in the transversely-polarized nucleon. The function $h_1(x, Q^2)$ is chirality-odd; i.e., the parton leaves the nucleon with one chirality and enters into it with the opposite chirality (or vice versa). This is the cause of the suppression of this function in DIS processes, where helicity is conserved.
• The transversity function $h_1(x, Q^2)$ appears from the bilocal generalization of the tensor operator $\bar{q}\sigma_{\mu\nu}i\gamma_5 q$.
• The chirality function $g_1(x, Q^2)$ is chirality-even (partons enter and leave the nucleon with the same chirality). This function is not suppressed in DIS processes and is well measurable in these processes. The function $g_1(x, Q^2)$ appears from the bilocal generalization of the axial charge operator $\bar{q}\gamma_\mu\gamma_5 q$.
• Sum rule:

(a) the tensor charge introduced through the relation

$$2s^i \delta q^a\left(Q^2\right) = \left\langle Ps\left|\bar{q}\sigma^{0i}i\gamma_5\frac{\lambda^a}{2}q\left(Q^2\right)\right|Ps\right\rangle, \tag{4.72}$$

where λ^a is the quark flavor matrix, can be represented as the integral of the transversity function:

$$\delta q^a\left(Q^2\right) = \int_0^1 dx\left[h_1^a\left(x, Q^2\right) - h_1^{\bar{a}}\left(x, Q^2\right)\right]. \tag{4.73}$$

Here, the superscripts a and $\bar{a}$ refer to the quarks and antiquarks, respectively;

(b) a similar sum rule for the axial charge

$$2s^i \Delta q^a\left(Q^2\right) = \left\langle Ps\left|\bar{q}\gamma^i\gamma_5\frac{\lambda^a}{2}q\left(Q^2\right)\right|Ps\right\rangle \tag{4.74}$$

has the form

$$\Delta q^a(Q^2) = \int_0^1 dx [g_1^a(x, Q^2) + g_1^{\bar{a}}(x, Q^2)].$$ (4.75)

The comparison of the sum rules for these two cases indicates not only the similarity of the formulas, but their differences. First, the antiquark contributions enter into formulas (4.73) and (4.75) with different signs, because the tensor charge is a charge-odd term, whereas the axial charge is a charge-even term. Second, the transversity function $h_1(x, Q^2)$ is not normalized to the tensor angular momentum and has not a simple physical interpretation in spin fractions carried by partons in contrast to $g_1(x, Q^2)$. Third, all components of the tensor charge have a nonzero anomalous dimension, but no one of these components mixes with the gluon operators under renormalization. On the contrary, the flavor-nonsinglet axial charge Δq^a (where $\Delta q^0 \propto \Sigma$) has anomalous dimension due to triangular anomaly.

• Model prediction. In the nonrelativistic quark model, $h_1(x)$ and $g_1(x)$ coincide with each other. In the relativistic bag model (Jaffe and Ji 1992b), they are slightly different.

• In the quark-parton model, the following sum rule for the angular momentum was recently proven (Backer et al. 2004):

$$\frac{1}{2} = \frac{1}{2} \sum_{a=q,\bar{q}} \int dx h_1^a(x) + \sum_{a=q,\bar{q},g} \langle L_T \rangle^a,$$ (4.76)

where the left-hand side is the nucleon spin, the first term on the right-hand side is the transversity contribution to the nucleon spin, and the second term corresponds to the contribution of the transverse component of the orbital angular momentum of partons to the spin. This formula is of interest because the transversity contribution to the nucleon spin decreases with an increase in Q^2, whereas the contribution from the orbital angular momentum increases. This provides a hope to separate these contributions in experiment. Transversity has been calculated for the region $Q^2 \leq 0.5 \text{ GeV}^2$ in various models. The tensor charges were also calculated on a lattice, as well as with the QCD sum rules (Barone 2004). These calculations gave the following results: $\delta u \sim 0.1$–0.7 and $\delta d \sim -0.1$–0.4 at $Q^2 = 10 \text{ GeV}^2$.

• Drell–Yan processes with the production of lepton pairs $(l\bar{l})$ are purest for determining the structure functions $g_1(x)$ and $h_1(x)$, although the cross sections for these processes are much smaller than the cross sections for DIS. One of such processes is

$$\bar{p}(\uparrow) + p(\uparrow) \rightarrow l\bar{l} + X.$$ (4.77)

Using a longitudinally polarized antinucleon beam and a longitudinally polarized nucleon target, we can measure the following asymmetry associated with helicity:

$$A_{LL} = \frac{\sum_a e_a^2 g_1^a(x) g_1^{\bar{a}}(y)}{\sum_a e_a^2 f_1^a(x) f_1^{\bar{a}}(y)}.$$ (4.78)

If both the beam and target are polarized transversely to the beam direction, we can measure the following two-spin transverse asymmetry associated with transversity:

$$A_{TT} = \frac{\sin^2 \theta \cos 2\phi}{1 + \cos^2 \theta} \frac{\sum_a e_a^2 h_1^a(x) h_1^{\bar{a}}(y)}{\sum_a e_a^2 f_1^a(x) f_1^{\bar{a}}(y)}. \tag{4.79}$$

Here, θ and ϕ are the polar and azimuth angles of the scattering of partons in their center-of-mass frame, respectively. The function $h_1(x)$ appears here as the leading (scaling) term due to the presence of the antiquark.

In the case of reactions with two polarized protons planned at RHIC (Saito 2004), the expected effect is (1–2) %. The smallness of the expected effect in this case is determined by two factors: (a) the number of antiquarks in the proton is small and their transverse polarization is also small, and (b) in the RHIC kinematics, when $\sqrt{s} = 200$ GeV, $M < 10$ GeV, and $x_1 x_2 = M^2/s \leq 3 \cdot 10^{-3}$, the experiment probes the region of very small x values, where small transversity is expected. Correspondingly, small asymmetry is expected. In this case, it is more appropriate to deal with a polarized antiproton beam as in reaction (4.77), but at moderate energies. In particular, the designed PAX experiment (Efremov 2004; PAX Collaboration 2005), when $30 \leq s$ (GeV^2) ≤ 45, $M \geq 2$ GeV, and $x_1 x_2 = M^2/s \geq 0.1$, will probe the region of large x values, where noticeable transversity is expected. Calculations for $M = 4$ GeV and $s = 30$ GeV^2 show that transversity is 0.3 and is almost constant in the range $x_F = x_1 - x_2 = 0$–0.3. At $M = 4$ GeV and $s = 45$ GeV^2, transversity is also 0.3, but in a wider range $x_F = 0$–0.5 (Rathmann 2004). In the PAX experiment, the measurement of transversity through the production of J/ψ particles in same process (4.77) can be very useful, because the cross section for this process is two orders of magnitude larger than the Drell–Yan cross sections, and asymmetry is also about 0.3. This illustrates the practicality of the use of polarized antiproton beams for measuring transversity.

Transversity in DIS appeared to be suppressed by a factor of m_q/Q, where m_q is the mass of the current quark. The masses of current quarks are very small. It would be appropriate to produce heavy quarks, but the cross sections for such DIS processes are very small. Thus, transversity makes a negligibly small contribution to DIS processes.

For completeness, we present the following expression for asymmetry, when one of the colliding particles is longitudinally polarized and the other particle is transversely polarized:

$$A_{LT} = \frac{2 \sin 2\theta \cos \phi}{1 + \cos^2 \theta} \frac{M}{\sqrt{Q^2}} \frac{\sum_a e_a^2 [g_1^a(x) y g_T^{\bar{a}}(y) - x h_L^a(x) h_1^{\bar{a}}(y)]}{\sum_a e_a^2 f_1^a(x) f_1^{\bar{a}}(y)}. \tag{4.80}$$

Here, twist-3 functions, namely, $g_T^a(x)$ and $h_L^a(x)$, appear for the first time; they, as expected, have an order of $M/|Q|$. A deeper insight in this problem can be found in Barone (2002).

• Other possibilities for measuring transversity. One of the possible channels of studying transversity is presented by semi-inclusive processes in the current fragmentation region (Jaffe and Ji 1992b):

$$e + p \rightarrow e' + H + X. \tag{4.81}$$

The spin and twist properties of such processes are determined by parton distribution and fragmentation functions. If the target is transversely polarized, the following asymmetry is measured:

$$A_T^H(x, z, Q^2) \propto \frac{\Lambda}{\sqrt{Q^2}} \frac{\sum_a e_a^2 [h_1^a(x, Q^2) \hat{e}^{a/H}(z, Q^2) + g_T^a(x, Q^2) \hat{f}_1^{a/H}(z, Q^2)]}{\sum_a e_a^2 f_1^a(x, Q^2) \hat{f}_1^{a/H}(z, Q^2)}. \tag{4.82}$$

This expression includes not only the transversity function, but also the function of the fragmentation of the quark a to the hadron H, $\hat{f}_1^{a/H}$, which is a twist-2 function and has even chirality, and the twist-3 functions g_T^a and $\hat{e}^{a/H}$ with even and odd chiralities, respectively. The factor in front of the fraction indicates the presence of twist-3 terms. In principle, this formula makes it possible to extract the transversity function if the other three functions have been determined in preliminary experiments with polarized and unpolarized targets. Even if such experiments are possible, they are very difficult.

Another semi-inclusive, but attractable process is the process of the production of polarized hyperons. They are interesting because their weak two-particle decays make it possible to determine their polarization with high efficiency.

• Transversity distribution functions depending on the quark transverse momentum k_T. With allowance for the transverse quark momentum in the nucleon, five additional distribution functions appear in addition to three functions (f_1, g_1, and h_1) described above. Some of them are directly associated with transversity, and the first experimental data on them appear. Let us discuss these functions.

We have four physical quantities: the momentum P and spin S_T (the subscript T means that it is perpendicular to the momentum P) of the nucleon, as well as the transverse momentum k_T of the quark and its spin S_{qT}. Using them, we can determine the following spin-dependent transverse asymmetries of quarks.

Case 1 Let the nucleon be transversely polarized. The difference between the numbers of quarks with spins parallel and antiparallel of the nucleon spin is given by the expression

$$\Delta N = N_{q\uparrow/p\uparrow}(x, k_T) - N_{q\downarrow/p\uparrow}(x, k_T)$$

$$= (S_T \cdot S_{qT}) h_1(x, k_T^2)$$

$$- \frac{1}{M^2} \cdot \left[(k_T \cdot S_T)(k_T \cdot S_{qT}) + \frac{1}{2} k_T^2 (S_T \cdot S_{qT}) \right] h_{1T}^\perp(x, k_T^2). \tag{4.83}$$

The function $h_1(x, k_T^2)$ after integration with respect to the quark transverse momentum is transformed to the known scaling transversity function, whereas the new function $h_{1T}^{\perp}(x, k_T^2)$ after such integration vanishes. In the absence of integration, the fragmentation of quarks with different directions of transverse polarization in hadrons leads to asymmetry that is called the Collins effect (Collins 1993). In this case, the functions $h_1(x, k_T^2)$ and $h_{1T}^{\perp}(x, k_T^2)$ lead to asymmetries of the form $\sin(\phi_h + \phi_S)$ and $\sin(3\phi_h - \phi_S)$, respectively. Here, the azimuth angles refer to the final hadron (the subscript h) and to the proton polarization (the subscript s). Asymmetries are measured in the reactions

$$e + p \uparrow \to e' + h + X. \tag{4.84}$$

Such reactions are called semi-inclusive DIS (SIDIS).

Case 2   Let us consider the polarized proton with unpolarized quarks inside it. In this case, the following asymmetry can be expected in the number of unpolarized quarks:

$$\Delta N_2 = N_{q/p\uparrow}(x, k_T) - N_{q/p\downarrow}(x, -k_T) = \frac{(\vec{k}_T \times \vec{P}) \cdot \vec{S}_T}{M} f_{1T}^{\perp}(x, k_T). \tag{4.85}$$

Such an appearance of asymmetry is called the Sivers effect, and the function $f_{1T}^{\perp}(x, k_T)$ is called the Sivers distribution function (Sivers 1990).

Case 3   Transverse-polarized quarks inside the unpolarized proton can give asymmetry of the form

$$\Delta N_3 = N_{q\uparrow/p}(x, k_T) - N_{q\downarrow/p}(x, k_T) = \frac{(\vec{k}_T \times \vec{P}) \cdot \vec{S}_{qT}}{M} h_1^{\perp}(x, k_T). \tag{4.86}$$

The function $h_1^{\perp}(x, k_T)$ is called the Boer–Mulders distribution function (Boer and Mulders 1998). The Sivers and Boer–Mulders distribution functions are odd under time reversal, because it contains the product of an odd number of vectors changing sign under time reversal. However, such a "time parity violation" at the parton level "is allowed" due to the presence of the final-state quark interaction, which corrects the situation.

Experimental determination of the transversity function is a key element of many current polarization programs. Investigations are conducted in the following fields.

1. The measurements of single-spin asymmetry in the SIDIS reaction

$$e + p \uparrow \to e' + \pi + X. \tag{4.87}$$

In this reaction, nonzero asymmetry can appear only in the case of the inclusion of the transverse quark momentum. However, the transverse momentum leads to the noncollinear kinematics of the scattering of a quark by another quark, and the application of the parton model requires the proof of the factorization theorem for this case. This proof of factorization in noncollinear kinematics was recently given in Ji et al. (2004). As a result, asymmetry can appear due to two causes. The first is the Collins effect, when the transversely-polarized final quark fragments to an unpolarized hadron:

$$\Delta N_C = N_{h/q\uparrow}(z, \vec{P}_{hT}) - N_{h/q\downarrow}(z, \vec{P}_{hT}) = \frac{(\vec{k}_T \times \vec{P}_{hT}) \cdot \vec{S}_{qT}}{zM_h} H_1^{\perp}(z, \vec{P}_{hT}^2). \quad (4.88)$$

Transversely-polarized quarks appear in the Collins effect only due to interaction with other quarks in the same nucleon. If the initial nucleon is polarized, its quarks can give asymmetry when forming hadrons (Sivers effect). In the general case, this asymmetry has the form

$$\Delta N_S = N_{h/p\uparrow}(z, \vec{P}_{hT}) - N_{h/p\downarrow}(z, \vec{P}_{hT})$$

$$\propto A(y) \Im\left[\frac{(\vec{k}_T \cdot \vec{P}_{hT})}{M_h} h_1 H_1^{\perp}\right] \sin(\phi_h + \phi_S)$$

$$+ B(y) \Im\left[\frac{(\vec{k}_T \cdot \vec{P}_{hT})}{M_h} f_{1T}^{\perp} D_1\right] \sin(\phi_h - \phi_S)$$

$$+ C(y) \Im\left[\lambda(\vec{k}_T, \vec{k}_T, \vec{P}_{hT}) h_{1T}^{\perp} H_1^{\perp}\right] \sin(3\phi_h - \phi_S). \quad (4.89)$$

Here, $\Im[\ldots]$ is the convolution integral with respect to the variables $\vec{k}_T$ and $\vec{k}_T$. Taking azimuthal moments, we can determine different asymmetries separately. For example, to determine the Collins asymmetry in experiment, it is necessary to calculate the moment

$$\langle\sin(\phi_h + \phi_S)\rangle = \frac{\int d\phi_h d\phi_S \sin(\phi_h + \phi_S)[\Delta N_S(\phi_h, \phi_S) - \Delta N_S(\phi_h, \phi_S + \pi)]}{\int d\phi_h d\phi_S[\Delta N_S(\phi_h, \phi_S) - \Delta N_S(\phi_h, \phi_S + \pi)]}.$$
$$(4.90)$$

The Sivers asymmetry $\langle\sin(\phi_h - \phi_S)\rangle$ is determined similarly.

The HERMES collaboration recently reported their preliminary data on these asymmetries measured in the kinematic region $0.02 < x < 0.4$ and $0.2 < z < 0.7$ for $\langle Q^2\rangle = 2.4$ GeV^2 (Airapetian et al. 2004). In particular, the Collins asymmetry is $A_T^{\pi^+} > 0$ and $A_T^{\pi^-} < 0$ in agreement with expectations, because transversity functions have the corresponding signs: $h_1^u > 0$ and $h_1^d < 0$. However, it was found that $|A_T^{\pi^+}| < |A_T^{\pi^-}|$, which contradicts the model expectation $|h_1^d| \ll |h_1^u|$. The real situation is more complex. The asymmetries measured for $\pi^{\pm}$ are the combinations of the following transversity functions:

$$A_T^{\pi^+} \propto 4h_1^u H_1^{\perp f} + h_1^d H_1^{\perp uf}, \qquad A_T^{\pi^-} \propto 4h_1^u H_1^{\perp uf} + h_1^d H_1^{\perp f}. \quad (4.91)$$

Here, the superscripts f and uf denote favorable and unfavorable (small as expected in the model) functions for the corresponding asymmetries, respectively. The data on asymmetry in the production of π^- implies the relation $H_1^{\perp uf} \approx -H_1^{\perp f}$. Additional independent experiments are obviously necessary for measuring the Collins transversity functions $H_1^{\perp uf}$ and $H_1^{\perp f}$. Another surprising result of the

HERMES experiment is that the asymmetry $A_T^{\pi^0}$ is approximately equal to $A_T^{\pi^-}$ and is also negative in contrast to the expectation according to isotopic invariance. However, similar measurements, but for small values $x \leq 0.1$, in the COMPASS experiment show that the asymmetries for both π^+ and π^0 are consistent with zero, as expected for small x, z, and p_T values (Pagano 2004). The HERMES collaboration has obtained the first indication that the Sivers function is positive and nonzero, but more precise measurements are necessary for final conclusions.

2. Transversity can also be determined in the two-spin SIDIS process

$$e + p \uparrow \rightarrow e' + \Lambda \uparrow + X, \tag{4.92}$$

where the initial nucleon is transversely polarized and the transverse polarization of the Λ hyperon is measured. This reaction can test the fragmentation function

$$H_1(z) = N_{h\uparrow/q\uparrow}(z) - N_{h\uparrow/q\downarrow}(z), \tag{4.93}$$

which is an analog of the transversity function h_1. However, it is difficult to predict the polarization of the Λ hyperon, because experimental data on the transversity fragmentation function $H_1(z)$ are absent.

3. An additional possibility of determining transversity in reactions of lepton-induced production of a hadron pair on a transversely-polarized target was pointed out in Collins (1994). This is the measurement of the azimuthal asymmetry of the plane of two hadrons with respect to the angle $\phi_{RS} = \phi_R + \phi_S - \pi$, where ϕ_R is the azimuth angle of the plane of two hadrons with respect to the lepton scattering plane and ϕ_S is the azimuth angle of the quark polarization (Bacchetta and Radici 2004). In 2002, the COMPASS collaboration made the first attempt to perform such measurements on the 160-GeV muon beam with the polarized ^{6}LiD target. Authors stated that asymmetry in the effective-mass region up to 1.5 GeV can be measured with an accuracy of several percent (Joosten 2004).

4. Asymmetry in the production of pions on polarized proton/antiproton beams or/and targets can appear due either to the presence of transverse momenta of quarks or to the contributions from high twists. This asymmetry also provides information on the transversity functions. Under the assumption of the applicability of factorization and in noncollinear geometry, asymmetry can be written in the form

$$d\sigma \uparrow - d\sigma \downarrow \propto \sum \left[h_1\left(x_a, k_T^2\right) + \left(k_T^2/M^2\right)h_{1T}^\perp\left(x_a, k_T^2\right) \right] \otimes f_1\left(x_{b+}, k_T'^2\right)$$

$$\otimes \Delta_{TT} \hat{\sigma}(a \uparrow b \rightarrow c \uparrow d) \otimes H_1^\perp\left(z, k_T^2\right) \tag{4.94}$$

for the case of the Collins effect and in the form

$$d\sigma \uparrow - d\sigma \downarrow \propto f_{1T}^\perp\left(x_a, k_T^2\right) \otimes f_1\left(x_b, k_T'^2\right) \otimes d\hat{\sigma}(ab \rightarrow cd) \otimes D_1\left(z, k_T^2\right) \tag{4.95}$$

for the case of the Sivers effect. The data of the E704 experiment are in better agreement with the Sivers model, whereas the Collins effects are noticeably suppressed. However, such data are yet scarce and more data, particularly for RHIC energies, should be collected. Single-spin asymmetries in processes with the production of direct photons and jets simultaneously in the final state can be very interesting. Such experiments are developed for the STAR setup.

References

Airapetian, A., et al.: (2004). arXiv:hep-ex/0408013

Alguard, M.J., et al.: Phys. Rev. Lett. **37**, 1281 (1976)

Altarelli, G., Parisi, G.: Nucl. Phys. B **126**, 298 (1977)

Anselmino, M., Gambino, P., Kalinowski, J.: Z. Phys. C **64**, 267 (1994)

Artru, X.: In: Proceedings of the Fifth Workshop on High Energy Spin Physics, Protvino, Russia, p. 152 (1993)

Artru, X., Mekhfi, M.: Z. Phys. C **45**, 669 (1990)

Ashman, J., et al., European Muon Collaboration: Phys. Lett. B **206**, 364 (1988)

Ashman, J., et al., European Muon Collaboration: Nucl. Phys. B **328**, 1 (1989)

Bacchetta, A., Radici, M.: In: Proceedings of DIS 2004 (2004). arXiv:hep-ph/0407345

Backer, B.L.G., et al.: HTP-PH/0406139 (2004)

Barone, V.: Phys. Rep. **359**, 1 (2002)

Barone, V.: In: Proceedings of the Sixteenth International Spin Physics Symposium, Trieste, Italy, p. 12 (2004)

Bjorken, J.: Phys. Rev. **170**, 1547 (1960)

Bjorken, J.D., Paschos, E.A.: Phys. Rev. **185**, 1975 (1969)

Blümlein, J., Kochelev, N.: Nucl. Phys. B **498**, 285 (1997)

Blumlein, J., Tkabladze, A.: Nucl. Phys. B **553**, 427 (1999)

Boer, D., Mulders, P.J.: Phys. Rev. D **57**, 5780 (1998)

Callan, C.G., Gross, D.J.: Phys. Rev. Lett. **22**, 156 (1969)

Close, F.E.: An Introduction to Quarks and Partons. Academic, London (1979) (Mir, Moscow, 1982)

Collins, J.C.: Nucl. Phys. B **396**, 161 (1993)

Collins, J.C.: Nucl. Phys. B **420**, 565 (1994)

Dicus, D.A.: Phys. Rev. D **5**, 1367 (1972)

Dokshitzer, Yu.L.: Sov. Phys. JETP **46**, 641 (1975)

Efremov, A.V.: In: Proceedings of the Sixteenth International Spin Physics Symposium, Trieste, Italy, p. 413 (2004)

Eidelman, S., et al.: Phys. Lett. B **592**, 166 (2004)

Feynman, R.P.: Photon Hadron Interactions. Benjamin, New York (1972)

Forte, S., Mangano, M.L., Ridolfi, G.: Nucl. Phys. B **602**, 585 (2001)

Goldstein, R.G., Moravcsik, M.J.: Ann. Phys. **98**, 128 (1976)

Goldstein, R.G., Moravcsik, M.J.: Ann. Phys. **142**, 219 (1982)

Goldstein, R.G., Moravcsik, M.J.: Ann. Phys. **195**, 213 (1989)

Gribov, V.N., Lipatov, L.N.: Sov. J. Nucl. Phys. **15**, 438 (1972)

Jaffe, R.: In: Proceedings of the Tenth International Spin Physics Symposium, Nagoya, Japan, p. 19 (1992a)

Jaffe, R.L., Ji, X.: Phys. Rev. Lett. **67**, 552 (1991)

Jaffe, R.L., Ji, X.: Nucl. Phys. B **375**, 527 (1992b)

Ji, X., Ma, P., Yuan, F.: Phys. Lett. B **597**, 299 (2004)

Joosten, R.: In: Proceedings of the Sixteenth International Spin Physics Symposium, Trieste, Italy, p. 381 (2004)

Klein, M., Reimann, T.: Z. Phys. C **24**, 151 (1984)

Leader, E.: Spin in Particle Physics. Cambridge Univ. Press, Cambridge (2001)

Lipatov, L.N.: Sov. J. Nucl. Phys. **20**, 95 (1975)

Martin, A.D., Roberts, R.G., Stirling, W.J., Thorne, R.S.: Eur. Phys. J. C **23**, 73 (2002)

Pagano, P.: In: Proceedings of the Sixteenth International Spin Physics Symposium, Trieste, Italy, p. 469 (2004)

Panofsky, W.: In: Proceedings of the International Symposium on High Energy Physics, Vienna (1968)

PAX Collaboration: Technical proposal for antiproton-proton scattering experiments with polarizations (2005). arXiv:hep-ex/0505054v1

Ralston, J., Sopper, D.E.: Nucl. Phys. B **152**, 109 (1979)

Rathmann, F.: In: Proceedings of the Sixteenth International Spin Physics Symposium, Trieste, Italy, p. 145 (2004)

Saito, N.: In: Proceedings of the Sixteenth International Spin Physics Symposium, Trieste, Italy, p. 58 (2004)

Sivers, D.: Phys. Rev. D **41**, 63 (1990)

Wandzura, S., Wilczek, F.: Phys. Rev. B **72**, 195 (1977)

Part II
Polarization Technology

Polarization technology is the field of polarization physics devoted to the development of methods and techniques for obtaining polarized beams and targets, as well as for measuring their parameters (polarimetry). As energies in experiments were increased, new ideas and methods have been developed. At the very beginning of its development (the early 1950s), polarization technology involved simple methods such as the use of nuclear reactions (elastic and inelastic), polarizing filters (passage of particles through ferromagnetic foils), and decays (e.g., the decay of a pion into a muon; the final muon is polarized). The necessity of obtaining polarized particles with intermediate and high energies required development of high-current sources of polarized particles for injection into an accelerator. The research program for spin correlations also stimulated developments of efficient polarized targets for experiments with fixed targets. In 2000, the unique RHIC collider was built, in which two polarized proton beams with a center-of-mass energy of 200 GeV collide; this energy will be soon increased to 500 GeV. With an increase in the energy of polarized beams, polarimetry, i.e., technique for measuring the polarizations of these beams and targets was actively developed. Correspondingly, the second part of this course of polarization physics consists of four chapters. The first chapter presents methods for obtaining high-energy polarized beams. Various methods for producing polarized proton, electron, and muon beams are discussed. The second chapter describes various polarized targets; the third chapter is devoted to polarized sources of atomic beams and optically pumped polarized ion source and the fourth chapter, to polarimetry. These are the foundations of modern polarization technology.

Chapter 5
Methods for Obtaining Polarized Beams

Methods for obtaining high-energy polarized beams are strongly different for different particles. The problem of obtaining polarized proton beams is most complicated. In this case, it is necessary to create high-current sources of hydrogen ions with a high polarization degree and to guide such a beam through a long chain of accelerating units for reaching the final energy. It is particularly difficult to guide beams through strong-focusing accelerators, where a high-accuracy device, so-called "Siberian snake," should be used to preserve polarization during the acceleration of the protons. This device will be described later. The polarimetry of high-energy proton beams is also difficult. The problem of obtaining polarized electron beams in circular accelerator is somewhat easier. This problem is simplified with the use of ring accelerators/colliders due to the effect of the synchrotron-radiation-induced self-polarization of electrons (so-called Sokolov–Ternov (ST) effect), which will be discussed in the section devoted to polarized electron/positron beams. The problem of obtaining polarized electron beams in linear accelerators is somewhat more complicated. In this case, it is necessary to create high-current sources of polarized electrons. It is particularly easy to obtain polarized muon beams. Muons appear being already polarized in the weak decay of pions. For this reason, many difficulties inherent in the production of polarized proton and even electron (linac) beams are absent in this case. These aspects will be discussed in more details later.

For completeness, we mention here three methods for producing polarized proton beams, which played an important role in the development of polarization physics. The first method, which is applicable for proton kinetic energies ≤ 1 GeV, was experimentally discovered in the early 1950s (Oxley et al. 1954; Chamberlain et al. 1954; Stoletov and Nurushev 1954; Meshcheriakov et al. 1956). In this method, a secondary proton beam was produced by scattering of the primary proton beam on the internal target of light nuclei (Be, C), and extracted from an accelerator. The polarization of such beams reaches 60 % and the intensity reaches $\sim 10^6$ polarized protons/s. However, the polarization of such beams decreases rapidly with an increase in the initial energy. As a result, this method is inapplicable for energies ≥ 1 GeV. Moreover, a low intensity of polarized beams forced physicists to build

S.B. Nurushev et al., *Introduction to Polarization Physics*,
Lecture Notes in Physics 859, DOI 10.1007/978-3-642-32163-4_5,
© Moskovski Inzhenerno-Fisitscheski Institute, Moscow, Russia 2013

the powerful polarized proton sources and use the method of the acceleration of initially polarized protons. At this intermediate energy region the specific technique of producing the longitudinally polarized proton beam was proposed in Kumekin et al. (1959) and was invented in Meshcheriakov et al. (1963). In this method the first scattering occurs in vertical plane (usually in horizontal). Therefore the polarization vector becomes in horizontal plane and you may use efficiently the transverse (vertical) magnetic field in order to rotate the polarization vector to the longitudinal direction. In such a way you are avoiding the application of the non efficient solenoidal fields.

The second method of producing the polarized beam of higher energy was proposed in Overseth (1969) for proton accelerators with energies ≥ 100 GeV. It is based on the fact that the decay of the $\Lambda(\bar{\Lambda})$ hyperon ($\Lambda \rightarrow p + \pi^-$) is caused by weak interaction; for this reason, protons (antiprotons) in these decays are polarized up to 65 %. This method was first implemented at the Fermilab Tevatron in Grosnick et al. (1990). A similar method in application to the SPS accelerator was proposed as early as in Dalpiaz et al. (1972), but it has not been implemented. In application to the U-70 accelerator at IHEP, this method was proposed in the late 1970s (Apokin et al. 1977; Nurushev et al. 1980), but was implemented only in Galyaev et al. (1992). At the Fermilab, the effective beam polarization is 45 % (of both beam signs) at a polarized beam energy of 200 GeV. In this case, the total intensity of the polarized beam at the experimental target is $\sim 9 \times 10^6$ polarized protons per spill when the 800-GeV/c primary beam with an incident flux of 10^{12} protons per 20 s spill is incident on the target. This polarized beam was efficiently used in the E704 polarization experiment. Similar efficient use of the polarized beam was done at U70 accelerator. But in both cases for obtaining the polarized beams the external production target was used. But in some cases such an approach is not suitable (for example at colliders).

For that reason the possibility of producing the polarized proton (antiproton) beam through hyperon (antihyperon) decay using the internal production target in the accelerators/colliders was studied in Chetvertkova and Nurushev (2007, 2008). The first estimates of the polarized beam parameters are made for U70 accelerator. The conclusion can be made that there is a possibility to obtain the polarized beam on the internal production target of the same quality as on the external target but avoiding several problems peculiar to the external production target scheme.

The acceleration of the polarized proton beam is very complicated and expensive technique. Though theoretically it is possible to accelerate the polarized protons in U-70 and at LHC, it's doubtful, that it will be done in the nearest future. In such situation the more simple method of producing the polarized proton beam at LHC on the internal carbon target 0.7 μm thick was also estimated. Such rough estimates show that the polarized proton beam energy of 3.6 TeV, intensity $I \sim 6 \cdot 10^9$ pp/s at primary proton beam energy 7 TeV, luminosity $L \sim 3 \cdot 10^{37}$ cm^{-2} s^{-1}. Polarization is expected to be about 30 %. The polarized antiproton beam intensity might be 1 % of the polarized proton beam keeping all other proton beam parameters.

At Paul Sherrer's Institute (Switzerland) for the first time the polarized neutron beam was obtained by the use of the third method, polarization transfer mechanism from the longitudinally polarized proton beam of 560 MeV and polarization 60 % to neutrons at zero neutron production angle. This beam was polarized up to 50 % and efficiently used by physicists in completion of the full set of polarization experiments in neutron-proton elastic scattering (Arnold et al. 1997). There are no the theoretical or experimental proof of the existence of such mechanism of spin transfer from polarized protons to antiprotons. For that reason Chetvertkov and Nurushev (2009) proposed to make the search experiment at BNL. The combination of the AGS as a source of the polarized protons and, for example, RHIC as the storage ring may become very optimal scheme for such search of the spin transfer mechanism from protons to antiprotons. Evidently antiproton beam channel, spin rotator and polarimeter should be built and implemented.

5.1 Acceleration of Polarized Protons

A quantitative test of QCD predictions is of current importance for polarization physics. This test requires experiments at high momentum transfers and measurements of small spin effects. For such experiments, polarized proton beams with very high intensities and high polarization degrees are necessary. Until recently, this seems to be impossible because only strong-focusing accelerators operate at high energies. For beam focusing, extremely inhomogeneous magnetic fields are used in these accelerators. This is seen in the name of such an accelerator: alternating gradient synchrotron (AGS). Such accelerators provide difficulties for accelerating polarized particles. The first attempt to accelerate polarized protons in the AGS showed that the known method of "jumping" of depolarizing resonances and local corrections of imperfection resonances have the limiting possibility up to 22–25 GeV. They do not solve the problem of polarization preservation at higher energies (Khiari et al. 1989). An idea for the fundamental solution of the problem of polarization preservation in high-energy accelerators was proposed by Derbenev and Kondratenko (1975). In the simple presentation, the idea is as follows. The depolarization of a proton beam occurs when the relative spin rotation frequency $v_S = \gamma G$ (γ is the Lorentz factor and G is the anomalous gyromagnetic ratio of the proton) is an integer or a multiple of the frequency of betatron oscillations of the beam. In this case, resonance appears and the beam is depolarized; i.e., the spins of protons are chaotically oriented in space. Methods for avoiding this depolarization were proposed. The following idea was revolutionary. Two complexes of magnets are mounted on the beam orbit in the ring at an angular distance of 180° from each other. Each complex rotates the spin about the axis lying in the orbit plane, and these two axes are perpendicular to each other. In this case, according to the spin motion equation, spin oscillation frequency is now $1/2$. This frequency is energy independent. As a result, the problem of depolarization suppression has been solved. This idea, which was called the Siberian snake technique, was first implemented at the AGS–RHIC accelerating storage complex (BNL, USA) (Mackay 2004).

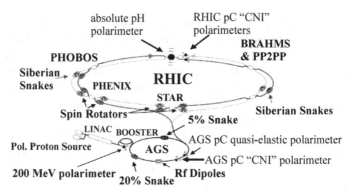

Fig. 5.1 Layout of the AGS–RHIC complex

5.1.1 BNL Accelerating Storage Complex

The BNL accelerating storage complex, whose layout is shown in Fig. 5.1, begins with the source of polarized negative hydrogen ions, which are obtained by optical pumping, OPPIS; this source will be described in Sect. 6.2. It provides a current of 0.5 mA and a polarization of 80 %. The polarized ions are accelerated by RF and linear accelerators to 200 MeV ($\gamma = 1.21$) and are stripped before injection to the booster. The beam is accelerated in the booster to an energy of 2.46 GeV ($\gamma = 2.62$). Polarized protons are injected into the AGS, where they are accelerated to an energy of 24.3 GeV ($\gamma = 25.94$). Accelerated polarized protons are transported from the AGS and are injected into the first and, then, to the second ring of the RHIC. Polarimeters are placed on this path: one polarimeter on the source (OPPIS), one at the exit of the linear accelerator, one $p + C$ polarimeter in the AGS, and two $p + C$ polarimeters in the collider (one in each ring). The polarimeter of quasi-elastic pp scattering, which was constructed for the preceding experiment (Khiari et al. 1989), is also used at the AGS. The alternating gradient synchrotron is equipped with two warm partial helical snakes for 5 and 20 % (percentages of the rotation angle of the complete snake, which is 180°). The RHIC collider is equipped with two pairs of full snakes and eight spin rotators; it is planned to use existing RF dipoles (one in each ring) for the fast flip of the beam polarization in the future. The polarized jet hydrogen target with a polarized-proton density of $1 \cdot 10^{12}$ protons/cm^2 and a polarization of (92 ± 1.8) % has been mounted and tested at RHIC.

Proton spin precession in the rest frame is described by the Thomas–Frenkel equation, which was previously called the BMT equation (Thomas 1927; Frenkel 1926; Bargman et al. 1953):

$$\frac{d\vec{S}}{dt} = \frac{e}{\gamma m}\vec{S} \times \left[(1 + G\gamma)\vec{B}_\perp + (1 + G)\vec{B}_l\right]. \tag{5.1}$$

Here, $G = (g - 2)/2 = 1.792817$ is the anomalous magnetic moment of the proton in the units of the nuclear magneton; m and γ are the mass and Lorentz factor of the proton, respectively; $B_\perp$ and B_1 are the transverse and longitudinal components of the magnetic field, respectively; t is time; and these parameters are defined in the laboratory frame. The equation of momentum rotation under the action of the Lorentz force has the form

$$\frac{d\vec{p}}{dt} = \frac{e}{\gamma m}\vec{p} \times \vec{B}. \tag{5.2}$$

Let us compare formulas (5.1) and (5.2) under the assumption that the magnetic field is static and uniform. Let us additionally assume that the longitudinal field component is zero. In this case, it is seen that the rate of spin rotation around the magnetic field is higher than the rate of velocity-vector rotation around the same field by a factor of $f = (1 + \gamma G)$. If $G = 0$ spin and velocity rotate with equal frequency. Both vectors for a positively charged particle rotate in the same direction. As it will be seen below, owing to the presence of the Lorentz factor, the number of depolarizing resonances increases with energy. On the contrary, a particle does not change the direction in the purely longitudinal magnetic field, whereas spin rotates with the frequency that is higher than the cyclotron frequency by a factor of G. However, the longitudinal field (solenoid) is inefficient as energy increases and is not used to control the polarization.

Thus, in planar geometry, when horizontal fields are absent and only the vertical field exists, the spin precession frequency in the particle rest frame is $\nu_S = \gamma G$. For simplicity, this frequency is called the spin tune or matching.

The radial magnetic field not only excites vertical betatron oscillations of the beam, but also shifts the spin direction from the vertical direction. If the particle radial field deviates by an angle φ by radial field, spin deviates by the angle $\varphi_s = (1 + \gamma G)\varphi$. The value $\nu_s = d\varphi_s/dt$ is called the spin tune. As known, the spin rotation operations around different coordinate axes do not commute with each other; for this reason, not only spin tune can change, but also the spin rotation axis can deviate from the vertical direction.

When moving in an actual ring accelerator, spin is primarily subjected to two types of depolarizing resonances. One of them is called intrinsic and occurs when spin tune is a multiple of the frequency of vertical betatron oscillations Q_v. Another type of depolarizing resonance is called imperfection or error resonance. It appears due to errors in the adjustment of the magnetic elements (under the assumption of their perfect identity). This resonance appears when spin tune is an integer. The minimum interval between imperfection resonances is 523 MeV and is determined from the equality $\gamma G = 1$. For most resonances of both types, the resonance appearance condition can be written in the form

$$\nu_S = n + n_V Q_V + n_h Q_h. \tag{5.3}$$

Here, n, n_V, and n_h are integers. The last term appears in the presence of solenoids or turned quadrupole lenses in the ring; as a result, couplings appear between vertical and horizontal oscillations of the particle. The notion of the resonance strength ε is introduced. It determines the deviation of the mean polarization from the vertical direction and is measured as the ratio of this angle to 2π. According to the

Froissart–Store formula (Froissart and Store 1960), polarization before input to resonance, P_{in}, and polarization after output from resonance, P_{out}, are related as

$$P_{out} = \left(2e^{-\frac{\pi|\varepsilon|^2}{2\alpha}} - 1\right) P_{in}. \tag{5.4}$$

Here, the parameter $\alpha = d(\gamma G)/d\theta$ specifies the rate of changing spin matching per unit angle of particle rotation in radians. According to this formula, the larger the coefficient α, the better the polarization preservation. If it is very small, polarization flop probability is large, and polarization can be preserved. This method was tested successfully at VEPP and not very successfully at Saturn and ZGS.

5.1.2 Calculation of Resonance Strength

The equation of spin motion in the external magnetic field was given above, see Eq. (5.1). A particle moving in cyclic accelerators under ideal conditions rotates about the vertical magnetic field $\vec{B} = \vec{B}_\perp = B_0 \vec{y}$ with a constant frequency ν along a closed planar orbit. Spin also rotates about the same axis with frequency $\nu_s' = \gamma G \nu$. In applications, the ratio $\nu_s = \nu_s'/\nu = \gamma G$ called spin matching is used more often. In existing accelerators with variable gradient, first, focusing and defocussing magnetic elements exist and, second, perturbation fields always exist owing to errors in the production of the magnetic elements or to errors in their adjustment. As a result, the total magnetic field acting on a particle is represented in the form

$$\vec{B} = B_0 \vec{y} + (B_y \vec{y} + B_x \vec{x}). \tag{5.5}$$

The horizontal field component B_x shifts the spin direction from the vertical direction. As a result, the vertical spin component decreases. Since the horizontal field is weak, the decrease in the vertical polarization is small. However, owing to the increase in such perturbations, spin matching can be in resonance with the frequencies of betatron oscillations of the particle. In this case, strong depolarization of the beam is possible.

The motion of the particle along the y axis has two components corresponding to the motion along the closed orbit Y_α and betatron oscillations Y_β. In the presence of an error in the adjustment of magnetic elements, Y_α is nonzero even for the ideal case of the monoenergetic beam ($\Delta p/p = 0$). As a result, spin imperfection resonance appears. The Fourier component of this resonance has the form $e^{\pm i K \theta}$, where K is an integer. Moreover, particles undergo betatron oscillations about the equilibrium orbit. The perturbating field initiating such a motion of the particle also has the Fourier component with the frequencies of betatron oscillations: $e^{i(kP \pm \nu_y)}$, where k, P, and ν_y are an integer number, the accelerator superperiod P, and the frequency ν_y of vertical betatron oscillations, respectively. As mentioned above, such a resonance is called intrinsic resonance. Note that, when spin matching $\nu_s = \gamma G$ is an integer k or $kP \pm \nu_y$, the coherent action of the field on spin occurs, and this leads to the depolarization of the beam. The resonance strength due to all listed factors is given by the formula

$$\varepsilon_{K_0} \propto \oint B_x e^{iK_0\theta} d\theta. \tag{5.6}$$

Using formula (5.1), we can show that spin motion in the magnetic field of the ring accelerator can be represented in the form

$$\frac{d\vec{S}}{d\theta} = \vec{S} \times \vec{\Omega}_a. \tag{5.7}$$

Here, the operator $\vec{\Omega}_a$ can be decomposed in three unit coordinate vectors:

$$\vec{\Omega}_a = -t\vec{x} + r\vec{s} - \kappa\vec{y}. \tag{5.8}$$

The unit vectors $\vec{s}$, $\vec{x}$, and $\vec{y}$ are defined as follows. The first two vectors lie in the orbit plane, and $\vec{s}$ is directed along the tangent to the orbit in the particle motion direction, $\vec{x}$ is perpendicular to $\vec{s}$ and directed outward, and $\vec{y}$ is directed along the upward normal to the orbit plane. The parameters in formula (5.8) are given by the expressions

$$\kappa \approx \gamma G, \quad r = (1+\gamma G)y' - \rho(1+G)\left(\frac{y}{\rho}\right)', \quad t = (1+\gamma G)\rho y''. \tag{5.9}$$

Prime in these formulae means the differentiation with respect to θ and ρ is the current curvature radius of the orbit. Then, in terms of the variables $S_\pm = S_1 \pm S_2$ in view of formulae (5.9), Eq. (5.7) can be represented in the form

$$\frac{dS_+}{d\theta} = i\kappa S_+ + i\varsigma^* S_3, \qquad \frac{dS_-}{d\theta} = -i\kappa S_- - i\varsigma S_3, \qquad \frac{dS_3}{d\theta} = i\left(\varsigma S_+ - \varsigma^* S_-\right). \tag{5.10}$$

Here, the coupling constant of transverse motions is given by the expression

$$\varsigma = -t - ir = -(1+\gamma G)\left(\rho y'' + iy'\right) + i\rho(1+G)\left(\frac{y}{\rho}\right)' = \sum_m \varepsilon_m e^{\pm iK_m\theta}. \tag{5.11}$$

Then, the formula for the resonance strength can be written in the form

$$\varepsilon_{K_0} = \frac{1}{2\pi} \oint \frac{\varsigma(\theta)}{\rho(s)} e^{iK_0\theta(s)} ds = -\frac{\gamma G}{2\pi} \oint y'' e^{iK_0\theta(s)} ds = -\frac{\gamma G}{2\pi} \oint G(s) e^{iK_0\theta(s)} ds. \tag{5.12}$$

Here, the function $G(s)$ is the focusing–defocussing accelerator strength. For an accelerator with the (focusing–flight–defocussing–flight) structure, the strength of the main intrinsic resonance is determined in the explicit form

$$\varepsilon_K = -\frac{\sqrt{\gamma G}}{2\pi}\left\{ G_F\sqrt{\varepsilon_N\beta_y(F)} + G_D\sqrt{\varepsilon_N\beta_y(D)}e^{i(\frac{K_0}{M}-\mu)\pi} \right\}$$
$$\cdot \frac{\sin[M(\frac{K_0}{M}-\mu)\pi]}{\sin[(\frac{K_0}{M}-\mu)\pi]} e^{i(M-\frac{1}{2})(\frac{K_0}{M}-\mu)\pi}. \tag{5.13}$$

Here, ε_N is the normalized emittance; G_F and G_D are the focusing and defocussing quadrupole strengths, respectively; μ is the phase increment on a cell; $\beta_y(F)$ and $\beta_y(D)$ are the vertical amplitude functions in focusing and defocussing quadrupoles, respectively; and M is the total number of cells in the accelerator.

Equation (5.13) provides a number of interesting conclusions. First, strongest resonances are spaced by a value of M, which is the total number of cells in the accelerator. Second, the resonance strength is proportional to M. Third, the strength of intrinsic resonances increases with energy as $\sqrt{\gamma}$, whereas the strength of the imperfection resonance increases linearly with γ.

The presentation in this subsection is based on Courant et al. (1986).

5.1.3 Calculations for the AGS

The application of this calculation technique to the AGS for the case $Q_v = 8.7$ and normalized emittance $\pi\varepsilon = 10\pi$ mm · mrad shows the presence of three strong depolarizing intrinsic resonances for spin matching values of $0 + Q_v$, $36 - Q_v$, and $36 + Q_v$. Recall that the periodicity of magnetic structure of the AGS is $P = 12$; therefore, resonances for which n are multiples of P are strong (see Eq. (5.3)). The proton Lorentz factors for these points are 4.85, 15.23, and 24.93, respectively ($n_V = \pm1$ in Eq. (5.3)). The relative strengths of these resonances are approximately 0.015 for the first two resonances and 0.027 for the third resonance. This third resonance is strongest, and injection energy for RHIC was chosen below this value. The remaining intrinsic resonances are much weaker.

The number of resonances and their strengths at RHIC increase rapidly according to the general laws. Similar calculations for this case at $Q_v = 29.212$ indicate the presence of four strong intrinsic resonances at energies above 100 GeV. Their strengths are from 0.35 to 0.45. Even below 100 GeV, the number of resonances with strengths an order of magnitude larger than at the AGS is so large that no methods except for the Siberian snake technique can overcome them.

To the end of 2004, the situation with the acceleration of polarized proton beams at RHIC was the following. The linear accelerator gave protons with a polarization of 80 % and an energy of 200 MeV, and the booster accelerated them up to 1.5 GeV. To suppress imperfection resonances at integers 3 and 4, the orbit was corrected by means of the corresponding harmonics of the magnetic field. The beam was extracted from the booster slightly earlier (at $\gamma G = 4.5$) since the first intrinsic resonance occurs at $v_s = \gamma G = 4.9$.

The beam in the AGS was accelerated from $\gamma G = 4$ to $\gamma G = 46.5$. Here, to suppress depolarization, a partial snake based on a warm solenoid with a spin strength of 5 % was first used. In 2003, polarization obtained at the exit of the AGS with this solenoid was only 28 %. In this case, the solenoid operated in combination with the alternating magnetic field of a dipole, which assisted to rotate spin in four strong intrinsic resonances mentioned above.

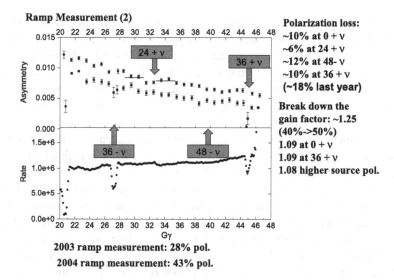

Fig. 5.2 Illustration of the improvement of the beam polarization at the exit of the AGS in 2004 as compared to 2003. The table to the right from the figure shows both losses of polarization and gains at certain resonances

A strong coupling was revealed between the vertical and horizontal motions of particles, which was caused by this solenoid. This solenoid changed to a warm helical magnet, which also served as a partial snake, but with a higher strength (8 % at the injection energy and 6 % at the energy of beam extraction from the AGS). As a result, a polarization of 50 % was obtained at the exit of the AGS in Huang (2004). The energy dependence of raw asymmetry measured in 2003 with the solenoid and in 2004 with the helical magnet is shown in the upper panel in Fig. 5.2 (the lower panel presents the collected statistics). The decrease in asymmetry with an increase in energy is clearly seen, but it is difficult to explain this behavior, because the energy dependence of the analyzing power of the polarimeter (elastic pC scattering in the region of Coulomb–nuclear interference) is unknown. This decrease in polarization can be due both to the analyzing power and to beam depolarization in the acceleration process. The certainly established fact, which is seen in Fig. 5.2, is that the partial dipole snake (upper circles) works better than the solenoid (lower squares). The analyzing power of the pC polarimeter at the energy of the beam extraction from the AGS was calibrated using another polarimeter based on a polarized jet target placed at RHIC. As a result, it was found that the beam extracted from the AGS in 2004 had a polarization of 50 %. Under the assumption that polarization is not lost in the booster, the beam at the entrance to the AGS would have a polarization of 50 % rather than 80 %, as found at the exit of the linear injector, i.e., almost at the entrance to the AGS. Therefore, the beam in the AGS is possibly depolarized from 80 % to 50 %, and this property should be further studied.

However, the problem of polarization preservation in the AGS remains unsolved.

Polarized protons should be accelerated at RHIC from the energy corresponding to $\gamma G = 46.5$ to the energy corresponding to $\gamma G = 478$ (250 GeV). For polarization preservation, two Siberian snakes spaced by 180° on the orbit are used. Spin rotation axes lie in the horizontal plane and are perpendicular to each other. Each snake consists of four superconducting helical dipoles. The winding of each dipole is a clockwise spiral. This winding generates a helical magnetic field, which is always perpendicular to the beam axis. As a result, spin matching is independent of the beam energy in contrast to the case of the solenoid. Beam swinging is also minimal. The main requirements on the snake are the following: first, it should rotate spin by 180° about the beam axis and, second, it should be transparent for the beam. This means that the beam parameters outside the snake must be unchanged. In addition, two spin rotators are mounted in pairs in each ring of RHIC around the beam intersection points for the STAR and PHENIX setups. They are used to convert the vertical polarization of the beam to the horizontal plane (orbit plane). Then, two dipoles of the accelerator transform the beam polarization to the longitudinal polarization. After the interaction of the beams, a similar pair of dipoles and magnets of the rotator return the beam polarization to the vertical position. This is necessary for the most important goals of the polarization program at RHIC associated with investigation of the spin structure of the nucleon.

The operation of the spin rotators was tested by measuring asymmetry in the absence and presence of supply of the rotators. Left–right asymmetry is observed in the first case and disappears when the rotators operate. Note only that the experimental measurement accuracies are insufficiently high for separating systematic errors. A polarization of 50 % was obtained at the exit of the AGS for $1 \cdot 10^{11}$ particles in a bunch. According to the same data, polarization at the exit of the linear injector was 80 %. Since information on polarization at the entrance of the AGS is absent, it is impossible to attribute all polarization losses to the AGS. Polarization is possibly lost in the booster. This is also a very important problem for the 1.5-GeV booster of the U-70 accelerator complex at IHEP (Protvino), because its parameters are close to the parameters of the booster for the AGS. The transportation channels for polarized beams obviously also require attention.

A luminosity of $4 \cdot 10^{30}$ cm^{-2}s^{-1} and a polarization of 40 ± 5 % at $\sqrt{s} = 200$ GeV were achieved at RHIC in 2004. To the next beam run, physicists are going to update the solenoid in the laser source of polarized protons in order to increase the polarization from 80 % to 85 %. It is also planned to implement a superconducting partial snake with 25 % spin strength (spin rotation by 45°) at the AGS. Large attention is focused on the possibility of improving the planarity of the orbits, including geodesy, beam position monitors, electronics, etc.

The nearest aim of physicists is to reach an energy of $\sqrt{s} = 500$ GeV, a luminosity of $1.5 \cdot 10^{32}$ cm^{-2}s^{-1}, and a polarization of at least 70 %. This will be the achievement of the main goal for beginning of the main polarization program on investigation of the spin structure of the nucleon.

5.2   Polarized Electron Beams

In this section, methods for producing polarized high-energy electron beams are discussed in application to the beams used at present at the largest operating accelerators. These are the ring accelerator HERA in Germany and the linear electron accelerator SLC at SLAC, USA. In these two accelerators, polarized electron beams are obtained using fundamentally different methods and we describe them separately.

5.2.1   Polarized Electron Beam of the HERA Ring Collider

The HERA collider with a length of 6.336 km has two rings. In one ring, protons are accelerated from 40 GeV to an energy of 820 GeV, and electrons in the other ring are accelerated from 14 GeV to 27.5 GeV. It is known that electrons in ring accelerators are self-polarized due to synchrotron radiation through the Sokolov–Ternov effect (Sokolov and Ternov 1963). Polarization in an ideally uniform static magnetic field is given by the expression

$$P = -P_{max}\left(1 - e^{-\frac{t}{\tau_{ST}}}\right). \tag{5.14}$$

According to this formula, the spins of electrons in time τ_{ST} are aligned oppositely to the direction of the magnetic field, and polarization reaches a maximum, which asymptotically $(t \to \infty)$ is

$$P_{max} = \frac{8}{5\sqrt{3}} = 92.4\,\%. \tag{5.15}$$

The physical cause of the spontaneous polarization of electrons is as follows. Electrons moving in a ring accelerator emit photons. In this case, the probability of emitting a photon depends on the mutual orientation of the electron spin and the external driving magnetic field. When emitting a photon, electron spin flop occurs. In this case, the state where spins are aligned against the external field is stable. Positron spins are favorably oriented along the field due to the positive charge. Let us assume that, when polarization P is reached, the number of electrons in the ring accelerator is $n = n_+ + n_-$, where n_+ and n_- are the numbers of electrons with spins directed along and against the magnetic field, respectively. By definition, polarization is $P = (n_+ - n_-)/(n_+ + n_-)$. According to these two relations,

$$n_+ = n(1 + P)/2, \qquad n_- = n(1 - P)/2. \tag{5.16}$$

For example, if the polarization of the positron (electron) beam with intensity n in the ring is 0.8, the spins of 90 % and 10 % of all positrons are directed along and against the field, respectively. Owing to the opposite charges, electrons are polarized oppositely to positrons.

However, it is difficult to preserve such a large polarization in existing accelerators. The presence of focusing and defocussing magnetic fields, as well as possible errors in the adjustment of the magnetic elements, can lead to depolarizing effects. Moreover, the Sokolov–Ternov process is slower than other competing processes in storage rings. This time of polarization "pumping" is given by the expression

$$\tau_{ST} = \frac{8\rho^3 m^2 c^2}{5\sqrt{3}\hbar e^2 \gamma^5}. \tag{5.17}$$

Here, m and e are the electron mass and charge, respectively; $\hbar$ is Planck's constant; c is the speed of light; and ρ is the current curvature radius of the particle trajectory in the magnets. The parameter τ_{ST} (the subscript ST means the Sokolov–Ternov process) is the polarization increase rate according to formula (5.14).

The parameters P_{max} and τ_{ST} depend on the accelerator parameters and vary in wide ranges. In particular, τ_{ST} is 40 min for energy of 27 GeV at HERA, is 300 min for energy of 46 GeV at LEP, and is as small as 2 min for energy of 29 GeV at TRISTAN. Polarizations really achievable in the accelerators will be presented below.

As in the case of the proton, the electron spin motion in the magnetic field is described by the FT–BMT equation (Frenkel 1926; Thomas 1927; Bargmann et al. 1959)

$$\frac{d\vec{S}}{dt} = \frac{e}{\gamma m}\vec{S} \times \left[(1+G\gamma)\vec{B}_\perp + (1+G)\vec{B}_\parallel\right]. \tag{5.18}$$

Here, $G = (g-2)/2 = 0.001159652$ is the anomalous magnetic moment of the electron in the units of the Bohr magneton; m and γ are the mass and Lorentz factor of the electron, respectively; $B_\perp$ and $B_\parallel$ are the transverse and longitudinal components of the magnetic field, respectively; and t is time. These parameters are defined in the laboratory frame, whereas spin is defined in the electron R frame. Electron momentum rotation induced by the Lorentz force is described by the equation

$$\frac{d\vec{p}}{dt} = \frac{e}{\gamma m}\vec{p} \times \vec{B}. \tag{5.19}$$

In the uniform static magnetic field, spin precesses with the frequency

$$\nu_s = \gamma G \nu_L, \tag{5.20}$$

where γ is the electron Lorentz factor and ν_L is the Larmor frequency of the rotation of a charged particle (electron in this case) about the same magnetic field. For simplicity, we introduce the quantity $f_s = \gamma G$ and call it spin matching (as above in the case of protons). Thus, the spin rotation frequency ν_s (or spin tune) differs from the Larmor frequency ν_L by spin factor f_s (or spin matching) and, at high energies, can be much higher than the latter. For example, spin matching for electron energy of 27.5 GeV is about 60. With noninteger ν_s values for electrons moving on a periodic (closed) orbit, Eq. (5.18) has a stable periodic solution usually denoted as $\vec{n}_0$. In ideal accelerators with an ideally planar orbit, the solution $\vec{n}_0$ is perpendicular to the orbit plane. The equilibrium polarization $\vec{P}_S$ has the same direction. If horizontal fields exist on the beam trajectory or the adjustment of magnetic elements is violated, the vector $\vec{n}_0$ is not vertical. This means that the original Sokolov–Ternov approach with the assumption of the ideal conditions of the electron motion (homogeneous magnetic field and the absence of horizontal fields) is inapplicable directly to existing accelerators. A possible modification of this method is the semiclassical method applied in Baier et al. (1970), and Derbenev and Kondratenko (1973). According to these works, the equilibrium value of the polarization vector $\vec{P}_{eq}$ is given by the expression

$$P_{eq} = \frac{8}{5\sqrt{3}} \frac{\langle |\rho|^{-3} \vec{b} \cdot \vec{n}_0 \rangle}{\langle |\rho|^{-3} [1 - \frac{2}{9} (\vec{n}_0 \cdot \vec{v})^2] \rangle}. \tag{5.21}$$

Here, the unit vector $\vec{b}$ specifies the direction of the magnetic field and the angular brackets stand for the averaging of quantities in the brackets both over the ring and the ensemble of particles.

As mentioned above, the polarization of electrons appears due to synchrotron emission of photons by electrons. In turn, this process leads to the electron recoil. Change in the electron motion direction according to relation (5.19) leads to change in the spin direction. That is, spin diffusion occurs. As a result, formula (5.21) is transformed as

$$P_{eq} = \frac{8}{5\sqrt{3}} \frac{\langle |\rho|^{-3} \vec{b} \cdot (\vec{n}_0 - \vec{d}) \rangle}{\langle |\rho|^{-3} [1 - \frac{2}{9} (\vec{n}_0 \cdot \vec{v})^2 + \frac{11}{10} |\vec{d}|^2] \rangle}, \qquad \vec{d} = \gamma \frac{\partial \vec{n}}{\partial \gamma}. \tag{5.22}$$

The vector $\vec{n}$ is a generalization of the vector $\vec{n}_0$ to the case of the presence of various perturbating factors. For example, the difference of the particle energy from the nominal value is presented in the form of the vector $\vec{d}$, which appears in the expression for polarization linearly (numerator) and quadratically (denominator). The presence of vertical betatron oscillations together with errors in the adjustment of the magnetic elements leads to the appearance of resonances under the condition

$$v_s = k + m_x Q_x + m_z Q_z + m_s Q_s, \tag{5.23}$$

where k, m_x, m_z, and m_s are integers and Q_x, Q_z, and Q_s are the betatron frequencies. Relation (5.23) can be used to find depolarizing resonances. A more careful analysis of spin diffusion shows that the depolarization rate is proportional to polarization "pumping" rate (5.17) multiplied by a polynomial of spin matching v_s. All strong resonances with pole expressions for them in the denominator should be taken into account. The number of resonances and their strengths increase with energy. So-called synchrotron side band resonances are considered as most dangerous. They appear as secondary resonances of strong parent resonances of the first order with the numbers $m_x = \pm 1$ or $m_z = \pm 1$; in this case, m_s is a small integer.

Although the polarization mechanism is weak and quite long-term, physicists demonstrated that it can be observable, enhanced, and used in electron storage rings. The radiative polarization effect was first observed almost simultaneously at the VEPP (USSR) and ALCO (Italy) machines. These effects have been measured with a high accuracy at the VEPP, VEPP-2, VEPP-3, and VEPP-4 machines in Novosibirsk. Somewhat later, the polarization of electron beams due to synchrotron radiation was observed at CESR, SPEAR, and DORIS setups (Potaux et al. 1971;

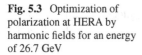

Fig. 5.3 Optimization of polarization at HERA by harmonic fields for an energy of 26.7 GeV

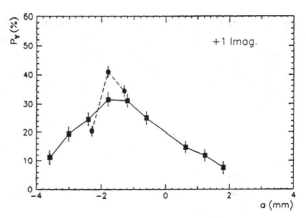

Shatunov 1990; Mackay et al. 1984; Barber et al. 1984). However, all these machines operated at relatively low energies of several GeVs. As known, depolarizing resonances become strong with an increase in energy. In view of this circumstance, the production of the 16.5-GeV polarized electron beam at PETRA in Bremer et al. (1982) is noticeable achievement. For this, a special device was designed and applied to suppress the field harmonics close to spin matching.

Beginning with 1990, electron polarization was observed at the TRISTAN, LEPP, and HERA accelerators. A 26.7-GeV electron beam with a polarization of 56 % was obtained at the HERA collider in 1992. The vertical polarization of the electron beam with an energy of about 46 GeV obtained at LEPP in 1993 by correcting the closed orbit was 57 %.

Special investigations with the artificial excitation of depolarizing resonances show that the fourth harmonics, $-1, 0$, and $+1$ and the second harmonic are dangerous at HERA. Since each resonance has a complex amplitude with the real and imaginary parts, there are eight parameters to be minimized. Eight short magnetic coils with horizontal magnetic fields (their slangy name is "bumps") were used to individually suppress all eight amplitudes. Figure 5.3 presents preliminary results of optimizing the orbit at HERA for an energy of 26.7 GeV ($\nu_s = 60.5$). In this case, the imaginary part of the amplitude for the $+1$ harmonic was fitted. When the imaginary part of the amplitude for this harmonic reaches 1.8 mm, polarization above 30 % was obtained (the solid line with points). An additional correction of the orbit provided a polarization close to 40 % (dashed line with experimental points).

A similar scheme for optimizing the closed orbit was also applied for TRISTAN at an energy of 28.86 GeV ($\nu_s = 66.5$). First, the standard deviation of the vertical closed orbit was corrected to a level of 0.3 mm; in this case, polarization did not change. Then, with the use of eight bumps, the slope of the vertical closed orbit at a harmonic close to spin matching was corrected. Figure 5.4 shows how the correction of the $+66$-th harmonic makes it possible to increase polarization from 7 % to (75 ± 15) %.

For the adjustment of an accelerator with a polarized beam, it is very important not only to rapidly polarize the beam, but also to rapidly depolarize it. The

Fig. 5.4 Correction of depolarization at TRISTAN by harmonic fields for an energy of 28.86 GeV

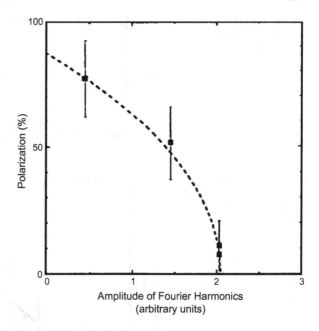

experience of works on the above machines shows that this can be made using the same correctors used to polarize the beam. After the depolarization of the beam, the correctors can be switched on, the polarization increase rate in an exponential can be measured, and the polarization increase time τ can be determined by the fitting method with expression (5.14). The time τ consists of two components:

$$\tau^{-1} = \tau_{ST}^{-1} + \tau_{dep}^{-1}. \tag{5.24}$$

Here, τ_{ST} is the theoretical polarization increase rate given by formula (5.17) and τ_{dep} is the beam depolarization rate. From relation (5.24), τ_{dep} can be expressed in terms of τ. Then, using the formula

$$P_{eq} = P_{ST} \cdot \frac{\tau_{dep}}{\tau_{ST} + \tau_{dep}} = P_{ST} \cdot \frac{\tau}{\tau_{ST}}, \tag{5.25}$$

we can determine polarization P_{eq} with a small systematic error. The results of the application of the described method to LEP for an energy of 46.5 GeV ($\nu_s = 105.5$) are shown in Fig. 5.5. With $\tau = (35 \pm 15)$ min, a polarization of $P_{eq} = (19.7 \pm 3.1)$ % was obtained. The measurements at TRISTAN were performed at an energy of 14.76 GeV ($\nu_s = 33.5$). Figure 5.6 shows that $P_{eq} = (69 \pm 24)$ % was obtained (after the recalculation of asymmetry to polarization) with $\tau = 68$ min. Finally, Fig. 5.7 shows the result of the measurements at HERA for an energy of 26.7 GeV ($\nu_s = 60.5$). A polarization of (46.5 ± 5) % was obtained. All listed polarization values are in good agreement with the values measured with polarimeters (Barber 1992).

From the very beginning, the HERA complex design implied the production of longitudinally-polarized electron beams for analyzing spin effects in electroweak interactions.

Fig. 5.5 Polarization pumping time at LEP for an energy of 46.5 GeV

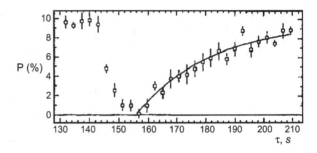

Fig. 5.6 Polarization pumping time at TRISTAN for an energy of 14.76 GeV

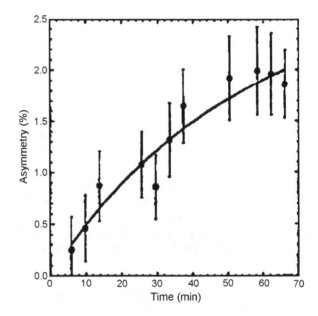

Fig. 5.7 Polarization pumping time at HERA for an energy of 26.7 GeV

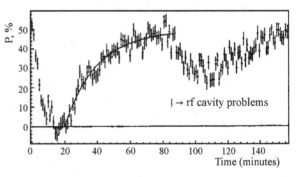

This problem formulated by physicists is that the electron polarization at the point of the collision of protons and electrons should be longitudinal, i.e., is directed along the momentum (or against the momentum) of the electron. There is an additional problem: the possibility of the fast reversal of the electron-beam polarization

Table 5.1
Longitudinally-polarized
electron beams

Machine	Energy (GeV)	Polarization (%)	Year
VEPP	0.65	80	1970
ACO	0.53	90	1970
VEPP-2M	0.65	90	1974
SPEAR	3.7	90	1975
VEPP-3	2	80	1976
VEPP-4	5	80	1982
PETRA	16.5	70	1982
CESR	5	30	1983
DORIS	5	80	1983
LEP	47	57	1993
HERA	26.7	60	1993
HERA (longitudinal polarization)	27.5	70	1994

direction should be ensured. This problem is simpler than the above two problems, but it should be solved in view of the conditions of the physical experiment. These requirements appear from the HERMES program proposed in HERMES (1990). To demonstrate the real possibility of obtaining, first, the transverse and, then, longitudinal polarizations of the electron beam, specialists performed a large amount of work, and the experiment began in 1994.

The necessity of obtaining the longitudinal polarization of the electron beam requires a special device called the spin rotator. The spin rotator is a set of radial magnetic fields that makes it possible to change the stable polarization direction from the vertical direction to the longitudinal direction. In this case, the spin rotator should not change the dynamics of electron motion. Owing to the introduction of radial fields, the closed orbit is not planar, the stable spin direction changes, and this leads to the emission of photons by electrons in view of the finite sizes of the beam. As a result, the spin rotator is a source of spin diffusion.

To obtain the longitudinal polarization of electrons, the term $\vec{\gamma}\vec{d} = \gamma \frac{\partial \vec{n}}{\partial \gamma}$ in formula (5.21) should be made zero at points where the term $|\rho|^{-3}$ is large. This was ensured by the so-called spin matching method, which includes two procedures: the optimization of the closed orbit and the correction of beam optics. There were many doubts that spin rotators operate as planned and that the spin matching method is efficient. Nevertheless, a longitudinal polarization of 65 % was obtained in May 1994 with the use of a pair of spin rotators and application of the two procedures listed above. This was the first case where the Sokolov–Ternov effect was used in a storage ring to obtain longitudinally-polarized electrons (Barber et al. 1995). Since that time, the very successful implementation of the HERMES polarization program begins. Brief chronology of obtaining polarized electron beams in electron storage rings is presented in Table 5.1.

All electron polarizations presented in Table 5.1 are transverse, except for the last row. Here, it is indicated that a longitudinal electron polarization of 70 % was obtained for the first time at the HERA accelerator. The lifetime of this polarization was about 10 h. The polarized positron beam with a polarization of 50 % was also obtained at the HERA accelerator.

5.2.2 Polarized Electron Beam of the SLC Linear Collider

The SLAC Linear Collider (SLC) at the Stanford Linear Accelerator Center (SLAC) is a linear collider with a length of 3 km intended for studying the production and decay of intermediate bosons. It consists of two parts: a linear accelerator (linac) and two arcs. The linac is intended for accelerating electrons and positrons to an energy of 46.6 GeV, and arcs, for organizing their collisions. From the very beginning, the collider design involved the possibility of the acceleration of longitudinally-polarized electrons. Since the polarization mechanism due to synchrotron radiation is absent in linear accelerators, a specially developed source of polarized electrons is used in them. The polarization currently achieved at the SLC in the normal regime is 80 %. The SLC is unique because it is the collider with highest energy and a longitudinally-polarized electron beam and allows in parallel experiments in the collider mode (SLD detector) and with fixed targets in a special region for the extracted electron beam (End Station A (ESA) region). These experiments with fixed targets were pioneering in the investigation of the spin structure of the nucleon, such as the E-80 (Alguard et al. 1976) and E130 (Baum et al. 1983) experiments. After the discovery of the "spin crisis" effect in 1987 at CERN in the ESA region the experiments listed in Table 5.2 were carried out. The parameters of polarized electron beams for these experiments were significantly improved as compared to the parameters of the beams for the first experiments. The aim of these experiments is to measure the structure functions of the proton and neutron and to test the Bjorken and Ellis–Jaffe sum rules. The results of these experiments with a high accuracy confirmed the conclusion of the EMC experiment (CERN) that quarks carry a very small fraction of the spin of the parent nucleon. They also confirm the validity of the Bjorken sum rule and the violation of the Ellis–Jaffe sum rule.

Below, we describe the method for obtaining longitudinally-polarized electron beams at the single largest electron–positron collider SLC. In this presentation, we follow primarily Woods (1994).

5.2.2.1 Source of Polarized Electrons

Polarized electrons are obtained by irradiating a GaAs photocathode by polarized laser beams (see Fig. 5.8).

In this scheme, two different laser generators are used, because experiments with fixed targets in the ESA region and at the SLC require beams of various time structures (see Tables 5.2 and 5.3). For experiments with fixed targets, the Ti-sapphire

Table 5.2 Beam parameters in the ESA region

Parameter	E142	E143	E154	E155	
N, stat.	$2 \cdot 10^{11}$	$4 \cdot 10^9$	$2 \cdot 10^{11}$	$4 \cdot 10^9$	N is the collected statistics, f is the
f, Hz	120	120	120	120	frequency of electron beam cycles,
τ, ns	1	2	100	100	T is the beam run duration, and
E, GeV	22.7	29.2	48.6	48.6	Year is the experiment date
P, %	40	84	80	85	
T, month	2	3	2	3	
Year	1992	1993	1995	1995	

Fig. 5.8 Source of polarized electrons at SLC

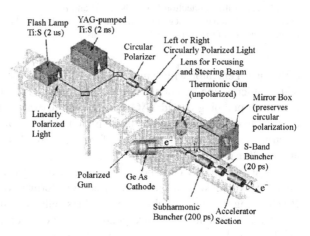

Table 5.3 SLC beam parameters

Parameter	1993	1994
N^+	$3.0 \cdot 10^{10}$	$3.5 \cdot 10^{10}$
N^-	$3.0 \cdot 10^{10}$	$3.5 \cdot 10^{10}$
f, Hz	120	120
σ_x, μm	0.8	0.5
σ_y, μm	2.6	2.4
Luminosity, $cm^{-2} s^{-1}$	$5 \cdot 10^{29}$	$1 \cdot 10^{30}$
Z/h (peak)	50	100
$\sqrt{s}$, GeV	91.26	91.26
P, %	63	80
Active time, %	70	70
T, month	6	7
Collected Z	60K	100K

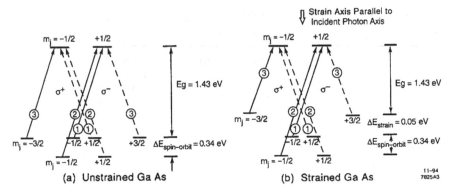

Fig. 5.9 Energy levels of (**a**) unstrained and (**b**) strained GaAs and allowed transitions from valence levels to the conduction band: the *solid* and *dashed lines* mark transitions stimulated by photons with positive and negative helicities, respectively; and the numbers in the *circles* indicate the probabilities of the corresponding transitions

(TiS) laser generator with lamp pumping is used to obtain pulses with a duration of two microseconds. For the SLD experiment at the SLC, two TiS laser generators with pumping on Nd:YAG (neodymium-activated yttrium–aluminum garnet) are used to obtain two pulses with a duration of 60 ns. One pulse is used to form an electron beam and the other pulse, to form a positron beam for the SLD experiment at the collider.

Linearly-polarized laser beams are guided through the system of mirrors to a quarter-wave Pockers cell, where they are transformed to circularly-polarized light. By changing the sign of voltages on Pockers cells, the direction of the circular polarization of photons and, correspondingly, the electron polarization direction can be changed. This change occurs randomly with a frequency of 120 Hz (the SLC frequency) and is ensured by a special generator. This operation is very important for minimizing systematic errors in the experiment.

Figure 5.9 shows the scheme of the transfer of electrons from valence levels to the conduction band in the GaAs photocathode. Figure 5.9a corresponds to the case of the unstrained GaAs photocathode. Photons with positive helicity and energy in the range 1.43 eV $< E <$ 1.77 eV can transfer valence electrons from two $j = 3/2$ levels to two $j = 1/2$ levels in the conduction band (the solid lines). According to the Clebsch–Gordan coefficients, the probabilities of these two transitions are related as 3:1. Therefore, the expected electron polarization is 50 %. In this case, the polarization direction for electrons emitted from the photocathode coincides with the photon polarization direction because they move in the opposite directions. As seen in Fig. 5.9a, the $j = 3/2$ levels are degenerate. If this degeneracy is lifted, the situation changes sharply as shown in Fig. 5.9b. In this case, only the transition from one level holds, and energy is insufficient for the transition from the other level. As a result, 100 % electron polarization can be achieved. The degeneracy of $j = 3/2$ levels can be really lifted by preparing the so-called stressed GaAs material. This is made as follows. Thin layers of this material are grown on a GaAsP substrate. The

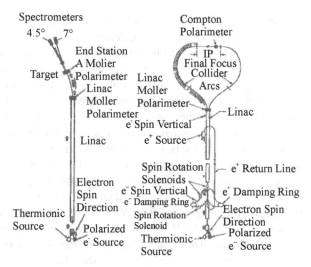

Fig. 5.10 Schemes of electron beam transport to the (*left*) ESA region and (*right*) SLD detector region

mixture of these two materials leads to the shift of the levels shown in Fig. 5.9b. Such materials are produced commercially. Their quantum efficiency defined as the number of photoelectrons per incident photon is about 0.2 %.

The further improvement of the operational parameters of the source was described in Clendenin et al. (2002).

5.2.2.2 Polarized Electrons in the ESA Region for Experiments with Fixed Targets

The remaining problems are the acceleration of the polarized electron beam from 60 keV to 46.6 GeV and its transport to a user. These problems are simplest for the ESA experimental region (Fig. 5.10).

Since the beam is longitudinally polarized, spin remains unchanged in the longitudinal accelerating electric field. However, at a finite energy, the beam deviates in the horizontal plane by the angle $\theta = 428$ mrad. In this case, spin advances the electron rotation angle by $\Delta\theta_s = \theta_s - \theta = \gamma G\theta$. When $\Delta\theta_s = n\pi$, the electron beam is longitudinally polarized. This occurs at energies $E = n \cdot 3.24$ GeV.

The beam polarization is measured by a Moller polarimeter described in the section "Polarimetry." Then, the beam interacts with a polarized target for the measurement of the spin structure functions. The parameters of polarized electron beams and the experiments performed in the ESA region with a fixed polarized target are listed in Table 5.2.

5.2.2.3 Polarized Electrons in the SLD Region for Collider Experiments

The next polarization experiment at SLAC was performed on the SLD instrument. The main goal of this detector is to check the predictions of the Standard Model, in particular, the Standard-Model predictions for spin effects. The longitudinally-polarized electron beam for this detector is obtained using a complicated procedure. The right scheme in Fig. 5.10 illustrates the formation of this beam. First, two bunches of polarized electrons are produced by an electron gun operating under a high voltage of 120 kV. Such a high voltage is necessary for the removal of the charge saturation effect in the photocathode at pulsed currents above 6 A. Two electron bunches are formed and accelerated up to 1.19 GeV. Then, they are injected into the storage ring, where the beam is cooled for 8 ms; i.e., its sizes are reduced in order to decrease the beam emittance. After that, they are transported to the linac and are driven into it by a kicker magnet. The first bunch is accelerated to the final energy and moves to the interaction point after the preceding positron bunch. The second electron bunch is accelerated only to 39 GeV and is used to obtain positrons. Positrons with energies 2–20 MeV are collected in the source and are accelerated to an energy of 200 MeV. Then, they are transported almost to the beginning of the linear accelerator. The beam emittance is decreased in the storage ring for 16 ms. Then, positrons are injected back into the SLC and are accelerated to a final energy of 46.6 GeV. After that, a separating magnet directs the electron and positron bunches to the north and south arcs, respectively. In these arcs, each electron loses an energy of 1 GeV on synchrotron radiation. As a result, the beam collision energy is 91.2 GeV, which corresponds to the Z boson mass.

When the electron beam is transported from the linear accelerator to the storage ring, spin is rotated by $450°$ immediately before entering the solenoid. The solenoid rotates spin from the horizontal position to the vertical plane. This spin transformation is necessary in order to avoid depolarization in the storage ring. With this vertical polarization, electrons return to the linear accelerator and their acceleration continues.

More detailed analysis shows that the polarization of electrons in this cycle is preserved at a level of 99 %; i.e., polarization loses 1 % because the electron beam directed to the storage ring has an energy of 1.19 GeV rather than 1.21 GeV as calculated. In front of the solenoid, the beam polarization appears to be rotated by $442°$ rather than by $450°$. For this reason, the solenoid rotates the polarization vector from the horizontal position to the direction deviated from the vertical direction by $8°$. This explains the 1 % loss of polarization.

The electron and positron beams leaving the linac are trapped by magnets for transporting to the north and south arcs. Each arc contains 23 achromatic lenses, and each achromatic lens consists of 20 magnets with combined functions. The spin of 46.6-GeV electrons is rotated by $1085°$ in each achromatic lens, whereas the betatron phase advance is $1080°$. Thus, the SLC has the operating point very close to spin matching. This means that small vertical oscillations of the beam in an achromatic lens, as well as errors in the adjustment of the magnetic elements, can lead to the deviation of the polarization from the vertical direction. Since this effect is cumulative, it can strongly change the longitudinal polarization component. As a result, having a certain polarization of the beam at the exit of the linac, it is impossible to predict the polarization direction at the interaction point. There are only two empirical methods. In the first case, solenoids existing at the SLC are used to ensure a certain orientation of the beam polarization at the exit of the linac. Then, the actual beam polarization at the interaction point can be measured using a Compton polarimeter. Three such measurements make it possible to determine three Euler angles specifying the beam polarization direction. These parameters are sufficient for reconstructing the spin transport matrix. Now, the relation between polarizations at the linac exit and interaction point can be determined using this matrix.

In the second method, two local spin perturbations ("bumps") are created by means of seven last achromatic lenses in the arc. The amplitudes of these bumps are selected empirically so as to obtain the maximum longitudinal polarization at the interaction point. These two methods provide the same polarization within 1 %. However, the method of two spin bumps is used as the convenient variant because it gives a high emittance for a plane beam (Table 5.3). Such a beam is obtained even in the storage ring. The use of solenoids for spin rotation leads to the relation of the dynamics of the beam in the horizontal and vertical planes, destroying the appropriate shape of the beam. Finally, necessary luminosity is not achieved. Therefore, this method is inappropriate for creating longitudinal polarization, and the method of spin bumps holds as the convenient variant. The beam parameters are given in Table 5.3.

5.3 Polarized Muon Beams

A polarized muon beam is naturally obtained through the weak decay of pions in flight

$$\pi \rightarrow \mu + \nu. \tag{5.26}$$

Since this decay proceeds with parity violation, muons are longitudinally polarized. The second feature of this pion decay channel is that its probability is larger than 99 %. The third feature of this channel is that it is two-particle, and the decay kinematics facilitates the selection of muons with appropriate energies and, correspondingly, appropriate polarization. In particular, if muons (μ^-) moving forward in the pion direction in the pion (π^-) rest frame are selected, these muons have left helicity and their polarization is close to 100 %. If we want to measure the helicity of muons, it is necessary to select muons moving backward in the rest frame. However, in view of the Lorentz transformation, these muons have other (lower) energies. However, this is inappropriate for experiments; moreover, they have a low intensity. Thus, there is no answer to the question how the polarization of a polarized muon beam can be reversed without changing the other parameters. We can recall the possibility of changing the polarized μ_L^+ beam from the decay of positive pions to the polarized μ_R^- beam from the decay of negative pions. However, changes in both the charge and helicity of the muon beam can lead to uncontrollable false asymmetries.

Fig. 5.11 (**a**) Polarization
and (**b**) flux of the 200-GeV/c
μ^+ beam versus the
momentum of the parent pion

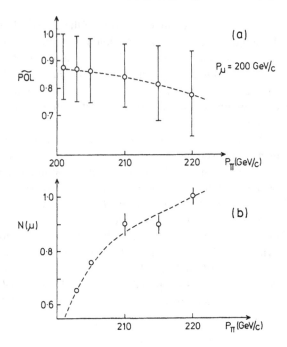

This method for obtaining polarized μ_L^+ beams was successfully applied in the famous EMC (<u>E</u>uropean <u>M</u>uon <u>C</u>ollaboration) experiment (Gabathuler 1984). Figure 5.11 taken from that work shows how the authors optimized the beam in polarization and intensity. Figure 5.11a presents the results of the measurements for optimizing the polarization. The magnetic fields in the muon channel were fixed so that only 200-GeV muons pass through the channel. The regime in the preceding, pion channel was changed so as to vary the pion energy. The polarization of the muon beam was measured by a polarimeter for six pion energies. Although the polarization measurement accuracy is low (about 10 %), it is seen that a large polarization is obtained for pion energies close to the muon energy. This conclusion is consistent with theoretical expectation. The polarization measurement accuracy was later increased to several percent (see section "Polarimetry").

Figure 5.11b shows the flux of 200-GeV/c muons as a function of the pion momentum. The measurements were carried out simultaneously with the measurements of muon polarization. In agreement with the expectations, the yield of the muons increases with the momentum. As known from the fundamental relations of polarimetry, the optimum of the working point should be at the point of the maximum of the quality factor $M = N \cdot P_\mu^2$, and this factor within the measurement accuracy has a plateau in the scanned pion momentum range.

In the SMC and COMPASS experiments that followed the EMC experiment (on the same channel), the parameters of the beam, polarimeters, and other facilities were somewhat improved, but the energy range of the muon beam remained almost the same.

At the end of 2004, the COMPASS setup operated with the muon beam whose parameters are: an energy of 160 GeV, an intensity of $2 \cdot 10^8$ muons/cycle, a cycle duration of 4.5 s, and a polarization of 76 % (Bressan 2004).

References

Alguard, M.J., et al.: Phys. Rev. Lett. **37**, 1261 (1976)

Apokin, V.D., et al.: CERN/SPS/77-61 (1977)

Arnold, J., et al.: Nucl. Instrum. Methods Phys. Res. A **386**, 211 (1997)

Baier, V.V., Katkov, V.M., Strakhovenko, V.M.: Zh. Èksp. Teor. Fiz. **31**, 908 (1970)

Barber, D.P.: In: Proceedings of the Tenth International Symposium on High Energy Spin Physics, Nagoya, Japan, p. 83 (1992)

Barber, D.P., et al.: Phys. Lett. B **35**, 498 (1984)

Barber, D.P., et al.: Phys. Lett. B **343**, 436 (1995)

Bargman, V., et al.: Phys. Rev. Lett. **2**, 435 (1953)

Bargmann, V., et al.: Phys. Rev. Lett. **2**, 435 (1959)

Baum, G., et al.: Phys. Rev. Lett. **51**, 1135 (1983)

Bremer, H.D., et al.: DESY report 82-026 (1982)

Bressan, A.: In: Proceedings of the Sixteenth International Spin Physics Symposium, Trieste, Italy, p. 48 (2004)

Chamberlain, O., et al.: Phys. Rev. **93**, 1430 (1954)

Chetvertkov, M.A., Nurushev, S.B.: In: Proceedings of the XIII Advanced Research Workshop on High Energy Spin Physics DSPIN-07, Dubna, p. 185 (2009)

Chetvertkova, V.A., Nurushev, S.B.: In: Proc. of XII Advanced Research Workshop on High Energy Spin Physics DSPIN-07, Dubna, p. 41 (2007)

Chetvertkova, V.A., Nurushev, S.B.: In: Proc. of the XIX International Baldin Seminar on High Energy Physics Problems, Dubna, vol. 2, p. 89 (2008)

Clendenin, J.E., et al.: In: Proceedings of the Fifteenth International Spin Physics Symposium, Upton, New York, USA, p. 1042 (2002)

Courant, E.D., Lee, S.Y., Tepikian, S.: AIP Conf. Proc. **145**, 174 (1986)

Dalpiaz, P., et al.: CERN/ECFA **1**, 284 (1972)

Derbenev, Ya.S., Kondratenko, A.M.: Zh. Èksp. Teor. Fiz. **37**, 968 (1973)

Derbenev, Ya.S., Kondratenko, A.M.: Dokl. Akad. Nauk SSSR **223**, 830 (1975)

Frenkel, J.: Z. Phys. **37**, 243 (1926)

Froissart, M., Store, R.: Nucl. Instrum. Methods **1**, 297 (1960)

Gabathuler, E.: In: Proceedings of the Seventh International Symposium on High Energy Spin Physics, Marselle, France, pp. 2–141 (1984)

Galyaev, N.A., et al.: IHEP Preprint No 92-159 ОП OMMC, Protvino (1992)

Grosnick, D.P., et al.: Nucl. Instrum. Methods Phys. Res. A **290**, 269 (1990)

HERMES collaboration: A Proposal to Measure Spin Dependent Structure Function at HERA (1990)

Huang, H.: In: Proceedings of the Sixteenth International Spin Physics Symposium, Trieste, Italy, p. 683 (2004)

Khiari, F.Z., et al.: Phys. Rev. D **39**, 45 (1989)

Kumekin, Yu.P., Marish, K.S., Nurushev, S.B., Stoletov, G.D.: JINR Preprint P-278, Dubna (1959)

Mackay, W.: In: Proceedings of the Sixteenth International Spin Physics Symposium, Trieste, Italy, p. 163 (2004)

Mackay, W.W., et al.: Phys. Rev. D **29**, 2483 (1984)

Meshcheriakov, M.G., et al.: Zh. Èksp. Teor. Fiz. **31**, 361 (1956)

Meshcheriakov, M.G., et al.: At. Energy **14**, 38 (1963) (in Russian)

Nurushev, S.B., et al.: In: Proceedings of the International Symposium on High Energy Physics
· with Polarized Beams and Targets, Lausanne, Switzerland, p. 501 (1980)

Overseth, O.E.: National Accelerator Laboratory, Summer Study Report SS-120, 1 (1969)

Oxley, C., et al.: Phys. Rev. **93**, 806 (1954)

Potaux, D., et al.: In: Proceedings of the Eighth International Conference on High Energy Accelerators, p. 127. CERN, Geneva (1971)

Shatunov, Yu.M.: Part. Accel. **32**, 139 (1990)

Sokolov, A.A., Ternov, I.M.: Dokl. Akad. Nauk SSSR **153**, 1053 (1963)

Stoletov, G.D., Nurushev, S.B.: Report of Inst. for Nucl. Problems, Dubna (1954)

Thomas, L.H.: Philos. Mag. **3**, 1 (1927)

Woods, M.: In: Proceedings of the Eleventh International Symposium on High Energy Spin Physics, Bloomington, USA, p. 230 (1994)

Chapter 6
Polarized Targets

The development of polarized proton targets for accelerator experiments became one of the important directions in polarization physics in the early 1950s. The necessity of such targets became obvious after theoretical works where it was shown that reconstruction of the (pion–nucleon or nucleon–nucleon) scattering matrix requires primarily the measurements of polarization parameters (see Sect. 2.2 "Reaction matrix"). Experiments with unpolarized particles allow the determination of only one observable, namely, differential cross section, whereas the other observables (more than three and eleven for the pion–nucleon and nucleon–nucleon scattering, respectively) should be measured in experiments with polarized initial and/or final particles. Such experiments obviously require both polarized beams and polarized targets.

The principle of developing polarized targets (solid and gas) is the same as that for sources of polarized ions: it is necessary to choose one of, e.g., four hyperfine states of the hydrogen atom. Let this state corresponds to positively polarized protons (line 1 in Fig. 6.1). For the production of proton beams with negative polarization, it is necessary to choose, for example, line 3 in the same figure. Another possible variant is the combination of two other lines.

In the external static magnetic field H, the levels of hydrogen atoms in the stationary state are occupied by both electrons and protons according to the Boltzmann distribution

$$n_\pm = e^{\pm \frac{g\beta H}{kT}}, \qquad (6.1)$$

where g is the Landé factor, β is the Bohr or nuclear magneton, k is the Boltzmann constant, and T is the target temperature. This formula shows that the population densities of lower energy levels (e.g., the ground level $|-\frac{1}{2}, +\frac{1}{2}\rangle$) is much larger than those of upper levels (e.g., the level $|+\frac{1}{2}, +\frac{1}{2}\rangle$). By definition, the target polarization is

$$P_T = \frac{n_+ - n_-}{n_+ + n_-} = \tanh \frac{g\beta H}{kT}. \qquad (6.2)$$

S.B. Nurushev et al., *Introduction to Polarization Physics*,
Lecture Notes in Physics 859, DOI 10.1007/978-3-642-32163-4_6,
© Moskovski Inzhenerno-Fisitscheski Institute, Moscow, Russia 2013

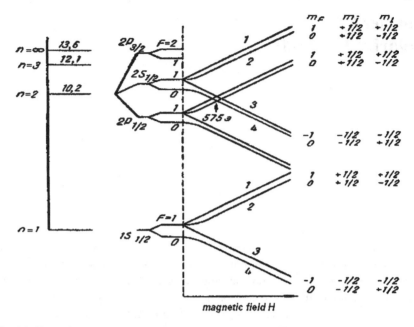

Fig. 6.1 Energy level scheme and Zeeman splittings of the hyperfine components of the hydrogen atom: (j, m_j), (I, m_I), and (F, m_F) are the spins and spin projections for the electron, proton, and total spin $\vec{F} = \vec{j} + \vec{I}$, respectively; n is the principal quantum number; and the abscissa axis is the ratio of the external magnetic field B to the so-called critical field B_C, which corresponds to the interaction between the magnetic moments of the electron and proton, leads to the singlet–triplet doublet with $F_S = 0$ and $F_T = 1$, and is $B_C = 507$ Oe for the ground state $1S_{1/2}$

Since $n_\pm$ are nonnegative integers, this formula indicates that the target polarization can vary in the range

$$-1 \le P_T \le 1. \tag{6.3}$$

The substitution of the experimental values $B = 2$ T and $T = 2$ K and the table values of the other parameters $g = g_n = 5.56$, $\beta = \beta_n = 3.152 \cdot 10^{-14}$ MeV T^{-1}, and $k = 8.617 \cdot 10^{-11}$ MeV K^{-1} into formula (6.2) provides the following estimate for the expected polarization of protons in a target:

$$P_T \approx 0.002. \tag{6.4}$$

This polarization of the proton target is too small for applications. At the same time, this method called the "brute force" method based only on the use of the strongest possible magnetic field and the lowest possible temperature is of applied interest for the production of polarized electrons. This is because the magnetic moment of the electron is larger than the magnetic moment of the proton by a factor of $\frac{g_e m_p}{g_p m_e} \approx 660$, and the electron polarization is almost complete under the same conditions as for protons. Polarized electron targets have recently attracted great interest in view of the proposals of physicists to use them to create antiproton polarizers (Rathmann et al. 2006; Meyer 1994).

The development of a polarized solid proton target became possible in 1953, when American physicist Overhauser proposed the idea of dynamic nuclear polarization (DNP) (Overhauser 1953a, 1953b, 1953c).

This idea was immediately tested and completely confirmed experimentally (Carver and Slichter 1953a, 1953b). On the basis of this idea, numerous polarized targets both with continuous polarization generation and with "frozen spins" have been developed. We will return later to polarized proton targets and discuss these mechanisms in more detail.

Now, we briefly present the essence of the DNP (Dynamic Nuclear Polarization) idea following Bilenky et al. (1964), Atsarkin (1980), Jeffries (1963), and Paul (1960).

Let us consider the ground state $1S_{1/2}$ of the hydrogen atom. The hydrogen atom consists of two particles with spin 1/2, a proton and an electron. Since they have magnetic moments, they interact with each other and this interaction leads to the formation of four energy levels. Three levels correspond to three projections of the total spin $F = 1$ of the (proton + electron) system (triplet states). The fourth state corresponds to $F = 0$ (singlet state). These states exist in the absence of a magnetic field, but only as a doublet with $F = 1$ and 0, because the triplet states are degenerate (see Sect. 1.5). Two additional interactions appear in the hydrogen atom in the external magnetic field H. The interaction of the electron magnetic moment with the field H lifts the degeneracy of the triplet state. It is split into three levels and, thus, four lines, the so-called fine structure lines appear. As a result of the interaction of the proton magnetic moment with the field H, each of the fine structure lines is split into two lines, which are called the hyperfine lines. These four spectral lines are designated by the Dirac brackets in Fig. 6.1 and also by numerals 1–4 from top to bottom. In Dirac brackets $|-\frac{1}{2}, \frac{1}{2}\rangle$ and $|-\frac{1}{2}, -\frac{1}{2}\rangle$, the first and second spin projections refer to the electron and proton, respectively. The upper spectral lines are correspondingly denoted as $|\frac{1}{2}, \frac{1}{2}\rangle$ and $|\frac{1}{2}, -\frac{1}{2}\rangle$ or as lines 1 and 2, respectively.

The difference between the fine structure energy levels is

$$\Delta E_e = \hbar \nu_e = g_e \mu_B H. \tag{6.5a}$$

At the same time, the difference between two neighboring hyperfine lines is

$$\Delta E_p = \hbar \nu_p = g_p \mu_n H, \tag{6.5b}$$

where the subscript p refers to the proton and μ_n is the nuclear magneton. Substituting the known values of the parameters (see above), we obtain

$$\nu_p \approx \nu_e / 660. \tag{6.6}$$

This relation allows us to estimate the difference between the frequencies of radiated waves from the transitions between different electronic levels and within a given electronic level.

Transitions with simultaneous flips of electron and proton spin projections are called forbidden, and transitions with only electron spin flip ($|M, m\rangle \leftrightarrow |M \pm 1, m\rangle$) are called allowed. By means of an RF magnetic field perpendicular to the main field H, transitions between the energy levels of the hydrogen atom can be performed in paramagnetic impurity ions with spin $1/2$. In this case, the relaxation time of the forbidden levels is much longer than the relaxation time of the allowed levels. As a result, saturation is achieved in a certain time when populations of two levels become the same. Let us consider the transfer between the $|\frac{1}{2}, \frac{1}{2}\rangle$ and $|-\frac{1}{2}, -\frac{1}{2}\rangle$ levels (lines 1 and 3). For brevity, we specify the first and second levels by the signs $-$ and $+$, respectively. When the populations of the forbidden levels become the same, we have

$$\frac{N_+}{N_-} = \frac{n_-}{n_+}, \tag{6.7}$$

where $N_\pm$ and $n_\pm$ are the numbers of electrons and protons with the spin projections $\pm\frac{1}{2}$, respectively. Since the relaxation time of the allowed transitions is much shorter than the relaxation time of the forbidden transitions in an alternating magnetic field, the distribution of electrons under these conditions is a nearly Boltzmann distribution. Then, it follows from relations (6.5a), (6.5b) and (6.7) that

$$\frac{n_+}{n_-} = e^{\frac{g_e \mu_B H}{kT}}. \tag{6.8}$$

Now, after the dynamical excitation, the polarization of protons is determined by the expression

$$P_T = \frac{n_+ - n_-}{n_+ + n_-} = \tanh \frac{g_e \mu_B H}{kT}. \tag{6.9}$$

The comparison of this expression with expression (6.2) shows that polarization increases after the dynamical excitation of the target. As mentioned above, this increase is a factor of about 660; i.e.,

$$P_T \approx \frac{g_e}{g_n} P_{stat} \approx 6.6 \cdot 10^2 P_{stat}. \tag{6.10}$$

This increase in polarization underlies dynamic nuclear polarization (DNP).

In this chapter, various types of polarized targets are successively discussed. Owing to various features, solid targets with continuous polarization generation and with frozen spins, jet gas targets and storage cell gas targets are considered separately. The parameters of polarized targets used in the experiments depend on the type of reactions. For example, polarized targets for electron beams should be sufficiently thin for a beam, and the target material should have a small radiation thickness. On the contrary, a target for muon beams should be very thick for a beam in order to compensate low yields of a reaction. The sizes of polarized targets for hadron beams should be intermediate between the sizes of polarized targets for electrons and muons.

In the absence of an external magnetic field ($H = 0$), the electron spin j and proton spin I is summed to the total spin F. It is conserved because the corresponding operator commutes with the Hamiltonian. It has two eigenvalues $F_T = 1$ (triplet) and $F_S = 0$ (singlet). As a result of the interaction between the magnetic moments of the electron and proton, these two levels are split even in the absence of an external magnetic field (see Fig. 6.1). In this case, the triplet level is above the singlet one by the energy $\Delta W = \hbar \cdot 1420.4$ MHz $\approx 5.8 \cdot 10^{-6}$ eV. The principal quantum numbers n are given in Fig. 6.1 to the right of the solid vertical line and the absolute energies of the levels of the hydrogen atom measured from the ground level $n = 1$ are presented to the left of this vertical line. In particular, the levels with $n = 2$ and 3 are spaced from the level with $n = 1$ by energies 10.2 and 12.1 eV, respectively. The ionization energy of the hydrogen atom is 13.6 eV. Lamb shift polarimetry is based on the $2S_{1/2}$ excited level.

Let us consider now changes in the levels in an external magnetic field B taking the ground state as an example. The dashed vertical line corresponds to the point $B = 0$. In the external magnetic field, the triplet state with the total spin projections $m_F = 1, 0$, and -1 is split into three hyperfine lines, and the energies of the first two lines with $m = 1$ and 0 (specified as 1 and 2 in Fig. 6.1, respectively) increase with magnetic field, whereas the energy of the line with $m = -1$ (line 3) decreases. The energy of the singlet line (line 4) also decreases. The right column shows the spin states of the electron and proton that enter into certain lines. For example, states 1 and 3 contain spins with positive and negative projections on the direction of the field H, respectively. In both cases, the electron and proton spins are parallel to each other. They are antiparallel in remaining states 2 and 4. Since the target polarization is determined only by proton spin orientations, the DNP process at low temperatures ($\sim$1 K) is used to align proton spins in a magnetic field range of 2–5 T. In this case, pump frequencies are in a range of 50–150 GHz. Polarization can be obtained without the use of the DNP method, but with the use of low temperatures and strong magnetic fields. This method was already mentioned above as the brute force method and began to be applied to ND (deuterated ammonia) since Kageya (2004).

In this chapter, we consider two types of polarized targets. Representatives of the first group are the so-called polarized solid targets. They have a high nucleon density comparable to the Avogadro number ($\sim 10^{23}$ nucleons/cm^3). These targets are usually used with the extracted primary and secondary particle beams. Polarized jet targets and polarized storage-cell gas targets belong to the second group of targets, are used in accelerators and colliders as internal targets, and have a nucleon density of 10^{12} and 10^{14} nucleons/cm^3, respectively. The possibility of using polarized jet targets and polarized storage-cell gas targets as internal targets is based on the fact that the effective density of such targets increases proportionally to the high frequency of beam circulation in accelerators/colliders (100 kHz–1 MHz). At colliders, an additional factor increasing the effective target thickness is associated with the beam lifetime ranging from 5 to 10 h. Hence, this factor is $\sim 10^4$ (in seconds). As a result, the effective densities of polarized jet targets and polarized storage-cell gas targets at colliders are about 10^{22} and 10^{24} nucleons/cm^3, respectively. Apparently, the effective densities of gas targets approach closely the density of polarized solid targets. Polarized gas targets have two important advantages over solid targets: first, they completely consist of polarized protons and, second, their polarization is frequently reversed. We emphasize that two types of targets, solid and gas, do not replace, but supplement each other. In particular, gas targets cannot be used on extracted or secondary beams, whereas polarized solid targets cannot be used as internal targets in accelerators.

The important parameter of any polarized target is the so-called target quality factor defined as

$$M = \kappa \cdot n \cdot (d \cdot P)^2. \tag{6.11}$$

Here, κ is the factor of the filling of a working cell by a useful substance (concerns polarized solid targets), n is the target density, d is the dilution factor defined as the ratio of polarized nuclei to the total number of nuclei (polarized + unpolarized), and P is the target polarization. The parameter M minimizes the time of achieving a given asymmetry measurement accuracy in an experiment with a given beam. Particular examples are given below.

We discuss only individual examples of polarized solid targets used in experiments with hadron, electron, and muon beams. These examples give the general representations about their properties such as used materials, magnets for the pumping and preservation of polarization, generators for pumping polarization, radiation resistance, factor of merit (FOM) of typical targets and their operational characteristics. These examples illustrate both large successes in the development of polarized targets and unsolved problems.

For the case of polarized gas targets, we consider only one polarized jet target and one polarized storage-cell gas target that are most advanced and used at the largest accelerators. We point to a proposal to use polarized gas targets with high densities to polarize antiprotons (Meyer 1994). There is also a proposal to experimentally test this idea (Rathmann et al. 2006).

Current promising developments aimed at improving the parameters of polarized targets are not discussed. We concern them in the context of the description of the chosen targets.

6.1 Polarized Solid Targets

In this section, polarized targets used on hadron, electron, and muon beams are described. We consider targets of widely known experimental facilities HERA, PROZA, E704, E143, EMC, SMC, and COMPASS. The last two targets are modernizations of the EMC target. Choosing the targets for description, we wish to present a variety of both targets and physical problems solved with them. Polarized solid targets can be conditionally divided into two groups. The first group includes targets with continuous polarization generation. Such targets are used with high-intensity beams (10^{10}–10^{12} particles/s). The second group includes frozen-spin targets. These targets are used at beam intensities $< 10^8$ particles/s, but provide a large useful solid angle and a relatively weak magnetic field in the target region. Both types of targets are presented in the experiments discussed below.

6.1.1 Polarized Target of the HERA Facility

In 1971–1975, in the HERA (High Energy Reaction Analysis) experiment at the 70-GeV IHEP proton accelerator, which was largest at that time, the polarization P and spin rotation parameter R were measured in the following reactions of the elastic scattering of particles and antiparticles by the polarized proton target developed in Saclay, France:

$$\pi^{\pm} + p \uparrow \rightarrow \pi^{\pm} + p, \qquad K^{\pm} + p \uparrow \rightarrow K^{\pm} + p, \qquad p^{\pm} + p \uparrow \rightarrow p^{\pm} + p. \tag{6.12}$$

Here, the proton and antiproton are denoted as $p^{\pm}$ for symmetry. These reactions are appropriate for using polarized solid targets, because completely definite kinematics of a process allows an almost complete suppression of the background from unpolarized nuclei in a target, where their number is almost an order of magnitude larger than the number of polarized nuclei (the dilution factor $d \sim 0.1$). In other words, in above reactions, the dilution factor in polarized solid targets is not already of special importance as a background source. The measurements were carried out at a momentum of 40 GeV/c with a negative beam and at a momentum of 45 GeV/c with a positive beam. The beams were obtained on internal targets and were transported on the polarized target of the experiment (Raoul et al. 1975; Bruneton et al. 1976). The experiment was unique both in the number of particles and antiparticles and in energy. The results still remain unique.

When designing a polarized target, it is necessary to take into account a number of circumstances, in particular, the physical observables that are measured. The described experiment was designed to jointly measure two parameters: polarization P and spin rotation parameter R. Taking into account reaction kinematics, this condition imposed strong requirements on the design of the magnet of the polarized target. Next requirements on the target design are determined by the beam parameters. They are presented in Table 6.1.

According to Table 6.1, the maximum intensity of a beam, in this case the π^--meson beam, did not exceed $6 \cdot 10^6$ π^-/cycle or 10^6 π^-/s. This means that frozen spin polarized targets made of organic hydrogenous materials can be used in the experiment. The use of such targets has a number of advantages. In particular, in this case it is easier to prepare the material of a target with the additive of paramagnetic pentavalent chromium atoms. Moreover, polarization can be held by a weak magnetic field of 0.5 T. The effect of such a field on the trajectory of the recoil proton is weaker than the effect of a polarization generating field of 2.5 T. Such a target can provide a larger useful solid angle for the experiment.

The following important parameter of a beam that should be taken into account when designing a target is its transverse size. According to Table 6.1, the sizes of the beam at the target are less than 2 cm. Therefore, the input diameter of a cell with a working substance should be ≥ 2 cm. The next parameter following from the experimental conditions is the useful solid angle necessary for detection of secondary particles. In the HERA experiment, it was determined by two requirements: first, to measure elastic polarization, the open polar angle should be in the range

Table 6.1 Parameters of the beams for the HERA experiment

Parameter	Negative beam	Positive beam
Momentum, GeV/c	40	45
Particle emission angles, mrad	0	27
Momentum band $\Delta p/p$, %	± 2	± 2.7
Beam composition, %	π^- (97.9); K^- (1.8); $\bar{p}$ (0.3)	p (94); π^+ (5); K^+ (1)
Beam intensity at the experimental target (ET) (for proton beam intensity of $5 \cdot 10^{12}$ protons/cycle on the U-70 internal target), particles/cycles	$3 \cdot 10^6$	$1 \cdot 10^6$
Beam sizes at ET (rms) $x \times y$, mm	10×15	15×18
Angular beam divergence at the experimental target $x' \times y'$, mrad	$(\pm 2.5) \times (\pm 1.5)$	$(\pm 1.5) \times (\pm 1.3)$
Momentum dispersion in, D, at the momentum collimator, $D = \Delta x/(\Delta p/p)$, mm/ %	6	4.5

30 mrad $< \theta <$ 60 mrad and the azimuth angle should be about this value and, second, the necessity of the simultaneous measurement of the spin rotation parameter R. Such a measurement is performed with the recoil particle; consequently, the useful polar angle should be close to 90° in the laboratory frame. For this reason, superconducting Helmholtz coils are calculated taking into account these conditions. They have a polar opening angle of $\pm 45°$ around a recoil angle of 90° in the laboratory frame and $\pm 7°$ in the azimuth angle. The opening angle for detecting recoil protons scattered in the vertical plane is the same (Fig. 6.2).

Propanediol ($C_3H_8O_2$) cooled by a flow of liquid ^{3}He was used as a target material. Liquid ^{3}He flowed in a tube. The same tube contained a precooler, a condenser, and devices for the expansion of ^{3}He. The tube was placed inside the cryostat of liquid ^{4}He coaxially with it (see Fig. 6.2). A microwave cavity for target excitation by a microwave field was placed at the end of this cylindrical tube inside it. This part of the target is coldest in a ^{3}He exhaust line. The total volume of used ^{3}He was 25 l, and it was always in the closed cycle. The consumption of liquid ^{4}He in the target was 2 l/h, and its general consumption taking into account the superconducting magnet was approximately 10 l/h. The superconducting coils of the magnet were later supplemented by additional windings in order to reduce the working current from 260 A to 237 A; in this case, the working field remained at a level of 2.5 T.

The thin-walled copper cavity 24 cm^3 in volume was filled with frozen propanediol balls with a diameter of about 2 mm. Polarization was reversed by small frequency adjustment near 70 GHz without change in the strength and sign of the magnetic field. Polarization in the excitation mode reached a maximum value of 85 % at an excitation power of 50 mW. Excitation was carried out at a temperature of 0.48 K.

Fig. 6.2 Polarized target of
the HERA facility (side
view): the magnetic field is
parallel to the horizontal
plane (perpendicularly to the
figure plane); the cryostat is
oriented upwards from the
beam axis

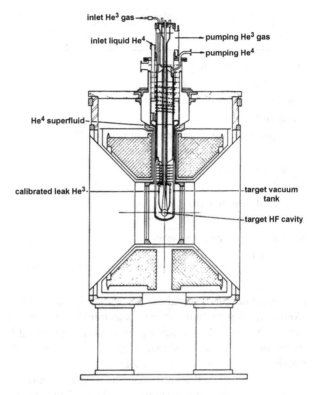

inlet He³ gas

inlet liquid He⁴

pumping He³ gas

pumping He⁴

He⁴ superfluid

calibrated leak He³

target vacuum
tank

target HF cavity

Table 6.2 presents the parameters of the polarized propanediol target. The diameter and length of the target cell were 2 and 8.3 cm, respectively. At the excitation time instant, the field was 2.5 T and the excitation frequency was 70 GHz. The uniformity of the magnetic field in the target volume under excitation was better than 10^{-4}. After excitation, the target polarization was measured using the nuclear magnetic resonance (NMR) technique. The coils were winded around of the cell with the target. These coils absorbed or emitted powers proportional to the target polarization. Measurements consisted of two stages.

At the first stage, the so-called NMR signal of natural (equilibrium) polarization was measured. Measurements of this equilibrium signal were performed under the same conditions as at excitation (a field of 2.5 T and a temperature of 0.48 K) only with the disconnected microwave generator. These measurements gave a calibration factor for determining the absolute target polarization after excitation. The level of this signal is about 0.19 %, and it is usually difficult to measure such a weak signal. At the second stage, the signal enhanced approximately by two orders of magnitude after excitation is measured. Usually, a time of 20 min is sufficient to increase polarization from zero to 0.8 of its maximum value, which is about 80 %. However, at least 2 h is required for increasing it to 80 %. In both cases, polarization is proportional to the area of the curves scanned in time. Processing of a large number of such curves taking into account such factors as the nonlinearity of a Q meter

Table 6.2 Parameters of the polarized propanediol target (Bruneton et al. 1976; Raoul et al. 1975)

Target parameters	Propanediol ($C_3H_8O_2$)
Frozen-propanediol density, g/cm^3	1.1
Free hydrogen density, g/cm^3	0.12
Bound-to-free proton number ratio	4.2
Radiation length, cm	45
Target polarization (average for a running period), %	80
Time of P increasing to $0.8 \cdot P_{max}$, min	20
Target diameter, cm	2
Target length, cm	8.3
Effective H_2 density, g/cm^3	0.085
Temperature, K	0.48
Microwave excitation power, mW	50
Frequency, GHz	70

and the scattering of signals in waveguides is a laborious problem. This information from an analog-to-digit converter was transferred to a computer and was processed. Polarization was determined by comparing the integral from the NMR signal with the natural polarization signal. Target polarization on the average over the entire running period was 77 % (see Table 6.2).

The accuracy of the measurement of NMR polarization by the instrument was ±5 %, where 3 % were due to the error in the determination of the equilibrium signal and 4 % were due to the uncertainty of the signal enhancement factor (Autones et al. 1972). The target operated successfully for several years.

6.1.2 Polarized Target of the PROZA Setup (IHEP)

The PROZA experimental setup (for polarization in charge exchange reactions) was created five years after the end of the HERA experiment and was mounted in the same beam line 14 of the U-70 accelerator. Therefore, the hadron beam parameters remained the same as those for the HERA experiment (see Table 6.1). The only innovation is the extraction of a proton beam from the U-70 accelerator by a bent single crystal (Aseev 1989). The beam energy was 70 GeV and the intensity was 10^7 protons per cycle. The uniqueness of this method is that it was applied for the first time to the strong-focusing accelerator. Another feature was that the crystal deflected the beam at a very large angle of 80 mrad, and there was no confidence in obtaining the necessary intensity under these conditions. This beam has almost no impurity of other particles, is monochromatic, and has small transverse sizes. It is now used in the PROZA experiment.

The main objective of the development of a new type of a polarized "frozen spin" target is to study polarization in charge exchange reactions. An example of two-

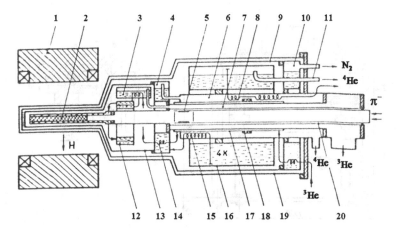

Fig. 6.3 Target of the PROZA setup (designations are explained in the text)

Table 6.3 Polarized target of the PROZA setup

Proton target parameter	Parameter value
Target size, mm	Diameter $= 19.6$, length $= 200$
Target material	Propanediol, $C_3H_8O_2$
Paramagnetic impurity (concentration), 10^{20} spin/cm^3	Complex Cr^5 ($1.8^{+0.1}_{-0.2}$)
Generation/holding field, T	2.08/0.4
Maximum polarization P_+/P_-, %	$+(90 \pm 3)/(94 \pm 3)$
DNP, T (mK)/M (mV)/HF (GHz)/n (mol/s)	$0.2/90/56/3 \cdot 10^{-2}$
Time of increase to $0.8\, P_{max}$, min	50
Freezing regime, T (K)/(mol/s)	$0.02/2 \cdot 10^{-3}$
Decay time P_+/P_-, h	1200/800

particle charge exchange reactions is the $\pi^- p(\uparrow) \to \pi^0 n$ reaction. In addition to elastic scattering, this reaction is the second successful example of using polarized solid targets because the reaction kinematics is completely determined.

The target has been developed by physicists at the Laboratory of Nuclear Problems (JINR, Dubna) with the assistance of physicists from IHEP. The target scheme is shown in Fig. 6.3, and its key parameters are given in Table 6.3. Its basic units will be briefly described below.

The PROZA experiment at the U-70 accelerator (IHEP) was conducted with a transversely polarized frozen target (Borisov et al. 1980). The idea of the development of such a target was proposed for the first time in Neganov (1966) and Hall et al. (1966). The target consists of a unique magnet with the combined function of the generation and preservation of polarization (Fig. 6.4), an economical horizontal dissolution refrigerator on the ^{3}He–^{4}He mixture (Fig. 6.3), a group of pumps for

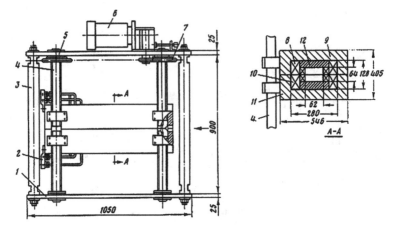

Fig. 6.4 Design of the Dzhin magnet of the PROZA setup: the arrow specifies the beam incoming direction; the target cryostat is inserted in the same direction; the AA-section cut of the magnet is shown on the right

vacuum production and the evacuation of flash ^{3}He and ^{4}He gases, and the systems of the generation and measurement of polarization.

The polarized-target cell had a diameter of 1.96 cm, a length of 20 cm, and a working volume of $V = 60$ cm^3 (Borisov 1986). The magnet should provide the necessary field (2.5 T) with uniformity $\Delta B/B \sim 10^{-4}$. The preliminary calculations of the magnetic field configuration were performed using the technique developed at IHEP (Daikovskii and Portugalov 1978). Such a magnet that was designed according to this model, was made at IHEP, and is in operation since 1980 is shown in Fig. 6.4 (Burkhin 1981).

Let us begin with the magnet design. Load-bearing units consist of two steel plates /1, 5/ fastened by four supports /3/. The magnet itself consists of the top /9/ and bottom /11/ III-shaped magnetic cores attached to the bearing plates by four screws /4/. Both magnetic cores can be budged or extended by means of a shift power system containing engine /6/, a worm gear reducer, and chain drive /7/. To increase the saturation induction, the working gap of the electromagnet is formed by two plates /12/ made of the 49-KF material. The magnetic core is made of SR-3 steel. The basic parameters of the magnetic core were chosen on the basis of the preliminary calculations. For a more precise correction of the field (to 0.1 %), the adjustable bend of polar plates /12/ was provided in the magnet design. The final correction of the field was carried out by shimming (placing permalloy foils in certain places along the magnet poles). Two coils /8, 10/ are incorporated in the magnetic core; each of them consists of 12 sections made of a $\varnothing 7 \times 1.5$-mm copper tube. The average number of turns in sections is 7. All sections are electrically connected in series, and water is supplied to them in parallel by means of collector /2/. A thermo-contact for switching-off of the magnet current at temperature >70 °C is mounted at the water-output end of each section. The cooling system is designed for desalted water (resistivity ≥ 10 kΩ/cm) at a pressure difference of 5.5 atm and an input water

temperature of 15 °C. The general water flow rate under the specified conditions is
4 m^3/h.

Let us shortly describe the magnet power supply. The field of the magnet was
generated by a thyristor source of stabilized current based on a KTU serial thyristor
rectifier ($I_{nom} = 800$ A and $V_{nom} = 240$ V). Updating was necessary for increasing
the long-term stability of the source to ± 0.01 % and for reducing the amplitude of
voltage pulsations on the magnet to 0.05 V. It has been experimentally determined
that heat release in the dissolution chamber of the refrigerator due to eddy currents
at this pulsation amplitude is $\leq 10^2$ erg/s at a temperature of 19 mK, which is ad-
missible.

The problems of magnet protection against switching-off of water or increase in
its temperature above 70 °C have been solved. The reached accuracy of the magnet
current setting is ≤ 0.01 %.

The magnetic field distribution in the target region has been measured by a Hall
sensor and an NMR device. For the closed magnetic core, the vertical and horizontal
distances between the poles were 64 and 62 mm, respectively (see the right part in
Fig. 6.4). The distance between the poles of the extended magnetic core was 26 cm.
The aim was that the field and its uniformity in the first state can be enough to
generate a sufficiently high polarization. In the second state, high uniformity, as well
as a high field, was not required. The idea of the creation of such a simple, cheap, and
reliable magnet was implemented by the simple mechanical moving of the magnet
poles without changing other parameters. The magnetic field in the first case was
2.18 T, and its uniformity was $\Delta B / B \sim 10^{-4}$ (Fig. 6.5). Curve 1 corresponds to the
optimum form of the poles selecting by means of bending screws. Curve 2 appeared
after the careful shimming of the fields using permalloy foils. This field is sufficient
for creating acceptable polarization in 2–3 h. In the case of the extended poles, the
field was 0.45 T and field uniformity in the target region was $\Delta B / B \sim 10^{-2}$. These
field parameters ensured the target characteristics presented in Table 6.3. Another
advantage of this magnet is that it provides a large solid angle. Only owing to these
characteristics, the extensive research program for charge exchange processes with
small cross-sections has been successfully completed.

The total weight of the magnet is ~ 1 t and its dimensions are $1.05 \times 0.86 \times$
0.95 m.

The improved version of this magnet designed for a target length of 400 mm was
created and tested in Grachev (1993).

The magnetic field of the Dzhin magnet has been measured in the target region.
Figure 6.5 shows the relative uniformity of the magnetic field along the target axis
beginning from its center.

The 60.3-cm^3 target cell (position /2/ in Fig. 6.3) was filled with frozen
propanediol-1,2 ($C_3H_8O_2$) balls about 2 mm in diameter; the filling factor of the
cell was 0.6. Propanediol contained additives of the EBA-Cr (V) paramagnetic material. Propanediol contains ten unpolarizable bound protons per approximately one
polarizable free proton. The effective dilution factor of target polarization given by
the expression $F = \frac{N_{bound}}{N_{bound} + N_{free}}$ was 0.9 taking into account liquid helium and foils
on target windows. The measured target density was 0.62 g/cm^3.

Fig. 6.5 Relative uniformity
of the magnetic field of the
Dzhin magnet along the
longitudinal target axis: curve
1 is obtained after the
optimization of the positions
of pole tips and curve *2*, after
shimming by permalloy foils;
the origin of the abscissa axis
is at the target center

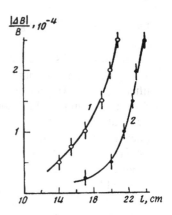

The thermal isolation of low-temperature units of the cryostat is ensured by vacuum jacket /19/, tight screen /9/ at a temperature of 77 K, and tight screen /13/ at
1 K. Gaseous ^{3}He is cooled by a cooling coil placed in a vessel with liquid nitrogen
/10/ and then enters the channel of gas heat exchanger /17/. Coal trap /18/ placed
in this ring channel adsorbs various impurities, including hydrogen. Bath /4/ with
liquid helium ^{4}He at a temperature of 1 K under vacuum provides the condensation
of ^{3}He in condenser /14/. Bath /4/ was supplied by liquid ^{4}He from reservoir /16/
through cooled coil /15/. After condensation, liquid ^{3}He enters the heat exchanger
located in ^{3}He evaporation bath /3/, and, then, counter flow heat exchanger /12/.
After dissolution, ^{3}He passes through the cell with the target substance along the
channels of the counter flow heat exchanger and enters an evaporation bath. The
evacuation of ^{3}He from the evaporation bath is performed through ring channel /7/,
the evacuation of ^{4}He from bath /4/ is carried out through ring channel /6/. Evaporating ^{4}He from helium reservoir /16/ can be removed either through tube /11/, or
through coil /8/ located in channel /6/. Thus, ^{3}He in channel /17/ is cooled simultaneously with ^{3}He evacuated from the evaporation bath, ^{4}He evacuated from bath /4/,
and ^{4}He evaporating from tank /16/. This design ensured economical consumption
of cryogenic liquids in the target.

Thin-walled stainless steel tube /20/ 43 mm in diameter provides the channel
for the input of the cell with the substance. The beam passes through the same
channel. After the insertion of the cell with the substance and its hermetic sealing
in the channel, hollow heat-insulating plug /5/ is inserted in the channel. The plug
contains thin-walled thermal screens made of a copper foil 20-μm-thick on the beam
path and the coal trap creating high vacuum in the channel. Such a device allows the
fast and simple insertion and removal of the cell with the substance.

A GDI-7 tunable microwave generator with a long-term frequency instability
of $\sim 10^{-4}$ is used to generate polarization. In view of the high stability of the microwave generator, automatic frequency tuning is not required; consequently, the
microwave system is considerably simplified. The high voltage of a power supply
unit is only controlled in microwave generator operation.

The equilibrium polarization signal was measured at a temperature of 0.8 K in
a magnetic field of 2.1 T. The measurement accuracy was 3 % and the absolute

accuracy of target polarization measurement was 5 %. The reached polarizations are presented in Table 6.3. Polarization is lower than that in similar facilities, because the reached field is somewhat weaker (2.1 T) than that in those facilities (2.65–2.7 T).

Polarization sign change takes 2–4 h and is performed once in two days.

6.1.3 Polarized Target of the E704 Experiment (FNAL)

The E704 experiment at Fermilab was carried out with a longitudinally polarized frozen target developed jointly by physicists from Saclay (France) and Argonne National Laboratory (USA) (Grosnick et al. 1997). The target consisted of the following basic units (Fig. 6.6): (a) a dissolution refrigerator on the ^{3}He–^{4}He mixture, (b) a superconducting solenoid, (c) a polarization generation system, and (d) a polarization measurement system (Chaumette et al. 1989, 1990, 1991). For the convenient service of the target, its basic elements were placed on mobile supports, which make it possible to release a place on the beam for the liquid-hydrogen target and for the elements of a Primakoff polarimeter and a Coulomb-nuclear interference polarimeter.

The cell of the polarized target had a diameter of 3 cm and a length of 20 cm. It was filled with frozen 1-pentanol ($C_5H_{12}O$) balls 2 mm in diameter. 1-Pentanol contained 6 wt % of water with additives of the EBA-Cr(V) paramagnetic material. 1-Pentanol contains one polarizable proton per approximately six unpolarizable bound nucleons. The effective dilution factor of target polarization was 8.4 taking into account liquid helium and foils on target windows. According to estimates, balls filled about 98 % of the total target volume and the packing factor was 0.65. The measured target density was 0.62 g/cm^3. The parameter $A = (N_A \rho l)^{-1}$, where N_A is the Avogadro number, ρ is the target density, and l is the target length, which is used when determining total cross sections, had a value of 1040 ± 38 mb for protons.

The superconducting solenoid (see Fig. 6.6) had a total length of 86 cm and an inner diameter of 9.4 cm. It consumed liquid helium with a rate of 1.5 l/h taking into account consumptions on siphons, gates, and other transitive elements. The solenoid could give a maximum field of 6.5 T at a supply current of 185 A. For polarization generation in this experiment, it operated with a field of 5 T. In this operating mode, the nonuniformity of the magnetic field over the entire target volume did not exceed $\Delta B/B \leq \pm 5 \cdot 10^{-5}$. After polarization build-up and transition to the frozen-spin regime, the solenoid was moved by 16 cm upwards along the beam in order to increase the useful polar angle in the laboratory frame up to 130 mrad. This was necessary for measuring asymmetry in the production of π^0 mesons in the central region, i.e., close to 90° in the center-of-mass frame. At an initial proton beam momentum of 200 GeV/c, this angle corresponds to an angle of about 100 mrad in the laboratory frame. After this motion of the solenoid, the magnetic field in the target volume remains ≥ 1.9 T. This field is quite enough for long-term preservation of polarization in the frozen spin mode.

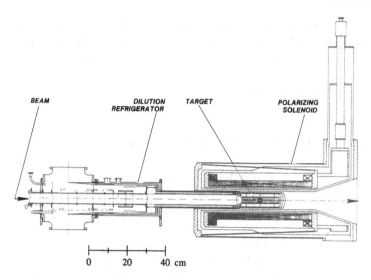

Fig. 6.6 Polarized target of the E704 experiment at Fermilab: the dissolution cryostat, target, and superconducting solenoid in the position for polarization generation are shown; the beam enters from the left

The ^{3}He–^{4}He dissolution refrigerator was a self-contained unit, had a horizontal design, and was coaxial with a beam. A free channel with a minimum extraneous substance was kept at the refrigerator center for beam passage through the target. A special gadget for the fast installation of the cell with the target at the workplace was inserted through the same channel. This installation was performed when the refrigerator was entirely cold and was in helium atmosphere. The pumping speed of the evacuation pumps was 5500 m^3/h. The temperature in the frozen mode was 60 mK and the ^{3}He flow was 4 mmol/s. This flow in the polarization build-up mode was 24 mmol/s.

Temperatures were measured by means of carbon resistors calibrated on germanium resistors.

All equipment was mounted on a support, which moved perpendicularly to the beam axis. During the collection of statistics, a leak appeared in the ^{3}He evacuation pump and the frequent reversal of the target polarization became impossible. The relaxation time of target polarization was 50 d at a target temperature of ≤80 mK. A microwave back-wave tube was used as a microwave generator for frequencies near 70 GHz. It ensures the possibility of fine frequency tuning to increase the population of different levels of the hydrogen atom if necessary. Thus, target polarization was reversed at a fixed magnetic field.

Target polarization was measured by the standard NMR method at a frequency of about 100 MHz. Signals were detected from three coils placed on the cell of the polarized target equidistantly at the beginning, middle, and end of the target. Owing to noise, the middle coil did not work during the collection of statistics. The signals from the coils were read out through the CAMAC system and were processed by a computer. In the frozen mode, the measurements were performed with a period of several hours. The absolute polarization was determined by comparing the excited (amplified) signal with the equilibrium signal. The equilibrium signal was measured at a temperature of 1 K and a field of 2.5 T. In 3–4 h of generation, the polarization of free protons usually reached $P_T = 77$ % and $P_T = -80$ %. These values were obtained separately from two extreme NMR coils in measurements in the generation and frozen modes. All these values coincided within the measurement accuracy. The polarization decay rate in the frozen mode was (1.51 ± 0.16) % per day.

The analysis of all polarization measurement data showed that the error in polarization at the 2σ level was ± 6.5 % (Hill 1992). This error included discrepancies in temperature and the statistical error in the measurements of the equilibrium NMR signal, NMR signal background, and linear and nonlinear instabilities, spatial uniformity in the polarization distribution, as well as extrapolation and interpolation errors. The contributions of the majority of these errors were symmetric and uncorrelated.

The positive direction of polarization in this experiment meant the high population of the low laying Zeeman level. In other words, positive target polarization means that it is directed along the solenoid field. Since the solenoid field in the polarization generation regime was directed against the beam momentum, positive polarization is also directed against the beam momentum. Polarization was reversed once a day in order to reduce the level of the systematic errors associated with the reversal of the beam polarization.

This polarized target was specially developed to measure the difference between the total cross sections in pure spin states in the interaction of longitudinally polarized 200-GeV proton and antiproton beams with the longitudinally polarized protons of the target. This experiment has been successfully carried out. In parallel with this experiment, two-spin asymmetry A_{LL} (π^0) has been measured in the inclusive production of π^0 mesons in the central area at collisions of the longitudinally polarized protons and antiprotons with the longitudinally polarized protons of the target (Grosnick et al. 1997).

6.1.4 Polarized Target of the E143 and E153 Experiments (SLAC)

This target is taken in order to illustrate polarized solid targets used with electron beams, intensities that can be withstood by these polarized targets, and arising problems (Crabb 1995).

The SLAC linear accelerator was taken as an example in the section devoted to polarized electron beams. In that section, the beam parameters in the E143 experiment are given in the table. The electron beam with an intensity of $5 \cdot 10^{11}$ s^{-1} was incident on the polarized solid cryogenic target with continuous polarization generation. Ammonia compounds $^{15}NH_3$ and $^{15}ND_3$ were used as target materials. To avoid the strong radiation damage of the target, the beam was extended at the input to the target to the sizes of the working cell. The simplified layout of the target is presented in Fig. 6.7.

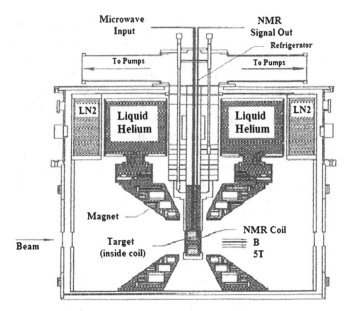

Fig. 6.7 Polarized ammonia target of the E143 experiment at SLAC

The cryostat of the target was oriented vertically and passed in the coils of the superconducting magnet. The refrigerator with continuous ^{4}He evacuation operated at a temperature of 1 K. This target has an original design: two working ammonia targets ^{15}NH$_3$ and ^{15}ND, as well as an empty target and a cell with a carbon target settled one over the other, are placed in the refrigerator. Such a design had an appreciable gain in statistics collection time. Any of the targets listed above could be inserted quickly into the beam by means of a mechanical drive and a remote-controlled engine. Waveguides for the microwave generator, signal cables from NMR coils, and the waveguide for the transfer of microwave power from the generator to the target also passed inside the probe.

The superconducting magnet created a required uniform magnetic field of 5 T in the target volume. The detailed description of the system can be found in Crabb and Day (1995).

Ammonia used in the target was irradiated by different electron beams in a vessel with liquid argon. The typical irradiation dose was 10^{17} electrons/cm^2. After irradiation, polarization is generated. Generation for protons gave a polarization of 95 % in agreement with the previous measurements (Crabb 1991). However, the maximum polarization reached in the E143 experiment with the beam was only 80 %. This occurred because of the specificity of the arrangement of the materials of the targets. Deuterated ammonium was placed above the cell with ammonia (^{15}NH$_3$), closer to the microwave generator. As a result, a portion of a wave from the generator, which was intended for ammonia, was absorbed in ^{15}ND$_3$ immediately behind the waveguide.

Fig. 6.8 Increase in deuteron polarization (in deuterated ammonia) in time; the effect of addition frequency modulation on this process is clearly seen

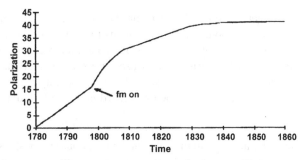

Deuteron polarization generated immediately after irradiation was as low as 13 %. However, it was found that generated polarization is at least doubled when the same material is irradiated at temperature ≤ 1 K (Boden et al. 1991). Under the conditions of the E143 experiment, the increase factor was 3, as seen in Fig. 6.8.

Deuteron polarization is additionally enhanced when using a frequency modulation signal. The time of the application of frequency modulation is marked in Fig. 6.8 by an arrow with the "FM switching" legend near a time (abscissa axis) of 1800. Around a time of 1805, frequency modulation was erroneously switched on and the polarization generation rate decreased.

During the experiment, target polarization decreased due to radiation damage. To restore polarization, the method of annealing at a temperature 85 K was used. A careful analysis of annealing and irradiation cycles has shown that this process is complex. For example, two cycles of a decrease in the polarization of protons were observed. The first cycle corresponds to a very fast polarization decrease at the beginning of irradiation by a high radiation dose. After a time, this cycle changes to a slower decrease. In the case of deuterons, there is only one slow cycle. The Teflon cell containing ammonia is also polarized under irradiation. The measured polarization of the cell reached 8 %, and a necessary correction was introduced in proton polarization.

Another important applied result of such investigations is the detection of the polarization of ^{15}N nitrogen atoms. Both polarizations increase during generation, but the behaviors of the curves are different in the presence and absence of microwave radiation. A thorough analysis is necessary to adequately correct the experimental data on nitrogen polarization.

According to these investigations, irradiated ammonia exposed to continuous microwave radiation is a radiation-resistant material for beam intensities up to 10^{11} particles/s. Hydrogenous organic materials do not withstand such irradiations.

6.1.5 Polarized Target of the EMC Experiment (CERN)

The EMC (European Muon Collaboration) experiment on deep inelastic scattering was performed on a muon beam with energy of about 200 GeV. The measurement of the nucleon structure function $g_1 (x, Q^2)$ requires longitudinally polarized beams and targets. A muon beam naturally has longitudinal polarization. Its large transverse sizes (about 5 cm), low intensity ($\sim 3 \cdot 10^7$ muons/cycle), and small cross section for its interaction with nucleons required the development of the largest longitudinally polarized target used in experiments (Gabathuler 1984). Moreover, experimentally measuring so-called raw asymmetry, asymmetry in deep inelastic scattering is expected to be low according to the formula

$$A_m = P_\mu P_p F D A. \tag{6.13}$$

Here, P_μ and P_p are the beam and target polarizations, respectively; D is the factor of polarization transfer from the muon to a virtual photon; and F is the target dilution factor; i.e. the ratio of the number of free protons in the target to the number of all nucleons. For comparison, $F = 0.13$ and 0.17 for 1-pentanol and ammonia, respectively. Experimenters have chosen ammonia. The parameter A is physical asymmetry from which the function g_1 (x, Q^2) is determined. The quantity A_m is estimated as about 1 %. The measurement of such a small asymmetry is very sensitive to changes in beam parameters in time and to equipment efficiency. Such systematic errors in experiments with polarized hadron and electron beams are strongly suppressed by frequent reversal of the beam polarization. This is impossible in the case of the muon beam. Owing to the large sizes of the target, target polarization cannot also be quickly reversed. In order to reduce false asymmetry in this case, the target was divided into two identical parts with opposite polarizations. In this case, the measurement is performed simultaneously with two opposite signs of polarization. Nevertheless, it is impossible to completely eliminate the difference between geometrical acceptances of instruments to these two parts of the target. As a result, the necessity of the reversal of the target polarization remained, and this reversal was performed, but not frequently.

The superconducting solenoid of the EMC target had a basic coil 1600 mm in length and a free gap at the center with a diameter of 190 mm. At a current of 180 A, it produces a field of 2.5 T (see Fig. 6.9).

Twelve correcting coils each 132 mm in length are winded over the basic coil. The correcting coils are distributed uniformly along the basic coil. The number of turns and current for each correcting coil are selected separately. The aim was to increase the edge fields of the basic coil and, thus, to expand the region of the uniform field. The magnet was cooled by liquid helium at a temperature of 4.2 K. The flash cold gas cooled radiating screens. Average heat incoming to the solenoid at its complete loading was about 6 W. The supply of liquid helium to the magnet was provided automatically from a 2000-l Dewar vessel placed near the target refrigerator. All gas was gathered and used again by the transfer system. The cryostat of the target was entirely welded with the use of aluminum and titanium alloys.

The basic coil and each of the correcting coils had individual supply and provided the uniformity of the magnetic field better than $\pm 5 \cdot 10^{-5}$ during 24 h. Time of activating the magnet to the working field was about 10 min.

The target material was cooled by the ^{3}He–^{4}He dissolution cryostat, which was united with the solenoid, but had an independent vacuum jacket. With the existing materials for the target, the dynamic polarization process requires a power of at

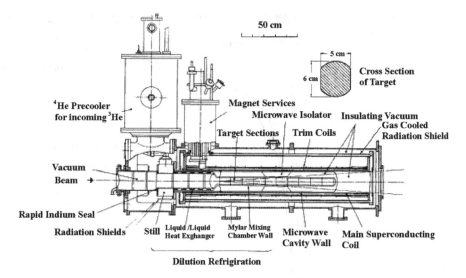

Fig. 6.9 Polarized target of the EMC experiment

least 1 mW per 1 g of the substance. For the entire volume of the EMC target, this corresponds to a necessary power of 1.1 W. The expected power of the designed cryostat at 0.5 K was 2.5 W. The test of the produced cryostat power gave a power of 2.0 W at a temperature of 0.5 K. A difference of 0.5 W between the design and actual cold productivities is entirely explained by the fact that the power of pumps is insufficient to ensure the necessary speed of ^{3}He circulation in the circuit. However, the reached power is sufficient to cool the target to the necessary temperature.

The usual technique of target loading was inapplicable because of the high weight of the target, and the original approach was implemented. In this method, direct access is ensured to the mixing bath for loading of a material in the horizontal plane. The target material stored in a special container at liquid nitrogen temperature (77 K) is overloaded to the mixing bath. Special indium vacuum seals are quickly closed, isolating the cell with the substance. The container remains near the cell, providing a vacuum volume for evacuation, and serves as a beam guide. The container has foils on two sides and minimum heat-insulating foils on the beam path.

Two parts of the target are placed in separate cavities. Each cavity has a diameter of 159 mm and a length of 400 mm. Microwave power is introduced separately into each cavity through WG22 square-section waveguides (8 mm). A VKE2401-N3 extended interaction oscillator (Varian) was used as a microwave generator. Two such generators provided the independent polarization of two parts of the target. Despite the careful screening of halves of the target, small mutual induction was revealed. However, induced polarization was measured and necessary corrections were introduced in order to determine the correct polarizations of the halves of the target.

For the measurement of target polarization, four coils for detecting NMR signals were mounted on each of two halves of the target. Signal processing, calibration, and

Fig. 6.10 Polarization generation on the EMC ammonia target

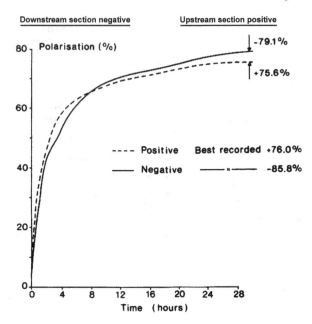

inclusion of corrections were performed by the standard method. The measurement results are presented in Fig. 6.10.

It is seen that a polarization of 70 % was generated in approximately 12 h. Polarizations +76 % and −79 % were reached by 24-h generation.

With the use of this target, the well-known, currently unexplained spin-crisis phenomenon was observed. In order to find a key to the explanation of this phenomenon, the EMC program has been consistently continued on the same muon beam by two groups, SMC and COMPASS. These groups have confirmed with a better accuracy the existence of this phenomenon, though cardinal solutions of the problem have not been given. However, they have improved the polarized target, which will be briefly described below.

6.1.6 Polarized Target of the SMC Experiment (CERN)

In this section, we will follow Kynäräinen (1994), and Adams et al. (1999). The target is also consisted of two halves. As the target material, 1-butanol containing 5 wt % of water and 4 wt % of EHBA-Cr (V) as a paramagnetic additive was used. This material has been chosen because it is free of the background from polarized nuclei, which is characteristic of ammonia where nitrogen nuclei are polarized. Butanol has a higher hydrogen content than propanediol. The method for preparing butanol balls is the same as in the case of propanediol (see Sect. 6.1.2 "Polarized target of the PROZA setup").

Each half of the target was placed in a cylindrical cell 50 mm in diameter and 650 mm in length (the initial length was 600 mm). The 100-GeV muon beam had the transverse sizes 16×15 mm. Therefore, the beam entirely kept within the target. For the separation of events from two halves of the target, the gap between them should be 200 mm (originally 300 mm). The total sizes of the target were $2 \times 12 \times 80$ cm and the packing factor was 0.65.

The loading of the target material was made using the same method as in the case of the EMC experiment. In the coldest part of the dissolution chamber, the refrigerator gave a temperature of 30 mK. The cooling power of the refrigerator was 1 mW at 50 mK, 15 mW at 100 mK, 400 mW at 300 mK, and 1.3 W at 0.5 K.

The magnetic system of the target consists of three independent magnets. The solenoid produces a basic field of 2.5 T with uniformity better than $6 \cdot 10^{-5}$ over the entire target volume. This field is obviously used to build-up polarization in the target. The following dipole magnet produces a field of 0.5 T with a uniformity of 10 %. The third magnet consisting of 16 coils is a solenoid for correcting the longitudinal field of the basic solenoid. The inner diameter of the basic solenoid over which the other magnets are winded is 265 mm and the free opening angle is $7.5°$.

A new feature of the magnetic system is the possibility of reversing polarization in 0.5 h by using the combination of the magnetic fields of the solenoid and dipole. For this purpose, the solenoid field is decreased to 0.5 T and, then, the dipole is switched on. Now, it is necessary to hold the total field at a level of 0.5 T, synchronously increasing the field of the dipole and reducing the solenoid field. When the solenoid field passes through zero, the dipole field should be 0.5 T and/or above. Then, the solenoid field is increased having the opposite sign, and the dipole field can be decreased. When the solenoid field reaches 0.5 T, the dipole field can already be switched off. The basic requirement is that the total field should always be above or equal to 0.5 T. The most dangerous situation is if this field becomes zero. In this case, polarization also vanishes. This procedure is controlled by a computer. This method for reversing polarization is very profitable. Polarization was usually reversed once a week, and this process took a day. Now, this procedure can be made once in 5 h, it takes only 0.5 h, and polarization is preserved.

Figure 6.11 illustrates polarization reached in the SMC target. The comparison shows that this result close to the results obtained with the EMC target (Fig. 6.10).

6.1.7   Polarized Target of the COMPASS Setup (CERN)

The main aim of the COMPASS experiment is to determine the gluon contribution to the nucleon spin. For this purpose, the physical program proposes investigation of two processes. The first process includes the production of open-charm mesons, and the second process includes the production of hadron pairs with a high transverse momentum. Both processes are studied on a longitudinally polarized muon beam with a momentum of 160 GeV/c. Deuterated lithium with longitudinal polarization is used as a target. The COMPASS setup is based on the SMC equipment

Fig. 6.11 Polarization
generation in the SMC
butanol target

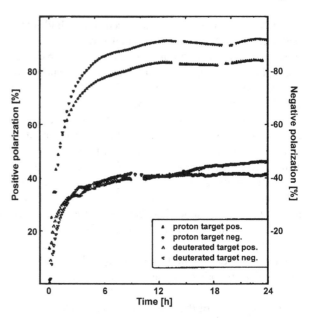

with two differences. First, the new superconducting solenoid with a larger aperture
is under development to increase the useful solid angle of the instruments from 70 to
180 mrad. Second, the COMPASS collaboration has great achievements in the development of ^{6}LiD materials for the polarized target (Gauthtron et al. 2004). These
achievements are presented below.

Two-spin asymmetry in the production of particles in the deep inelastic scattering
of polarized leptons by the polarized target is defined by the expression

$$A_m = \frac{N(+) - N(-)}{N(+) + N(-)} = P_B P_T F A_t. \tag{6.14}$$

Here, $N(+)$ and $N(-)$ are the normalized count rates of detectors for the parallel
and antiparallel orientations of the polarizations of initial colliding particles, respectively; P_B and P_T are the beam and target polarizations, respectively; A_m and A_t are
the measured (raw) and true asymmetries, respectively; and F is the dilution factor
defined as the ratio of free (polarizable) nucleons to the total number of nucleons
in the target. One of the important criteria in selecting materials for the target is the
so-called factor of merit M defined by the expression

$$M = \rho \kappa (F P_T)^2. \tag{6.15}$$

Here, ρ is the target density and κ is the packing factor of the target with the
working substance. For a given asymmetry measurement accuracy, measurement
time is minimal at the maximum value of the M factor. The SMC and COMPASS
targets are compared in Table 6.4, which illustrates the achievements of the COM-
PASS collaboration. These achievements are the highest dilution and the quality
factors currently received in ^{6}LiD. Polarization reversal by means of the combination of the solenoid and dipole is used in the COMPASS experiment, as well as in

Table 6.4 Comparison of some parameters of the SMC and COMPASS targets

Parameter	SMC	SMC	COMPASS
Target material	NH_3	D-butanol	^6LiD
Density, g/cm^3, ρ	0.85	1.10	0.84
Polarization, %	H: 90	D: 50	D: 50
Packing factor, κ	0.60	0.60	0.55
Dilution factor, F	0.176	0.238	0.50
Quality factor, M	10.3	6.7	16.0

the SMC experiment. In the COMPASS experiment, the computer-controlled reversal of polarization is performed once in eight hours and takes only 33 min. As a result, appreciable economy of the beam usage time is achieved.

The above review of solid targets shows that frozen targets are profitable for using with low-intensity beams and when the large solid angle is required. Targets with continuous generation and with radiation-resistant materials are used with high-intensity beams. Organic materials (butanol, propanediol, 1-pentanol, ethanol, etc.) have an order of magnitude lower resistance to radiation damage than ammonia NH_3 or ^7LiH in which the paramagnetic centers are created through irradiation by high doses of intense beams of electrons, protons, or other charged particles. There are the following unsolved problems:

- Production of pure polarized proton targets,
- Fast reversal of the target polarization,
- Fast turn of polarization in the necessary direction (longitudinal or transverse to the beam).

6.2 Polarized Gas Targets

The solid targets considered in the preceding section have a number of significant disadvantages. Some of them are as follows. The first is a large inverse dilution factor; i.e., a target material includes many unpolarizable nucleons. The contribution of these nucleons is sometimes an order of magnitude larger than the contribution from polarizable nucleons. For this reason, the effective target polarization appears to be an order of magnitude lower than the generated polarization of free protons. As a result, it is difficult to study polarization effects in inclusive reactions in the absence of strong kinematic constraints, which exist for exclusive processes. Second, the generation of polarization and, correspondingly, its reversal takes a lot of time, several hours. For this reason, it is impossible to frequently reverse polarization to substantially suppress the effects of false asymmetry. These two main difficulties are absent in polarized gas targets. However, the density of the polarized nucleons in gas targets is much lower than that in polarized solid targets. In particular, this density in polarized solid targets is about $5 \cdot 10^{22}$ protons/cm^3, whereas the highest density of polarized protons in the RHIC jet target is only $1.3 \cdot 10^{12}$ protons/cm^3 (Wise 2004). The next qualitative jump in developing pure polarized proton targets was the development of polarized storage-cell gas targets. The highest density reached in the HERMES experiment is $1 \cdot 10^{14}$ protons/cm^3 (Ackerstaff et al. 1998). Nevertheless, the difference between the densities in polarized solid targets and polarized gas targets still remains large. Therefore, the most reasonable solution is to use polarized gas targets in colliders. In this case, the effective thickness of the target increases due to two factors. The first factor is the rotation frequency of the internal accelerator beam, which is $7.8 \cdot 10^4$ and $2 \cdot 10^5$ Hz in the RHIC and U-70 accelerators, respectively. The second positive factor for a collider is the beam lifetime. The beam lifetime in RHIC is about 10 h, which gives an additional factor of $3.6 \cdot 10^4$. As a result, the gain factor for RHIC is $2.5 \cdot 10^9$. This is a factor by which the effective thickness of the gas target increases. In particular, the effective thickness of a jet target is $\sim 10^{17}$ polarized protons/cm^3, and the effective density of polarized storage-cell gas targets is 10^{19} polarized protons/cm^3. This density is already comparable with the density of polarized solid targets. Thus, such targets are very attractive owing to, first, almost the same density as that of polarized solid targets, second, unit dilution factor (pure proton target), and, third, fast polarization reversal and the almost point size of the target (in the case of a gas jet target).

The basic characteristics of polarized gas targets for two largest facilities, RHIC and HERMES, are described below.

6.2.1 RHIC Polarized Jet Target (BNL)

The choice of a polarimeter for measuring the RHIC beam polarization was discussed in detail for a long time. The special working group was organized; after the detailed analysis of the problem, it recommended to choose the polarimeter on the basis of elastic proton–proton scattering in the Coulomb–nuclear interference region. In addition to other advantages, this polarimeter is a self-calibrated device; i.e., its analyzing power is equal to the polarization arising in the same process at collision of unpolarized protons (details see in Sect. 8.1 "Polarimetry"). This polarimeter was implemented for the first time in the E704 experiment at Fermilab in 1990 (Akchurin et al. 1993). In January 2000, the workshop on polarimetry at Brookhaven National Laboratory (BNL) approved the recommendation of the working group. The variant of the creation of the jet polarized hydrogen target was also recognized as optimal. In June 2000, design works began. On October 8, 2003, the polarized atomic hydrogen source was mounted at the testbench at BNL. The polarized atomic hydrogen source was placed at the standard position in the RHIC ring in spring 2004. The first physical results on the measurement of the polarization of a 100-GeV proton beam were presented on the 16th International Spin Physics Symposium in Trieste in October 2004 (Bravar 2004).

The general view of the RHIC polarized jet target is given in Fig. 6.12 (Wise 2004). All sources of the polarized nuclear beams are based on the same principle

RF dissociator with adjustable frequency

Preliminary cooling of the gas
Magnets with six conic poles

HF generators with highly uniform
magnetic field

Target's magnet consisting of two Helmholtz coils

On-line Breit–Rabi polarimeter

Detector for a calibration ion beam

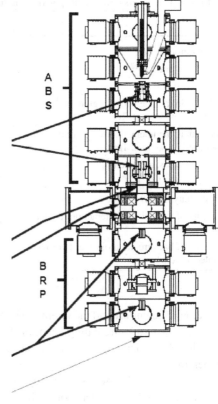

Fig. 6.12 RHIC polarized jet target

and have the same configuration. Variations appear in applications to individual accelerators. The given polarized atomic hydrogen source is situated in the vertical
position. Its sizes above and below the beam level are 230 and 120 cm, respectively. The diameter of the largest chamber in the projection on the orbit plane is
80 cm. This chamber (at the beam level) contains the detectors for recoil protons.
The atomic beam source (designated as ABS) is in the top part of Fig. 6.12. It begins
with a tank for molecular hydrogen, which enters a dissociator, where molecules are
dissociated into hydrogen atoms by a 250-W RF discharge. The dissociator tube is
cooled by water. The tube ends with an aluminum nozzle cooled to 30–100 K. In the
latest version, an adapter 12 cm in length cooled to 70–140 K has been inserted between the tube and nozzle. All these tools reduce the recombination of atoms on the
walls, reduce the velocity of atomic hydrogen, and improve the temperature stability
of hydrogen.

Hydrogen atoms are separated by the Stern–Gerlach method using sextupole
magnets in electron spin states; i.e., states with spin $+1/2$ (hyperfine states $|1\rangle$ and
$|2\rangle$) and $-1/2$ ($|3\rangle$ and $|4\rangle$) are spatially separated. These sextupole magnets also fo-

Table 6.5 Parameters of the RHIC polarized atomic hydrogen beam

Parameter	Value	Comment
Polarization, %	92.4 ± 2	
Density p/cm^2, 10^{12}	1.3 ± 0.2	
Beam size, mm	5.8	FWHM
Vacuum in the interaction region (with the switched-off source), Torr	$4 \cdot 10^{-9}$	
Vacuum in the interaction region (with the switched-on source), Torr	$1.4 \cdot 10^{-8}$	
Flow of atoms, atoms/cm^2	$1.24 \cdot 10^{17}$	
Magnetic field B, T	0.12	Vertical field
Uniformity $\Delta B/B$	$5 \cdot 10^{-3}$	
Density of atoms in the interaction region, atoms/cm^3	$1.0 \cdot 10^{12}$	
Density of background molecular hydrogen, molecules/cm^3	$1.5 \cdot 10^{10}$	These are molecules from the dissociator

cus atomic beams around the interaction point (the point of the interaction of atomic hydrogen with the circulating beam of the accelerator). Then, nuclear polarization is induced by changing the population of the hyperfine levels by means of the RF generator.

The polarized beam passes through the interaction region, and one third of the beam is focused on the detector of the Breit–Rabi polarimeter (designated as BRP in Fig. 6.12).

The polarized atomic hydrogen source is divided into sections for evacuation: six sections are provided for the atomic beam source, three sections, for the Breit–Rabi polarimeter, and one section, for the target (the place of the intersection of the accelerator beam and jet). Each section is equipped with two pumps, except for the first section where three pumps are used. As a result, vacuum in the interaction region reaches $4 \cdot 10^{-9}$ and $1.4 \cdot 10^{-8}$ Torr in the absence and presence of the jet, respectively. These values and some parameters of the polarized jet target are presented in Table 6.5.

The importance of high vacuum in the interaction region is seen on the following example. Vacuum in RHIC is $1.4 \cdot 10^{-8}$ Torr $\sim 2 \cdot 10^{-11}$ atm. Taking into account the Loschmidt number $2.7 \cdot 10^{19}$ molecules/cm^3, we find that the RHIC ring contains $5.4 \cdot 10^8$ molecules/cm^3. These are primarily N$_2$ nitrogen molecules containing 28 nucleons. Hence, the density of background nucleons from the residual gas is $1.5 \cdot 10^{10}$ nucleons/cm^3. The density of polarized protons in the jet is $1.0 \cdot 10^{12}$ protons/cm^3. Therefore, the expected background is about 1 %. If vacuum in the ring appears to be 10^{-6} Torr, i.e., two orders of magnitude lower than in RHIC, the expected background-to-signal ratio is 1:1. Under such conditions, the direct use of the RHIC source is difficult and maybe even impossible.

The next important element of the polarized atomic hydrogen source is the magnetic field. To preserve nuclear polarization, atomic hydrogen should meet an adiabatically varying magnetic field on its path; otherwise, it can be depolarized. The adiabaticity condition quantitatively means that the direction of the applied magnetic field in the rest frame of atomic hydrogen should vary with a frequency of less than $10^{-3}\omega_0$, where ω_0 is the Larmor frequency ($\gamma_0 B$). This adiabaticity condition was violated by the magnetic field of two coaxial coils placed near the interaction point. Their functions are, first, to preserve the polarization of atomic hydrogen and, second, to eliminate the field on the path of recoil protons with low (<10 MeV) energy. However, these coils eliminated the magnetic field at two places on the path of atomic hydrogen. To correct fields at these places, iron magnetic screening was used. Finally, the residual depolarization was reduced to <0.4 %. It is necessary to notice that the source in RHIC is on a rectilinear section far enough from ring magnets, and these fields were disregarded.

The polarization of atomic hydrogen was measured by the Breit–Rabi polarimeter in one minute. For this purpose, strong- and weak-field high frequency transitions were tuned separately. These systems were placed both upstream and downstream of the interaction point. The equipment for detecting atomic hydrogen had the large amplification of signals at low noise and good grounding. It operated with one third of the atomic beam passed through the interaction region. Let us denote the ratio of the numbers of transitions to states $|2\rangle$ and $|1\rangle$ as $\kappa = N_2/N_1$ and the efficiencies of weak- and strong-field transitions in the sextupole magnets of the Breit–Rabi polarimeter upstream of the interaction point as $(1-\varepsilon_{1-3})$ and $(1-\varepsilon_{2-4})$, respectively. Then, the polarizations of atomic hydrogen beams are given by the expressions

$$P^+ = \frac{1 + \kappa \cos\theta - 2\kappa\varepsilon_{2-4}\cos\theta}{1+\kappa},$$

$$(6.16)$$

$$P^- = \frac{-1 - \kappa \cos\theta + 2\kappa\varepsilon_{2-4}\cos\theta}{1+\kappa}.$$

Here, $\tan\theta = B_c/B$, where B_c is the critical field ($B_c = 507$ G for the hydrogen atom) and B is the polarization-holding magnetic field (Helmholtz coils in Fig. 6.12). The values measured in 2004 are $P^+ = (95.7 \pm 0.1)$ % and $P^- = -(95.7 \pm 0.1)$ %.

The usual determination of the efficiencies of high frequency transitions involves measurements with all possible high frequency transitions. Then, the populations of the corresponding levels are obtained by solving the resulting equations. Another approach was used for this polarimeter. One of two spin states is usually transmitted to the interaction region. The sextupole magnets of the polarimeter can bend both nuclear spin states from the detector. If both states are transmitted to the interaction region and the magnets are switched on, only atoms that are not deviated enter the detector. Their occurrence frequency was 52 Hz. If the high frequency transitions are switched off, unpolarized atoms enter the detector with a frequency of 20900 Hz. Hence, the efficiency of high frequency transitions is 99.7 %. The measured average target polarization was (92 ± 2) % (Nass et al. 2004). The basic correction to polarization came from molecular hydrogen that was passed through the interaction region and amounted 1.5 wt % of atomic hydrogen. Table 6.5 presents polarization including this correction (Zelenski et al. 2004).

6.2.2 Polarized Storage-Cell Gas Target of the HERMES Experiment (DESY)

In the preceding section, polarized jet targets used as internal targets in accelerators and colliders have been considered. We pointed to a number of advantages of polarized jet targets over polarized solid targets. However, some important physical problems require denser polarized gas targets. Two examples of such problems are as follows. The measurements of the proton and neutron spin structure functions at the HERMES experiment at the HERA collider (Coulter et al. 1990) required polarized hydrogen, deuterium, and helium-3 gas targets with densities $\geq 10^{14}$ particles/cm^2, whereas jet targets have densities below 10^{12} particles/cm^2. The second problem is the development of an antiproton polarizer or polarizing filters (Dobeling et al. 1985), where targets with high densities are required to reach the necessary luminosity in an acceptable time.

The design principle of a polarized storage-cell gas target differs only slightly from the design principle of the jet polarized target described above in application to RHIC. The basic constructive difference is that a storage cell, which is a tube about 40 cm in length where atomic hydrogen is stored in time, is placed in the region of the intersection of the polarized atomic beam with the circulating beam (interaction region, IR). The main engineering problem is to avoid the loss of the target polarization because an increase in the number of collisions of hydrogen atoms with the walls of the storage tube. Another problem arises because the density of atoms is two orders of magnitude higher than that of the jet target. For this reason, it is necessary to carry out good evacuation of the IR in order to avoid the deterioration of vacuum in the accelerator (Nass et al. 2003). The scheme of the polarized target (with the storage cell) currently used in the HERMES experiment is presented in Fig. 6.13.

Unpolarized atomic hydrogen leaves the dissociator nozzle and, moving from left to right, enters the sextupole magnets (Baumgarten et al. 2002a). The first magnet separated atoms in electron spin, and the second magnet focuses the chosen nuclear-polarized beam to the interaction region. The transfer of electron polarization in nuclear polarization is ensured by RF generators in combination with the magnets. Nuclear-spin-polarized atomic hydrogen passes through a tube and enters the storage cell. It is a thin-walled aluminum tube with the 21×8.9-mm elliptic section (see insert in Fig. 6.13) and a length of 40 cm. The inner surface of the tube is covered with a special composite (Drifilm) to reduce the effect of collisions of atoms with the wall and, correspondingly, the depolarization effect (ΔP_{WD}, the subscript WD means wall depolarization). The other sources of depolarization are spin-exchange interactions between atoms (ΔP_{SE}) and the interaction of the magnetic moment of an atom (electron) with the induced field of the beam (ΔP_{BI}). The

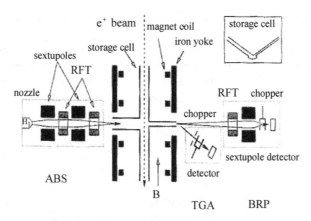

Fig. 6.13 Polarized storage-cell gas target of the HERMES experiment. The following units are arranged from left to right in the direction of the motion of hydrogen atoms: the atomic beam source (denoted as ABS) consisting of the nozzle, sextupoles with conic poles, and RF generators; the T-shaped storage cell; the polarized electron beam with an energy of about 27 GeV enters from above; the coils of the magnet producing the polarization holding magnetic field are shown; the magnetic screen around the coils is given; the mass spectrometer (TGA), Breit–Rabi polarimeter (BRP), and beam chopper are presented

cell is in a longitudinal or transverse magnetic field. In 1997–2000, a longitudinal magnetic field of 3.3 kG produced by the superconducting solenoid was applied. Since 2001, the transverse holding field produced by a standard warm dipole was used. An electron beam with an energy of about 27 GeV and a polarization of about 60 %, which was described in detail in Sect. 5.2 "Polarized electron beams," enters the storage cell from above (see Fig. 6.13). The products of the interaction of two beams, electron and hydrogen, are detected by the detectors located along the beam. The primary goal of the HERMES experiment is the study of the nucleon spin structure functions. For this purpose, in addition to deep inelastically scattered electrons, other charged particles, for example, a pair of charged hadrons is simultaneously detected. Such reactions are called semi-inclusive. All detected particles are identified in kind, energy, and emission angle.

A small-diameter tube is attached to the storage cell from the right (see insert in Fig. 6.13). This tube allows one to take a portion of the polarized atomic beam for the analysis of its mass composition by means of the target gas analyzer (TAG in Fig. 6.13) (Baumgarten et al. 2003). The same portion of the beam is another time taken to measure its polarization using the Breit–Rabi polarimeter (BRP in Fig. 6.13) (Baumgarten et al. 2002b).

There are some parameters relating the average target polarization P_T to the atomic hydrogen polarization P_a (Lenisa 2004). These parameters are the fraction of atomic hydrogen injected into the cell, α_a; the fraction of atoms surviving collisions with the storage-cell walls, α_r; and the ratio of the polarization of hydrogen molecules to the polarization of the atomic beam, β, which was recently determined as $\beta = 0.64 \pm 0.19$ (Airapetian et al. 2004) from the data of the 1997 running period

Table 6.6 Parameters of the polarized storage-cell gas targets of the HERMES experiment used in the 1997, 2000, and 2002 running periods

Parameters	$H_{\parallel}$ (1997)	$H_{\perp}$ (2002)	$D_{\parallel}$ (2000)
α_a	0.960 ± 0.010	0.918 ± 0.032	0.919 ± 0.026
α_r	0.945 ± 0.035	0.979 ± 0.023	0.997 ± 0.017
P_{z+}	0.908 ± 0.016	0.859 ± 0.032	0.927 ± 0.017
$-P_{z-}$	0.908 ± 0.016	0.859 ± 0.032	0.915 ± 0.010
$-\Delta P_{SE}$	0.035	0.055	≤ 0.001
$-\Delta P_{WD}$	0.02	0.055	No
$-\Delta P_{BI}$	No	0.015	No
P_+	0.851 ± 0.031	0.783 ± 0.041	0.851 ± 0.029
$-P_-$	0.851 ± 0.031	0.783 ± 0.041	0.840 ± 0.026
T (10^{14} nucleons/cm^2)	0.7	1.1	2.1
FOM ($P^2 t$), 10^{14} nucl./cm^2	0.5	0.67	1.5

with the longitudinally polarized target has been determined. The corresponding relation has the form

$$P_T = \alpha_a \big[\alpha_r + (1 - \alpha_r \beta) \big] P_a. \qquad (6.17)$$

Table 6.6 presents the parameters of targets in various running periods of the HERMES experiment (FOM is the factor of merit).

The comparison of longitudinally polarized hydrogen and deuterium targets leads to the following conclusions. In the same holding magnetic fields, spin-exchange processes and processes of collisions with the storage-cell walls for deuterium are suppressed as compared to hydrogen by the ratio $(B_c^H / B_c^D) \approx 20$. The second conclusion is that the positive and negative polarizations P_z almost coincide for hydrogen and are significantly different for deuterium. This difference is due to the fact that deuterium has a larger number of spin degrees of freedom and, correspondingly, the number of parameters is larger in this case.

Table 6.6 also shows that the optimum working condition of a target was observed with the deuterium target in 2000, when conditions ΔP_{WD} (wall-collision depolarization) $= \Delta P_{SE}$ (spin-exchange depolarization) $= \Delta P_{BI}$ (beam-field-induced depolarization) $= 0$ were satisfied. This was explained by two reasons. The first reason is the low critical magnetic field $B_c^D = 117$ G, in contrast to the proton case, where $B_c^H = 507$ G. The second reason is the better covering of the storage-cell surface. This procedure also improved the work of the target with $H_{\perp}$ in the 2002 running period in comparison with the 1997 running period with $H_{\parallel}$, as seen from the comparison of the depolarization effects in these two running periods presented in Table 6.6.

At the same time, a problem was outlined. According to Table 6.6, the target used in 2002 is inferior to the target used in 1997 in both the quality factor and polarization. The cause of this problem is that the target density in 2002 was higher and, correspondingly, depolarization effects increase due to a larger number of collisions with walls and to spin-exchange processes. The solution of the problem seems to increase the holding magnetic field proportionally to the increase in the target density.

6.3 Conclusions

The above presentation shows that enormous jump in the development of polarized gas targets has been made in the past decade. In combination with gas targets, any modern collider can provide two types of experiments in parallel: collider experiments and fixed-target experiments. This feature expands the energy range accessible to experiments at colliders. Another important result of these developments is that almost all one- and two-spin fixed-target inclusive experiments become possible due to purity (in the content of polarized protons). These two main achievements in the development of polarized gas targets (not speaking about other achievements) are remarkable achievements in polarization physics.

References

Ackerstaff, K., et al.: Nucl. Instrum. Methods Phys. Res. A **417**, 230 (1998)

Adams, D., et al.: Nucl. Instrum. Methods Phys. Res. A **437**, 23 (1999)

Airapetian, A., et al.: Eur. Phys. J. D **29**, 21 (2004)

Akchurin, N., et al.: Phys. Rev. D **48**, 3026 (1993)

Aseev, A.A.: Preprint No. 89-57, IHEP, Serpukhov (1989)

Atsarkin, V.A.: Dynamic Nuclear Polarization in Solid Dielectrics. Nauka, Moscow (1980)

Autones, P., et al.: Nucl. Instrum. Methods **103**, 211 (1972)

Baumgarten, C., et al.: Nucl. Instrum. Methods Phys. Res. A **496**, 263 (2002a)

Baumgarten, C., et al.: Nucl. Instrum. Methods Phys. Res. A **482**, 606 (2002b)

Baumgarten, C., et al.: Nucl. Instrum. Methods Phys. Res. A **508**, 268 (2003)

Bilenky, S.M., Lapidus, L.I., Ryndin, R.M.: Usp. Fiz. Nauk **84**, 243 (1964)

Boden, B., et al.: Z. Phys. C, Part. Fields **40**, 175 (1991)

Borisov, N.S.: In: Proceedings of the 7th International Symposium on High Energy Spin Physics, Protvino, USSR, p. 236 (1986)

Borisov, N.S., et al.: Preprint No. I-80-98, JINR, Dubna (1980)

Bravar, A.: In: Proceedings of 16th International Spin Physics Symposium, Trieste, Italy, p. 700 (2004)

Bruneton, K., et al.: Prib. Tekh. Eksp. **5**, 46 (1976)

Burkhin, M.M., et al.: Prib. Tekh. Eksp. **1**, 30 (1981)

Carver, T.R., Slichter, C.P.: Phys. Rev. **92**, 212 (1953a)

Carver, T.R., Slichter, C.P.: Probl. Sovrem. Fiz. **6**, 182 (1953b)

Chaumette, P., et al.: In: Proceedings of the 8th International Symposium on High Energy Spin Physics, Minneapolis, USA, 1988. AIP Conference Proc., vol. 187, p. 1331. AIP, New York (1989)

Chaumette, P., et al.: In: Advances in Cryogenic Engineering, vol. 35. Plenum, New York (1990)

Chaumette, P., et al.: In: Proceedings of the 9th International Symposium on High Energy Spin Physics, Bonne, 1990, vol. 2, p. 237. Springer, Berlin (1991)

Coulter, K., et al. (HERMES Collaboration): Proposal DESY PRC-90/01 (1990)

Crabb, D.G.: In: Proceedings of the 9th International Symposium on High Energy Spin Physics, Bonne, 1990, vol. 2, p. 289. Springer, Berlin (1991)

Crabb, D.G.: In: Proceedings of the 6th Workshop on Spin Phenomena in High Energy Physics, Protvino, Russia, p. 152 (1995)

Crabb, D.G., Day, D.B.B.: Nucl. Instrum. Methods Phys. Res. A **356**, 9 (1995)

Daikovskii, A.G., Portugalov, Yu.I.: Preprint No. OMVT 78-68B, IHEP, Serpukhov (1978)

Dobeling, H., et al.: Proposal cernpssc/85/80 (1985)

Gabathuler, E.: In: Proceedings of the 6th International Symposium on High Energy Spin Physics, Marseille, France, vol. C2, p. 141 (1984)

Gauthtron, F., et al.: In: Proceedings of the 16th International Spin Physics Symposium, Trieste, Italy, p. 791 (2004)

Grachev, O.A., et al.: Prib. Tekh. Eksp. **3**, 189 (1993)

Grosnick, D.P., et al.: Phys. Rev. D **55**, 1159 (1997)

Hall, H.E., et al.: Cryogenics **6**, 8 (1966)

Hill, D.: Argonne National Laboratory Report No. ANL-HEP-TR-92-68 (1992)

Jeffries, C.D.: Dynamical Nuclear Orientation. Interscience, New York (1963) (Mir, Moscow, 1965)

Kageya, T.: In: Proceedings of the 16th International Spin Physics Symposium, Trieste, Italy, p. 812 (2004)

Kynäräinen, J.: Nucl. Instrum. Methods Phys. Res. A **356**, 47 (1994)

Lenisa, P.: In: Proceedings of the 16th International Spin Physics Symposium, Trieste, Italy, p. 808 (2004)

Meyer, H.O.: Phys. Rev. E **50**, 1485 (1994)

Nass, A., et al.: Nucl. Instrum. Methods Phys. Res. A **505**, 633 (2003)

Nass, A., et al.: In: Proceedings of the 16th International Spin Physics Symposium, Trieste, Italy, p. 776 (2004)

Neganov, B.S.: Zh. Eksp. Teor. Fiz. **50**, 1445 (1966)

Overhauser, A.: Phys. Rev. **89**, 689 (1953a)

Overhauser, A.: Phys. Rev. **91**, 476 (1953b)

Overhauser, A.: Phys. Rev. **92**, 411 (1953c)

Paul, W.: In: Proceedings of the International Symposium on the Polarization Phenomena in Nuclei, Basel, Switzerland (1960)

Raoul, J.C., et al.: Nucl. Instrum. Methods **125**, 585 (1975)

Rathmann, F., et al.: Phys. Rev. Lett. **94**, 014801-1/4 (2006)

Wise, T.: In: Proceedings of the 16th International Spin Physics Symposium, Trieste, Italy, p. 757 (2004)

Zelenski, A., et al.: In: Proceedings of the 16th International Spin Physics Symposium, Trieste, Italy, p. 761 (2004)

Chapter 7
Polarized Hydrogen Ion Sources for Accelerators/Colliders

In this chapter, we consider two types of commonly accepted polarized hydrogen ion sources. The first source is called polarized atomic beam source (PABS). In this source, atoms polarized in electron spin using the Stern–Gerlach method with the subsequent microwave pumping of nuclear polarization are first obtained. The other source is called optically pumped polarized ion source (OPPIS). This method involves the capture of polarized electrons from optically polarized nuclei by protons and the subsequent application of spin- and charge-exchange processes to obtain polarized protons or polarized negative hydrogen ions.

A substantial stage in the development of these sources was initiated by two proposals by Zavoiskii (1957a, 1957b). The first proposal based on the transfer of the polarization of an electron at its capture by a proton in a charge-exchange target has led to the development of modern OPPISs. The second proposal concerning the use of the Lamb shift has led to the development of the first PABSs of the polarized H^- and D^- ions for most tandem accelerators. First, they were also used as sources at meson factories, but were later replaced by OPPISs (Zelenskii 2003).

Both PABS and OPPIS have three identical devices for obtaining a polarized beam. The first device provides electron-spin-polarized atoms. The second device is necessary for the transfer of the electron polarization to a nucleus, a proton in this case. Finally, the third device provides either the ionization of nuclear-spin-polarized atoms for the production of a polarized proton beam or charge exchange for the production of a polarized negative hydrogen ion beam. Then, polarized protons or ions are injected into an accelerator. The final result on the production of the polarized beam obviously depends on the effective operation of each of three devices.

A significant difference between two sources is that a PABS deals with thermal atoms (velocities of $\sim 10^5$ cm/s), while an OPPIS provides quite fast beams (energies 3–8 keV and velocities $\sim 10^8$ cm/s). Therefore, the final ion beams can differ both in intensity and in polarization.

Below, we describe the PABS and OPPIS one after another. These two sources provide the best parameters of polarized beams and are competing.

S.B. Nurushev et al., *Introduction to Polarization Physics*,
Lecture Notes in Physics 859, DOI 10.1007/978-3-642-32163-4_7,
© Moskovski Inzhenerno-Fisitscheski Institute, Moscow, Russia 2013

7.1 Polarized Atomic Beam Sources

At present, either polarized protons or polarized negative hydrogen ions H^- are used for injection into accelerators. Correspondingly, sources are equipped with additional units, such as charge exchange units. Below, we will briefly discuss the parameters of the following latest sources:

(i) the polarized ion source at the Institute for Nuclear Research, Russian Academy of Sciences (INR RAS) (Belov et al. 1995).
(ii) the IUCF–CIPIOS source (Derenchuk 2002).

7.1.1 Polarized Ion Source at the Institute for Nuclear Research, Russian Academy of Sciences

In the mid-1990s, a polarized negative hydrogen ion source with a pulse current up to 1 mA, a current duration of 180 µs, a polarization of 87 ± 2 %, and a normalized emittance of 1.8π mm · mrad, which contains 90 % of the beam, was developed at the Moscow INR RAS. This achievement was based on a number of the important engineering developments, which were widely applied later. This source seems to be very promising and we will briefly describe its features primarily following Belov et al. (1995).

The production of highly polarized proton beams with intensities of about the intensities of unpolarized beams obtained at the same accelerators was a dream of researchers in polarization physics for a long time. Estimates show that this achievement requires a pulsed source of polarized negative hydrogen ions with a current of about 10 mA and a pulse duration of ~200 µs. The necessity of polarized H^- ions is explained by the possibility of using multiturn injection to increase the intensity of the circulating beam and to improve its emittance. In combination with a booster, this method allows the storage of the desired intensity. The optimum stripping of negative ions is certainly necessary before injection into the booster.

The polarized atomic beam source (PABS) at the INR RAS provides pulsed polarized positive and negative hydrogen ion beams. The charge exchange reaction between polarized thermal hydrogen atoms and deuterium ions in the deuterium plasma was used for the first time to obtain polarized hydrogen ions in this source. The source provides a polarized H^+ ion beam with a pulse current of 6 mA and a polarization of 85 % (Belov et al. 1987, 1990). To obtain a polarized H^- beam in the same source, the deuterium plasma was enriched in D^- ions using a specially designed device with a cesium-vapor surface-plasma converter.

The layout of the polarized hydrogen ion source is shown in Fig. 7.1. The beam of hydrogen molecules enters a dissociator from the left. Under the action of a microwave discharge in the dissociator, molecules dissociate into neutral hydrogen atoms H^0. Then, H^0 atoms pass through a nitrogen-cooled nozzle to a collimator and enter a sextupole magnet separating atoms in electron spin. A mass spectrometer detecting the composition of the beam is located downstream of the magnet.

Fig. 7.1 Layout of the
polarized atomic beam source
at the INR RAS

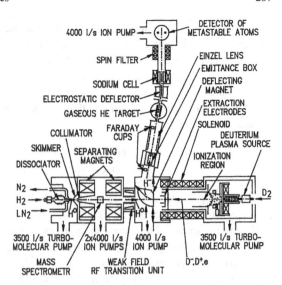

The focusing of atoms occurs in the next sextupole magnet. After that, electron-spin-polarized atoms arrive at the region of a weak magnetic field with the microwave transfer of electron polarization to nuclear polarization. This section of the source was described in detail in Belov et al. (1987). Then, atoms enter an ionizer. The charge-exchange region is in a solenoid producing a magnetic field of 1.3 kG. Nuclear-spin-polarized atomic hydrogen and negative deuterium ions D^- enter the solenoid from the left and right, respectively, and are involved in the charge exchange reaction

$$D^- + \vec{H}^0 \rightarrow D^0 + \vec{H}^-. \tag{7.1}$$

The use of this reaction to produce polarized negative ions H^- was proposed in Haeberli (1968). The cross section for this reaction at D^- energy $\sim$10 eV is 10^{-14} cm^2 (Hummer et al. 1960). The radial motion of polarized H^- ions is bounded by the solenoid field. They are drawn back by the electric field of the extraction system (extracting electrodes are indicated in Fig. 7.1) and are accelerated to 20 keV. Together with these ions, D^- ions and electrons are also drawn. In a deflection magnetic field, they are separated owing to the difference between their momenta. The magnetic field deflects H^- ions at an angle of 100° and extracts them from the source region. The intensity of the H^- beam was measured by the Faraday cylinder. The beam emittance was measured by the two-slit method described in Belov et al. (1994). The polarization of H^- ions was measured by a low energy polarimeter based on the Lamb shift effect. In this method, polarized protons are neutralized and L_α and L_β metastable states appear. The L_β state has a very short lifetime, and the L_α state is long-lived and its population is measured. In this case, H^- ions should be converted to protons. This is performed using a helium cell placed upstream of the polarimeter (Fig. 7.1). The helium density in the cell is as low as 10^{14} cm^{-2}. The probability of double charge exchange $H^- \rightarrow H^+$ at such density is about 0.5 %, but it is enough for the measurements. After the production of polarized protons using the helium cell, protons are deflected by an electrostatic deflector to the polarimeter. The measured polarization appeared to be 87 ± 2 %. It is assumed that polarization is not lost in the charge exchange process.

A feature of the pulsed polarized hydrogen ion source at the INR RAS is the production of the polarized ions by the charge exchange of polarized atoms in the plasma:

$$H^0 \uparrow + D^+ \rightarrow H^+ \uparrow + D^0; \qquad (7.2)$$

$$H^0 \uparrow + D^- \rightarrow H^- \uparrow + D^0. \qquad (7.3)$$

The energy of colliding particles in the plasma is ~ 10 eV; for such energies, the cross sections for reactions (7.2) and (7.3) are $5 \cdot 10^{-15}$ and 10^{-14} cm^2, respectively. Owing to such large cross sections for the charge exchange reactions, a high efficiency of the charge exchange of neutral hydrogen atoms and a high intensity of polarized hydrogen ions can be reached.

At INR RAS in 1986 the polarized proton source with the following parameters was developed (see Figs. 7.2 and 7.3):

- the pulse current of the polarized proton beam is 6 mA with a free atomic beam and 11 mA with a storage cell in the ionizer;
- the normalized emittance is 1.7π mm mrad with a free atomic beam and 1π mm mrad with the storage cell in the ionizer;
- the pulse duration is 100 µs;
- pulse repetition rate is 1–10 Hz;
- polarization is 80–90 %.

The following parameters of the polarized hydrogen atom beam have been reached:

- the pulsed beam intensity is $2 \cdot 10^{17}$ atoms/(cm^2 s);
- the most probable velocity of atoms is $2 \cdot 10^5$ cm/s (cooling by liquid nitrogen).

Polarized negative hydrogen ions are very useful at injection into accelerators because they can increase the intensity of a circulating beam by several times.

To obtain polarized negative ions, plasma enriched in unpolarized negative ions D^- should be generated in the plasma charge-exchange ionizer (Belov et al. 1987).

An unpolarized D^- ion source with a current of 2 µA based on the plasma source without the converter was developed in 1990 at the INR RAS. In 1993, the current of unpolarized D^- ions was increased to 1.2 mA and the plasma converter with the use of Cs vapors was developed for the application of the surface-plasma method for producing negative ions. Then, the current of unpolarized D^- ions reached 11 mA in 1996 due to placing of the plasma converter at the input of the ionizer solenoid. The development of a two-stage arc converter made it possible to reach a current of more than 60 (up to 90) mA in 2001. On the basis of these achievements, the parameters of the negative polarized hydrogen ion source have been improved:

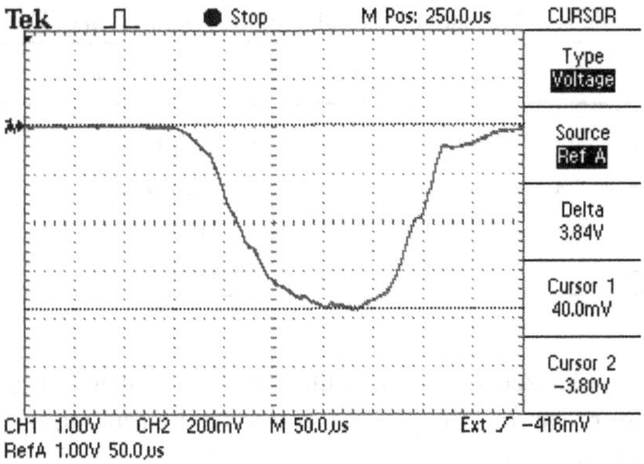

Fig. 7.2 Oscillogram of the
$H^- \uparrow$ ion current pulse from
the polarized ion source at the
INR RAS: the vertical and
horizontal scale units are
1 mA and 50 µs, respectively

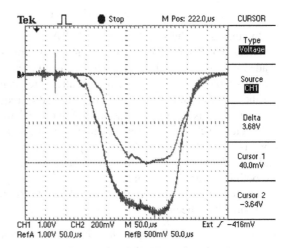

Fig. 7.3 Oscillogram of the
$H^- \uparrow$ ion current pulse and
(lower pulse) unpolarized D^-
ion current pulse from the
polarized ion source at the
INR RAS: the horizontal
scale unit is 50 µs

- the pulse current of the $H^- \uparrow$ ion beam is 3.8 mA;
- the pulse current of the unpolarized D^- ion beam is 60 mA (up to 90 mA);
- the polarization is 85–90 %;
- the normalized emittance is 1.7π mm mrad;
- the pulse duration is 170 µs;
- the pulse repetition rate is 1–10 Hz.

7.1.2 IUCF–CIPIOS Polarized Ion Source

The cooler injector polarized ion source (CIPIOS) is intended for producing polarized and unpolarized H^- and D^- ions. It was developed in 1999 in cooperation between the Indiana University Cooling Facility (IUCF) and the INR RAS.

Unpolarized H^- and D^- ions are produced in the plasma ionizer when hydrogen and deuterium are supplied to the plasma source, respectively. For the production

Table 7.1 Measurement results for the polarization of D^- in IUCF

State	Expected P_z	Measured P_z	Expected P_{zz}	Measured P_{zz}
+ vector	+1	0.909	+1	0.891
− vector	−1	−0.684	+1	0.695
+ tensor	0	0.003	+1	0.875
− tensor	0	−0.020	−2	−1.591

of polarized D^- ions, deuterium and hydrogen are supplied to the dissociator of the source and to the plasma source of the ionizer, respectively, and are involved in the reaction

$$D^0 \uparrow + H^- \rightarrow D^- \uparrow + H^0. \tag{7.4}$$

In contrast to the source at the INR RAS the CIPIOS involves sextupole magnets (sextupoles) made of permanent magnets with a magnetic field of 1.4 T. This made it possible to achieve the better focusing of an atomic beam and to reduce the emittance of the polarized beam; the latter circumstance is particularly important for the injection of polarized ions into the radio frequency quadrupole (RFQ). The cryogenerator is used to cool atoms in the dissociator. In the CIPIOS, the deuteron polarization scheme with two sextupole magnets and three high frequency transitions is used; it can provide vector polarizations ± 1 and tensor polarizations $+1$ and -2. The polarization direction in the beam extracted from the source is vertical for both $H^- \uparrow$ and $D^- \uparrow$ ions.

The following characteristics of the IUCF (CIPIOS) negative polarized hydrogen ion source have been obtained:

- the pulse current of the polarized $H^- \uparrow$ ($D^- \uparrow$) ion beam is 1.8 (2) mA.
- the pulse current of the unpolarized H^- (D^-) ion beam is 40 (30) mA.
- the polarization of $H^- \uparrow$ is 80–85 %;
- the polarization of D^- see in Table 7.1;
- the normalized emittance of $H^- \uparrow$ ($D^- \uparrow$) is 1.2π mm mrad;
- the pulse duration is up to 500 μs;
- the pulse repetition frequency is 1–4 Hz;
- the CIPIOS provides the long-term stability of intensity and polarization, polarization was constant during the accelerator operation runs with a duration of $\sim$1000 h;
- the CIPIOS operates with high reliability: it operates without staff with an automated control system during $\sim$1000 h.

7.2 Optically Pumped Polarized Ion Source (OPPIS)

The worldwide largest, yet single polarized proton collider RHIC (Relativistic Heavy Ion Collider) with the design center-of-mass energy $\sqrt{s} = 500$ GeV was commissioned in 2000 at Brookhaven National Laboratory, BNL (USA). The collider at this energy should provide a luminosity of $L = 2 \cdot 10^{32}$ cm^{-2}s^{-1} and a polarization of 70 %. The collider with the design parameters should deliver 120 bunches. The polarization of every bunch is set by a certain program. The optically pumped polarized ion source (OPPIS) developed by the Collaboration consisting of physicists from BNL; INR RAS; LAMPF; TRIUMF; and KEK. OPPIS is used as an injector of polarized particles (Zelenski et al. 2002). In order to reach the desired luminosity, every bunch should contain $2 \cdot 10^{11}$ polarized protons. Taking into account losses of the beam on its transportation, the source should produce at least 10^{12} polarized H$^-$ ions, which corresponds to a current pulse integral of $\sim$150 mA μs. The source with a current of 0.5 mA and a duration of 300 μs satisfies this condition.

To increase the phase density of the beam, the multiturn injection of the beam into the booster is now used. In this case, it is possible to compress $5 \cdot 10^{11}$ protons in one bunch, to accelerate in the booster to 1.5 GeV, and then to inject into the alternating gradient synchrotron (AGS). The AGS cycle lasts 3–5 s, whereas the source operates at a frequency of 1 Hz; additional cycles are used for the diagnostic purposes, for example, for the measurement of the beam polarization.

The layout of the RHIC OPPIS is shown in Fig. 7.4.

The superconducting solenoid consists of three windings having independent power supplies. The aim of this design is to create the desired configuration of the longitudinal magnetic field along the beam line. The electron cyclotron resonance (ECR) ionizer operates at a frequency of 29.2 GHz and requires a magnetic field of $\sim$10 kG. The hydrogen plasma formed in an ECR discharge contains a large fraction of protons ($\sim$90 %). The extraction of protons from the ECR region and their formation are performed using three-grid multiaperture ion-optical system. The ion-optical system is located at the beginning of a table with a maximum magnetic field of 27 kG. The length of this magnetic table is about 30 cm. A cell with optically polarized rubidium vapors completely occupies this table. The distance between the grids of the ion-optical system and the beginning of the rubidium cell is about 3 cm. The superconducting solenoid can be moved relative to the ECR and Rb cell in order to optimize the parameters of the source, as well as for repair and preventive works. Deflecting plates consisting of four stainless steel pipes 5 mm in diameter are mounted at the Rb-cell output. This improved vacuum pumping and allowed one to refuse water cooling because protons through the deflecting system arrived directly at the water-cooled screen.

After the deflecting plates, neutral electron-spin-polarized atomic hydrogen enters the so-called Sona-transition region (named after the inventor of the method). In this region, polarization is transferred from an electron to a proton. The careful adjustment of the transition point position and the magnetic field gradient in this region can ensure 100 % efficiency of polarization transfer. Change in the sign of the field should be performed at small gradients of the field. For this reason, the Sona-transition region is placed inside the screen in the form of a soft iron tube 50 mm in diameter and 155 mm in length. A coil 600 mm in diameter is placed outside the screen. The coil field is directed oppositely to the field of the superconducting solenoid. This is made for two reasons. First, it is possible to displace the point of

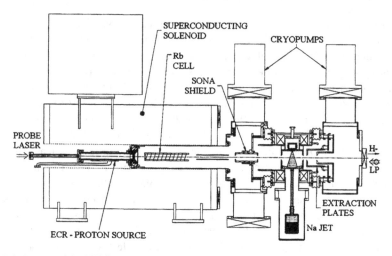

Fig. 7.4 Layout of the RHIC optically pumped polarized H$^-$ ion source. The plasma proton source is denoted as ECR (Electron Cyclotron Resonance), the laser probe is used to measure the thickness and polarization of Rb vapors from the Faraday rotation of the polarization plane (the beam enters from the left through a window in the ECR); the superconducting solenoid contains the ECR, Rb cell, and Sona-transition region; the sodium jet ionizer is in the field; and LP is the laser for the optical pumping of the vapor in a separate solenoid

the transition of the magnetic field through zero. Second, the magnetic field gradient in the Sona-transition region can be reduced by means of the coil field. For example, at a coil current of 100 A, the transition point is shifted by 8 cm against the atomic beam and the magnetic field gradient (taking into account the effect of the magnetic screen) is reduced to $dB/dz < 0.2$ G/cm. These conditions theoretically ensure 100 % polarization transfer from the electron to the proton. The correction coil simultaneously suppressed the residual field of the superconducting solenoid on the ionizer; for this reason, a discharge current in the high-voltage field of an extractor could be significantly reduced.

After the Sona region, hydrogen atoms already basically polarized in nuclear spin enter the ionizer where they are scattered on sodium vapors and are transformed to H$^-$ ions. The ionizer is placed in a magnetic field of 0.15 T in order to avoid the depolarization of the atoms owing to hyperfine interaction. At the chosen field, such a depolarization should be less than 2.5 %. However, such a field increases the emittance of the polarized H$^-$ beam at the ionizer output. The increase in emittance can be estimated by the formula

$$\Delta\varepsilon_n \approx 1.6\pi B R^2, \tag{7.5}$$

where B is the magnetic field in Teslas and R is the radius of the ionizer cell in centimeters. Unnormalized emittance is calculated for atomic-beam energy of 3 keV. The substitution of the numerical values of the parameters into formula (7.5) yields

$$\Delta\varepsilon_n \approx 0.2\pi \text{ cm mrad.} \tag{7.6}$$

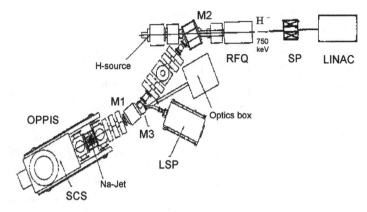

Fig. 7.5 Layout of the injection of the polarized H⁻ ions into the 200-MeV LINAC: OPPIS is the optically pumped polarized ion source; SCS is the superconducting solenoid, M_1 and M_2 are the deflecting magnets, LSP is the Lamb-shift polarimeter, optics box is a room for a laser unit providing the pumping of Rb vapors; a laser beam from this unit is adjusted on the beam axis; SP is the solenoid for turning the polarization direction from horizontal in vertical; H-source is the high-current unpolarized H⁻ ion source, RFQ is the RF quadrupole, and Na-Jet is the sodium jet ionizer

Fig. 7.6 Sodium ionizer with vertical geometry. A voltage of ~35 kV is applied to the ionizer for the acceleration of H⁻ ions to 35 keV and their subsequent injection into the microwave quadrupole

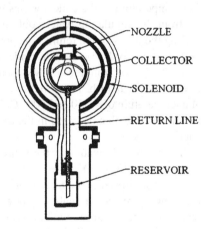

The emittance of the polarized atomic hydrogen beam entering the ionizer cell with a diameter of 2 cm is as small as 0.02π cm mrad and almost does not affect the final emittance given by formula (7.6). After the ionizer, the H⁻ beam enters a microwave quadrupole having an acceptance of 0.2π cm mrad (see Fig. 7.5).

Hence, these parameters are nearly matched.

Note that a significant part of the neutral beam (about 70 %) at the ionizer input is lost because of its large size.

In order to increase reliability at long-term operation, the geometry of the jet target–ionizer has been changed from horizontal to vertical (see Fig. 7.6).

The nozzle in this geometry was mounted on the collector cover. New Inconel (special alloy) heaters 2.5 mm in diameter were used. Such a heater in the bottom

Fig. 7.7 Arrangement of the elements of the Lamb-shift polarimeter intended to measure the polarization of the H⁻ ion beam at energies 2–35 keV

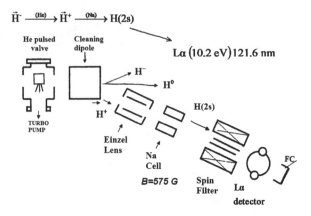

part is connected to a sleeve through a thick nickel wire 25-cm long. As a result, the working temperature of the Inconel heaters decreased to 120 °C; correspondingly, they could be simply wrapped by a copper foil instead of soldering used previously. Thus, the reliability of their operation was increased. At a working temperature of ~500 °C, the temperature of the hottest points of the heaters did not exceed 120 °C. The temperature of the collector was maintained within 140 ± 10 °C by usual water from the circulating water cooling system of the source. Special measures for sealing the joints of the return transport tube with the collector on the one side and with the tank for Na on the other side strongly reduced the heating of this tube. The yield of H⁻ ions from the cell was measured as a function of the temperature of the ionizer for Na and Rb. These measurements showed that the H⁻ yield was saturated at a temperature of ~350 and 500 °C for Rb and Na, respectively. The H⁻ yield for Rb increases with a decrease in the energy of the atomic beam and is ~16 % at an energy of 1 keV.

The beam polarization at the source output is measured by the Lamb-shift polarimeter (see Fig. 7.7).

This polarimeter is optimized for the work with pulsed polarized ion sources. Its advantages are a high analyzing power and a high count rate. As a result, polarization can be measured with an accuracy of 5–10 % in 2 min. The measurement accuracy is primarily limited by systematic errors. Nevertheless, as a fast, absolute, and online polarimeter, it is convenient for source adjustment and polarization monitoring. The Lamb-shift polarimeter in the RHIC OPPIS is used in two possible configurations. The polarization of H⁻ ions from the source is measured in the first case. To this end, the longitudinally polarized H⁻ ion beam from the source with an energy of ~3 keV is focused by an electrostatic lens and is deflected by a magnet at an angle of 47.5° (see Fig. 7.7). In this case, the polarization of ions is turned by 180° and remains longitudinal, as necessary for the Lamb-shift polarimeter. To obtain metastable hydrogen atoms, H⁻ should be recharged first to H⁰ (hydrogen atom) and then to H⁺ (metastable hydrogen atom). For this purpose, H⁻ ions are passed through a pulsed helium ionizer placed in a magnetic field of ~1 kG to avoid the depolarization in these charge exchange processes. The proton beam is deflected

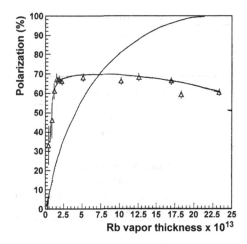

Fig. 7.8 (*Left scale, triangles*) Polarization and (*right scale*) output current of H⁻ ions versus the thickness of optically polarized rubidium vapors

by a magnetic dipole by an angle of ∼5° and enters the polarimeter (see Fig. 7.7). The polarimeter includes an electrostatic lens, a Na cell, a spin filter (a combination of a solenoid with a magnetic field of 575 G and a capacitor), a photon detector, and a Faraday cylinder.

In the second configuration of the polarimeter, protons are selected from a sodium target–ionizer. Though their fraction is only 0.3 % of the H⁻ yield, it is quite enough for the measurements. These polarized protons are deflected by a magnet by ∼100°, and polarization remains longitudinal. Then, they enter the same Lamb-shift polarimeter where their polarization is measured.

The measured proton polarization is shown in Fig. 7.8 as a function of the thickness of the optically polarized Rb target.

As seen in the figure, polarization is almost constant in the Rb vapor density range $(2–12) \cdot 10^{13}$ atoms/cm². At low Rb densities, polarization decreases owing to the neutralization of protons. In the working range $((5–10) \cdot 10^{13}$ atoms/cm²), this process leads a depolarization of 2–3 % and average polarization is ∼65 %. At high densities, polarization decreases due to optical pumping attenuation.

Any ion plasma source contains an impurity of H_2^+ charged molecules. The OPPIS produces also molecules H_2^- whose energy is half of the energy of the main H⁻ ions. They constitute a polarized background; consequently, it is necessary to numerically determine both the background and its contribution to the polarization of the main beam. Figure 7.9 shows the measured current of the negative ion yield as a function of the voltage applied to the ECR device.

It is seen that the basic peak is observed at a voltage of 2 kV, and the bump corresponding to the background molecules is observed at ∼4 kV. This bump with pulsed feed (small squares) is lower than that with continuous feed. Their contribution ranges within 5–10 % and polarization is diluted by ∼2–3 %.

This effect is also observed in the RHIC polarized jet target (Zelenski et al. 2004).

To conclude the chapter, we present the comparative characteristics of pulsed polarized ion sources (see Table 7.2).

Table 7.2 Comparative characteristics of the polarized ion sources

Laboratory	Source type	Bunch current (mA)	Polarization	Pulse duration, µs	Pulse repetition rate, Hz
INR	Atomic beam with a quasiresonance charge exchange plasma resonator	11 for H$^+$ 3 for H$^-$	0.8 0.9	200	10
BNL	Optical pumping	1 for H$^-$	0.7–0.8	500	1
JINR	Atomic beam ionizer with a Penning gauge	0.4 for D$^+$	0.6	400	0.2
Juelich	Atomic beam with a Cs-beam ionizer	0.01 for H$^-$	0.9	20	1

Fig. 7.9 Yield of negative ions from the ECR versus the voltage applied to it

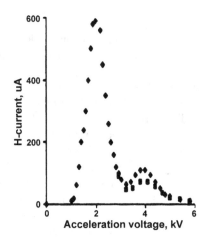

7.3 Conclusions

Two polarized ion sources successfully representing two competing directions, namely, the polarized atomic beam source and the optically pumped polarized ion source have been considered in detail. Both methods provide polarized sources meeting modern requirements. Moreover, both directions have good reserves for the further improvement of the beam parameters and the satisfaction of continuously increasing requirements of researchers in polarization physics.

References

Belov, A.S., et al.: Nucl. Instrum. Methods Phys. Res. A **255**, 442 (1987)
Belov, A.S., et al.: In: Proceedings of the International Workshop on Polarized Ion Sources and Polarized Gas Jets, KEK Report 90-15, p. 69 (1990)

Belov, A.S., et al.: Instrum. Exp. Tech. **37**, 131 (1994)

Belov, A.S., et al.: In: Proceedings of the 6th Workshop on Spin Phenomena in High Energy Physics, Protvino, Russia, p. 115 (1995)

Derenchuk, V.P.: In: Proceedings of the 15th International Spin Physics Symposium, Upton, New York, 2002. AIP Conf. Proc., vol. 675, p. 887 (2002)

Haeberli, W.: Nucl. Instrum. Methods **62**, 355 (1968)

Hummer, D.G., et al.: Phys. Rev. **119**, 668 (1960)

Zavoiskii, E.K.: Zh. Eksp. Teor. Fiz. **32**, 408 (1957a)

Zavoiskii, E.K.: Zh. Eksp. Teor. Fiz. **32**, 731 (1957b)

Zelenskii, A.N.: Doctoral Dissertation in Mathematical Physics, Moscow (2003)

Zelenski, A., et al.: In: Proceedings of the 15th International Spin Physics Symposium, Upton, New York, 2002. AIP Conf. Proc., vol. 675, p. 881 (2002)

Zelenski, A., et al.: In: Proceedings of the 16th International Spin Physics Symposium, Trieste, Italy, p. 761 (2004)

Chapter 8
Beam Polarimetry

Polarization investigations with polarized proton, photon, electron (positron), and muon beams are now conducted in many laboratories throughout the world. Investigations with polarized antiproton beams obtained from antilambda decay (Carey et al. 1990) were also recently performed. Accordingly, physicists have developed various devices to measure the polarizations of these beams. Such devices are called polarimeters.

In this chapter, we define some terms particularly those that have not yet been settled and formulate the general requirements on polarimeters and on necessary accuracy in beam polarization measurements. Then, the classification of polarimeters is given and the physical foundations of polarimetry are briefly presented.

Examples of polarimeters are given beginning with the origin of polarization physics at accelerators covering a wide energy range. We also present the latest results on polarimetry obtained at the AGS accelerator and Relativistic Heavy Ion Collider (RHIC), Brookhaven National Laboratory (BNL, USA). Schemes for calculating polarimeters are illustrated on an example of the expected acceleration of the polarized proton beam at the U-70 accelerator.

Examples of polarimeters currently used on polarized lepton beams are also briefly reviewed.

Polarimetry is an actively developing field of polarization technology and is devoted to the research and development (R&D) of the methods for measuring the polarization of beams and targets. Let us describe the procedure for measuring polarization. The scattering of the initial beam by the polarimeter substance can be characterized by three parameters: the initial beam polarization, the polarimeter response function usually called the analyzing power, and the asymmetry of particles scattered in the polarimeter. Measuring the asymmetry of scattered particles and knowing one of the remaining parameters from other measurements, one can determine the last desired parameter. Some terms in polarimetry have not yet been settled. For this reason, it is useful to take some definitions and terms from optics and, specially, from the section of optics devoted to the polarization of light (Prokhorov 1984). In particular, we use the term polarimetry for the set of the regular investigations for the development of polarimeters, i.e., devices for measuring

S.B. Nurushev et al., *Introduction to Polarization Physics*,
Lecture Notes in Physics 859, DOI 10.1007/978-3-642-32163-4_8,
© Moskovski Inzhenerno-Fisitscheski Institute, Moscow, Russia 2013

the polarization of beams and targets. Both methods and devices applied in polarimetry obviously depend on many factors, for example, on interaction (strong, electromagnetic, or weak) underlying polarimeter operation. All polarimeters based on strong interactions are relative because neither the magnitude nor the sign of their analyzing power can be predicted a priori. Polarimeters based on nuclear reactions of elastic scattering, inclusive formations of pions, etc. are relative polarimeters. On the contrary, all polarimeters based on electroweak interactions are absolute polarimeters, because both the magnitude and the sign of their analyzing power can be predicted a priori. Møller, Mott, and Compton polarimeters are absolute polarimeters. Polarimeters based on weak interactions (decays of muons, hyperons, etc.) are also absolute. There are polarimeters based on the interference between strong and electroweak interactions. We call such polarimeters mixed or interference (by analogy with optics). Depending on the relative contributions of these interactions, they can be closer to absolute or relative polarimeters, but cannot approach any of these classes, because mixed polarimeters are based on the interference effect. Mixed polarimeters cannot be implemented without its calibration. The pp polarimeter based on the interference between Coulomb and nuclear interactions is a typical example of mixed polarimeters. According to the above classification, it cannot be called absolute (though it is often called absolute), because the calculation of the analyzing power of this polarimeter involves many hadron amplitudes a priori unknown. Therefore, this instrument should be calibrated in advance, as any relative polarimeter. At the same time, an absolute polarimeter does not require such calibration. The characteristics of polarimeters also depend on the kinematic variables of a reaction such as the initial energy, scattering angles, and energy of secondary particles. Since polarization is a vector, all three its components should be measured to determine its direction in space. Such a measurement requires an absolute polarimeter rather than a relative polarimeter that determines only the value of polarization without its sign. The detailed classification of the polarimeters will be given below.

8.1 Basic Relations in Polarimetry

Left–right, "raw," or experimental asymmetry is the directly measured difference $\Delta n = n_+ - n_-$ between the particle yields n_+ and n_- for the different orientations of the polarization vector of the beam (target) $\vec{P}_B$ ($\vec{P}_T$) divided by their sum $n = n_+ + n_-$:

$$\varepsilon(x_F, p_T, s) = (n_+ - n_-)/(n_+ + n_-) = \Delta n/n, \qquad (8.1)$$

or

$$\varepsilon(x_F, p_T, s) = A_N(x_F, p_T, s) P_B \cdot \vec{n} \cdot \vec{n}_P. \qquad (8.2)$$

Here, $\vec{n}$ is the unit vector perpendicular to the reaction plane and $\vec{n}_P$ is the unit vector along the polarization vector. The angle between them is designated as ϕ.

Analyzing power or physical asymmetry $A_N(x_F, p_T, s)$ in this formula is the raw asymmetry ε of a particular process reduced to the 100 % polarization of the beam P_B (or target P_T) and x_F, p_T, and s are the Feynman parameter, momentum transfer, and the square of the total energy of the colliding particles in their center-of-mass frame, respectively. For brevity, these variables will often be omitted, as we did already for n_+ and n_-.

Polarimetry involves an important parameter called **the dilution factor** $D = $ signal/(signal + noise) or the inverse parameter

$$R = D^{-1}. \tag{8.3}$$

Unfortunately, there is a confusion on notation for the dilution factor in the literature. In particular, the dilution factor is denoted as $D = d$ in Sect. 6.1, in the subsection devoted to the COMPASS target. This circumstance should be taken into account when reading original articles. From general point of view the factor D should relate the polarization of the target $\vec{P}_T$ to the polarization $\vec{P}_n$ of the free or polarizable nucleon in the target by relation $\vec{P}_n = D\vec{P}_T$. For example, the polarized propanediol target has a polarization $|\vec{P}_T| \sim 70\,\%$ and $D \sim 0.1$, so $|\vec{P}_n| \sim 7\,\%$.

The factor R quantitatively determines the contribution of background processes and characterizes the imperfection of a **polarized target** (**or beam**). For a perfect target $R = 1$, but such a target does not yet exist. Taking into account this factor, relation (8.1) is represented in the form

$$\varepsilon(x_F, p_T, s) = D(n_+ - n_-)/(n_+ + n_-) = D \cdot A_N. \tag{8.4}$$

Here, $n_\pm$ are yields of the polarized nucleon in the target and A_N is the analyzing power taken as the foundation of polarimetry.

Therefore, the variance of raw asymmetry is given by the expression

$$\sigma^2(\varepsilon) = R/n. \tag{8.5}$$

Equation (8.2) provides the other useful relation

$$\frac{\sigma^2(\varepsilon)}{\varepsilon^2} = \frac{\sigma^2(P_B)}{P_B^2} + \frac{\sigma^2(A_N)}{A_N^2}. \tag{8.6}$$

Relations (8.5) and (8.6) are fundamental ones and widely used in polarimetry. Let us consider a particular example of what statistics is required to determine beam polarization with a given relative accuracy δP_B. It is assumed that the analyzing power is measured in a separate experiment with high accuracy. Neglecting the term containing the analyzing power in relations (8.5) and (8.6), we find

$$N = \frac{R}{(\varepsilon \delta P_B)^2}. \tag{8.7}$$

We emphasize two important circumstances. First, experimentally measured asymmetry ε is D times smaller than physical asymmetry (see Eq. (8.4)). Second, time required to achieve the necessary statistical accuracy δP_B in the measurement of the beam polarization is D times smaller in the case of the absence of the dilution effect. Both these factors are present, in particular, in the experiments with solid polarized targets made of organic materials, where $R \approx 8$–10. Ammonium NH_3 has the lowest dilution factor, $R = 4.2$, among the currently available inorganic materials.

Equation (8.7) can be represented as an expression for the beam polarization variance $D(P_B)$:

$$D(P_B) = \frac{R}{N \cdot A_N^2}. \tag{8.8}$$

The count rate N depends on luminosity L, differential cross section I for the reaction used in polarimetry, and detector acceptance $\Delta\Omega$:

$$N = L \cdot I \cdot \Delta\Omega. \tag{8.9}$$

From (8.8) and (8.9), it follows that

$$D(P_B) = \frac{R}{L \cdot I \cdot A_N^2 \cdot \Delta\Omega}. \tag{8.10}$$

One of the important parameters of the polarimeter is the factor of merit (FOM)

$$M = A_N^2 \cdot I. \tag{8.11}$$

According to Eq. (8.10), a polarimeter with a higher FOM, M (at fixed luminosity, dilution, and acceptance parameters) provides a higher accuracy in the determination of the beam polarization. The FOM, M allows the comparison of different polarimeters: the higher the M value, the better the polarimeter.

8.2 Classification of Polarimeters

The first classification of polarimeters mentioned above is based on the type of interaction used in polarimeters. A polarimeter based on an electroweak process is absolute; i.e., the magnitude and sign of the analyzing power of this polarimeter can be predicted a priori. Such a polarimeter can be used to determine an unknown polarization of the beam (target) without additional calibration. Polarimeters based on strong interaction are relative. Since there is no quantitative theory of strong interaction, the parameters of such polarimeters cannot be calculated a priori. These polarimeters should be selected empirically ("empirical polarimeters"), i.e., experimentally. They should be calibrated using absolute polarimeters.

The general classification of polarimeters is shown in Fig. 8.1 (SPIN Collaboration 1992). This scheme obviously misses the type of polarimeters based on the interference effect between the strong and electroweak interactions. An example of mixed polarimeters is the Coulomb–nuclear interference (CNI) polarimeter. This polarimeter requires calibration and, correspondingly, is classified as a relative polarimeter. After calibration, it can possibly be transferred to the class of absolute polarimeters. As seen in Fig. 8.1, polarimeters can be on-line, i.e., providing information on the beam (target) polarization immediately during an experiment and off-line, i.e., providing information after the collection of statistics because of the slow collection of statistics or the complexity of a polarimeter, for example, a large amount of information or because of the necessity of the careful processing and introduction of various corrections, etc.

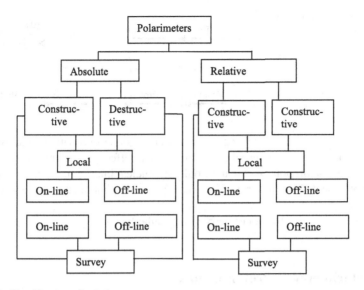

Fig. 8.1 Classification of polarimeters

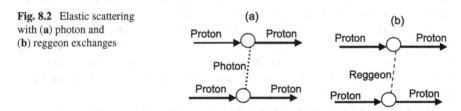

Fig. 8.2 Elastic scattering with (**a**) photon and (**b**) reggeon exchanges

On lepton beams, absolute polarimeters are usually used and provide the highest accuracy (≈ 1 %). Polarimeters that can operate without interrupting the basic experiment are called constructive. Polarimeters, interrupting the basic experiment in order to measure the beam polarization are called destructive. Polarimeters that measure polarization in the region of the interaction between the beam and target (experiment target) in the case of experiments with the fixed target or in the region of the collision of beams (in colliders) are called local.

General-purpose polarimeters used, for example, for the adjustment of an accelerator or its separate units and beam lines are called survey polarimeters.

Examples of all types of these polarimeters will be considered below in the corresponding sections.

Diagrams of elastic and inelastic processes are presented in Figs. 8.2 and 8.3, respectively.

They represent the Feynman diagrams for all high-energy polarimeters currently used in experiments, except maybe a Compton scattering polarimeter. For the Compton scattering polarimeter, it is sufficient to replace one of the proton lines by a photon line, the exchange lines in one diagram by a photon line, and the exchange lines in the other diagram by a boson line. Almost all absolute polarimeters are cal-

Fig. 8.3 Inelastic scattering with (**a**) photon and (**b**) reggeon exchanges

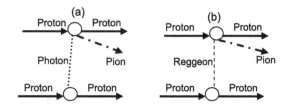

culated in the one-photon exchange approximation sometimes with the inclusion of higher order diagrams in unusual cases. Interference effects in such mixed polarimeters as the Primakoff polarimeter and Coulomb–nuclear interference polarimeter can also be represented in the form of diagrams with one-photon and one-reggeon exchanges.

Let us discuss now the polarimetry of particular beams.

8.3 Polarimetry of Proton Beams

To cover a wide energy range, we will consider possible proton polarimeters in application to the 70-GeV U-70 accelerator at the Institute for High Energy Physics (IHEP, Protvino, Russia) and to RHIC (BNL, USA). By the example of the first accelerator whose parameters are known, we present the scheme for calculating the characteristics of polarimeters. By the example of the second accelerator, the correctness of these calculations will be illustrated taking into account that the basic polarimeters are almost identical in both cases.

It is supposed that a polarized proton beam is accelerated at the U-70 accelerator complex from 25 keV to 70 GeV (see Chap. 5). The source provides 25-keV polarized negative hydrogen ions that are accelerated to 100 keV and injected into the Ural-30 linear accelerator. Then, the charge exchange injection of the 30-MeV proton beam into the 1.5-GeV booster occurs and protons from the output of the 1.5-GeV booster are transported and injected into the U-70 accelerator. After the acceleration of the polarized proton beam to 70 GeV, it is extracted from the accelerator by fast or/and slow beam extraction systems and is distributed over several experimental setups with fixed targets. The main objective of this section is to analyze the possible methods for measuring the beam polarization in the corresponding regions of the acceleration or transportation chain.

In 1981, it was proposed to accelerate the polarized proton beam at the U-70 accelerator using the Siberian snake method (Ado et al. 1983). At that time, this proposal was not accepted. At present, it seems appropriate to return to this program especially after the successful commissioning of the 200-GeV polarized beam at RHIC and the promising advance to the final RHIC energy $\sqrt{s} = 500$ GeV. This subject was discussed in detail at the 11th Workshop on High Energy Spin Physics (Dubna, Russia, 2005). The reasons concerning the possibility of accelerating the polarized proton beam in the U-70 synchrotron were presented by Shatunov (2005), Vasiliev (2005) presented the possible polarization program with the use of this

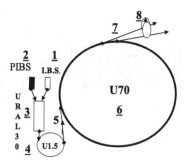

Fig. 8.4 U-70 accelerator complex: (*1*) unpolarized ion source, (*2*) polarized ion source, (*3*) Ural-30 RF linear accelerator, (*4*) 1.5-GeV booster, (*5*) beam transportation channel from U1.5 to U70, (*6*) U70 accelerator, (*7*) beam extraction and transportation system, (*8*) general and local polarimeters

beam, and the report by Nurushev (2005) concerned polarimetry. The general tone of reports and discussions was positive.

Polarimeters necessary for the measurement of the polarizations of protons in the energy range 30 MeV–70 GeV are discussed below on the basis of work (Nurushev 2005).

8.4 U-70 Accelerator Complex at IHEP

The layout of an accelerator complex for the 70-GeV polarized proton beam is presented in Fig. 8.4.

Figure 8.4 presents all elements of the accelerating complex; the measurements of the beam polarization are planned at the outputs of these elements and they will be successively discussed below.

8.4.1 Polarimeter on 25 keV for an Atomic Beam Source

Polarized atomic beam source (PABS) and optically pumped polarized ion source (OPPIS) have been described in detail in Chap. 7. In both cases, polarimeters based on the Lamb shift of the hydrogen atom levels were applied to measure the polarization of atomic beams with energies 10–30 keV. Such a polarimeter operates as follows (see Fig. 8.5) (Belov et al. 1987).

The polarized proton beam is focused by electrostatic lens /1/ on a cell filled with sodium vapors /4/. The beam position in the vertical plane is varied by electrostatic corrector /2/. Vacuum valves /3/ control the proton flux. Some polarized protons in the cell with sodium are transformed to metastable hydrogen atoms $\vec{H}^0$ in the states α and β ($2S_{1/2}$). Then, the crossed static magnetic fields of solenoid /5/ and

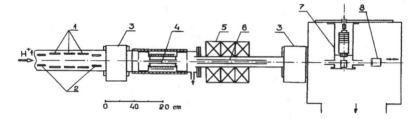

Fig. 8.5 Lamb-shift polarimeter for measuring the polarization of positive and negative hydrogen ions

the electric field of capacitor /6/ eliminate the β state of the atom. The beam of metastable atoms mainly in the α state is detected by special photon detector /7/ with a multiplier of secondary electrons. The beam passed through the polarimeter is detected by Faraday cylinder /8/. Measuring the count rates for various initial polarizations of the proton beam, it is possible to determine the beam polarization. It has appeared to be 76 ± 2 %. This accuracy was achieved in about 1 min. Therefore, such an instrument can be used as an absolute polarimeter in the on-line mode. In the present configuration, this polarimeter is destructive, because the polarized ion beam should be taken away from the injection line for the period of its operation (1 min); i.e., it interrupts the experiment.

8.4.2 Relative Polarimeter for an Energy of 30 MeV

A rich experience of physicists throughout the world can be used when developing a relative polarimeter for measuring the beam polarization at the output of the Ural-30 linear accelerator. The analyzing power for the elastic proton–carbon (pC) scattering at this energy has been measured with an accuracy of about 1 %. According to these results, the relative pC polarimeter can be designed for energy of 30 MeV using the scattering of protons at an angle of 65° in the laboratory frame. In this case, the analyzing power is expected to be $A_N = (57.4 \pm 0.9)$ %. Let us assume that this instrument is directly used in the 1:1 scale. Since the Ural-30 accelerator provides a current of $I = 3$ mA and a bunch duration of 40 µs, the number of protons in each bunch is $7.5 \cdot 10^{11}$. Let the thickness and width of a target be 106.9 mg/cm^2 and 3 mm, respectively (the beam size is approximately 30 mm). For the useful solid angle of the scintillator telescope $\Omega = 10^{-5}$ srad, 60 events per bunch are expected. Taking into account 16 bunches and 6 cycles per minute, we obtain $6 \cdot 10^4$ events in 10 min. This statistics is sufficient for measuring the beam polarization with an accuracy of 5 %.

A possible variant of the polarimeter for energy of 30 MeV shown in Fig. 8.6 was successfully used at Argonne National Laboratory to measure the polarization of 50-MeV protons in the injector (Ratner 1974). The polarimeter is based on the elastic scattering of protons on a carbon target at an angle of 55°, where the analyzing

Fig. 8.6 Possible variant of a
polarimeter at the output of
the Ural-30 accelerator

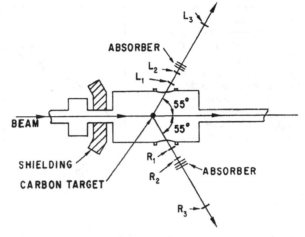

power is about 85 % and the accepted by recoil detector differential cross section is
10 mb. The width and height of the target were 0.05 and 7.5 cm, respectively, and its
thickness was 0.08 g/cm². Each of two symmetrically located telescopes consisted
of three scintillation counters. The thicknesses of the first two scintillators were
chosen so that scattered protons left the most part of energy in them. Then, counts
from inelastic processes are suppressed by means of thin polyethylene plates in front
of the third scintillator. The width, height, and thickness of the third scintillator were
4.04, 8.08, and 0.16 cm, respectively. This counter was determining, was placed at
a distance of 66 cm from the target, and covered a solid angle of 6.7 msrad. The
measurements of the mean free path, ionization losses, and time of flight ensured a
sufficiently reliable separation of the elastic process.

The same method can be used for an energy of 30 MeV, but it is necessary to
select the thickness of the filters in front of the last counter and to move to an angle
of 65°.

8.4.3 Polarimeter for the 1.5-GeV Booster

The analyzing power of pp scattering has been measured in a wide energy range.
The experiment for energy closest to the higher energy of the booster has been
executed at the PS accelerator at CERN (Albrow et al. 1970). The analyzing power
for protons with a kinetic energy of $T = 1.34$ GeV at a laboratory angle of $\Theta_L = 12°$
is $A_N = (37 \pm 2)$ % (see Fig. 8.7). The points in Fig. 8.7 are taken from McNaughton
and Chamberlin (1981), McNaughton et al. (1981a, 1981b), Bevington et al. (1978),
Besset et al. (1980), Bugg et al. (1978), Greeniaus et al. (1979), and Cheng et al.
(1967).

It has been found that the differential cross section is 66 mb/srad with an accuracy
of about 5 %. Let us assume that two scintillator telescopes each with a solid angle

Fig. 8.7 Analyzing power of *pp* elastic scattering at a laboratory angle of 17° versus the initial kinetic energy of the proton (the most accurate measurements performed in Los Alamos); it covers the lower energy range of the booster of the U-70 accelerator; the curve is calculated with the Arndt's phase shifts

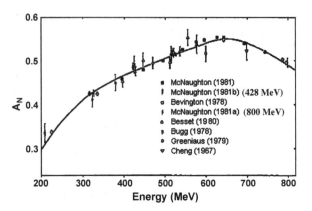

of 0.7 msrad will be jointly used. The booster can provide approximately 30 bunches each containing $2 \cdot 10^{11}$ protons. The bunch duration is very short, approximately 80–100 ns. This condition imposes significant constraints on the data collection rate. For safety, it is supposed that only one event can be detected in the passage of each bunch through a target. This can be implemented using a polyethylene target with a thickness of about 50–100 μm. For a beam polarization of 70 %, it is expected that an accuracy of 5 % will be achieved in 2 h.

A possible variant of the polarimeter for the 1.5-GeV booster presented in Fig. 8.8 is successfully used at the Dubna synchrophasotron to measure the polarization of proton and deuteron beams in the booster energy range Azhgirei et al. (2002). Asymmetry in proton–proton elastic scattering is measured with the use of a polyethylene target. To subtract the background coming from *pp* quasielastic scattering, the carbon target is used. The data for a laboratory scattering angle of 14° are more statistically valid. For a better selection of desired events, it is necessary to add one scintillation counter to each forward arm. It is suggested to calculate and install a protection around the equipment.

The analyzing power of a polyethylene target measured on the setup presented in Fig. 8.8 is shown in Fig. 8.9 along with the results of the previous measurements. According to this figure, the analyzing power of the polyethylene target at the highest booster energy is close to 35 %.

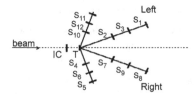

Fig. 8.8 Possible variant of a polarimeter for the booster: T is the target (CH_2 or C), S_{1-8} are the scintillation counters, IC is the ionization chamber; the polarimeter is based on *pp* elastic scattering

Fig. 8.9 Effective analyzing power A_N (CH$_2$) for scattering at a laboratory angle of 14° versus the kinetic energy of protons

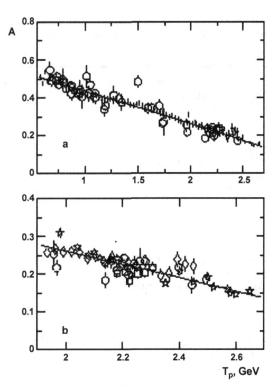

8.4.4 Polarimeters for an Energy of 70 GeV

The absolute and relative polarimeters developed for the main accelerator should operate from the injection energy 1.3 GeV to the highest energy 70 GeV and measure the polarization of the internal beam at all stages of its circulation.

Before to discuss particular polarimeters, it is necessary to specify the key parameters of the U-70 accelerator that are substantial for the choice of a polarimeter. Most these parameters are presented in Table 8.1 (Tarasov 1964).

Using these data and involving additional information, we can estimate the necessary characteristics of a beam. First, we will estimate the transverse sizes of the beam. Taking the beam emittance from Table 8.1 and the amplitude function from additional sources, the transverse sizes of the 70-GeV beam can be determined from the expressions

$$\sigma_i = \sqrt{\frac{\varepsilon_i \beta_i}{6\pi}}, \quad \dot{\sigma}_i = \sqrt{\frac{\varepsilon_i}{6\pi \beta_i}}. \tag{8.12}$$

Here, σ_i $(i = x, y)$ is the transverse size of the beam in the corresponding direction, $\dot{\sigma}_i$ is the angular divergence of the beam, and ε_i and β_i, $i = x, y$ are the emittance (contains 95 % of the beam) and amplitude function, respectively. The jet target is placed on a rectilinear section of the ring between truncated blocks D^* and F^*. Here, the amplitude functions are β_x, $\beta_y = 25$ m. Thus, the transverse sizes

Table 8.1 Comparative characteristics of polarimeters for the proton beams at RHIC and the U-70 accelerator (PPJT is the polarized proton jet target)

Reaction	A_N, %	$\Delta\sigma$, µb	M, µb	Setup	Comment
1. $p_\uparrow + p \to p + p$, 100 GeV/c $10^{-3} \le |t|$ (GeV/c)$^2 \le 2\cdot10^{-2}$, $\Delta\phi/\phi = 8.6\cdot10^{-2}$	2 ± 0.3	100	$4\cdot10^{-2}$	PPJT, $n_T = 10^{12}$ p/cm^2; Si detectors	Based on the polarimeter at RHIC. Recoil protons are detected
2. $p_\uparrow + C \to p + C$, 100 GeV/c $7\cdot10^{-3} \le |t|$ (GeV/c)$^2 \le 3\cdot10^{-2}$, $\Delta\phi/\phi = 0.13$	1	9530	0.953	Carbon foil, Si detectors	There are no data on the differential cross sections. Details of estimates see the text
3. $p_\uparrow + A \to \pi^- + X$, 70 GeV/c $x_F \approx 0.5$, $p_T \approx 0.7$ GeV/c; $\Delta p/p \approx \pm1$ %, $\Delta\Omega \approx 8$ µsrad	20	0.06	$2\cdot10^{-3}$	Internal target, extraction of π^- to beam line no. 2	It is necessary to calibrate the polarimeter
4. $p_\uparrow + p \to p + p$, $p + p_\uparrow \to p + p$, 70 GeV $0.2 \le -t$ (GeV/c)$^2 \le 0.3$	2.23 ± 0.15	120	$3\cdot10^{-2}$	PPJT of the PROZA setup and scintillator hodoscopes	Absolute polarimeter for internal and external beams
5. $p_\uparrow + p \to p + p$, 1.34 GeV, $\theta_{lab} = 12°$, $\Delta\Omega \approx 0.7$ msrad	37 ± 2	46	6	CH$_2$ target and two telescopes	Relative polarimeter in line
6. $p_\uparrow + C \to p + C$, 30 MeV, $\theta_{a\,lab} = 65°$ $\Delta\Omega \approx 10$ µsrad	57.4 ± 0.9	0.15	0.05	C target, scintillation counters	Relative polarimeter in line
7. $H^+ + Na^- \to H^0 + Na^0$, 10 keV	100	–	–	Microwave generator, γ detector	Absolute polarimeter in line

of the 70-GeV beam are $\sigma_x = 2.4$ mm and $\sigma_y = 2.1$ mm and the angular divergences are $\dot\sigma_x \approx 0.1$ mrad and $\dot\sigma_y \approx 0.1$ mrad. The parameters at the injection energy are $\sigma_x = 17.6$ mm, $\sigma_y = 15.4$ mm, $\dot\sigma_x \approx 0.7$ mrad, and $\dot\sigma_y \approx 0.7$ mrad. Running ahead, we note that the jet (with FWHM = 6 mm) completely covers the 70-GeV beam and only 1/6 of the beam at the injection energy. As a result, polarization average over the entire 70-GeV beam can be measured. The polarization of only 1/6 of the 1.3-GeV beam is measured for once. This is insufficient and it is necessary to seek the possibility of scanning the beam or the target in order to measure the polarization of the entire beam.

Among known polarimeters, only the Coulomb–nuclear interference (CNI) polarimeter has weak energy dependence. This polarimeter also satisfies another important criterion: it has the highest quality factor (FOM). The third important fact is

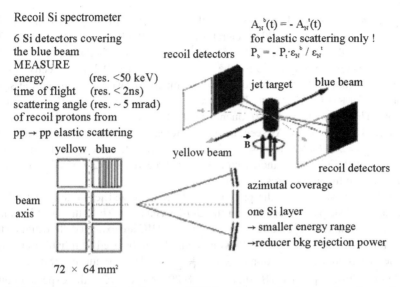

Recoil Si spectrometer

6 Si detectors covering
the blue beam
MEASURE
energy   (res. <50 keV)
time of flight   (res. < 2ns)
scattering angle (res. ~ 5 mrad)
of recoil protons from
pp → pp elastic scattering
    yellow blue

$A_N^b(t) = - A_N^t(t)$
for elastic scattering only !
$P_b = - P_t \cdot \varepsilon_N^b / \varepsilon_N^t$

recoil detectors

jet target   blue beam

yellow beam

recoil detectors
azimutal coverage

one Si layer
→ smaller energy range
→reducer bkg rejection power

beam
axis

72 × 64 mm²

Fig. 8.10 Layout of the CNI polarimeter at RHIC based on *pp* elastic scattering; the resolution in the recoil proton energy is <50 keV, the resolution in the time-of-flight is <2 ns, the angular resolution is ~5 mrad; indices b and t designate the beam and target, respectively

that the CNI polarimeter has been successfully implemented at RHIC for energy up to 100 GeV. These discussions lead us to the choice of *pp* and *p*C polarimeters in the CNI region, those that have been implemented at AGS and RHIC (Bravar et al. 2004; Jinnouchi et al. 2004).

For this reason, we will briefly describe two internal absolute CNI polarimeters and two possible external polarimeters: the relative inclusive polarimeter on charged pions and the absolute polarimeter based on *pp* elastic scattering.

8.4.4.1 *pp* CNI Polarimeter on the Polarized Jet Target with the Detection of the Recoil Proton

The *pp* CNI polarimeter for 70 GeV has a large differential cross section in the CNI peak region:

$$\frac{d\sigma}{dt}(-t \approx 3 \cdot 10^{-3} (\text{GeV}/c)^2) \approx 100 \text{ mb}/(\text{GeV}/c)^2. \qquad (8.13)$$

Its average analyzing power is about 2 %. Therefore, the quality factor is FOM ≈ $4 \cdot 10^{-2}$ (mb/(GeV/c)²). The layout of this polarimeter is presented in Fig. 8.10.

The same polarimeter will probably be used at the U-70 synchrotron. The polarized jet target has a jet surface density of $10^{12} p \, \text{cm}^{-2}$, a polarization of $P_T = (92 \pm 1.8)$ %, and a jet size of 5 mm (FWHM) and operates in the direct current mode. Two blocks of silicon strip detectors each with overall dimensions $72 \times 64 \text{ mm}^2$ are placed on the left and right of the beam at a distance of about 80 cm from the polarized jet target. First, to measure the analyzing power of *pp* elastic scattering, it is necessary to scatter the 70-GeV unpolarized proton beam on the polarized jet target. Let us estimate the expected intensity per second for the unpolarized beam of the U-70 accelerator accepting the following parameters. The internal beam flux per second is $I = 5 \cdot 10^{12} \cdot f$, where $f = 200$ kHz, i.e., $I = 10^{18}$ protons/s. Since the duty cycle for the U-70 accelerator is 0.2, the effective intensity of the beam is $2 \cdot 10^{17}$ protons/s. For comparison, we can estimate the same parameter for RHIC. At present, RHIC provides 55 bunches with $5 \cdot 10^{10}$ protons in a bunch. The beam circulation frequency is 78 kHz. Thus, the RHIC beam intensity is also $2 \cdot 10^{17}$ protons/s. As seen, the beam intensities in the U-70 accelerator and RHIC ($\sqrt{s} = 200$ GeV) are now the same. As known, the energy and intensity in each ring of RHIC will be increased in two or three years to 250 GeV and at least by an order of magnitude, respectively.

For further estimations, the parameters of the polarized jet target at RHIC are taken as the basic parameters also for the U-70 accelerator. Thus, the same luminosity and approximately the same count rate as for RHIC are expected. To estimate the count rate, one (or both) of the following methods can be used. If reliable data exist for the differential cross sections $\frac{d\sigma}{d\Omega}$ or $\frac{d\sigma}{d\Omega dE}$ for elastic scattering (Smiths et al. 1981) or inclusive process (Bozhko et al. 1980), respectively, the expected count rate of the polarimeter can be estimated by the formula

$$N_e = L \cdot \Delta\sigma. \tag{8.14}$$

Here, L is the luminosity and the effective cross section $\Delta\sigma$ corresponding to the useful acceptance of the polarimeter is given by the expressions

$$\Delta\sigma = \frac{d\sigma}{d\Omega} \Delta\Omega \cdot \kappa \tag{8.15}$$

for elastic scattering and

$$\Delta\sigma = \Delta\Omega \cdot \Delta E \cdot \kappa \frac{d\sigma}{d\Omega dE} \tag{8.16}$$

for the inclusive reaction. Here, $\Delta\Omega$, ΔE, and κ are the solid angle, energy covering band, and efficiency of the polarimeter, respectively. To determine luminosity, it is also necessary to know the target surface density N_T. Then, luminosity is given by the expression

$$L = N_B \text{ (protons/s)} \cdot N_T \text{ (protons/cm}^2\text{).} \tag{8.17}$$

We made the following assumptions. An exponential dependence on t is taken for pp elastic differential cross section at 70 GeV/c. The slope parameter is known from the experimental data: $B = 11.3$ (GeV/c)$^{-2}$. The differential cross section at $t = 0$ is calculated from the optical theorem, neglecting the real part of the amplitude. The measurement range has been accepted the same as for RHIC, namely, $0.002 \leq |t| \leq 0.02$ (GeV/c)^2. The azimuthal covering was accepted as $\frac{\delta\phi}{\phi} = 0.086$. It corresponds to the overall dimensions of the blocks of the detectors specified above. Thus, the effective cross section is 100 μb. Multiplying this value by luminosity, we obtain $N = 20$ events/s. To achieve an accuracy of 5 % in the calibration of the analyzing power, approximately $1.2 \cdot 10^6$ events are required. This number of

Jet-Target Holding Magnetic Field

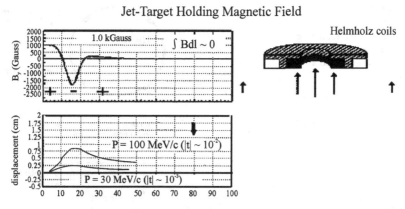

Fig. 8.11 Distribution of the polarization holding magnetic field along the recoil proton path

events can be collected in approximately 17 h of the operation of the U-70 accelerator. Thus, knowing the polarization of the polarized jet target P_T, we determine the analyzing power A_N of pp elastic scattering from the relation $\varepsilon_T = A_N P_T$, where ε_T is the raw asymmetry measured on the polarized jet target with the unpolarized proton beam. At the second stage, beam polarization is measured using an unpolarized jet. Raw asymmetry in this case is determined by relation $\varepsilon_B = A_N P_B$. From these two measurements, the beam polarization is determined by the expression $P_B = \frac{\varepsilon_B}{\varepsilon_T} P_T$. This completes the calibration of the beam polarization. The beam polarization can be measured with an accuracy of 5 % approximately in 28 h.

There is the second possibility of determining the beam polarization simultaneously with the measurement of the spin correlation parameter. In this case, both the target and beam should be polarized.

The absolute calibration of polarization apparently takes a long time and is inappropriate for such on-line goals as the accelerator adjustment for work with the polarized beam. This problem requires though a relative, but much faster polarimeter. Such a polarimeter will be described below.

One feature of this polarimeter is presented in Fig. 8.11. Recoil protons have a very low kinetic energy of about 1–10 MeV. Their trajectories are considerably bent by a polarization holding magnetic field (internal Helmholtz coil, the field is shown by upward arrows) at the output of the target. To compensate this field, the external Helmholtz coil generates a magnetic field of the opposite polarity. The total field distribution along the path of protons is shown in the upper panel. The idea is that the integral of the field becomes zero. This aim is generally achieved but incompletely. The lower panel shows the deviations from the path in the absence of a magnetic field (the abscissa axis). Two cases are considered. In the first case, protons have a low momentum of 30 MeV/c (the upper line).

In the second case, protons have a momentum of 100 MeV/c (the lower line). As seen, the distortion of trajectories is strongest for protons with very low energies. However, these protons are most necessary in polarimetry, because the maximum analyzing power is achieved at a recoil proton momentum of about 30 MeV/c. This

Fig. 8.12 Polarimeter based on the elastic Coulomb–nuclear scattering of protons on carbon nuclei; this device is now used at AGS

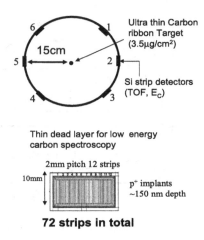

indicates the necessity of the careful screening of the recoil proton path from magnetic fields.

8.4.4.2 *p*C Recoil Polarimeter for the CNI Region

The *p*C polarimeter is useful due to very high luminosity and large cross section. Let us assume that the same equipment as at AGS is used (see Fig. 8.12; the diameter of the "detector" at RHIC (the circle on which the detector is located) is 80 cm).

In this case, the carbon strip target with a surface density of 3.5 μg/cm^2 (Jinnouchi et al. 2004) contains $1.75 \cdot 10^{17}$ carbon atoms/cm^2, which makes it possible to reach a luminosity of $L = 3.5 \cdot 10^{34}$ cm^{-2} s^{-1} at the U-70 accelerator. Taking a value of 9.5 mb for the cross section for the setup, the count rate is expected to be $3.3 \cdot 10^8$ events/s. This count rate should be naturally reduced by a factor of 10^4–10^5. This can be achieved in several ways. For example, one way is to use a target whose width is much smaller than the beam size. In this method, it is necessary to scan the beam to obtain the average, rather than local, beam polarization. At RHIC, the strip target has a width of 5 μm, whereas the beam size is about 1 mm. Therefore, it is expected that the count rate will decrease by a factor of 200. The count rate reduction factor for the same target in the U-70 accelerator with a beam size of about 20 mm is $4 \cdot 10^3$. Thus, the count rate becomes $8 \cdot 10^4$ events/s, which is acceptable. When the average analyzing power in the *p*C reaction is 1 %, the measurement of the beam polarization with an accuracy of 5 % requires a statistics of about $8 \cdot 10^6$ events. Therefore, the measurement at one point will take 100 s. For the highest energy of the U-70 accelerator with a plateau duration of 2 s, the result can be obtained approximately in 50 s. At higher energy, measurements at 10 points can be desired. Then, such measurements will take approximately 1000 s. Thus, the polarimeter on the *p*C reaction is the fastest polarimeter. Such a fast polarimeter could be very useful in accelerator adjustment, in measures against depolarizing resonances, etc.

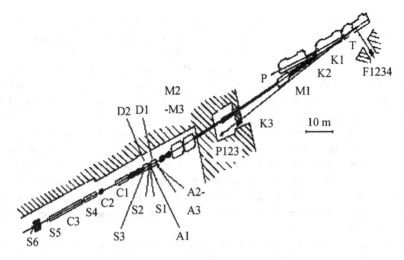

Fig. 8.13 Layout of beam line no. 2 of the U-70 accelerator, which can be used with the specified experimental setup as a relative polarimeter for the internal polarized beam of the U-70 accelerator: (P) is the circulating proton beam; (T) is the internal target; (K_1, K_2, K_3) are the collimators; (M_1, M_2, M_3) are the rotation magnets; (A_1–A_3, S_1–S_6) are the scintillation counters; (C_1–C_3) are the threshold Cherenkov detectors; (D_1, D_2) are the differential Cherenkov detectors; and (P_{123}) and (F_{1234}) are the scintillator telescopes for the control of the intensity of the internal proton beam incident on the target

There is another way to reduce the count rate directly by reducing the beam intensity. In this case, the control of the low-intensity circulating beam is a significant problem. The optimum can be found by combining these two methods.

8.4.4.3 External Inclusive Pion Polarimeter

After the commissioning of the U-70 accelerator, the yields of particles in inclusive reactions were measured with the use of internal targets. The particle emission angle varied from 0 to 15 mrad and the range of the momenta of secondary particles was 10–60 GeV/c. The particles were guided to beam line no. 2 (see Fig. 8.13). The effective solid angle and the range of momenta in the beam line were specified by collimators and magnetic dipoles and quadrupoles. The selection and identification of the particles were carried out using scintillation counters, as well as threshold and differential Cherenkov detectors. The particle momentum was measured also by Cherenkov detectors. The initial beam intensity was controlled by two scintillator telescopes (Gorin et al. 1971).

Now, it is proposed to use beam line no. 2 (or no. 14) and advanced equipment for measuring the polarization of the circulating beam using the analyzing power in the inclusive production of pions. The polarized proton beam with upward or downward polarization collides with the internal carbon target in the form of a foil with the sizes 50 μm (width) × 5 cm (height) × 20 μm (thickness). In this case, luminosity is 10^{35} cm^{-2}s^{-1}. Since a significant analyzing power is expected in the regions $p_T \geq 1.0$ GeV/c and $x_F \geq 0.5$, the corresponding production angle and momentum of secondary particles should be selected according to this requirement.

The differential cross section for the $p + $ Be reaction at 67 GeV/c was measured in Bozhko et al. (1980). The largest reached production angle was 20 mrad and the momentum of negative pions was 34 GeV/c. The differential cross section in the laboratory frame was $\frac{d\sigma}{d\omega \cdot dp} = 11 \frac{\text{mb}}{\text{sr} \cdot \text{GeV}}$. A count rate of $3 \cdot 10^4$ events/s is expected at a beam momentum spread of 2 % and an effective solid angle of 8 µsrad. Assuming that negative pions with higher momentum (e.g., 50 or 60 GeV/c) can be observed at the same angle and that the yield of such particles decreases by two orders of magnitude, the beam polarization can still be measured with a desired accuracy of 5 % in approximately 20 min. It will be a fast relative polarimeter operating on-line with a computer. It can be very useful for the adjustment of the accelerator with the polarized beam.

Note that the analyzing power of positive and negative pions at 70 GeV/c has not been directly measured. The authors took the interpolated values of the analyzing power between 22 and 200 GeV/c. However, the accuracy of such an interpolation is doubtful. Moreover, experimental data do not allow the measurement of the beam polarization with a necessary accuracy of 5 %. Therefore, measurements with the required accuracy should be performed at the U-70 accelerator. For this reason, an absolute polarimeter based on pp elastic scattering will be discussed below.

At present the magnetic bump system is used at U70 for directing the circulating proton beam to the internal target placed at distances ± 40 mm. The produced on this target the secondary charged particles are extracted by using the basic magnetic elements of U70. This technique could be also used in the case of the primary polarized proton beam.

8.4.4.4 External Absolute Polarimeter Based on pp Elastic Scattering

Polarization in pp elastic scattering at 70 GeV/c has been measured only at one point at Fermilab. The data show that $P = (2 \pm 1.3)$ % at $t = -0.3$ (GeV/c)^2. A more accurate measurement of polarization in pp elastic scattering at 45 GeV/c has been performed by the HERA collaboration (Gaidot et al. 1976). The average polarization in the range $0.2 \leq -t$ (GeV/c)$^2 \leq 0.3$ was found to be $P = (2.23 \pm 0.15)$ %. The cross section for this reaction has also been measured with a high accuracy. Therefore, the unpolarized beam can first be extracted by a bent crystal at this energy (the U-70 accelerator should be adjusted to energy of 45 GeV) and our equipment can be calibrated with the polarized target. Then, detailed measurements of the analyzing power should be carried out at energy of 70 GeV. If the analyzing power is close to 2 % as shown by the Fermilab experiment, and we know the target polarization can be measured with accuracy better than 5 %. Thus, we will have an absolute polarimeter for 70 and 45 GeV with accuracy better than 10 %. Then, it is possible to use the same target without polarization for scattering of the 70-GeV polarized beam, and the beam polarization at this energy can be

Fig. 8.14 Possible variant of
an external absolute proton
polarimeter for general
purposes; the elastic
scattering of protons on the
polarized proton target is
used; the polarimeter is based
on the 14th beam line at the
PROZA setup with the
improvement of the
equipment for the detection
of elastic scattering

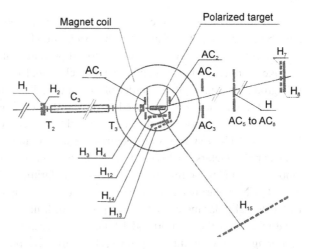

measured with desired accuracy. Moreover, if the extraction system and beam line
allow, this technique can be used in the entire energy range of the U-70 accelerator.
This technique becomes particularly favorable with a decrease in the beam energy,
because the analyzing power of pp elastic scattering increases in this case.

The count rate can be estimated under the assumption that we use the equipment
of the HERA collaboration (Gaidot et al. 1976) and the differential cross section
measured by them at $-t = 0.3$ (GeV/c)^2. An accuracy of 5 % can be achieved
approximately in 10 h if the beam intensity is $2 \cdot 10^7$ polarized protons per cycle.

Figure 8.14 shows the layout of a possible variant of the external absolute polarimeter for general purposes including the adjustment of the U-70 accelerator. It
has been successfully applied to measure the polarization of particles and antiparticles in the HERA experiment at IHEP (Gaidot et al. 1975).

The basic results of the discussion of proton polarimeters are summarized in
Table 8.1.

The vacuum in the U-70 accelerator is 10^{-6} Torr on average, whereas it is $2 \cdot 10^{-7}$ Torr in AGS and better than 10^{-8} Torr in RHIC. Therefore, it is necessary
to estimate the expected background on recoil detectors from the interaction of the
circulating beam with the residual gas in the ring.

For such a background, the beam chamber region "visible" for the detector is
dangerous. Since the recoil detector of the polarimeter identifies the interaction vertex only from the time of flight, its time resolution $\Delta\tau$ gives the length of this region
as $l = c \cdot \Delta\tau$ (for the background of relativistic particles). With a usual reserve, we
take this region as $3l$. Taking into account $\Delta\tau = \pm2$ ns, this region is found to be
±180 cm with respect to the center of the polarized jet target. Thus, the total length
of the "background" region is 360 cm. It is necessary to determine the number and
type of the nuclei in this region and to compare with the density of nuclei (protons)
in the working target.

The number of such atoms is $N(A) = (10^{-6}/760)n_L$, where $n_L = 2.68 \cdot 10^{19} A$/cm^3 is the Loschmidt number. Thus, $N(A) = 3.53 \cdot 10^{10} A$/cm^3. Hence, the
surface density will be $n_S = 1.27 \cdot 10^{13} A$/cm^2. This number should be compared

with the surface density of the working substance of the polarized target, $10^{12} p/cm^2$. Apparently, the number of nuclei of the background material is an order of magnitude larger than the number of the polarized protons. Taking into account that the average atomic number of nuclei in the background is $A = 28$, the excess of the background has an additional order of magnitude. Therefore, vacuum in the region of the jet target should be improved by three orders of magnitude, i.e., to 10^{-9} Torr.

In summary, we note that the program of the acceleration of the polarized proton beam at the U-70 synchrotron requires the development of a set of absolute and relative polarimeters. In this section, we analyze the achievements of other research laboratories and try to find effective tools for polarimetry. Two internal absolute pp and pC CNI polarimeters have been selected, following experience at RHIC. Two new external polarimeters have been proposed. It has been shown that appropriate polarimeters for the Ural-30 and 1.5-GeV booster are based on pC and pp elastic scattering. This is the first step for seeking effective polarimetry for the polarized beam at the U-70 accelerator. Investigations will be continued.

8.5 Polarimetry of Electron Beams

Three types of polarimeters (Mott, Møller, and Compton polarimeters) for electron beams, which are well studied and widely applied in polarimetry, are considered in this section.

8.5.1 Mott Polarimeter

The Mott polarimeter (Wood 1994) is based on the electron-polarization dependence of the differential cross section for the scattering of transversely polarized electrons on the electrons of heavy nuclei. This subject was reviewed in Gay and Dunning (1992). Three such polarimeters were constructed at SLAC, and all of them are used either in the development of polarized electron sources (Hopster and Abraham 1988) or in experiments on the test of the space or time parity (Haeberly 1992). This is because the analyzing power of the Mott polarimeter is large at low energies and decreases rapidly with increasing energy. In particular, the analyzing power near a scattering angle of 90° is 29 % at 3 MeV and decreases to 5 % at 15 MeV (Haeberly 1992). When adjusting the polarized electron sources at SLAC, the Mott polarimeter made it possible to achieve an accuracy of 2 % in the measurement of the polarization of an electron beam with energy of about 200 keV.

8.5.2 Møller Polarimeter

This polarimeter (Møller 1932) is absolute, because it is based on the elastic scattering of two longitudinally polarized electrons:

$$e(\rightarrow) + e(\rightarrow) = e + e, \tag{8.18}$$

where the horizontal arrows in parentheses indicate that the electrons are longitudinally polarized.

These polarimeters were widely used in experiments with polarized electron beams at SLAC. This is explained by two reasons. First, the reaction kinematics is very simple and simplifies the equipment design. Second, the differential cross section for this reaction is precisely calculated; in the center-of-mass frame, it is given by the expression (Band 1994)

$$\frac{d\sigma}{d\Omega} = \frac{\alpha^2}{s} \frac{(3 + \cos^2 \theta)^2}{\sin^4 \theta} \left[1 - P_z^B \cdot P_z^T \cdot A_z(\theta)\right]. \tag{8.19}$$

Here, s is the square of the total energy of the initial two-electron system; θ is the center-of-mass electron scattering angle; α is the fine structure constant; P_z^B and P_z^T are the longitudinal beam and target polarizations, respectively; and the Møller asymmetry $A_z(\theta)$ is also theoretically calculated by the formula

$$A_z(\theta) = \frac{(7 + \cos^2 \theta) \sin^2 \theta}{(3 + \cos^2 \theta)^2}. \tag{8.20}$$

According to formula (8.20), the analyzing power is maximal near $\theta = 90°$. Thus, both electrons in the laboratory frame are scattered at identical angles with identical energies. In the general case of the elastic scattering of particles with identical masses, their scattering angles in the laboratory frame ϑ_1 and ϑ_2 are related as

$$\cot \vartheta_1 \cdot \cot \vartheta_2 = \frac{E + m}{2m}, \tag{8.21}$$

where E and m are the total energy of an initial electron in the laboratory frame and its mass, respectively. It is easy to estimate that $\vartheta_1 = \vartheta_2 = \vartheta$ for a center-of-mass scattering angle of 90° and the laboratory scattering angle ϑ is determined from the relation

$$\tan \vartheta = \sqrt{\frac{2m}{E + m}}. \tag{8.22}$$

For an energy of 50 GeV, this angle is 4.4 mrad. The energy of each final electron is 25 GeV.

A thin foil of an iron–cobalt alloy with a small vanadium addition is usually used as the polarized electron target. From the beginning of 1992 to 1996, five such polarimeters were constructed. Two of them were used for the problems of the SLC collider itself, whereas the other three polarimeters have been intended for experiments E142, E143, and E154 on studying the spin structure functions of the nucleon. The polarimeter of experiment E154 will be described below. This polarimeter was also more recently used in experiment E155 (Band 1996).

The 48.3-GeV longitudinally polarized electron beam is scattered on a target consisting of six foils with thicknesses 20–154 µm. The target plane is inclined to the beam axis at an angle of 20.7°. Scattered electrons are detected within the laboratory polar angles 3.59–8.96 mrad (94°–105° in the center-of-mass frame). A special system of collimators (called mask by the authors) separates a constant azimuth angle range of 0.20–0.22 rad. Separated recoil electrons are bent by a septum magnet with a magnetic field of 1.1 T in the horizontal plane and are detected by strip silicon detectors. Thus, both electrons enter the same detectors owing to relativistic two-particle kinematics.

The electron target is magnetized by two Helmholtz coils. The same coils reverse the target polarization direction before each new statistics collection run. The polarization of the electrons of the target is given by the expression

$$P_T = \frac{M}{n_e \mu_B} \cdot \left(\frac{g'-1}{g'}\right) \cdot \left(\frac{g_e-1}{g_e}\right). \tag{8.23}$$

Here, M is the magnetization of the electron target, n_e is the electron density, μ_B is the Bohr magneton, g_e is the gyromagnetic ratio of the free electron, and g' is the so-called magnetomechanical ratio of the target nucleus. All quantities except g' and M are known from the standard tables. The parameter g' was measured in Scott and Sturner (1969) for an alloy containing 50 % of iron and 50 % of cobalt (without vanadium). It appeared to be $g' = 1.916 \pm 0.002$.

To determine the polarization of the electrons of the target, it is necessary to find the magnetization M, which is related to the magnetic induction B and magnetic field H as

$$4\pi M = B - H. \tag{8.24}$$

The magnetization M has been measured specially by introducing a ferromagnetic foil between the so-called pickup coils. Voltages are induced in pickup coils when the foil is scanned by the magnetic field from -100 G to $+100$ G. According to the Faraday law, the magnetic field appearing in the foil is equal to the difference between the integrals of the induced voltages in the presence and absence of the foil in the coil. Taking into account all factors, the total relative error in the measurement of the polarization of the electron target was found to be 1.7 % (Band and Prepost 1996).

Polarization was measured with six permendur foils 3 cm in width and 39 cm in length with different thicknesses. Polarization measured with the 20-µm foil was 8.03 % and polarization measured with 30-, 40-, and 154-µm foils was 8.14 %. The systematic measurement error was 1.7 %.

For the first experiments on the measurement of the spin structure functions, a single-arm spectrometer was used; i.e., one recoil particle was detected. Two-arm spectrometers are used beginning with experiment E143. In this case, both scattered electrons were selected in the vertical plane using collimators. Then, they were analyzed by the magnetic field (septum magnet) in the horizontal plane. Silicon cell detectors were applied to detect particles, and 10.2 × 10.2-cm lead glass counters were placed behind the detectors to determine the total energy of each electron. The detection of both electrons in coincidence (a resolution time of 1 ns) made it possible to suppress corrections on the Levchuk effect (Levchuk 1994) to a level of ≤ 1 %. This effect is caused by the contribution to asymmetry from the electrons from low-lying atomic levels (primarily from the K shell). These electrons are unpolarized, and corrections on their contribution should be introduced. In the double coincidence method, it was necessary to introduce small corrections for the dead time and acceptance of the equipment to raw asymmetry.

The Møller polarimeter is not a continuously working polarimeter. It is used once a day and measurements with it take about 40 min. The polarization of the electron beam measured with this polarimeter depended both on the applied foil and on the quantum efficiency of the polarized electron source (SLAC polarized gun). The measured polarization of the electron beam was in the range 83–86 % during experiment E143. The total systematic error of measurements is estimated to be no more than 2 %. These results were in agreement with the measurements with other SLAC polarimeters.

8.5.3 Compton Polarimeter

The Compton polarimeter is an absolute, local, on-line, constructive, and fast polarimeter. This polarimeter is based on the elastic scattering of longitudinally polarized electrons on circularly polarized photons (Gunst and Page 1953). The circularly polarized photons with a wavelength of 532 nm are produced in a Nd:YAG (neodymium-activated yttrium aluminum garnet) laser generator. The system of reflecting mirrors directs photons with the conservation of their circular polarization to the so-called Compton interaction point. Electrons backscattered on photons (in the center-of-mass frame) pass a deflecting magnet and enter drift tubes and Cherenkov counters from nine beam lines. Each of these lines can serve as a polarimeter. However, the seventh beam line had the largest analyzing power and its indications were used for determination of the longitudinal polarization of the electron beam. This polarimeter allowed the measurement of electron beam polarization with a relative accuracy of 1 % in 3 min. The total systematic error did not exceed 1.3 %, and the greatest contribution to this error came from discrepancy in the measurement of the polarization of laser photons.

8.6 Polarimetry of Muon Beams

A polarized muon beam was obtained at the CERN SPS due to the weak decay of pions. Since this process is weak, it is sufficiently well calculated. Nevertheless, two experimental methods for measuring muon beam polarization have been developed. Polarimeters for both cases are described in detail in Adeva et al. (1994). In the first case, the elastic scattering of longitudinally polarized muons by longitudinally polarized electrons is used for an absolute polarimeter. A magnetized ferromagnetic material (49 % Fe + 49 % Co + 2 % V) is used for a polarized electron target with a thickness of about 2 mm. This target is described in detail in de Botton et al. (1992). The special measurements have shown that the polarization of electrons in the target is $P_z^e = (8.34 \pm 0.1) \cos \theta$ %, where θ is the angle between electron target polarization and the muon beam direction (z direction). This angle is usually as small as $\approx 20°$. The systematic error is determined primarily by discrepancy in the determination of the magnetomechanical ratio in the used alloy. The polarization of electrons was reversed in each SPS cycle. Moreover, the target orientation was reversed every 2 h to suppress false asymmetry.

Both final particles, muon and electron, are detected by a two-arm spectrometer. In addition to the track detectors, an absorber and a lead glass wall are placed in the muon and electron shoulders, respectively. This polarimeter was slow and required several months for collecting statistics. The result was found to be $P_\mu^{\mu-e} = -(79.4 \pm 1.7)$ % in good agreement with the Monte Carlo calculation of muon beam polarization $P_\mu = -(78 \pm 5)$ % for a μ^+ beam energy of $E_\mu = 190 \, \text{GeV}$ (Medved 1998).

Another method for determining μ^+ beam polarization is the measurement of the energy spectrum of positrons from the decay of μ^+ muons in flight (Marie 1995). Owing to the weak decay of the muon, where parity is not conserved, positrons from the decay are distributed anisotropically in the muon rest frame. As a result, the energy spectrum of positrons in the laboratory frame obviously depends on muon polarization (Lifshitz and Pitaevskii 1971). The spectral distribution of positrons in the laboratory frame is given by the expression

$$\frac{dN}{dy} = N_0 \left[\frac{5}{3} - 3y^2 + \frac{4}{3}y^3 - P_\mu \left(\frac{1}{3} - 3y^2 + \frac{8}{3}y^3 \right) \right]. \tag{8.25}$$

Here, $y = E_e/E_\mu$ is the Michel parameter, where E_e and E_μ are the energies of positrons and muons in the laboratory frame, respectively; N_0 is the number of decayed muons, and P_μ is the muon beam polarization.

The equipment for measuring the spectrum of positrons was almost the same as in the preceding polarimeter. Only measures were taken to reduce the background from the positrons produced outside the 30-m decay region. The field in the magnet was reversed. The positron momentum resolution was improved due to an additional proportional chamber in front of the lead glass counters. In addition, its energy was measured by an electromagnetic calorimeter. Knowing the muon and positron energies E_μ and E_e, we can construct the Michel distribution in the parameter $y = E_e/E_\mu$. The parent muon polarization determined by comparing the measured and calculated Michel spectra appeared to be $P_\mu^{\mu \to e \nu \bar{\nu}} = -(80.6 \pm 2.9)$ % in a good agreement with the expected theoretical value.

The second method provides shorter measurement time (an accuracy of 3 % is achieved in 24 h (Marie 1995)) than the first polarimeter. However, it requires more difficult calculations and the inclusion of many factors to correct systematic errors. As a result, according to the presented data, the accuracy of the second polarimeter is lower than that of the first polarimeter.

References

Adeva, B., et al.: Nucl. Instrum. Methods Phys. Res. A **343**, 363 (1994)

Ado, Yu.M., et al.: In: Proceedings of the 1st Workshop on Polarization Phenomena in High Energy Physics, Protvino, USSR, p. 121 (1983)

Albrow, M.G., et al.: Nucl. Phys. B **23**, 445 (1970)

Azhgirei, L.S., et al.: Pis'ma Fiz. Elem. Chastits At. Yadra **113**(4), 51 (2002)

Band, H.R.: AIP Conf. Proc. **343**, 245 (1994)

Band, H.R.: In: Proceedings of the 12th International Symposium on High Energy Spin Physics, Amsterdam, The Netherlands, p. 765 (1996)

Band, H.R., Prepost, R.: E-103 Technical Note 110 (1996)

Belov, A.S., et al.: Nucl. Instrum. Methods Phys. Res. A **255**, 442 (1987)

Besset, D., et al.: Nucl. Phys. A **345**, 435 (1980)

Bevington, P.R., et al.: Phys. Rev. Lett. **41**, 384 (1978)

Bozhko, N.I., et al.: Yad. Fiz. **31**, 1494 (1980)

Bravar, A., et al.: In: Proceedings of the 16th International Spin Physics Symposium, Trieste, Italy, p. 507 (2004)

Bugg, D.V., et al.: J. Phys. G **4**, 1025 (1978)

Carey, D., et al.: Nucl. Instrum. Methods Phys. Res. A **290**, 269 (1990)

Cheng, D., et al.: Phys. Rev. **163**, 1470 (1967)

de Botton, N., et al.: IEEE Trans. Magn. **32**, 2447 (1992)

Gaidot, A., et al.: Phys. Rev. Lett. **57**, 389 (1975)

Gaidot, A., et al.: Phys. Lett. **61**, 103 (1976)

Gay, T.J., Dunning, F.B.: Rev. Sci. Instrum. **63**, 1635 (1992)

Gorin, Yu.P., et al.: Yad. Fiz. **14**, 994 (1971)

Greeniaus, L.G., et al.: Nucl. Phys. A **322**, 308 (1979)

Gunst, S.B., Page, L.A.: Phys. Rev. **92**, 970 (1953)

Haeberly, W.: In: Proceedings of the 10th International Symposium on High Energy Spin Physics, Nagoya, Japan, p. 923 (1992)

Hopster, H., Abraham, D.L.: Rev. Sci. Instrum. **59**, 49 (1988)

Jinnouchi, O., et al.: In: Proceedings of the 16th International Spin Physics Symposium, Trieste, Italy, p. 515 (2004)

Levchuk, L.G.: Nucl. Instrum. Methods Phys. Res. A **345**, 496 (1994)

Lifshitz, E.M., Pitaevskii, A.P.: Relativistic Quantum Theory. Nauka, Moscow (1971) (Pergamon, Oxford, 1974), Part 2

Marie, F.: In: Proceedings of the 6th Workshop on Spin Phenomena in High Energy Physics, Protvino, Russia, p. 183 (1995)

McNaughton, M.W., Chamberlin, E.P.: Phys. Rev. C **24**, 1778 (1981)

McNaughton, M.W., et al.: Phys. Rev. C **23**, 838 (1981a)

McNaughton, M.W., et al.: Phys. Rev. C **23**, 1128 (1981b)

Medved, K.S.: In: Proceedings of the 13th International Symposium on High Energy Spin Physics, Protvino, Russia, p. 538 (1998)

Møller, C.: Ann. Phys. **14**, 532 (1932)

Nurushev, S.B.: In: Proceedings of the 11th Workshop on High Energy Spin Physics, Dubna, Russia, p. 517 (2005)

Prokhorov, A. (ed.): Encyclopedic Dictionary of Physics, p. 572. Sovetskaya Entsiclopediya, Moscow (1984)

Ratner, L.: In: Proceedings of the Symposium on High Energy Physics with Polarized Beams, Argonne, IL, vol. 1, p. 1 (1974)

Scott, G.G., Sturner, H.W.: Phys. Rev. **184**, 184 (1969)

Shatunov, Yu.M.: In: Proceedings of the 11th Workshop on High Energy Spin Physics, Dubna, Russia, p. 531 (2005)

Smiths, A.A., et al.: Yad. Fiz. **23**, 142 (1981)

SPIN Collaboration: Acceleration of polarized protons to 120 and 150 GeV in Fermilab Main Injector. University of Michigan Report no. UM-HE 92-05 (1992)

Tarasov, E.K.: Preprint No. 232, ITEP, Moscow (1964)

Vasiliev, A.N.: In: Proceedings of the 11th Workshop on High Energy Spin Physics, Dubna, Russia, p. 449 (2005)

Wood, M.: AIP Conf. Proc. **343**, 230 (1994)

Part III
Polarization Experiments and Their Results

In the last decade, impressing engineering advance has been achieved in many directions of polarization technology; this subject was considered in the second part of this book. The quality of the polarized electron beam has been considerably improved at SLC. Beginning with a polarization of 22 % in 1992, physicists achieved a polarization of 63 % in 1993 and routinely work with a polarization of 80 % and a luminosity of $2 \cdot 10^{30}$ cm^{-2} s^{-1} since 2000. Continuous improvement of the parameters of the polarized electron beam provided favorable conditions for the performance of four experiments with a fixed target (E142, E143, E154, and E155) in which the nucleon spin structure functions were measured with a high accuracy. At the same time, the SMC Collaboration at CERN has successfully completed the scientific program on a muon beam and gave a place for the COMPASS Collaboration. The main aim of the COMPASS Collaboration is to measure the distribution of polarized gluons in a nucleon. The HERMES program on the polarized electron beam with energy of about 27 GeV has been successfully implemented since 1998 and now it's completed. The first polarized proton beam with energy of 100 GeV and a polarization of about 30 % was obtained in 2000 at the RHIC collider. A polarization of 50 % was achieved in 2004 and 65 % in 2009. Works on an increase in polarization to 70 % and energy to designed 250 GeV are continuing.

Investigation of spin phenomena in the production of hyperons remains an active research field. The E704 Collaboration has published data on the left–right asymmetry in the inclusive production of Λ hyperons and on the depolarization parameter D_{NN} in the $p \uparrow + p \to \Lambda + X$ reaction at 200 GeV/c. The first quantitative proof that the analyzing power of Λ is nonzero and is not equal to the polarization of Λ at such a high energy is of noticeable interest. The measurement of D_{NN} in the production of Λ hyperons opened a direct way to separate the quantitative contribution of the spin transfer mechanism, though the existence of this mechanism has qualitatively been shown by the Fermilab measurements of the polarization of the Ω hyperon. Hyperon polarization was so far studied mainly on unpolarized hadron beams with zero strangeness. The WA89 Collaboration at CERN has published data on hyperon polarization measured on the 320-GeV Σ^- hyperon beam opening new possibilities for studying hyperon polarization.

Single-spin asymmetry represents a strict test for nonperturbative approaches in quantum chromodynamics (QCD). The E704 Collaboration data on single-spin asymmetry in the inclusive production of pions in the region of the fragmentation of the polarized beam attracted great interest in spin scientific community owing to the observation of significant polarization effects in such reactions (Adams et al. 1996; Nurushev 1997, 2002). Similar measurements have been recently performed at STAR for inclusive π^0 mesons at $\sqrt{s} = 200$ GeV. Appreciable spin effects (Adams et al. 2004) close to those observed in the E704 experiment were found in the region of the fragmentation of the polarized proton beam. In the region of the fragmentation of the unpolarized beam, asymmetry appeared to be zero within the measurement accuracy. Zero asymmetry in the central region was found in the E704 experiment (Adams et al. 1996). Recent PHENIX measurements of the asymmetry of π^0 mesons in the central region provided the same conclusions (Astier et al. 2001).

Double-spin asymmetry A_{LL} is another very attractive quantity for perturbative quantum chromodynamics, because it is sensitive to gluon polarization. The A_{LL} values have long been predicted for pions, the direct production of photons, J/ψ, and heavier objects. These predictions have not yet been tested experimentally. The experimental data on double-spin asymmetry $A_{LL}(\pi^0)$ were obtained for the first time by the E704 Collaboration in 1991 (Adams et al. 1991). Asymmetry appeared to be consistent with zero. This result has been discussed in a number of theoretical works. The basic conclusion of these discussions was that gluon polarization is insignificant. The PHENIX data obtained ten years later at energy of $\sqrt{s} = 200$ GeV (an order of magnitude larger than the energy in the E704 experiment) provide the same conclusion (Adler et al. 2004).

Scientific community always emphasized the importance of polarization investigations on neutrino beams. More recently, the NOMAD Collaboration published data on the polarization of Λ and $\bar{\Lambda}$ hyperons obtained on a muon neutrino beam (Astier et al. 2000, 2001). These experiments are of certain interest and, probably, are the beginning of neutrino polarization physics.

Elastic and diffraction processes are of special interest for polarization physics. Such fundamental problems as the role of the pomeron in spin-flip interactions, relations between the polarizations of particles and antiparticles, search for odderon, asymptotic behavior of helicity amplitudes, scattering and asymptotic relations can be studied in detail in these processes.

The current theoretical efforts are primarily concentrated on the interpretation of the experimental data on the spin structure functions and single- and double-spin asymmetries. The nature of the nucleon spin is the central point for theoretical understanding of the internal structure of nucleons, and the deep understanding of this problem has not yet been achieved. Theoretical interest in single-spin measurements follows from the possibility of developing nonperturbative models as perturbative quantum chromodynamics cannot explain these data. Double-spin asymmetries provide a good way for determining parton distributions in nucleons from experimental data.

The future of polarization physics is rather enticing. A wide variety of spin measurements (the E154 and E158 experiments on fixed targets, precise measurements of the parameters of the electroweak theory using the modernized SLD detector on colliding electron–positron beams) are carried out at the Stanford linear accelerator (SLAC). The development of the polarization program for RHIC begins. Investigations of the spin phenomena by COMPASS collaboration are planned for many years. The largest polarization accelerators such as eRHIC, J-PARK, U-70, and PAX at FAIR are at different development stages. Thus, the prospect of polarization investigations is promising.

We will discuss below the basic experimental results from various fields of high energy spin physics. These results will be compared with the predictions of various theoretical models. Finally, the current representation of the nucleon spin structure will be briefly presented.

A considerable advance has been achieved in high energy spin physics. During a very short time, a number of new phenomena concerning the nucleon spin structure, hyperon polarization, and single- and double-spin asymmetries in the production of hadrons have been revealed owing to considerable successes in the development of polarization technology. For example, the polarized proton and antiproton beams at the Tevatron have been obtained, the Siberian snake technique and spin rotators have been developed and applied in the acceleration of polarized electron and proton beams to very high energies (at SLC, HERA, AGS and RHIC). Significant methodical innovations have also been implemented in the construction of polarized solid and gas targets. All these advances made it possible to begin to study fine effects such as the dependence of spin phenomena on the quark flavor, the spin transfer mechanism, to comparatively study the polarization of hyperons and antihyperons, the analyzing power in the inclusive production of hadrons by proton, antiproton, and lepton beams, etc. These and other news in high energy spin physics will be discussed in the third part of the book following primarily review (Nurushev 1997) with additions of more recent results.

References

Adams, D.L., et al.: Phys. Lett. B **261**, 197 (1991)

Adams, D.L., et al.: Phys. Rev. D **53**, 4747 (1996)

Adams, D.L., et al.: Phys. Rev. Lett. **92**, 171801 (2004)

Adler, S.S., et al.: Phys. Rev. Lett. **93**, 202002 (2004)

Astier, P., et al., NOMAD Collaboration: Nucl. Phys. B **588**, 3 (2000)

Astier, P., et al.: Nucl. Phys. B **605**, 3 (2001)

Nurushev, S.B.: Int. J. Mod. Phys. A **12**, 3433 (1997)

Nurushev, S.B.: In: Proceedings of the 16th International Baldin Seminar on High Energy Physics Problems, Dubna, Russia, vol. 1, p. 147 (2002)

Chapter 9
Investigation of the Nucleon Structure Functions

9.1 Polarization as a Precision Tool for Measuring the Parameters of the Standard Model

Many experimental tests of the Standard Model at LEP and SLC are based on the measurement of asymmetries following from parity violation in the vertex

$$Z \to f\bar{f}. \tag{9.1}$$

Here, $f, \bar{f}$ is a fermion–antifermion pair (e.g., quark–antiquark, electron–positron).

The so-called forward–backward asymmetry A_{FB}^f in the yield of fermion f is defined as the difference between the yields of quarks in the reaction

$$e^- + e^+ \to f + \bar{f} \tag{9.2}$$

in the forward and backward hemispheres in the center-of-mass frame. Such a test at LEP was performed with unpolarized electron and positron beams. The following asymmetry was measured:

$$A_{FB}^f = \frac{\sigma_F^f - \sigma_B^f}{\sigma_F^f + \sigma_B^f} = \frac{3}{4} A_e A_f. \tag{9.3}$$

Here, σ_F^f and σ_B^f are the cross sections for the production of fermions f (quarks) in the forward and backward hemispheres with respect to the electron motion direction, respectively. Asymmetries A_e and A_f are caused by parity violation at the $Z \to f\bar{f}$ vertex and are given by the expression

$$A_f = \frac{2v_f a_f}{v_f^2 + a_f^2} = \frac{g_L^2 - g_R^2}{g_L^2 + g_R^2}. \tag{9.4}$$

S.B. Nurushev et al., *Introduction to Polarization Physics*,
Lecture Notes in Physics 859, DOI 10.1007/978-3-642-32163-4_9,
© Moskovski Inzhenerno-Fisitscheski Institute, Moscow, Russia 2013

Here, v_f and a_f are the vector and axial vector coupling constants, respectively; and g_L and g_R are the coupling constants with the left and right helicities, respectively.

Two kinds of asymmetry can be additionally measured with a polarized electron beam. The first asymmetry is also forward–backward asymmetry defined by the expression

$$\tilde{A}_{FB}^f = \frac{[\sigma_F^f(L) - \sigma_B^f(L)] - [-\sigma_F^f(R) - \sigma_B^f(R)]}{[\sigma_F^f(L) + \sigma_B^f(L)] + [\sigma_F^f(R) + \sigma_B^f(R)]} = \frac{3}{4} P_e A_f. \tag{9.5}$$

A comparison of formula (9.5) with formula (9.3) shows that the collection of statistics in the measurement of forward–backward asymmetry is more efficient by a factor of 25 (as the ratio $(P_e/A_e)^2$) if an electron beam with polarization P_e ($\approx 80\,\%$) is used. Another advantage of formula (9.5) is that it does not contain A_e and includes only the electron beam polarization. This allows the determination of asymmetry in the production of heavy quarks irrespectively of the asymmetry parameter A_e.

The following asymmetry appearing with the polarized electron beam is called the left–right asymmetry and is defined as

$$A_{LR} = A_e = \frac{\sigma_L - \sigma_R}{\sigma_L + \sigma_R} = \frac{2(1 - 4\sin^2\theta_W^{eff})}{1 + (1 - 4\sin^2\theta_W^{eff})^2}. \tag{9.6}$$

Here, σ_L and σ_R are the total pole cross sections for the production of the Z boson for the case of the left and right polarizations of the electron beam, respectively. The electroweak mixing angle θ_W^{eff} is defined by the relation

$$\sin^2\theta_W^{eff} = \frac{1}{4}\left(1 - \frac{v_e}{a_e}\right). \tag{9.7}$$

The best accuracy in the mixing angle is achieved in single measurements with asymmetry A_{LR}.

The SLD experiment showed that a setup equipped with the polarized electron beam for the exact determination of the parameters of the Standard Model is a powerful setup. In the further discussion, this statement is illustrated on some examples from Hertzbach (1995a, 1995b).

The measurement of left–right asymmetry on the SLD detector from 1992 to 1995 gave the value

$$A_{LR} = A_e = 0.1551 \pm 0.0040, \tag{9.8}$$

which provided the following accuracy in the determination of the effective mixing angle (Weinberg–Salam angle):

$$\sin^2\theta_W = 0.23049 \pm 0.00050. \tag{9.9}$$

For comparison, all four experiments at LEP (ALEPH, DELPHI, L3, and OPAL) during the entire operation time jointly achieved the following accuracy in the measurement of this angle:

$$\sin^2 \theta_W = 0.23186 \pm 0.00034. \tag{9.10}$$

Combining the weak mixing angles measured in the SLD and LEP experiments, we obtain the world average value

$$\sin^2 \theta_W = 0.23143 \pm 0.00028. \tag{9.11}$$

The latest value of this angle at the mass M_Z (Z boson) is (Yao et al. 2006)

$$\sin^2 \theta_W = 0.23122 \pm 0.00015. \tag{9.11a}$$

It is necessary to note that statistics in the SLD experiment is much less than that in experiments at LEP. As mentioned above, this gain is due to the use of the highly polarized electron beam in the measurement of asymmetry.

The estimate of the top quark mass from Eq. (9.11) coincides with the value measured at Fermilab. The Higgs boson mass was also estimated, but this estimate does not provide strong constraints because of a large measurement error.

The next example of the utility of the polarized electron beam was the direct determination of left–right asymmetry for heavy quarks in the SLD experiment with the use of Z bosons produced in the annihilation of longitudinally polarized electrons and unpolarized positrons. Asymmetry A_b in the production of the "beauty" quark b is sensitive to the contributions from the top quark, intermediate charged bosons, probably, Higgs particles, and new physics. In the SLD experiment, the quantity A_b is determined using several measurement methods supplementing each other.

These methods simultaneously applied to c and b quarks are as follows.

1. The determination of momentum-weighed charges.
   The result is $A_b = 0.843 \pm 0.046$ (stat.) ± 0.051 (syst.).
2. Identification by charged K mesons.
   The result is $A_b = 0.91 \pm 0.09$ (stat.) ± 0.09 (syst.).
3. Identification by leptons with high p_T values.
   The result is $A_b = 0.87 \pm 0.07$ (stat.) ± 0.08 (syst.); $A_c = 0.44 \pm 0.11$ (stat.) ± 0.13 (syst.).
4. The reconstruction of charmed mesons D$^+$ and D^{*+}.
   The result is $A_c = 0.64 \pm 0.11$ (stat.) ± 0.06 (syst.).

The result averaged over all four measurements is

$$A_b = 0.858 \pm 0.054. \tag{9.12}$$

This value is in agreement with the LEP average value

$$A_b = 0.883 \pm 0.031. \tag{9.13}$$

Both these results are consistent with the theoretical value

$$A_b(\text{theory}) = 0.935, \tag{9.14}$$

demonstrating a good agreement with the Standard Model. Such an approach gives the following average result for the asymmetry of the charmed quark

$$A_c(\text{muons}) = 0.577 \pm 0.097, \tag{9.15}$$

which is in agreement with the average LEP value

$$A_c = 0.65 \pm 0.05 \tag{9.16}$$

and with the Standard Model prediction

$$A_c(\text{theory}) = 0.667. \tag{9.17}$$

The fitting of the parameters of the Standard Model to the SLD and LEP data imposes constraints on the t quark mass: it should be in the range 168 GeV $< m_t <$ 192 GeV. In the agreement with this expectation, the top quark with the mass $m_t = 174$ GeV has been discovered at Fermilab on the CDF and D0 detectors. The top quark mass was determined under the assumption that the Higgs boson mass is in the range 60 GeV $< m_H < 1000$ GeV. In 1992 and 1993, the same SLD Collaboration measured the Bhabha scattering cross sections with polarized electrons:

$$e^- + e^+ \rightarrow e^- + e^+. \tag{9.18}$$

The set of the data provides the following value of the effective vector coupling constant of the electron:

$$v_e = -0.0414 \pm 0.0020 \tag{9.19}$$

with the best accuracy. Thus, polarization investigations exhibit obvious advantages in the test of the Standard Model.

The test of QCD by measuring the polarization of quarks is the important direction of polarization investigations. According to QCD, a left (right) polarized electron forms mainly a left (right) "twisted" jet. The SLD collaboration studied the production of the jet in order to determine the polarization of the quark producing this jet. In this case, the following parameter was used:

$$\Omega = \vec{t} \cdot (\vec{k}_1 \times \vec{k}_2), \tag{9.20}$$

where $\vec{t}$ is along the jet axis and $\vec{k}_1$ and $\vec{k}_2$ are the momenta of two leading particles in the jet. Then, all jet events can be classified into two groups with $\Omega > 0$ and with $\Omega < 0$, and asymmetry can be composed of their difference and sum. Such asymmetry allows the determination of the "handedness" of the jet. This jet handedness H is asymmetry in the numbers of jets with positive and negative $\vec{\Omega}$ values.

The analyzing power α and polarization P of the quark are combined in the quantity H

$$H = \alpha \cdot P. \tag{9.21}$$

The SLD Collaboration did not find a noticeably nonzero handedness of the jet and provided the upper bound $\alpha < 0.053$ at a 95 % C.L. for jets of light quarks.

Another subject studied by the SLD Collaboration is the correlation β between the polarization $\vec{S}$ of Z^0 and the orientation of the plane of three jets. The triple product $\vec{S} \cdot (\vec{k}_1 \times \vec{k}_2)$, where $(\vec{k}_1)$ and $(\vec{k}_2)$ are the momenta of two jets with the highest energies, can be sensitive to physics outside of the Standard Model. The correlation appeared to be in the range $-0.022 < \beta < 0.039$; i.e., it is consistent with zero and, at a 95 % C.L. excludes any indications of new physics beyond the Standard Model at the existing accuracy level.

The SLD Collaboration observed differences between the spectra of baryons and antibaryons for $p/\bar{p}$ and $\Lambda/\bar{\Lambda}$ (in events identified as light quarks). The polarization of the electron beam was used to identify jets induced by quarks q or antiquarks $\bar{q}$, and the presences of baryons and antibaryons in quark jets are compared. Any difference between these identified events was not observed for momenta ≤ 12 GeV/c. However, increasing correlation between quark jets $q(\bar{q})$ and the presence of baryon (antibaryon) in jets was observed at momenta above 12 GeV/c. This indicates that faster particles capture an initial quark (antiquark) with a larger probability.

The JINR group has analyzed the longitudinal correlation between the directions of the jets in the decay $Z^0 \to$ two jets (Efremov et al. 1995). The correlation sign is opposite to the Standard Model prediction obtained under the assumption of the factorization of the q and $\bar{q}$ distribution functions. This fact and small directivity value $H_q = (1.22 \pm 0.67)$ % can be treated as the prevalence of the vacuum chromomagnetic fields over the own field of the quark. A further experimental test of this hypothesis is necessary to finally ascertain the presence or absence of the new phenomenon.

9.2 Spin Structure Functions

The measurement of the spin structure functions remains the most important problem in spin physics. Despite the enormous efforts of physicists, many important problems such as the origin of the nucleon spin, the behavior of the spin structure functions at small and large factorization values, collision noncollinearity, dependence on Q^2, and exact criteria of various sum rules have not yet been solved (Ellis and Karliner 1995, 1996). The latest results on the spin structure functions of the proton, neutron, and deuteron published by Collaborations SMC (Adams et al. 1997a, 1997b), SLAC E142 (Anthony et al. 1993, 1996), and SLAC E143 (Abe et al. 1995a, 1995b; Crabb 1995) are shown in Figs. 9.1 and 9.2.

The SMC (<u>S</u>pin <u>M</u>uon <u>C</u>ollaboration) experiment has significantly advanced in achieving the minimum Bjorken parameter values x (Fig. 9.1), where there are indications of an increase in $g_1^p(x)$ with decreasing x (Fig. 9.1a). The region of small x

Fig. 9.1 Spin structure
functions (**a**) $g_1^p(x)$,
(**b**) $g_1^d(x)$, (**c**) $g_1^n(x)$, and
(**d**) $g_2^p(x)$ measured in the
SMC experiment

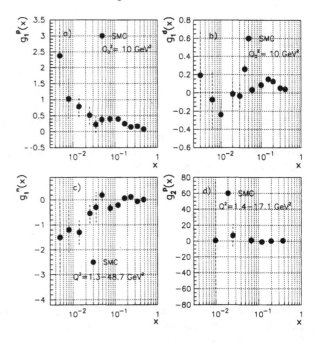

Fig. 9.2 Spin structure
functions (**a**) $g_1^p(x)$,
(**b**) $g_1^d(x)$, (**c**) $g_1^n(x)$, and
(**d**) $g_2^p(x)$ measured in the
E143 experiment

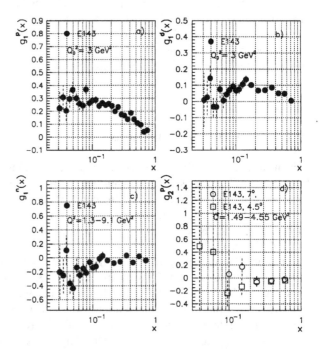

values attracts a lot of attention, because it is the only region where significant spin transfer is theoretically expected (Kaur 1977). The previous data on $g_1^d(x)$ (Adams et al. 1995) seemingly indicate appreciable negative values of this observable in the region $x < 0.02$ (see Fig. 9.1b). Accordingly, the neutron spin structure function $g_1^n(x)$ is also negative in almost the same x region. A more detailed presentation of the final results of the SMC experiment was given in Savin (1998).

Only from the SMC proton data, the following contributions of separate quarks to the nucleon spin have been obtained at $Q^2 = 10$ GeV^2:

$$\Delta\Sigma = 0.28 \pm 0.16, \qquad \Delta u = 0.82 \pm 0.05,$$
$$\Delta d = -0.44 \pm 0.05, \qquad \Delta s = -0.10 \pm 0.05. \tag{9.22}$$

Here, Δq, where $q = u, d, s$, are the contributions of the indicated quarks to the nucleon spin, the numbers refer to the proton spin, and $\Delta\Sigma$ is the sum of all contributions Δq.

Thus, only a small fraction of the nucleon spin, $\Delta\Sigma = 0.28 \pm 0.16$, is determined by the helicity of quarks, whereas the nonrelativistic quark model provides $\Delta\Sigma = 1$. This problem opened for the first time by the EMC Collaboration in 1987 was called "spin crisis" and has not yet been solved. Another surprising conclusion is that the strange quark sea is negatively polarized. It was generally expected that sea quarks should not be polarized. Many models have been developed under this assumption, but experiments appeared to be much richer than the expectations.

The next spin structure function measured by the SMC Collaboration was the so-called transverse spin function $g_2(x)$, which tests the combination of the transverse and longitudinal distributions of parton polarization in a nucleon. This distribution was measured in the scattering of a longitudinally polarized lepton on a transversely polarized nucleon. The function $g_2(x)$ consists of three components. The first component is determined by the leading twist-2 term $g_2^{WW}(x, Q^2)$ appearing from the same set of operators contributing to $g_1(x)$. The second component originates from twist-2 and is due to the transverse polarization distribution of quarks. The third contribution of the twist-3 dimension appears from the quark–gluon interactions. The result of the SMC experiment on the measurement of the spin structure function $g_2^p(x)$ is presented in Fig. 9.1d. The data on $g_2^p(x)$ are consistent with zero in the entire measured x region. A comparison of this result with theoretical calculations shows that the twist-2 term $g_2^{WW}(x, Q^2)$ well describes the data on $g_2^p(x)$ within the experimental accuracy and that the twist-3 term can be neglected. The accuracy of the measurement of $g_2^p(x)$ and the absence of a theoretical procedure for extrapolating this function to limiting x values do not allow the accurate test of the Burkhard–Cottingham sum rule

$$\Gamma_2^p = \int_0^1 g_2^p(x)dx = 0. \tag{9.23}$$

The E143 results (SLAC) (see Fig. 9.2) on the spin structure functions have a better accuracy in the measured x region and are in agreement with the SMC data in the overlapping x range ($> 4 \cdot 10^{-2}$). The E143 data were extrapolated to $x = 0$ using the (i) Regge model function $g_1^p(x) = C \cdot x^{-\alpha}$ and (ii) function $C \cdot \ln(1/x)$. The contribution from small unmeasured x values to the first moment is 0.006 ± 0.006 and 0.013 ± 0.003 in the first and second cases, respectively. These two values demonstrate the model dependence of the extrapolation to small x values. It has been found that the total contribution from quarks to the proton helicity is (27 ± 10) %, whereas the contribution from sea quarks to the proton helicity is $-(10 \pm 4)$ %, which is in agreement with the SMC results discussed above.

The results of the measurements of the spin structure function $g_1(x)$ for the deuteron and neutron are in agreement with the SMC data (see Figs. 9.2b and 9.2c), and they also show the tendency of change in the sign near $x \approx 0.05$. Some models predict such a change in the sign (Kaur 1977). The results of the measurements of the transverse spin structure function $g_2^p(x)$ in the E143 experiment are presented in Fig. 9.2d (Rondon 1995).

The data from two spectrometers placed at laboratory angles $4.5°$ and $7°$ are much below the bound following from the requirement of the positive sign of the distribution function.

They are consistent with the assumption that the transverse spin structure function $g_2^p(x)$ is small, but can be positive at $x < 0.1$ and becomes negative at $x > 0.1$. Such a behavior corresponds to the twist-2 contribution g_2^{WW} for the kinematics of this experiment. Large errors of the experimental data on g_2^{WW} do not exclude the possible twist-3 contribution of the same order. The authors made an estimate for the Burkhard–Cottingham sum rule. This integral is

$$\int_{0.03}^1 g_2^p(x)dx = -0.013 \pm 0.028 \qquad (9.24a)$$

for the proton and

$$\int_{0.03}^1 g_2^d(x)dx = -0.033 \pm 0.082, \qquad (9.24b)$$

for the deuteron. These results are consistent with zero values.

The experimental data make it possible to test two known sum rules. These sum rules provide relations between the first moments of the spin structure functions both symmetric and antisymmetric and the weak $SU(3)_f$ coupling constants F and D (vector and axial vector coupling constants, respectively). The Ellis–Jaffe sum rule has the form

$$\Gamma_1^{p(n)} = \int g_1^{p(n)}(x)dx = \pm\frac{1}{12}(F + D) + \frac{5}{36}(3F - D), \qquad (9.25)$$

and the Bjorken sum rule is given by the relation

$$\Gamma_1(Q^2) = \Gamma_1^p(Q^2) - \Gamma_1^n(Q^2) = \frac{1}{6}\left|\frac{g_A}{g_V}\right|\left[1 - \frac{\alpha_s(Q^2)}{\pi} - \cdots\right]. \qquad (9.26)$$

Fig. 9.3 Test of the sum
rules: (a) the Ellis–Jaffe sum
rule, (b) the Staude plot,
(c) $\Delta\Sigma$ vs. Δs, and (d) the
decomposition of the proton
spin into components

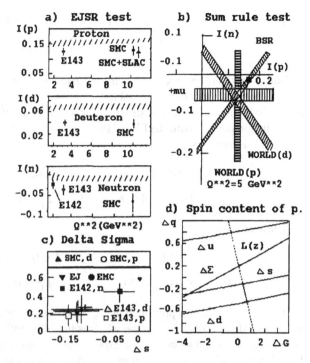

Tests of the Ellis–Jaffe sum rule are shown in Fig. 9.3a. The shaded bands are the
theoretical predictions for the first moments of the spin structure functions for the
proton, deuteron, and neutron. The data for the proton and deuteron are inconsistent
with the Ellis–Jaffe sum rule (the discrepancy is $\approx (4\text{–}6)\sigma$). The discrepancy for the
neutron is $\leq 3\sigma$ for the SMC and E143 data and is only 1σ for the E142 experiment.
In this case, a question concerning a correct extrapolation of the data to the point
$x = 0$ can arise.

The Ellis–Jaffe sum rule was tested by parameterizing lepton–proton asymmetry
A_1^p with the boundary conditions $A_1^p(x = 0) = 0$ and $A_1^p(x = 1) = 1$ (Nagaitsev
et al. 1996). The separate and joint analyses of the SMC and E143 data show the
mutual consistency of the data and indicate that the first moment Γ_1^p is more than
7σ below the predictions of the Ellis–Jaffe sum rule.

For the Bjorken sum rule, the E143 Collaboration provides

$$\Gamma_1^p - \Gamma_1^n = 0.163 \pm 0.010 \text{ (stat.)} \pm 0.016 \text{ (syst.)} \tag{9.27}$$

at $Q^2 = 3$ GeV^2. This result should be compared to the predicted value

$$\Gamma_1^p - \Gamma_1^n = 0.171 \pm 0.008. \tag{9.28}$$

The recent SMC results give

$$\Gamma_1^p - \Gamma_1^n = 0.183 \pm 0.034 \tag{9.29}$$

at $Q^2 = 10 \text{ GeV}^2$. This result should be compared to the predicted value

$$\Gamma_1^p - \Gamma_1^n = 0.186 \pm 0.002. \tag{9.30}$$

Both results are in agreement with the Bjorken sum rule at corresponding Q^2 values. The integrated experimental data (E80, E130, EMC, SMC, E142 and E143) on the first moment of the spin structure functions Γ_1 at $Q^2 = 5 \text{ GeV}^2$ are presented below in comparison with the Ellis–Jaffe sum rule and Bjorken sum rule.

The measurements:

$$\Gamma_1^p = 0.142 \pm 0.011, \qquad \Gamma_1^d = 0.038 \pm 0.006, \tag{9.31}$$

$$\Gamma_1^n = -0.061 \pm 0.016, \qquad \Gamma_1^p - \Gamma_1^n = 0.203 \pm 0.022. \tag{9.32}$$

The sum rules:

$$\Gamma_1^p = 0.164 \pm 0.005, \qquad \Gamma_1^d = 0.07 \pm 0.004, \tag{9.33}$$

$$\Gamma_1^n = -0.015 \pm 0.005, \qquad \Gamma_1^p - \Gamma_1^n = 0.181 \pm 0.003. \tag{9.34}$$

These results are also presented in Fig. 9.3b (Crabb 1995). The Bjorken sum rule is confirmed with a high accuracy, whereas the Ellis–Jaffe sum rule is inconsistent with the data (closed squares). The experimental results on the nucleon spin carried by valence quarks, $\Delta\Sigma$, and sea quarks, Δs, are presented in Fig. 9.3c for $Q^2 = 5 \text{ GeV}^2$. The $\Delta\Sigma$ value is close to 1/3 for all data, but the E142 data provide $\Delta\Sigma \approx 0.45$ for the neutron. On the average, 1/3 and 1/10 of the nucleon spin are carried by quarks and sea quarks, respectively, and the rest spin should be attributed to gluons and orbital angular momentum. The united experimental data provide the following decomposition of the nucleon spin at $Q^2 = 5 \text{ GeV}^2$:

$$\Delta u = 0.82 \pm 0.02, \qquad \Delta d = -0.43 \pm 0.02, \qquad \Delta s = -0.10 \pm 0.02. \tag{9.35}$$

Figure 9.3d helps to answer a question as how the nucleon spin could be shared between various parton components if the gluon contribution were known from independent measurements. It is seen that Δu provides the significant fraction of the spin, though it is considerably compensated by the negative contributions Δd and Δs to the total nucleon spin. An important objective is the real determination of the distribution of polarized gluons, which has not yet been measured precisely experimentally. This is the main problem both for the conducted experiments (COM-PASS, HERMES) and for the future programs (RHIC, U-70, J-PARK). The SMC, E142, and E143 data were also analyzed (Ramsey and Goshtasbpour 1995). This analysis differs from the approach of the experimental groups. In this analysis, the authors use the sum rules together with single experimental data in order to extract spin information, whereas the experimental groups use the data from many experiments to test the sum rules. Moreover, experimenters assume that the sea is symmetric in quark flavor and ignore the axially anomalous contribution. The authors discuss two models differing in gluon polarization. In the first case, they assume that

$\Delta G(x) = xG(x)$. In the second case, they choose $\Delta G(x) = 0$. They conclude that the naive quark model is insufficient for an explanation of the spin characteristics of the proton. They also notice that the total contribution from quarks to the proton spin Δq varies from 10 to 50 % in various experiments and the average value of this contribution is near $1/3$. Gluon polarization is very stable in various experiments ranging only from 0.44 to 0.46. In this case, the average contribution of the orbital angular momentum varies from -0.04 to -0.20. It is in agreement with Fig. 9.3d taken from Voss (1995). The model with $\Delta G = 0$ shows actually the same change in Δq as above with the positive contribution of the orbital angular momentum near 0.35. The sea contribution is negative in both variants of this model and is near -0.1. Such an appreciable polarization of the sea is the main cause of the violation of the Ellis–Jaffe sum rule. The authors conclude that further experimental investigations are necessary for the determination of the relative contributions from gluons and various flavors of sea quarks to the nucleon spin.

Aforementioned experiments determined the contributions Δu and Δd to the proton spin with a high accuracy, but they provided scarce information on the distributions of the polarizations of sea quarks and gluons. Since the region of small x values is very sensitive to the distributions of sea quarks and gluons (Ladinsky 1995), it is necessary to accurately measure the spin structure functions at small x values. This can be performed only at colliders of the HERA or RHIC type with the longitudinally polarized colliding electron and proton beams. Physicists already made the corresponding proposals, but they have not yet been accepted at HERA, since HERA is closed and are under development at RHIC (the eRHIC program).

It was proposed to use the Bjorken sum rule to determine the running strong coupling constant $\alpha_s(Q^2)$ (Ellis and Karliner 1995, 1996) instead of the experimental test of the Bjorken sum rule. According to the set of the experimental data,

$$\Gamma_1^p - \Gamma_1^n = 0.164 \pm 0.011. \tag{9.36}$$

Using this result and the Bjorken formula, one obtains

$$\alpha_s\left(M_z^2\right) = 0.116^{+0.003}_{-0.005}. \tag{9.37}$$

This value should be compared to the world average

$$\alpha_s\left(M_z^2\right) = 0.117 \pm 0.005. \tag{9.38}$$

A much higher accuracy is expected in forthcoming experimental data on the spin structure functions; this will open a new way for determining the running strong coupling constant by means of the Bjorken sum rule. The spin crisis becomes not so sharp as before, but it exists. Only $1/3$ of the nucleon spin is carried by quarks, and its other part has not yet been explained experimentally. It is possible to assume that the spin crisis concerns more likely theory rather than experiment. To justify this statement, we mention two theoretical works. Ma (1990) presented two statements clarifying the spin crisis problem: (i) the deep inelastic scattering of leptons serves as a probe for light (current) quarks rather than instanton (constituent) quarks, (ii) the proton spin in light cone dynamics is not simply the sum of the spins of individual quarks, but is the sum of the Melosh-rotating spins of light cone quarks. Using the credibility model, the author obtained the estimate $\Delta \Sigma = 0.227$, which is close to the experimental value $\Delta \Sigma = 0.29 \pm 0.06$. According Anselm and Ryskin (1995), a valence quark in the first stage of the evolution of the structure function can emit pseudoscalar mesons similar to π, η, and K and inverts the spin. As a result, the total chirality of quarks ($\Delta \Sigma$) decreases and a part of the spin is transformed to the orbital angular momentum of partons. Using a simple Hamiltonian with symmetric properties, the authors wrote the equation of the evolution for the structure function of the polarized quarks and could explain the spin crisis. According to their calculation, a part of the nucleon spin carried by quarks is 0.39 in agreement with the experimental data. If independent sources confirm the first or second theoretical work, this will mean that the spin crisis invented by theorists is closed by them.

The urgent problem in the research program for the nucleon spin structure is the necessity of a more accurate measurement of the gluon spin distributions. Physicists at the largest operating polarization setups such as COMPASS as well as at RHIC beginning a serious polarization program, work on this problem. In distant prospect, similar investigations are planned at the U-70 accelerator (Protvino, Russia) and J-PARC accelerator (Japan).

References

Abe, K., et al.: Phys. Rev. Lett. **74**, 346 (1995a)

Abe, K., et al.: Phys. Rev. Lett. **75**, 25 (1995b)

Adams, D., et al.: Preprint CERN-PPE/95-97 (1995)

Adams, D., et al.: Phys. Rev. D **56**, 5330 (1997a)

Adams, D., et al.: Phys. Lett. B **396**, 338 (1997b)

Anselm, A.A., Ryskin, M.G.: Z. Phys. C **68**, 297 (1995)

Anthony, P.L., et al.: Phys. Rev. Lett. **71**, 959 (1993)

Anthony, P.L., et al.: Phys. Rev. D **54**, 6620 (1996)

Crabb, D.G.: In: Proceedings of the 6th Workshop on High Energy Spin Physics, Protvino, Russia, vol. 1, p. 25 (1995)

Efremov, A., et al.: In: Proceedings of the 6th Workshop on High Energy Spin Physics, Protvino, Russia, vol. 2, p. 113 (1995)

Ellis, J., Karliner, M.: Phys. Lett. B **341**, 397 (1995)

Ellis, J., Karliner, M.: Phys. Lett. B **366**, 268 (1996)

Hertzbach, S.: In: Proceedings of the 6th Workshop on High Energy Spin Physics, Protvino, Russia, vol. 2, p. 5 (1995a)

Hertzbach, S.: In: Proceedings of the 6th Workshop on High Energy Spin Physics, Protvino, Russia, vol. 2, p. 219 (1995b)

Kaur, J.: Nucl. Phys. B **128**, 219 (1977)

Ladinsky, G.: In: Proceedings of the Workshop on the Prospect of Spin Physics at HERA, Zeuthen, p. 285 (1995)

Ma, B.-Q.: Preprint BIHEP-TH-90-36, Beijing (1990)

Nagaitsev, A.P., et al.: JINR Rapid Commun. No. 3, p. 59, Dubna (1996)

Ramsey, G.P., Goshtasbpour, M.: In: Proceedings of the 6th Workshop on High Energy Spin Physics, Protvino, Russia, vol. 1, p. 55 (1995)

Rondon, O.A.: In: Proceedings of the 6th Workshop on High Energy Spin Physics, Protvino, Russia, vol. 1, p. 15 (1995)

Savin, I.A.: In: Proceedings of the 13th International Symposium on High Energy Spin Physics, Protvino, Russia, p. 78 (1998)

Voss, R.: In: Proceedings of the Workshop on the Prospect of Spin Physics at HERA, Zeuthen, p. 25 (1995)

Yao, W.-M., et al.: J. Phys. G: Nucl. Part. Phys. **33**, 97 (2006)

Chapter 10
Hyperon Polarization

10.1 Dependence of Hyperon Polarization on Reaction Parameters

There are several detailed reviews on hyperon polarization (Pondrom 1985; Heller 1985; Lach 1994; Panagiotou 1990). These works discussed the dependences of hyperon polarizations on the initial energy, momentum transfer, Feynman parameter, and flavors of initial and final quarks. The above reviews contain the following conclusions:

- *the energy dependence of polarization*: it was certainly stated that Λ hyperon polarization is practically independent of the beam energy;
- *the dependence of polarization on p_T*: Λ hyperon polarization at a fixed Feynman parameter x_F increases linearly with p_T to $p_T \approx 1$ GeV/c and then becomes constant, i.e., exhibits a plateau whose height depends on x_F;
- *the dependence of polarization on x_F*: polarization increases almost linearly with x_F at a fixed momentum transfer p_T. The last two statements have been represented by the useful analytical expression (Pondrom 1985):

$$P(x, p_T) = -\left(a \cdot x_F + b \cdot x_F^3\right) \cdot \left[1 - \exp\left(-c \cdot p_T^2\right)\right]. \tag{10.1}$$

This function is an empirical function, because it has no strict theoretical proof. However, this function meets some general requirements. For example, it vanishes at $p_T = 0$ in agreement with the conservation of the angular momentum. It is an odd function of x_F as should be owing to the identity of two interacting protons. The parameters a, b, and c in the resulting formula have been determined by fitting it to the experimental data on Λ hyperon polarization at 400 GeV/c. These fitting parameters are as follows (Lundberg et al. 1989): $a = -0.268 \pm 0.003$, $b = -0.338 \pm 0.015$, and $c = (4.5 \pm 0.6)$ (GeV/c)$^{-2}$ with χ^2/DOF $= 109.4/69$ (DOF is the number of degrees of freedom). These parameters depend on the initial energy and flavor of produced hyperons (Λ, Σ, Ξ, etc.). The above formula corresponds to the factorization of the dependence of the polarization on x_F and p_T. The dependence on x_F reflects the identity of the initial particles (pp collisions), but is also used to describe the interaction of protons with nuclei. In view of the first item in the above list, P_Λ is factorized as

$$P_\Lambda = f_1(s) \cdot f_2(p_T) \cdot f_3(x_F). \tag{10.2}$$

The known DeGrand–Miettinen (DGM) model (DeGrand and Miettinen 1981a; DeGrand et al. 1985) does not indicate the factorization in the variables x_F and p_T (see below) and polarization decreases at large momentum transfers p_T. Formula (10.1) is very useful for comparison of various experimental data on hyperon polarization. Though all experimentally measured hyperon polarizations have simple dependences on all kinematic parameters (s, p_T, and x_F), theoretical models do not provide simple analytical expressions similar to the above empirical expression. The only known exception is the DGM model discussed below.

The program of the complete experiment for the inclusive production of hyperons is formulated in Swallow (1974). This program is similar to "a complete set of experiments" for nucleon–nucleon elastic scattering, which was formulated in the mid-1950s by Wolfenstein (1954) and Puzikov et al. (1957). According to this program, most experiments in the complete set are the measurements with a polarized proton beam and a polarized target. Such a program on the measurements of polarization observables in the inclusive production of hyperons began with the experiments at ZGS (Swallow 1974), AGS (Nessi 1988; Bonner et al. 1989), and Tevatron (Bravar 1996; Penzo 1995).

10.1.1 Energy Dependence of Λ Hyperon Polarization

In this section, we primarily discuss the energy dependence of Λ hyperon polarization (function $f_1(s)$ presented above), because it was extensively investigated in a wide energy range. The importance of these investigations follows from the fact that any polarization is proportional to the product

$$P \propto \frac{|F_{sf}|}{|F_{nf}|} \sin\alpha, \tag{10.3}$$

where F_{sf} is the spin-flip amplitude, F_{nf} is the spin-nonflip amplitude, and α is the phase difference between these two amplitudes. The amplitudes and phase are functions of the initial energy, Feynman parameter x_F, and transverse momentum transfer p_T. For polarization to be nonzero at any energy, F_{sf} and phase should be nonzero. The most popular theoretical models such as the Regge model or QCD predict a fast decrease in polarization with increasing energy. We do not know a model indicating an increase in polarization with energy. Polarization can seemingly be energy independent if, for example, interaction is carried only through a pomeron and the pomeron ensures spin flip. Such a possibility is indicated by the experimental data on polarization in the elastic scattering of hadrons at high energies (Nurushev 1990). The conclusion on the presence of the constant spin contribution to the differential section for the unpolarized pp elastic scattering in the energy range 9–500 GeV was previously made in Soffer and Wray (1973). The asymptotic behavior of the spin-flip amplitude can be determined, in particular, from the experimental data on hyperon polarization owing to a wide covering energy range. A brief review of the data on the energy dependence of Λ hyperon polarization will be given below.

There are some approaches to this subject. The first approach is associated with the results of the measurements of Λ hyperon polarization at the collider (Smith et al. 1987). The measurements are carried out at four ISR energies $\sqrt{s} = 31, 44, 53$, and 62 GeV. These data were approximated by the linear function

$$-P_\Lambda = a_1 + a_2 \cdot \left(\sqrt{s} - 62.4\right). \tag{10.4}$$

The fit of this function jointly to three data sets: (i) $\bar{x}_F = 0.39$, $p_T = 0.56$ GeV/c; (ii) $x_F = 0.58$, $p_T = 0.81$ GeV/c; and (iii) $x_F = 0.77$, $p_T = 0.92$ GeV/c gives the slope parameter $a_2 = (0.027 \pm 0.055)$ %/GeV. Therefore, pure change in polarization in an energy range from 31 to 62 GeV is (0.8 ± 1.7) %. This result shows that polarization is constant over the entire ISR energy range. The second approach is based on the results of the experiments with a fixed target at momenta from 12 to 400 GeV/c (Panagiotou 1990). The x_F and p_T values were taken in the polarization data in the regions $0.43 < x_F < 0.58$ and 1.15 GeV/c $< p_T < 1.58$ GeV/c, respectively. The data at four energies were fitted by the function

$$-P_\Lambda = a_1 + a_2 \cdot \ln p_L, \tag{10.5}$$

where p_L is the laboratory momentum of the incident beam. The χ^2 fit gives the parameters $a_1 = -26.3 \pm 2.0$ and $a_2 = 0.7 \pm 0.1$. Note that the slope parameter is positive (while the sign of the constant term is opposite), which reduces asymptotic polarization. Pure polarization change in the momentum range from 12 to 400 GeV/c is (2.45 ± 0.35) %. The slope parameter is nonzero, but demonstrates a very weak dependence on the initial energy.

It is difficult to compare these two results as they have no common kinematic regions. The unification of the results for the fixed target and collider is complicated because most works contain primarily only plots rather than tables of experimental data on polarization. Plots are obviously inappropriate sources of polarization values for a quantitative analysis. However, these sources have no alternative in some cases. The data thus collected are presented in Tables 10.1 and 10.2 (note that the negative sign of Λ polarization is included in the column heading, $-P_\Lambda$ and this quantity is given in percent). The last column is intended for comments and the previous column gives references to original works. Three functions were used for fitting over the entire energy range: two of them coincide with the functions given in Eqs. (10.4) and (10.5), and the third function appears in the Regge pole model:

$$-P_\Lambda = a_1 \cdot p_L^{a_2}. \tag{10.6}$$

Here, the parameter a_2 is the effective Regge trajectory. In the case of elastic scattering, it varies between -0.5 and -1 for small values of $-t$ ($-t \approx 0.1$ (GeV/c)^2)

Table 10.1 Data on the energy dependence of Λ hyperon polarization divided into x_F and p_T bins and fitting results

No.	p_L, GeV/c	$\sqrt{s}$, GeV	$-P_\Lambda$, %	Reference	Comment
\multicolumn{6}{l}{$\bar{x}_F = 0.39$, $\bar{p}_T = 0.56$ GeV/c}					
1	12.0	4.74	4.85 ± 2.05	(Abe et al. 1986)	2 points, $\bar{x}_F = 0.37$
2	300.0	23.7	0.0 ± 2.0	(Scubic et al. 1978)	$p_T = 0.58$ GeV/c
3	512.3	31.0	6.4 ± 2.9	(Smith et al. 1987)	From plots
4	1032.0	44.0	8.1 ± 2.6	(Smith et al. 1987)	From plots
5	1497.0	53.0	5.3 ± 2.4	(Smith et al. 1987)	From plots
6	2049.0	62.0	7.5 ± 0.8	(Smith et al. 1987)	From plots

1 $\quad -P_\Lambda = a_1 + a_2 \cdot \ln p_L$, fitting parameters $a_l = 0.87 \pm 2.88$, $a_2 = 0.8 \pm 0.41$; $\chi^2/\text{DOF} = 2.4$, $NP = 6$

2 $\quad -P_\Lambda = a_l + a_2 \cdot \sqrt{p_L}$, fitting parameters $a_l = 2.23 \pm 1.74$, $a_2 = 0.11 \pm 0.04$; $\chi^2/\text{DOF} = 1.8$, $NP = 6$

3 $\quad -P_\Lambda = a_1 \cdot p_L^{a_2}$, fitting parameters $a_l = 0.81 \pm 1.24$, $a_2 = 0.29 \pm 0.21$; $\chi^2/\text{DOF} = 2.04$, $NP = 6$

$\bar{x}_F = 0.58$, $\bar{p}_T = 0.81$ GeV/c

No.	p_L, GeV/c	$\sqrt{s}$, GeV	$-P_\Lambda$, %	Reference	Comment
1	12.0	4.74	14.9 ± 1.3	(Abe et al. 1986)	Aver. over 2 points
2	176.0	18.2	7.9 ± 5.5	(Gourlay et al. 1986)	New data
3	300.0	23.7	12.2 ± 1.2	(Scubic et al. 1978)	Aver. over 2 points
4	512.3	31.0	15.1 ± 2.6	(Smith et al. 1987)	From plots
5	1032.0	44.0	17.4 ± 3.5	(Smith et al. 1987)	From plots
6	1497.0	53.0	18.0 ± 3.2	(Smith et al. 1987)	From plots
7	2049.0	62.0	17.4 ± 1.2	(Smith et al. 1987)	From plots

1 $\quad -P_\Lambda = a_1 + a_2 \cdot \ln p_L$, fitting parameters $a_l = 12.38 \pm 1.99$, $a_2 = 0.46 \pm 0.33$; $\chi^2/\text{DOF} = 2.1$, $NP = 7$

2 $\quad -P_\Lambda = a_l + a_2 \cdot \sqrt{p_L}$, fitting parameters $a_l = 12.77 \pm 1.17$, $a_2 = 0.09 \pm 0.04$; $\chi^2/\text{DOF} = 1.4$, $NP = 7$

3 $\quad -P_\Lambda = a_1 \cdot p_L^{a_2}$, fitting parameters $a_l = 12.27 \pm 2.22$, $a_2 = 0.04 \pm 0.03$; $\chi^2/\text{DOF} = 2.0$, $NP = 7$

(Nurushev 1990). The fitting results for inclusive polarization are shown in Fig. 10.1 and presented in Tables 10.1 and 10.2. The inclusion of new experimental data with high accuracies actually worsens some earlier fits (see χ^2/DOF values in the tables, where NP is the number of experimental points and DOF means degrees of freedom).

The basic conclusions of the above analysis are as follows.

(i) All functions provide similar descriptions of the experimental data (the dashed line is the function $\sqrt{s}$, the dotted line is $\ln p_L$, and the solid line is the Regge

Table 10.2 Data on the energy dependence of Λ hyperon polarization divided into x_F and p_T bins and fitting results

No.	p_L, GeV/c	$\sqrt{s}$, GeV	$-P_\Lambda$, %	Reference	Comment
$\bar{x}_F = 0.77$, $\bar{p}_T = 0.92$ GeV/c					
1	12.0	4.7	21.0 ± 2.7	(Abe et al. 1986)	New data
2	300.0	23.7	$16.0 \pm 9.$	(Scubic et al. 1978)	New data
3	512.3	31.0	32.0 ± 4.1	(Smith et al. 1987)	From plots
4	1032.0	44.0	27.9 ± 6.4	(Smith et al. 1987)	From plots
5	1497.0	53.0	33.1 ± 5.8	(Smith et al. 1987)	From plots
6	2049.0	62.0	29.7 ± 4.3	(Smith et al. 1987)	From plots

1. $-P_\Lambda = a_1 + a_2 \cdot \ln p_L$, fitting parameters $a_1 = 16.0 \pm 4.3$, $a_2 = 2.0 \pm 0.8$; $\chi^2/\text{DOF} = 0.7$, $NP = 6$

2. $-P_\Lambda = a_1 + a_2 \cdot \sqrt{p_L}$, fitting parameters $a_1 = 20.8 \pm 2.8$, $a_2 = 0.26 \pm 0.11$; $\chi^2/\text{DOF} = 0.9$, $NP = 6$

3. $-P_\Lambda = a_1 \cdot p_L^{a_2}$, fitting parameters $a_1 = 17.4 \pm 3.8$, $a_2 = 0.08 \pm 0.04$; $\chi^2/\text{DOF} = 0.7$, $NP = 6$

$0.43 \leq x_F \leq 0.58$, 1.15 GeV/c $\leq p_T \leq 1.58$ GeV/c

No.	p_L, GeV/c	$\sqrt{s}$, GeV	$-P_\Lambda$, %	Reference	Comment
1	12.0	4.7	18.2 ± 1.85	(Abe et al. 1986)	Aver. over 4 points
2	24.0	6.7	25.7 ± 8.1	(Panagiotou 1990)	From plots
3	300.0	23.7	16.0 ± 1.8	(Scubic et al. 1978)	Aver. over 3 points
4	400.0	27.4	21.1 ± 0.8	(Lundberg et al. 1989)	Aver. over 2 point
5	2049.0	62.0	25.4 ± 2.2	(Smith et al. 1987)	From plots

1. $-P_\Lambda = a_1 + a_2 \cdot \ln p_L$, fitting parameters $a_1 = 14.6 \pm 2.9$, $a_2 = 1.0 \pm 0.5$; $\chi^2/\text{DOF} = 3.12$, $NP = 5$

2. $-P_\Lambda = a_1 + a_2 \cdot \sqrt{p_L}$, fitting parameters $a_1 = 16.9 \pm 1.5$, $a_2 = 0.18 \pm 0.07$; $\chi^2/\text{DOF} = 2.2$, $NP = 5$

3. $-P_\Lambda = a_1 \cdot p_L^{a_2}$, fitting parameters $a_1 = 14.7 \pm 3.5$, $a_2 = 0.06 \pm 0.04$; $\chi^2/\text{DOF} = 3.0$, $NP = 5$

pole dependence). The dash–dotted line in Fig. 10.1d corresponds to the fit from Panagiotou (1990). In this case, the slope is positive, whereas the slope of the new fit (with the Regge pole) is negative (see the solid line in Fig. 10.1d). This difference is insignificant now, but can lead to an essential discrepancy at asymptotic energy.

(ii) The energy dependence of Λ hyperon polarization is very weak, though there is some indication that polarization can vary with energy.

(iii) The existing data are scarce; therefore, additional measurements of polarization in a wide region of kinematic variables are necessary.

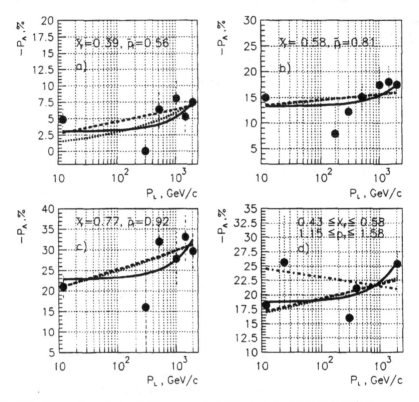

Fig. 10.1 Energy dependence of Λ hyperon polarization in various kinematic regions

Most theoretical models assume that hyperon polarization is energy independent. However, an interesting variety of hyperon polarizations was recently revealed. Figure 10.2a shows that polarization at ISR (marked by asterisks) at $\sqrt{s} = 62$ GeV (corresponding to $p_L = 2049$ GeV/c in experiments with the fixed target) is seemingly slightly larger than that at 800 GeV/c in experiments with the fixed target (closed circles) (Smith et al. 1987). The experiment reported in Smith et al. (1987) was performed on a beryllium target and a nuclear effect was undoubtedly presented in this experiment.

This nuclear effect can reduce Λ hyperon polarization only by ≈ 10 %, and this reduction is insufficient to explain the difference between polarizations. Hence, we can believe that there is a weak indication of an increase in the spin effect with energy. If this is confirmed, this phenomenon can become the important and surprising discovery.

The polarization of Σ^+ decreases with increasing energy (Cooper et al. 1983; Morelos et al. 1993) (see Fig. 10.2b), whereas the polarization of Ξ^- increases with energy (see Fig.10.2c) (Rameika et al. 1986; Duryea et al. 1991). This variety in the energy dependences of hyperon polarizations can give an important key to the detection of the source of hyperon polarization.

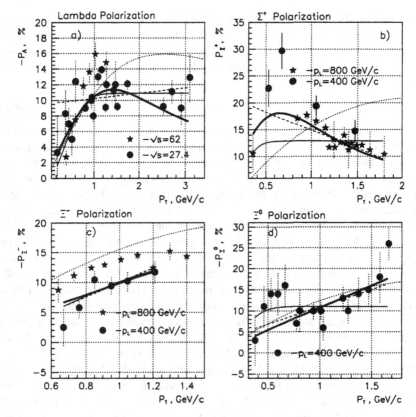

Fig. 10.2 Polarizations of the (**a**) Λ, (**b**) Σ^+, (**c**) Ξ^-, and (**d**) Ξ^0 hyperons produced by the proton beam versus p_T

10.1.2 Dependence of Hyperon Polarization on p_T

Another important subject is the dependence of hyperon polarizations on the momentum transfer p_T. The spin-flip amplitude F_{sf} in Eq. (10.3) should decrease as p_T at $p_T \to 0$ according to the conservation of the total angular momentum, whereas such a constraint does not exist for the nonspin-flip amplitude F_{nf}. Therefore, Λ hyperon polarization should begin to increase from zero linearly with p_T and, as data indicate, this dependence becomes flat near $p_T \approx 1$ GeV/c. Such a feature has been described by the parametrization (Lundberg et al. 1989):

$$-P_\Lambda = f_2(p_T) = a_1 \cdot \left[1 - \exp\left(-a_2 \cdot p_T^2\right) \right]. \tag{10.7}$$

The function $f_2(p_T)$ provides a good analytical description of the experimental data, which is very useful for comparison of various measurements (see examples below). However, this function has not two important features: its behavior at low transverse momenta is p_T^2 rather than p_T, and it is constant at high transverse momenta in contrast to the p_T^{-1} behavior with increasing p_T predicted by perturbative

QCD (pQCD) (Efremov 1978). We recall that both these features are present in the DGM recombination model mentioned above. These two features could be taken into account by multiplying function (10.7) by the term $1/p_T$ without introducing any new free parameter. Thus, new function $f_2(p_T)$ originating from pQCD has the form

$$f_2(p_T) = a_1 \cdot \left[1 - \exp\left(-a_2 \cdot p_T^2\right)\right] \cdot 1/p_T. \tag{10.8}$$

Two above functions and the function linear in p_T were fitted to the experimental data on P_Λ in the bin $0.3 < x_F \le 0.4$ at an initial momentum of 400 GeV/c reported in Heller et al. (1978). The fitting results are presented in Fig. 10.2b and Table 10.3. The thin solid line is the function $f_2(p_T)$ from Lundberg et al. (1989) given by Eq. (10.7) (the first line of the fitting parameters in Table 10.3), the dashed line is the linear approximation (the second line of the fitting parameters in Table 10.3), and the thick solid line is the pQCD stimulated function given by Eq. (10.8) (the third row in Table 10.3). It is seen in this figure that a good quantitative description of the data at a momentum of 400 GeV/c ($\sqrt{s} = 27.4$ GeV) is absent. Since the experimental points are strongly scattered, all three functions provide equally good descriptions in the measured p_T region. It is only possible to state that pQCD predictions for high p_T values do not contradict the existing experimental data. To confirm or reject this statement, it is necessary to expand the measurement region to higher p_T values. The data at the ISR energy $\sqrt{s} = 62$ GeV (see Fig. 10.2a) are better described by dependence (10.7) than by the other functions (see χ^2 values in Table 10.3). The DGM model predictions are presented in Fig. 10.2 by the dotted lines. The prediction is in good agreement with the experimental data for the polarization of Λ up to $p_T \le 1.5$ GeV/c, and some discrepancy is then observed. Since the calculations have been performed for the fixed value $x_F = 0.35$ and the experimental data have been averaged over the range $0.3 < x_F < 0.4$, it is difficult to explain such discrepancy only by the dependence of polarization on x_F. As the model has no free parameters, such an agreement is surprising. Very good agreement is observed for the polarization of Ξ^0 (see Fig. 10.2d). An appreciable discrepancy appears in the polarization of Σ^+ (see Fig. 10.2b). First, according to the DGM model, hyperon polarization should be independent of the initial beam energy. This is not the case: $P_{\Sigma^+}(800$ GeV/$c)$ is smaller than $P_{\Sigma^+}(400$ GeV/$c)$. Second, p_t dependence has a peak at $p_T \cong 1$ GeV/c, which is not predicted by the model. Finally, the model does not provide the quantitative description of the p_T dependence of Σ^+ polarization.

The data in Fig. 10.2b are better described by the pQCD function (the thick solid line) at both energies (see Table 10.3), though the p_T values are insufficiently high for the applicability of pQCD. The data for Ξ^- (see Fig. 10.2c) are well described by all three functions at 400 GeV/c (data from Rameika et al. 1986) and at 800 GeV/c (data from Duryea et al. 1991). There are some attributes of the beginning of a plateau in the data near $p_T \cong 1$ GeV/c, but new data at high transverse momenta are necessary to confirm such a conclusion. The polarization of Ξ^0 (Heller et al. 1978) (see Fig. 10.2d) exhibits the same features as the polarization of Ξ^- though the experimental points have wide dispersion.

Table 10.3 Dependence of the polarization of hyperons produced by a proton beam on p_T (comment to Fig. 10.2)

Reaction	a_1 (%)	a_2 (%)	χ^2/DOF	NP
$p \to \Lambda$, 400 GeV/c	10.9 ± 0.2	3.3 ± 0.4	5.9	23
$0.3 < x_F < 0.4$	9.6 ± 0.5	0.7 ± 0.4	6.7	23
$0.2 < p_T < 3.1$ GeV/c	22.4 ± 1.7	0.63 ± 0.07	6.1	23
$p \to \Lambda$, 2049 GeV/c	17.9 ± 2.0	1.6 ± 0.3	1.4	6
$x_F = 0.38$	-2.6 ± 1.4	16.6 ± 1.8	3.2	6
$0.4 < p_T < 1.3$ GeV/c	37.0 ± 21.0	0.04 ± 0.02	4.1	6
$p \to \Sigma^+$, 400 GeV/c	19.6 ± 1.2	40.2 ± 302.0	7.6	4
$x_F = 0.53$	32.9 ± 4.1	-12.4 ± 3.7	1.8	4
$0.5 < p_T < 1.5$ GeV/c	21.5 ± 1.9	3.5 ± 0.9	1.2	4
$p+ \to \Sigma$, 800 GeV/c	13.0 ± 0.2	14.5 ± 9.3	10.0	15
$0.44 < x_F < 0.52$	21.7 ± 1.1	-6.8 ± 0.9	5.2	15
$0.3 < p_T < 2$ GeV/c	17.0 ± 0.3	2.9 ± 0.4	4.4	15
$p \to \Xi^-$, 400 GeV/c	14.5 ± 5.6	1.2 ± 0.8	1.0	6
$0.29 < x_F < 0.61$	-2.0 ± 4.3	11.9 ± 4.5	1.2	6
$0.6 < p_T < 1.2$ GeV/c	276.0 ± 265.0	0.04 ± 0.04	1.2	6
$p \to \Xi^-$, 800 GeV/c	14.4 ± 0.5	2.8 ± 0.4	1.0	9
$0.32 < x_F < 0.7$	7.7 ± 1.5	5.6 ± 1.5	0.2	9
$0.6 < p_T < 1.4$ GeV/c	26.0 ± 6.5	0.7 ± 0.3	1.1	9
$p \to \Xi^0$, 400 GeV/c	11.0 ± 0.8	11.0 ± 9.0	2.0	16
$0.26 < x_F < 0.58$	2.7 ± 2.8	8.2 ± 2.77	1.4	16
$0.3 < p_T < 1.7$ GeV/c	634.0 ± 417.0	0.02 ± 0.01	1.5	16

Summarizing this subsection, we emphasize that the p_T dependences of polarization are different for different quark flavors. Such a difference cannot be explained in the popular DGM recombination model or the Lund model (Gustafson 1984). This fact implies the difference between the mechanisms of the interaction of partons with different flavors. The existing experimental data also do not contradict the pQCD predictions for high momentum transfers p_T.

10.1.3 Dependence of Hyperon Polarization on x_F

The DGM recombination model based on the $SU(6)$ symmetry and the Thomas spin precession mechanism predicts an almost linear dependence of Λ hyperon polarization on x_F with small corrections to higher x_F powers and to p_T. The polarization of the Λ hyperon is described in this model (in a certain approximation) by the analytical function

$$P_\Lambda = -A(x_F, p_T) \cdot p_T, \quad A(x_F, p_T) = f(x_F) \cdot g(x_F, p_T). \quad (10.9)$$

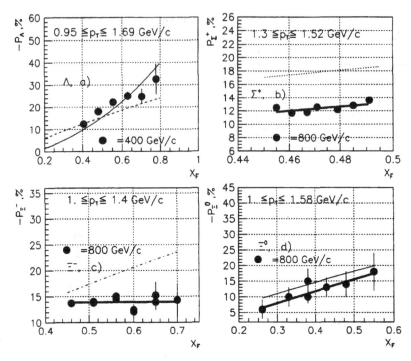

Fig. 10.3 Polarization of the (**a**) Λ, (**b**) Σ^+, (**c**) Ξ^-, and (**d**) Ξ^0 hyperons versus x_F

After the substitution of the numerical values for the intermediate parameters, the above functions can be presented as (DeGrand and Miettinen 1981a; DeGrand et al. 1985)

$$f(x_F) = \frac{0.12 \cdot x_F \cdot (1 - 0.31 \cdot x_F - 0.18 \cdot x_F^2)}{(1 - 0.35 \cdot x_F^2)}, \tag{10.10}$$

$$g(x_F, p_T) = \frac{1}{0.38 + (0.03 + 0.068 \cdot x_F + 0.04 \cdot x_F^2) \cdot p_T^2}. \tag{10.11}$$

Some experimental data for hyperon polarization as a function of x_F are presented in Fig. 10.3. The linear approximation is given by thick solid lines in Figs. 10.3b–10.3d. The behavior of the experimental data for hyperon polarization as a function of x_F depends on the quark flavor. The predictions of the DGM model are shown in these figures by dashed lines (the thin solid line in Fig. 10.3d). This comparison shows that the DeGrand–Miettinen model without any free parameters well describes the data for the polarization of Λ hyperons (see Fig. 10.3a); the calculations were performed with $p_T = 1$ GeV/c. The basic prediction of this model that the x_F dependences of the polarizations of some hyperons are the same is valid maybe only for the Ξ^0 hyperon (see Fig. 10.3d) and is incorrect for the Σ^+ (Fig. 10.3b) and Ξ^- (Fig. 10.3c) hyperons. Thus, the DeGrand–Miettinen model cannot explain the x_F dependence of the polarization of all hyperons. It is another

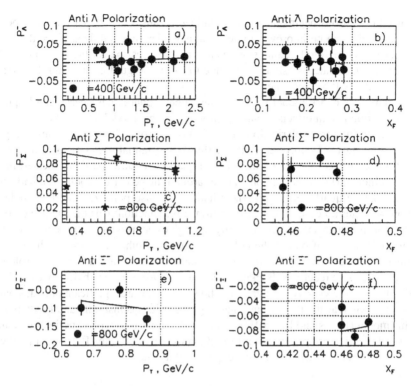

Fig. 10.4 Polarization of the $\bar{\Lambda}$, $\bar{\Sigma}^-$, and $\bar{\Xi}^-$ hyperons produced by the proton beam versus (**a, c,** and **e**) p_T and (**b, d,** and **f**) x_F, respectively

defect of this model. There are several critical remarks on this model in Fujita and Matsuyama (1988) and Magnin and Simao (1995), where the authors stated that the DGM model cannot describe polarization when hyperons are produced at small x_F values. This is due to the strong dependence of model predictions on the masses of sea quarks.

There is a more obvious disadvantage of this model: the model requires zero polarizations of the hyperons that do not have common quarks with the initial particles. According to this postulate, all antihyperons should be unpolarized, but this is not the case. For example, Fig. 10.4 shows some data on antihyperon polarizations. The $\bar{\Lambda}$ hyperon is unpolarized (see Figs. 10.4a and 10.4b), whereas the antisigma hyperon is polarized (see Figs. 10.4c and 10.4d); moreover, the sign and magnitude of polarization are the same as for the sigma hyperon. Another example is the polarization of the negative anticascade hyperon (see Figs. 10.4e and 10.4f), which has a magnitude of about 10 % and the negative sign as for the cascade hyperon. The solid lines in all figures are linear fits to the data. Thus, the DGM model has some problems, which will be solved with the appearance of new experimental data. However, this model for a long time gave correct guiding lines in many directions such as hyperon polarizations and relations between them, it predicts hyperon polarizations produced on various beams including polarized beams. For this reason, we expect that this model can be improved to describe new experimental data. Since there is no other model giving the same simple analytical expression for the description of the experimental data, we should continue to use it for guiding lines until a more perfect model will be developed.

A relativistic quark model with rotating valence quarks has been recently proposed to explain spin effects in hadron processes. This model has been applied, in particular, to interpret polarization in the inclusive production of hyperons at high energies (Meng 1991; Boros et al. 1993, 1996). Substantial foundations of the proposed model are as follows: (i) the valence quarks are Dirac particles located in a bag (confinement) and rotating on an orbit around the center of a polarized hadron; (ii) the valence quarks in the polarized hadron are also polarized and their polarization are determined by the $SU(6)$ hadron wave function; (iii) the inclusive production of final hadrons occurs primarily through direct production processes when the valence quarks of one of the colliding hadrons annihilate or join with the sea antiquarks of the other hadron; and (iv) the surface effect plays a significant role in one-spin inclusive reactions. Backgrounds reducing asymmetry obviously exist, but the analysis of the yields of Λ hyperons shows that they are significant only at small x_F values, whereas their contribution at large x_F values is insignificant. Predictions of this model (Boros and Liang 1996) and their comparison with experimental data on Λ polarization are presented in Fig. 10.3a by the solid line. The model predictions are in good agreement with the experimental data.

10.1.4 Polarization of Hyperons Produced by a Σ^- Beam

The results of the WA-89 experiment at CERN were published (Adamovich et al. 1995) (see Fig. 10.5). They were obtained on a 320-GeV/c Σ^- hyperon beam. All experimental data were described by the linear dependence on both variables x_F and p_T (thin solid lines in Fig. 10.5). Figure 10.5a shows the polarization P_Λ of Λ hyperons at $\bar{x}_F = 0.3$ as a function of p_T. The thick solid line is the calculation of the polarization P_Λ of Λ hyperons produced by a 300-GeV/c proton beam at the same x_F value by formula (10.1). The polarization of Λ hyperons produced by the Σ^- beam is obviously much smaller than the polarization produced by the proton beam. According to the DGM model (De-Grand and Miettinen 1981a; DeGrand et al. 1985), polarization P_Λ generated by the Σ^- beam should be about half the polarization P_Λ generated by the proton beam. This prediction seems approximately correct (cf. the thin and thick solid lines in Fig. 10.5a). Similar comparison was made for other measurements. For example, as seen in Fig. 10.5b, $\bar{\Lambda}$ hyperons are unpolarized in both cases (cf. the thin and thick solid lines representing the results of fitting the experimental data for Σ^- hyperon and proton beams, respectively). A very interesting feature is seen in Fig. 10.5c: the polarization of Σ^+ hyperons produced by the Σ^- hyperon beam (the thin solid line) is much smaller in magnitude and has

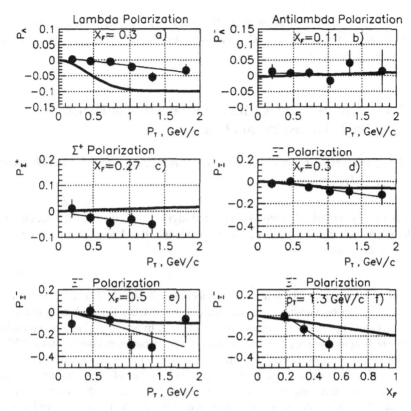

Fig. 10.5 Polarization of hyperons produced by the 320-GeV/c Σ^- beam: (**a**) P_Λ versus p_T, (**b**) $P_{\bar\Lambda}$ versus p_T, (**c**) P_Σ^+ versus p_T, (**d**) P_{Ξ}^- at $x_F = 0.3$ versus p_T, (**e**) P_{Ξ}^- at $x_F = 0.5$ versus p_T, and (**f**) P_{Ξ}^- versus x_F

the opposite sign as compared to that produced by the proton beam (the solid line, which was estimated by formula (10.1) as the corresponding data are absent).

According to the DGM model, the polarization of Σ^+ hyperons produced by the Σ^- beam should be the same in magnitude, but opposite in sign as the polarization of Λ hyperons produced by the proton beam. This prediction is seemingly confirmed. The polarizations of Ξ^- hyperons produced by both beams are close to each other (see Figs. 10.5e and 10.5f) in agreement with the model prediction. The authors of experimental work Adamovich et al. (1995) have made some conclusions. First, the polarization of Λ hyperons inclusively produced by the Σ^- hyperon beam is much smaller than that of Λ hyperons produced by the proton beam with the same initial energies. Second, $\bar\Lambda$ hyperons are also unpolarized as in the case of the proton beam. Third, Ξ^- hyperons have the same polarizations on both beams. Thus, hyperon polarization strongly depends on the flavors of the initial and final quarks. The same flavor dependence was previously observed for the polarizations of Σ^+ and Ξ^- hyperons. The dependences of their polarization on energy and p_T are contrary to those of the polarization of Λ (Nurushev 1993). More accurate data on the polarizations of hyperons (and other hadrons) produced by the Σ^- hyperon beam are obviously important.

10.2   Inclusive Reactions with the Production of Hyperons by the Polarized Proton Beam

10.2.1   Analyzing Power of Inclusive Reactions with the Production of Hyperons

The use of the polarized proton beam for studying spin effects in the production of hyperons can significantly enrich our knowledge. It is known that the transition matrix of the reaction

$$p \uparrow + p \to \Lambda + X \tag{10.12}$$

is a function of eight real parameters (Doncel and Mendez 1972). Therefore, eight observables can be measured. In this section, we discuss the analyzing power in reaction (10.12) of the inclusive production of the Λ hyperon, A_N, and the so-called depolarization tensor D_{NN} (one of the Wolfenstein parameters (Wolfenstein 1954)). In 1995, the E704 Collaboration published the data on the left–right asymmetry in reaction (10.12) (Bravar et al. 1997). The measurement of the asymmetry of the Λ hyperon is interesting for many reasons. First, we deal with a heavier quark, the s quark. Second, this reaction is sensitive to the quark flavor. The third reason is the possibility of the comparison of the analyzing power with the polarization of the Λ hyperon that appears in the same reaction when the initial proton beam is unpolarized. It is known that the analyzing power A_N and polarization P for a binary reaction are equivalent owing to time invariance. It is also well known that such a relation is not proved for inclusive reactions. The experimental data on asymmetry A_N for the inclusive production of Λ hyperons at 200 GeV/c are presented by closed circles in Figs. 10.6a and 10.6b as functions of p_T and x_F, respectively.

Some conclusions can be deduced from these figures. First, the BNL data at 18.5 GeV/c and the ZGS data are consistent with zero (Lesnik et al. 1975; Bonner et al. 1987) (data for 18.5 GeV/c are shown by stars and linear fits are presented by solid and dashed lines for 200 and 18.5 GeV/c, respectively). Second, the analyzing power A_N at 200 GeV/c is negative and its absolute value increases with both arguments from zero near $p_T \cong 0.5$ GeV/c and $x_F \cong 0.4$ to 10 % near $p_T \cong 1$ GeV/c and $x_F \cong 0.7$. The linear approximation provides a good description (see solid lines). In both cases, at 18.5 and 200 GeV/c, the analyzing power is not equal to the polarization: the polarization is much larger in magnitude than the analyzing power though their signs are the same. This result is a strong challenge for the DGM model according to which $A_N(\Lambda) = P_N(\Lambda)$ in inclusive production. At the same time, the relativistic quark model of rotating valence quarks (Boros

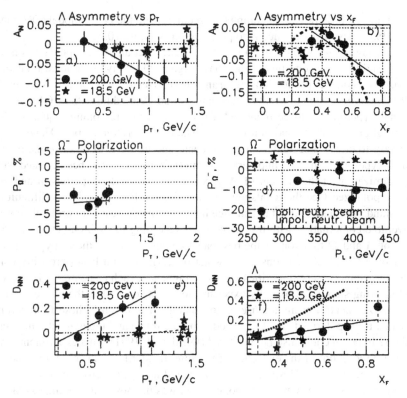

Fig. 10.6 Analyzing power A_N for Λ versus (**a**) p_T and (**b**) x_F, polarization P_{Ω^-} for (**c**) proton beam and (**d**) neutral beams, and the spin transfer tensor D_{NN} for Λ versus (**e**) p_T and (**f**) x_F

et al. 1996) provides a quantitative description of the dependence of the analyzing power of Λ on x_F (Fig. 10.6b, the dash-dotted line). According to this model, an appreciable negative asymmetry at large x_F values is caused by the recombination of $(u_v d_v)$ valence diquarks from the polarized projectile with the sea quarks of the unpolarized target. At the same time, zero asymmetry at small x_F values is due to indirect production processes (for example, the disintegration of heavy hyperons). According to the similarity mechanism, $A_N(\Lambda)$ in the intermediate region is similar to π^0 asymmetry. More precise data are necessary for testing such statements.

10.2.2 Spin Transfer Process in the Inclusive Production of Hyperons

The important methodical invention has been made in studying the polarization of the Ω hyperon. At the first stage, physicists tried to obtain polarized Ω hyperons traditionally by bombarding nuclear targets by the proton beam extracted with extension pulse and the separation of Ω hyperons at some small production angles

(Luk et al. 1993). However, in complete agreement with the DGM model (DeGrand et al. 1985), they have obtained zero polarization (see Fig. 10.6c).

Following the recommendation given in DeGrand et al. (1985), physicists have then passed to the second phase: they have organized two types of neutral beams, polarized (at nonzero production angle) and unpolarized (at zero angle) (Wood et al. 1996). Such succession of events can be understood. According to the DGM model, the final hyperon should be polarized if it has one or more common quarks with the initial proton. This condition is not satisfied for a proton–Ω pair. Hence, it is necessary to create a secondary hyperon beam and to use it to produce the polarized Ω hyperons. A neutral unpolarized beam (at zero angle) has been created with the optimization of the content of Λ hyperons. However, Ω hyperons turned out to be unpolarized (see the data shown by asterisks in Fig. 10.6d). It is difficult to understand the cause of this fact: strange hyperons in this beam are probably not in a favorable combination with the Ω particle. According to DeGrand et al. (1985), a pair of Λ and Ω hyperons is the best pair for spin transfer to the Ω hyperon and a pair of Σ^- and Ω hyperons is the worst pair. The neutral beam probably contains many unnecessary impurities without strangeness such as neutrons. However, the following step undertaken by experimenters was successful: they have created the polarized neutral beam and used it to generate Ω hyperons. These Ω hyperons appeared to be polarized (see Fig. 10.6d, closed circles) and the magnetic moment of the Ω hyperon could also be measured (Diehl et al. 1991). This experiment gave the first direct proof of the existence of the spin transfer mechanism in the production of the Ω hyperon. However, as the polarized neutral hyperon beam contained many types of hyperons and their polarization was unknown, it was impossible to quantitatively determine the fraction of initial polarization transferred to the final hyperons. To obtain such quantitative data, it is necessary to measure the so-called Wolfenstein parameters (Wolfenstein 1954) or spin tensors introduced in the beginning of the 1950s for the description of elastic scattering.

These parameters are also applicable for the inclusive production of hyperons (strictly speaking, for hyperons with spin 1/2 such as Λ and Σ hyperons, but are inapplicable directly to Ω^- hyperons having spin 3/2). If the primary beam, for example, protons has polarization P_B, the polarization of the final hyperon is given by the expression

$$P_Y = \frac{P_Y^0 + D_{NN} \cdot P_B}{1 + P_Y^0 \cdot P_B}, \tag{10.13}$$

where P_Y^0 is the polarization of hyperons produced by the unpolarized proton beam. This is the hyperon polarization discussed above.

It is assumed that all polarization vectors are directed perpendicularly to the hyperon production plane. This circumstance is represented by the subscripts of the depolarization parameter. If this assumption is invalid, there are simpler formulas for the determination of polarization (spin) transfer. The Wolfenstein parameter D_{NN} characterizes the fraction of the polarization of the initial beam transferred to the final hyperons. It provides the quantitative description of the spin transfer process and the tensor D_{NN} is well known for the nucleon–nucleon interactions at low energies. Moreover, this parameter has been measured for the Σ^0 hyperon at 18.5 GeV/c (Bonner et al. 1989). This result is interesting, because the experiment gave $D_{NN} = 0.26 \pm 0.16$, whereas the DGM model predicted $D_{NN} = 0.67$, which significantly contradicts the experimental data. A similar contradiction has been found and for the analyzing power: A_N (exper.) $= 0.02 \pm 0.03$ versus A_N (theory) $= 0.20$. To eliminate such discrepancies, experimenters proposed to introduce two additional parameters into the DGM model that describe spin flip and could give the good description of the data. More data are undoubtedly required to improve the theoretical model.

In 1996, the interesting results on spin transfer tensor D_{NN} measured by the E704 Collaboration were published (Bravar 1996). The results of the measurements of D_{NN} are presented in Figs. 10.6e and 10.6f as functions of p_T and x_F, respectively. It is clearly seen that this parameter becomes nonzero at large x_F and p_T values showing that approximately 20–30 % of the polarization of the initial proton are transferred to the final Λ hyperon. This value obtained for the polarization of the Λ hyperon cannot be compared to polarization transfer in the inclusive production of the Ω hyperon for several reasons. First, Λ and Ω hyperons have different spins (1/2 and 3/2, respectively). Second, any formula similar to expression (10.13) has not yet been proposed for the Ω hyperon. Third, the kinematic regions are different: the transverse momentum was zero in the case of the polarization of the Ω hyperon, whereas the tensor D_{NN} measured for the case of P_Λ is nonzero at $p_T \cong 1$ GeV/c and $x_F \cong 0.8$. The part of the polarization transferred by the tensor D_{NN} should be added to the direct polarization of the Λ hyperon, P_Λ^0, taking into account the sign of D_{NN} as seen in Eq. (10.13) for the final polarization of the Λ hyperon. The results for D_{NN} at 18.5 GeV/c are also presented in these figures. They show that D_{NN} is consistent with zero (Bonner et al. 1988). Thus, the E704 Collaboration has presented the first direct quantitative data on the spin transfer effect in the inclusive production of the Λ hyperon at high energies. This result differs from the prediction of the DGM model. According to this model, D_{NN} should be zero (DeGrand and Miettinen 1981b, 1985). However, the rotating quark model also can describe these data (Boros et al. 1996) without the introduction of any free parameters (see Fig. 10.6f, the dotted line). The sign and shape of the depolarization tensor D_{NN} are predicted correctly though the good quantitative description has not yet been achieved.

The dynamical mechanism providing a good description of the features of the observed spectrum of Λ hyperons including their polarization was proposed in Soffer and Tornquist (1991). This model is based on the one-pion exchange diagram in the Regge model. All features of the inclusive production of Λ hyperons are attributed to the $\pi + p \rightarrow K + \Lambda$ binary reaction with kinematics appropriately taken into account. The model is promising for the explanation of many other spin effects in various reactions.

A more complete investigation of the spin transfer mechanism with various polarized beams is obviously required. Data on the spin tensors at high energies are very scarce yet.

It is particularly important to supplement the RHIC polarization program with studying the spin transfer tensors (Wolfenstein parameters). This requires a polarimeter-type instrument for analyzing the polarization of scattered protons (or any stable baryons or antibaryons). Unfortunately, such a program for RHIC is absent today. But spin transfer tensors may be studied through hyperon decays.

References

Abe, F., et al.: Phys. Rev. D **34**, 1950 (1986)
Adamovich, M.I., et al.: Z. Phys. A **350**, 379 (1995)
Bonner, B.E., et al.: Phys. Rev. Lett. **58**, 447 (1987)
Bonner, B.E., et al.: Phys. Rev. D **38**, 729 (1988)
Bonner, B.E., et al.: Phys. Rev. Lett. **62**, 1591 (1989)
Boros, C., Liang, Z.-T.: Phys. Rev. D **53**, R2279 (1996)
Boros, C., Liang, Z.-T., Meng, T.-C.: Phys. Rev. Lett. **70**, 1751 (1993)
Boros, C., Liang, Z.-T., Meng, T.-C.: Preprint no. FUB-HEP 96-9, Berlin Free University (1996)
Bravar, A.: In: Proceedings of the 12th International Symposium on High Energy Spin Physics. Amsterdam, The Netherlands, p. 244 (1996)
Bravar, A., et al.: Phys. Rev. Lett. **78**, 4003 (1997)
Cooper, P.S., et al.: Phys. Rev. Lett. **51**, 863 (1983)
DeGrand, T.A., Miettinen, H.I.: Phys. Rev. D **23**, 1227 (1981a)
DeGrand, T.A., Miettinen, H.I.: Phys. Rev. D **24**, 2419 (1981b)
DeGrand, T.A., Miettinen, H.I.: Phys. Rev. D **31**, 661 (1985)
DeGrand, T.A., Markkanen, J., Miettinen, H.I.: Phys. Rev. D **32**, 2445 (1985)
Diehl, H.T., et al.: Fermilab-Pub-91/165, Fermilab (1991)
Doncel, M., Mendez, A.: Phys. Lett. B **41**, 83 (1972)
Duryea, J., et al.: Phys. Rev. Lett. **67**, 1193 (1991)
Efremov, A.V.: Sov. J. Nucl. Phys. **28**, 166 (1978)
Fujita, T., Matsuyama, T.: Phys. Rev. D **38**, 401 (1988)
Gourlay, S.A., et al.: Phys. Rev. Lett. **56**, 2244 (1986)
Gustafson, G.: In: Proceedings of the 2nd International Workshop on High Energy Spin Physics, Protvino, USSR, p. 212 (1984)
Heller, K.: J. Phys., Colloq. **C2**(Suppl. No. 2), 121 (1985)
Heller, K., et al.: Phys. Rev. Lett. **51**, 2025 (1978)
Lach, J.: Fermilab-Conf-94/031, Fermilab (1994)
Lesnik, A., et al.: Phys. Rev. Lett. **35**, 770 (1975)
Luk, K.B., et al.: Phys. Rev. Lett. **70**, 900 (1993)
Lundberg, B., et al.: Phys. Rev. D **40**, 3557 (1989)
Magnin, J., Simao, F.R.A.: CBPF-NF-071/95 (1995)
Meng, T.-C.: In: Proceedings of the 4th International Workshop on High Energy Spin Physics, Protvino, USSR, p. 112 (1991)
Morelos, A., et al.: Fermilab-Pub-93/331-E (1993)
Nessi, M.: In: Proceedings of the 8th International Symposium on High Energy Spin Physics, Minneapolis, MN, USA, vol. 1, p. 66 (1988)
Nurushev, S.B.: In: Proceedings of the 9th International Symposium on High Energy Spin Physics, Bonn, Germany, p. 34 (1990)
Nurushev, S.B.: In: Proceedings of the 5th International Workshop on High Energy Spin Physics, Protvino, Russia, p. 5 (1993)
Panagiotou, A.D.: Int. J. Mod. Phys. A **5**, 1197 (1990)
Penzo, A.: In: Proc. 6th International Workshop on High Energy Spin Physics, Protvino, Russia, vol. 2, p. 34 (1995)

Pondrom, L.G.: Phys. Rep. **122**, 57 (1985)

Puzikov, L.D., Ryndin, R.M., Smorodinsky, Ya.A.: Nucl. Phys. **3**, 436 (1957)

Rameika, G., et al.: Phys. Rev. D **33**, 3172 (1986)

Scubic, P., et al.: Phys. Rev. D **18**, 3115 (1978)

Smith, A.M., et al.: Phys. Lett. B **185**, 209 (1987)

Soffer, J., Tornquist, N.A.: Preprint CPT-91/P.2559, HU-TFT-91-30, CNRS Luminy-Case 907-CPT-F13288, Marselle Cedex 9, France (1991)

Soffer, J., Wray, D.: Phys. Lett. B **43**, 514 (1973)

Swallow, E.C.: In: Proceedings of the Symposium on High Energy Physics with Polarized Beams, Argonne, IL, vol. XI, p. 1 (1974)

Wolfenstein, L.: Phys. Rev. **96**, 1654 (1954)

Wood, D.M., et al.: Phys. Rev. D **54**, 6610 (1996)

Chapter 11
Inclusive Hadron Production

11.1 Single-Spin Asymmetry in Inclusive Hadron Production

The initial stage of the investigation of single-spin asymmetry in inclusive hadron production was performed in the 1960s with the appearance of polarized proton beams with a kinetic energy of 400 MeV (March 1960; Mcllwan et al. 1962) and 650 MeV (Borisov et al. 1967). The first and second experiments were carried out with emulsions and bubble chambers, respectively; the third experiment was a purely electronic experiment (scintillation counters) with high statistics. In all these experiments, nonzero asymmetry of pions was observed. To interpret asymmetry in inclusive pion production at 650 MeV, the known Mandelstam isobar model (Mandelstam 1958) was successfully used, whereas the one-pion exchange (OPE) model (Ferrari and Selleri 1963) appeared to be unsatisfactory (Nurushev and Solovyanov 1965). These data were also useful for the phase shift analysis. Physicists made huge efforts to advance polarization investigations from energies about 100 MeV to energies $\sim 10^5$ GeV (in the laboratory frame) for 50 years. At the same time, the prevailing opinion is still that the spin phenomena insignificant and can disappear with increasing energy. However, already the first results from the RHIC collider showed that polarization effects survive even at energies of 10^5 GeV. Now, it is possible to hope that polarization physics will bring many further surprising results. Single-spin asymmetries at high energies (> 10 GeV) have been recently considered in several works where one can find particular results and details of discussions (Soffer 1995; Nurushev 1995a, 1995b).

We will focus on the discussion of new and significant experimental data on the single-spin asymmetry at high energies (Nurushev and Ryskin 2006).

The E581/E704 Collaboration (Fermilab) created the polarized 200-GeV beam of protons (antiprotons) from the decay of $\Lambda(\bar{\Lambda})$ hyperons using the extracted 800-GeV proton beam from the Tevatron. With such a beam having the highest energy at that time (1990), the Collaboration carried out a series of the measurements of single-spin asymmetries (results on hyperons were discussed above). The data can be systematized as follows.

S.B. Nurushev et al., *Introduction to Polarization Physics*,
Lecture Notes in Physics 859, DOI 10.1007/978-3-642-32163-4_11,
© Moskovski Inzhenerno-Fisitscheski Institute, Moscow, Russia 2013

(a) Asymmetry in the polarized-beam fragmentation region (intermediate transverse momenta) in the inclusive production of π^0 and $\pi^\pm$ mesons in pp and $\bar{p}p$ collisions.
(b) Asymmetry in the Coulomb-nuclear interference region in elastic pp scattering.
(c) Asymmetry in the central region (high transverse momenta) in the production of direct photons and in the inclusive production of π^0 mesons with accompanying charged particles and without them.

11.1.1 Asymmetry in the Inclusive Production of π^0 Mesons by Polarized Proton and Antiproton Beams at Intermediate Transverse Momentum (Beam Fragmentation Region)

Asymmetry at large x_F values (the beam fragmentation region) has been measured for the inclusive production of π^0 in the reactions

$$p\uparrow + p \to \pi^0 + X, \qquad \bar{p}\uparrow + p \to \pi^0 + X. \tag{11.1}$$

The region of the kinematic variables was 0.5 GeV/c < p_T < 2.0 GeV/c, 0 < x_F < 0.8 (Adams et al. 1992; Nurushev 1991). As seen in Figs. 11.1a and 11.1c (data for π^0 are shown by stars in both figures), asymmetry is close to zero in the range 0 < x_F < 0.3 and then starts to increase almost linearly with x_F, approaching $A_N = 0.15 \pm 0.03$ in the pp reaction and $A_N = 0.072 \pm 0.037$ in the $\bar{p}p$ reaction in the range 0.6 < x_F < 0.8.

Note that the asymmetry sign is the same for both reactions, but $A_N(p\uparrow + p \to \pi^0 + X) > A_N(\bar{p}\uparrow + p \to \pi^0 + X)$ throughout the measured x_F range. The dashed lines on Figs. 11.1a and 11.1c are linear fits to the experimental data. Fitting results for π^0 are the following: $A_N = -(0.02 \pm 0.008) + (0.22 \pm 0.03) \cdot x_F$ for the proton beam with $\chi^2 = 2$/DOF, NP (the number of experimental points) = 7 and $A_N = -(0.001 \pm 0.009) + (0.11 \pm 0.03) \cdot x_F$ for the antiproton beam with $\chi^2 = 0.5$/DOF for seven points. Hence, it is shown that information on the transverse polarization of a primary quark passes to the final hadron (π^0 in this case), and the effect increases with x_F. The comparison of these two fits shows that the slope parameter in the asymmetry of π^0 for the antiproton beam is half the value for the proton beam; this fact has not yet been explained by any theoretical model.

11.1.2 Asymmetry in the Inclusive Production of $\pi^\pm$ Mesons by the Polarized Proton and Antiproton Beams in the Beam Fragmentation Region

Single-spin asymmetries in the inclusive production of $\pi^\pm$ mesons have been measured by the E704 Collaboration in the reactions

$$p\uparrow + p \to \pi^+ + X, \qquad \bar{p}\uparrow + p \to \pi^- + X \tag{11.2}$$

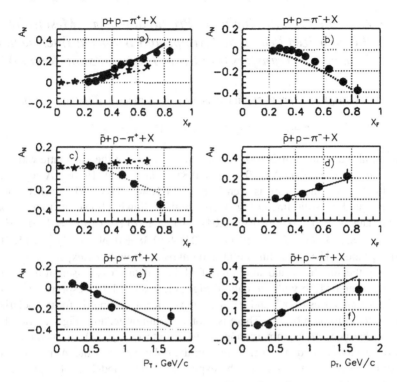

Fig. 11.1 Inclusive asymmetries of pions produced by the (**a**) and (**b**) proton and (**c**)–(**f**) antiproton beams with a momentum of 200 GeV/c; asymmetries for π^0 are shown by stars

at an initial momentum of 200 GeV/c and in the region of the kinematic variables 0.2 GeV/c $< p_T <$ 2.0 GeV/c, $0 < x_F < 0.85$ (Adams et al. 1991a). According to Figs. 11.1a and 11.1b, asymmetry $A_N^{\pm}$ is quite large for both π^+ and π^- mesons. However, the sign of asymmetry A_N^+ is opposite to the sign of asymmetry A_N^-. These opposite signs can be interpreted at the quark level under the assumption that u quarks are polarized in the same direction as the initial proton, but the d quark has the opposite polarization orientation. The asymmetries increase with x_F, indicating that the leading partons remember the spin states of the initial particles. The linear fits to the experimental data in Figs. 11.1a and 11.1b provided the following results: $A_N^+(x_F) = -(0.14 \pm 0.02) + (0.60 \pm 0.04) \cdot x_F$ and $A_N^-(x_F) = (0.13 \pm 0.02) - (0.43 \pm 0.03) \cdot x_F$ with $\chi^2 = 1.1$/DOF and $\chi^2 = 7.1$/DOF for π^+ and π^-, respectively. These fits are not presented in Figs. 11.1a and 11.1b in order to avoid the complication of comparison with the model predictions. Comparison with these two fits provides the conclusions: (a) the slope parameter for $A_N^+(x_F)$ is larger than that for $A_N^-(x_F)$, (b) both asymmetries become zero near $x_F \cong 0.2$–0.3, and (c) both asymmetries are larger in magnitude than asymmetry for π^0.

Since it is known that perturbative QCD is inapplicable at small p_T values, phenomenological models should be applied to explain such spin effects.

11.1.3 Asymmetry in the Inclusive Production of $\pi^{\pm}$ Mesons by the Polarized Proton and Antiproton Beams at Intermediate Momentum Transfers (the Beam Fragmentation Region)

The inclusive asymmetries of $\pi^{\pm}$ mesons have been measured in the reactions

$$p \uparrow + p \rightarrow \pi^{+} + X, \qquad p \uparrow + p \rightarrow \pi^{-} + X, \qquad (11.3)$$

at an initial momentum of 200 GeV/c and in the region of the kinematic variables: 0.2 GeV/c < p_T < 2.0 GeV/c, 0 < x_F < 0.85 (Adams et al. 1996a). A significant spin effect depending on x_F is seen in Figs. 11.1c and 11.1d. The dashed line in Fig. 11.1c and the solid line in Fig. 11.1d are the following linear fits to the experimental data: $A_N^+(x_F, \bar{p}) = (0.18 \pm 0.04) - (0.55 \pm 0.11) \cdot x_F$ ($\chi^2 = 1.0$/DOF for five points) and $A_N^-(x_F, \bar{p}) = -(0.11 \pm 0.04) + (0.38 \pm 0.09) \cdot x_F$ ($\chi^2 = 0.5$/DOF for five points). The linear fits are in good agreement with the experimental data, indicating the simplicity and similarity of all these data. Comparison of the data in Figs. 11.1a and 11.1b with the data in Figs. 11.1c and 11.1d, respectively, shows that the asymmetries of π^+ and π^- are almost mirror symmetric for the polarized proton and antiproton beams. There is another common feature: Figs. 11.1a–11.1d indicate that nonzero asymmetry appears in the interval $x_F \cong 0.2$–0.3 and then increases. A similar tendency is also seen in the polarization of the Λ hyperon. It is a very interesting fact and should be explained in any model for the quantitative description of single-spin asymmetry.

The asymmetries of in $\pi^{\pm}$ in the $\bar{p}p$ interaction are presented in Figs. 11.1e and 11.1f as functions of x_T. The asymmetry A_N increases almost linearly with p_T and a "threshold" behavior is observed at $p_T \cong 0.3$–0.4 GeV/c. The solid lines are the following linear fits to the experimental data: $A_N^+(p_T, \bar{p}) = (0.10 \pm 0.03) - (0.28 \pm 0.05) \cdot p_T$ ($\chi^2 = 2$/DOF for five points) and $A_N^-(p_T, \bar{p}) = -(0.06 \pm 0.03) + (0.23 \pm 0.04) \cdot p_T$ (χ^2/DOF = 2.6 for five points). The slope parameters are very close to each other and zero crossing points are also close to each other; hence, the same mechanism could be responsible for single-spin asymmetries in the inclusive production of π^+ and π^- mesons by a polarized antiproton beam. Data on the inclusive asymmetries of pions in the beam fragmentation region became very important not only due to theoretical interest (phenomenology of soft interactions), but also for the practical reason: these reactions can be effectively used for polarimetry of high-energy polarized proton and antiproton beams owing to their large cross sections and large analyzing power.

11.1.4 Theoretical Models

Any inclusive reaction can be represented at the parton level as the convolution of parton distributions in the initial hadron with the parton interaction cross section and the final parton fragmentation function. Transverse spin asymmetry can appear at any of these three stages. Correspondingly, theoretical models involve asymmetric parton distributions over k_T (parton intrinsic transverse momentum) in the polarized initial protons (Sivers effect), or in the parton–parton interaction (the so-called Szwed effect), or in the dependence of the fragmentation function of a final parton on its polarization (Collins effect). All three types of models have been implemented in practical calculations. We will discuss some of these models.

In Artru (1993) and Artru et al. (1994), the Collins-type string model was proposed to calculate asymmetry. The actual calculations were performed under the assumption that the polarization of a primary quark can be completely transferred to a final quark in the parton scattering subprocess ($D_{NN} = 1$). The quark–diquark composition of the polarized proton was postulated. Two strings (q)–(qq) decay according to the simple Lund rule. After kinematic calculations and some assumptions, A_N is numerically estimated for various variants of quark polarization. If quark polarization follows the $SU(6)$ wave function, we have $P_u = +2/3$ and $P_d = -1/3$ and the resulting asymmetry is significantly inconsistent with the data for negative pions. The maximum possible transverse polarizations of the quarks in a proton, $P_u = 1$ and $P_d = -1$, were chosen. The results are shown in Fig. 11.1a by the solid line (similar calculations for π^- and π^0 are not shown in Fig. 11.1). Constant quark polarization independent of the Bjorken parameter x leads to asymmetry that is larger than the measured value, whereas the x^2 dependence of quark polarization leads to the better description of the experimental data.

Transverse momentum distribution function of partons in a polarized hadron can be asymmetric (Sivers effect). A model providing a sufficiently good description of asymmetry in inclusive pion production was developed in Anselmino et al. (1994) on the basis of the relativistic quark–parton approach. Six free parameters of the model were determined by the best fit to the data on the asymmetry of pions. The asymmetry of π^- at fixed $p_T = 1.5$ GeV/c is shown by the dotted line in Fig. 11.1b as a function of x_F. Similar calculations were also performed for π^+ and π^0 (they are not shown in Fig. 11.1). The model determines the sign of the asymmetry of π^0 as positive.

Szwed model Its predictions will be discussed in the corresponding section, because it was applied only to data for large momentum transfers p_T, whereas relatively small momentum transfers are considered here.

11.1.5 Single-Spin Asymmetry at Very Small Transverse Momentum

The results of the E704 experiment on the measurement of the analyzing power of elastic pp scattering in the Coulomb–nuclear interference region are presented in Fig. 11.2a (Akchurin et al. 1993a).

We can ask a question: Are such data valuable? Answers are the following.

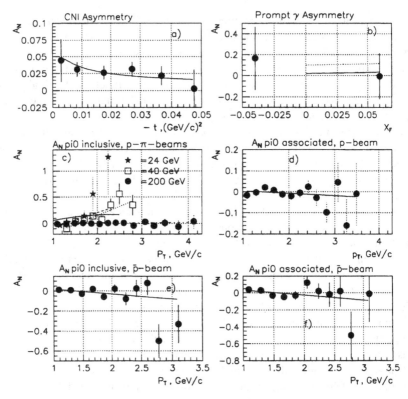

Fig. 11.2 Analyzing power A_N for (**a**) pp–pp, (**b**) direct photons, and (**c**)–(**f**) inclusive production of π^0 mesons by 200-GeV/c proton and antiproton beams; data for lower energies are also presented in panel (**c**)

- The analyzing power A_N is really nonzero. The value and shape of A_N correspond to previous theoretical calculations (Kopeliovich and Lapidus 1974; Bourrely and Soffer 1977; Buttimore et al. 1978). The formula used for fitting is

$$A_N = a \cdot \frac{z^{1.5}}{1 + z^2}, \quad z = \frac{t}{t_0}, \quad t_0 = \sqrt{3} \cdot \left(\frac{8\pi\alpha}{\sigma_T} \right) \cong 3.12 \cdot 10^{-3} \ (\text{GeV}/c)^2. \quad (11.4)$$

Here, σ_T is the total pp cross section, α is the fine structure constant, and the parameter a is defined below. Fitting gives $a = (4.73 \pm 0.92)$ % with $\chi^2/\text{DOF} = 0.31$ for six experimental points (see the solid line in Fig. 11.2a). Theoretical expectation for $a = \frac{\sqrt{3}(\mu-1)\sqrt{t_p}}{4 \cdot m}$ is 4.6 % in good agreement with the experimental value. This is the first experimental confirmation of the so-called Coulomb–nuclear interference phenomenon in polarization investigations at high energies.

- These data at such small $|t|$ values are also interesting for the direct determination of the spin-flip hadron amplitude from experiments. It is known that there is an indication of the pomeron contribution to the spin-flip amplitude. Though the statistics of these data is insufficient for a definite answer to the question on the presence of spin flip (experiment was not devoted to such physics), some estimates can be made. According to Akchurin et al. (1993b), these data provide the limits $\beta = 0.16 \pm 0.06$ and $\rho = 0.02 \pm 0.01$ for the imaginary and real parts of the ratio of the spin-flip amplitude to the spin-nonflip amplitude, respectively. Fitting was performed by including the data beyond the Coulomb–nuclear interference region. It is the best existing estimate of the spin-flip amplitude for elastic pp scattering. Theoretical estimate with the inclusion of the dynamically enhanced compact diquark and the experimental data at 6 and 10 GeV/c is 0.05–0.1 (Kopeliovich and Zakharov 1989).

11.1.6 Single-Spin Asymmetry in the Inclusive Production of Direct Photons in the Central Region

Single transverse-spin asymmetry in the production of direct photons was measured for the first time by the E704 Collaboration (Adams et al. 1995). A direct photon was separated on the basis of the careful measurements of yields of π^0 and η mesons, which are the basic sources of the background. Since the cross section for the direct production of photons is small and is determined by the difference of two large numbers, it is necessary to apply various methods for suppressing the background. In particular, the solid angle of a γ detector was doubled in the center-of-mass frame in order to reduce the background of single photons from the decays of π^0 and η mesons. In front of the γ detector, proportional chambers were placed to cut the background from the charged particles, primarily electrons. Large amount of works was performed for the simulation of the experimental setup, the appropriate choice of modes of its operation, and the determination of the efficiencies of the equipment. The measurement of polarization asymmetry is much more difficult than the measurement of cross sections, because polarization asymmetry is the difference of two almost equal numbers of scattering events to the left and right.

The single-spin asymmetry of the direct photons produced in the inclusive reaction $p \uparrow + p \rightarrow \gamma + X$ in the central region at a momentum of 200 GeV/c is presented in Fig. 11.2b as a function of x_F. The data were averaged over the measured p_T range. Asymmetry in the inclusive production of direct photons is consistent with zero within the experimental accuracy. The results of this first experiment on the measurement of the asymmetry of direct photons indicate a rather small gluon polarization ΔG. The dotted and solid lines in Fig. 11.2b illustrate the scales of expected twist-3 contributions for two parameterizations of the x dependence in the so-called "gluon pole" approximation (Qiu and Sterman 1991). The "fermion pole" approximation provides the opposite sign of the effect (Korotkiyan and Teryaev 1995). Experimental data with the existing measurement accuracy cannot distinguish between two models based on higher twist effects.

11.1.7 Single-Spin Asymmetry for π^0 Mesons at High Transverse Momenta p_T

Single-spin asymmetry at high transverse momenta in the central region was measured in the E704 experiment only for π^0 mesons (Adams et al. 1996b). The results
for single-spin π^0 inclusive asymmetry A_N for the 200-GeV/c polarized proton
beam are shown in Fig. 11.2c by closed circles. In the measured kinematic region,
A_N is consistent with zero within the measurement accuracy. Figure 11.2d shows
asymmetry for the case where the π^0 meson is accompanied by at least one charged
particle emitted in the direction opposite to the π^0 meson emission direction in the
center-of-mass frame. Such selection can enrich jet events, i.e., the parton hard collision process. The measured single-spin asymmetry is also consistent with zero.
Similar measurements were also performed with a polarized antiproton beam (see
Figs. 11.2e and 11.2f). The results are also consistent with zero.

Small single-spin asymmetries at high p_T values in the inclusive production of
π^0 mesons by 200-GeV/c proton and antiproton beams (see Figs. 11.2c and 11.2d)
put the problem of large asymmetries observed at 24 GeV/c (CERN) (Antille et al.
1980) and at 40 GeV/c by the PROZA Collaboration (IHEP) (Apokin et al. 1990).
These data are also presented in Fig. 11.2c (results for 24 and 40 GeV/c are shown
by stars and squares, respectively); hence, these data can be compared. The linear
fit $A_N(p_T) = a(1)(p_T - p_T^0)$ provides the following parameters:

- $p + p \uparrow = \pi^0 + X$, 24 GeV/$c$, $A_N(p_T) = (0.28 \pm 0.11) \cdot [p_T - (1.25 \pm 0.06)]$,
  χ^2/DOF $= 1.2$ for 6 points;
- $\pi^- + p \uparrow = \pi^0 + X$, 40 GeV/$c$, $A_N(p_T) = (0.33 \pm 0.08) \cdot [p_T - (1.5 \pm 0.10)]$,
  χ^2/DOF $= 1.2$ for 8 points;
- $p + p \uparrow = \pi^0 + X$, 200 GeV/$c$, $A_N(p_T) = (0.007 \pm 0.005) \cdot [p_T - (1.72 \pm 0.35)]$,
  χ^2/DOF $= 1.4$ for 15 points.

Analysis of these results and Fig. 11.2c indicates that asymmetry for the pp
reaction apparently decreases with increasing energy. This is confirmed by the
slope parameter $a(1)$ varying from 0.28 ± 0.11 at 24 GeV/c to $\cong 0.007 \pm 0.005$
at 200 GeV/c. Small slope parameters were also obtained at 200 GeV/c for the inclusive asymmetry of π^0 mesons produced by the polarized antiproton beam (see
Fig. 11.2e) and in charge conjugation reactions (see Figs. 11.2d–11.2f). The second
conclusion concerns a point where nonzero asymmetry appears. The corresponding parameter is $p_T^0 = 1.25 \pm 0.06$ GeV/c at 24 GeV/c, $p_T^0 = 1.5 \pm 0.1$ GeV/c at
40 GeV/c, and $p_T^0 = 1.72 \pm 0.35$ GeV/c at 200 GeV/c. Thus, the parameter p_T^0 tends
to increase with energy. To compare these conclusions to model predictions, we take
the Szwed model (Szwed 1990), which is based on two approximations: (a) on the
quark level, the partons of the incident particles are scattered in the external gluon
field and (b) hadronization proceeds through the recombination model. Polarization
in this model appears in the second order of perturbation theory. The result of calculations in this model for 24 GeV/c is shown by the solid line in Fig. 11.2c. With an
increase in energy, asymmetry decreases in agreement with the first our conclusion.

The Szwed model does not predict any changes in the sign of the asymmetry, and this fact contradicts the second above conclusion.

In addition, we will briefly mention several other theoretical models concerning discussed data.

- A simple model similar to the known Fermi relation (Fermi 1954) was proposed in Ryskin (1989) for asymmetry in inclusive processes. Under the assumption that the quark spin interacts with the chromomagnetic field of the color string, it was shown that the single-spin asymmetry is related to the inclusive cross section as

$$A_N \sim \delta p_T [d/dp_T][d\sigma/d^3 p]/[d\sigma/d^3 p], \qquad (11.5)$$

where δp_T is the additional transverse momentum acquired by the quark when a string is broken. It is assumed that this additional "kick" is then transferred to the final hadron with different signs depending on the spin orientation of the polarized quark. This relation was also used for the kaon and for the direct production of photons. This model has been successfully applied to the PROZA experiment data on single-spin asymmetry in the inclusive production of π^0 mesons at high p_T values. This model has been recently modified and successfully used to describe most experimental data presented above (Nurushev and Ryskin 2006). The model provides a simple analytical dependence for single-spin inclusive asymmetry and is applicable over almost the entire kinematic region. This model is presented in detail in Section "Theoretical Models" in the first part of the book.
- Zero single-spin asymmetry for π^0 mesons with high p_T values is expected according to perturbative QCD. The E704 experiment data (see the closed circles in Figs. 11.2c–11.2f) confirm this expectation.
- According to the rotating quark model, single-spin asymmetry should be zero in the central collisions (Meng 1991; Boros et al. 1993). These statements are valid for the data at 200 GeV/c.
- The model for describing single-spin asymmetry in inclusive hadron production at high transverse momenta was proposed (Troshin and Tyurin 1995). The main idea of this model is to explain the spin structure of a hadron by the spin structure of the constituent quarks: the constituent quarks are treated as quasiparticles consisting of current quarks and a surrounding cloud of quark–antiquark pairs of various flavors. Single-spin asymmetry in hadron production is proportional to the orbital angular momentum of current quarks inside a constituent quark. After the introduction of a certain phenomenological parameterization of quark distribution functions, the authors could calculate the analyzing power of the inclusive production of π^0 mesons. When they used the $SU(6)$ wave function for the polarized proton, the predicted asymmetry became regularly above the experimental values obtained in the E704 experiment. A more consistent description of the experimental results was achieved under the assumption that the polarizations of constituent quarks are maximal. In this sense, such a result supports the above conclusion, which is based on the Collins-type model, that the wave function of the polarized proton does not precisely follow the $SU(6)$ predictions. The other conclusions are

as follows: asymmetry becomes nonzero at $p_T > 1$ GeV/c, asymmetry depends slightly on the initial energy, and asymmetries for the charged pions are larger than that for the neutral pion. These predictions deserve to be tested.

The BNL Collaboration measured single-spin asymmetries in the inclusive production of $\pi^{\pm}$ (Saroff et al. 1990) for 13.3 and 18.5 GeV/c and proposed a hypothesis about the possible scaling law in the dependence of asymmetry on x_T or x_F. This hypothesis was based only on the data for π^+ (see Fig. 11.3a) because the asymmetry of π^- was insignificant for high x_T values. Since the energy range was narrow, it was difficult to justify any scaling law of the energy independence of asymmetry. The FODS-2 Collaboration has recently published the results of the measurements of inclusive asymmetries in the production of charged particles (pions, kaons, p, and $\bar{p}$) (Abramov et al. 1996). The 40-GeV/c polarized proton beam from the decay of Λ hyperons was used. The result for the asymmetry of π^+ is shown by open circles in Fig. 11.3a.

The new FODS-2 data confirm to some extent the scaling law for x_F. The data in Fig. 11.3a are fitted by the following linear functions:

13.3 GeV/c

$$A_N^+(x_T) = (0.32 \pm 0.09) \cdot \left[x_T - (0.32 \pm 0.04)\right], \quad \chi^2/\text{DOF} = 1.9 \text{ for 10 points;}$$

18.5 GeV/c

$$A_N^+(x_T) = (0.58 \pm 0.14) \cdot \left[x_T - (0.39 \pm 0.02)\right], \quad \chi^2/\text{DOF} = 0.51 \text{ for 9 points;}$$

40 GeV/c

$$A_N^+(x_T) = (0.33 \pm 0.08) \cdot \left[x_T - (0.38 \pm 0.02)\right], \quad \chi^2/\text{DOF} = 0.8 \text{ for 12 points.}$$

The data on A_N^+ immediately provide two conclusions: (a) the slope parameter is approximately constant and (b) the same is true for the zero crossing point x_T^0. It means that the asymmetry of π^+ exhibits energy scaling. For completeness, the experimental data on the inclusive asymmetry of π^0 shown in Fig. 11.2c are fitted by similar linear dependences:

24 GeV/c

$$A_N^0(x_T) = (0.94 \pm 0.37) \cdot \left[x_T - (0.38 \pm 0.02)\right], \quad \chi^2/\text{DOF} = 1.2 \text{ for 6 points;}$$

40 GeV/c

$$A_N^0(x_T) = (1.43 \pm 0.35) \cdot \left[x_T - (0.35 \pm 0.02)\right], \quad \chi^2/\text{DOF} = 1.2 \text{ for 6 points;}$$

200 GeV/c

$$A_N^0(x_T) = (0.06 \pm 0.05) \cdot \left[x_T - (0.18 \pm 0.04)\right], \quad \chi^2/\text{DOF} = 1.4 \text{ for 15 points.}$$

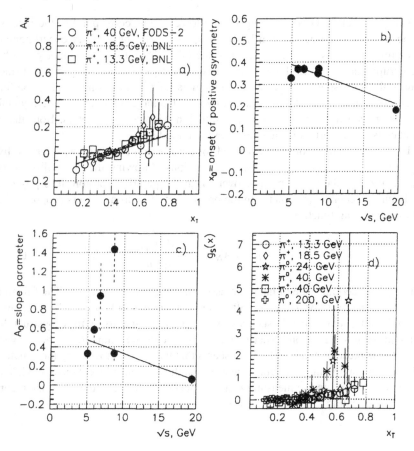

Fig. 11.3 (a) Analyzing power A_N for π^+ mesons, (b) zero crossing points x_0, (c) slope parameter A_0, and (d) scaling function $g_s(x_T)$

The energy dependences of two free parameters, which are the slope parameter A_0 and zero asymmetry crossing point x_T^0, can be extracted from the data of six experiments. Their energy dependences were parameterized as linear functions of $\sqrt{s}$ with the parameters determined by fitting to the above-listed data. The results are as follows:

$$x_T^0 = (0.013 \pm 0.005) \cdot \left[(36 \pm 12) - \sqrt{s} \right], \quad \chi^2/\text{DOF} = 1.2 \text{ for six points,}$$

$$A^0 = (0.03 \pm 0.006) \cdot \left[(22 \pm 3) - \sqrt{s} \right], \quad \chi^2/\text{DOF} = 3.5 \text{ for six points.}$$

The energy dependence of x_T^0 is presented in Fig. 11.3b. There are indications that the zero asymmetry crossing point x_T^0 decreases with increasing energy. This conclusion is based only on one point for 200 GeV/c, which can be unstable to zero asymmetry; for this reason, new data are necessary. The energy dependence of the slope parameter (see Fig. 11.3c) changes abruptly near 40 GeV/c (caution: there are two points at 40 GeV/c; the upper and lower points correspond to the results on the π^- and polarized proton beams, respectively). Three of six experimental points refer to the asymmetry of π^+ and the other three, to the asymmetry of π^0. The abrupt change in the slope parameter with energy concerns only π^0 data (the value at the highest energy follows from the data for π^0). Since the χ^2 criterion for the description of the distribution is not too good, a larger number of experimental data for π^0 are required to clarify the situation with two parameters. Both these parameters are important for testing theoretical models.

The form of the scaling law expected for hard parton scattering was derived by Sivers (1991) from general kinematic restrictions. Allowing the possibility of an asymmetric distribution of partons over k_T in the polarized proton, it is possible to obtain appreciable asymmetry in the production of hadrons. Sivers proposed the following scaling function:

$$g(x_T, \mu) = A_N(x_T) \cdot \frac{p_T^2 + \mu^2}{\mu \cdot p_t}, \tag{11.6}$$

where μ is a certain scale of hadron masses such that $m_q \ll \mu \ll p_T$. The function $g(x_T, \mu)$ contains information on soft coherent dynamics and it cannot be calculated in rigid scattering models. In such a situation, it is possible to try to reconstruct the behavior of $g(x_T, \mu)$ from experimental data. The desired function $g(x_T, \mu)$ with $\mu^2 = 0.5$ is presented in Fig. 11.3d. It seems that the data do not clarify the situation with scaling taking into account large errors in the data on π^0 for 24 and 40 GeV/c. Obviously, more precise measurements are necessary to make the conclusion concerning the scaling law proposed by Sivers.

11.2 Double-Spin Asymmetry in Inclusive Hadron Production

Double-spin asymmetry is a good quantity for theoretical models, because it is sensitive to the distributions of polarized partons. Double-spin asymmetry has long been predicted for the inclusive production of hadrons and jets in the collision of the longitudinally polarized proton beams with the longitudinally polarized proton targets (Babcock et al. 1979). It was emphasized that spin–spin asymmetry at high transverse momenta is more sensitive to the basic parton subprocesses than unpolarized observables. In this model, it was also accepted that single asymmetry at high p_T values is immaterial and this is apparently in agreement with the experimental results (see the preceding section). The first results on double-spin asymmetry in the inclusive production of neutral pions and jet-like photons were obtained by the E704 Collaboration with the polarized proton beam from the Tevatron at Fermilab (Adams et al. 1991a, 1994). These measurements were conducted in parallel with the basic experiment on the measurement of the difference between the total cross sections in pure spin states in the proton–proton and antiproton–proton interactions at 200 GeV/c (Grosnick et al. 1997). These results will be briefly discussed below.

11.2.1 *Double-Spin Asymmetry in the Inclusive Production of π^0 Mesons in the Central Region in the Collision of Longitudinally Polarized Proton and Antiproton Beams with the Longitudinally Polarized Target*

Double-spin asymmetry A_{LL} for the inclusive production of π^0 mesons by polarized proton and antiproton beams is consistent with zero (see Figs. 11.4a and 11.4b) (Adams et al. 1991a). Predictions of the aforementioned model for the Cralitz–Kaur version of the distributions of constituent quarks are presented by the dashed lines in Figs. 11.4a and 11.4b for $A_{LL}(pp \to \pi^0 + X)$ and $A_{LL}(\bar{p}p \to \pi^0 + X)$, respectively (the solid lines are linear fits). It is worth noting that valence quarks in the Cralitz–Kaur model "forget" the spin orientation of the initial proton through interaction with the sea and that only the 11.6 % fraction of proton polarization is determined by gluons. These predictions agree with the experimental data in a narrow experimentally measured kinematic region. High x_T values more interesting for the model were not reached in the E704 experiment.

11.2.2 *Double-Spin Asymmetry in the Inclusive Multiphoton Production of Jet Pairs*

For theory, it is certainly important to separate parton subprocesses in the purest form. For this aim, the use of the kinematics of a subprocess in interest is most accessible for experimenters. As is known, subprocesses end with the production of primarily two jets scattering in the opposite directions in the center-of-mass frame. The location of the detectors at $90°$ in the center-of-mass frame symmetrically with respect to the beam provides a chance to detect particles from two jets in coincidence. Then, the selection of the subprocess can be improved by imposing the usual coplanarity requirements and choosing the threshold energy for the selection of events with high transverse momenta. Such a procedure was used to select photons in the E704 experiment. In this case, it was required that more than one photon were detected in each calorimeter and they were detected in coincidence. Such events are called multiphoton pairs. The data obtained for double-spin asymmetry A_{LL} were published in Adams et al. (1994). The pseudo-mass M' and transverse momentum p_T^γ of a pair in each event are determined as

$$M' = \left| p_T^{\gamma 1} \right| + \left| p_T^{\gamma 2} \right| \quad \text{and} \quad p'_{T\gamma} = \left| p_T^{\gamma 1} \right| - \left| p_T^{\gamma 2} \right|, \tag{11.7}$$

where $p_T^{\gamma 1}$ and $p_T^{\gamma 2}$ are the transverse momenta of each multiphoton event. Most multiphoton pairs originate from two-jet events. In the measurements of the inclusive production of jets, parton distributions are smeared owing to the integration over the undetected jet. At the same time, no serious smearing problem appears when both jets are detected. Thus, two-jet production directly provides information

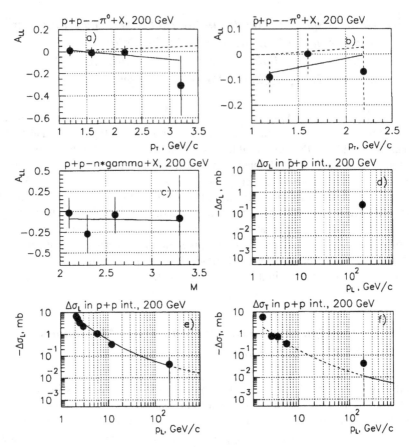

Fig. 11.4 Double-spin asymmetry A_{LL} at 200 GeV/c for (**a**) π^0 produced by the proton beam, (**b**) π^0 produced by the antiproton beam, and (**c**) multiphoton pairs; and the difference $\Delta\sigma_L$ between the total cross sections for (**d**) the longitudinally polarized antiproton beam and proton target, (**e**) the longitudinally polarized proton beam and proton target, and (**f**) the transversely polarized proton beam and transversely polarized proton target

about parton processes, particularly about the gluon contribution. Therefore, A_{LL} for the production of multiphoton pairs is also sensitive to gluon polarization. The data on A_{LL} are presented in Fig. 11.4c. According to this figure, A_{LL} do not significantly differ from zero within statistical uncertainty. The solid line in this figure is a linear fit to the experimental points; it shows, first, that double-spin asymmetry is consistent with zero, and, second, that asymmetry is independent of the parameter M'. The Monte Carlo simulation indicates that 93 % and 6 % of detected γ-ray photons were the products of the decays of π^0 and η mesons, respectively. The contributions from gluon–gluon, quark–gluon, and quark–quark scatterings to multiphoton-pair events in the range 2.0 GeV/c < $|M'|$ < 4.0 GeV/c are 45.5 %, 45.5 %, and 9.0 %, respectively. The Monte Carlo simulation taking into account the experimental conditions shows that experimental data preferably indicate gluon distribution functions with small gluon polarization.

11.2.3 Spin-Dependent Total Cross Sections

Interest in the measurements of total cross sections in pure spin states is caused by several reasons. First, as we discussed in the first part of this book, such measurements allow the complete reconstruction of the imaginary parts of three nonzero forward nucleon–nucleon scattering amplitudes. Second, having a set of such cross sections and using dispersion relations, one can reconstruct the real parts of these amplitudes, thereby completing the reconstruction of the forward nucleon scattering matrix. Third, such measurements can completely clarify the asymptotic behavior of forward spin-flip amplitudes.

The difference $\Delta\sigma_L$ between the total pp and $\bar{p}p$ cross sections in purely longitudinal spin states at 200 GeV/c have been measured by the E704 Collaboration (Grosnick et al. 1997). The previous measurements of this observable were limited by momenta ≤ 12 GeV/c for pp scattering and $\Delta\sigma_L(\bar{p}p)$ was not measured. The imaginary parts of forward elastic pp scattering amplitudes have been reconstructed from the measured total cross sections σ_T, $\Delta\sigma_L$, and $\Delta\sigma_T$ using the procedure proposed in Bilen'kij and Ryndin (1963). The real parts of these amplitudes can be determined by the measurement of the corresponding observables in the Coulomb–nuclear interference region or, as mentioned above, with the use of dispersion relations. However, such a program has been never implemented at high energies though data at 200 GeV/c are more advanced in this direction. The second aim is to test theoretical models. The third aim is to obtain an estimate of the possible contribution from spin-dependent interactions to the increase in the total cross section at high energies. The result for $\Delta\sigma_L(\bar{p}p)$ is presented in Fig. 11.4d and we cannot discuss the energy behavior of this cross section in the absence of other measurements. There are theoretical estimates for $\Delta\sigma_T$ and $\Delta\sigma_L$ at 200 GeV/c (Miettinen 1990). Applying simple reasonings based on the additive quark model, $SU(6)$ symmetry, helicity conservation, and the assumption that $\Delta\sigma_L(pp)$ is small, the author has predicted considerable value $\Delta\sigma_L(\bar{p}p) \cong 2$ mb at $p_{\text{lab}} = 200$ GeV/c. The experimental value $\Delta\sigma_L(\bar{p}p) = [-254 \pm 124 \text{ (stat.)} \pm 107 \text{ (syst.)}]$ μb presented in Fig. 11.4d is an order of magnitude smaller than the above estimate. This means that some important preconditions in the model are missed. At the same time, the jet production model (Ramsey and Sivers 1991) predicts $\Delta\sigma_L(pp) \sim 1$ μb, which is consistent with the experimental value $\Delta\sigma_L(pp) = [-40 \pm 48 \text{ (stat.)} \pm 52 \text{ (syst.)}]$ μb.

Figures 11.4e and 11.4f show the energy dependences of $\Delta\sigma_L(pp)$ and $\Delta\sigma_T(pp)$. The $\Delta\sigma_T(pp)$ value at 200 GeV/c (which has not been measured) has been obtained by extrapolating the data at low energies and its error has been estimated as 100 %. According to Dunne (1967), and Lapidus (1976), the Regge model with cuts provides the following expression for $\Delta\sigma_L(pp)$ and $\Delta\sigma_T(pp)$:

$$\Delta\sigma_L(pp) = \Delta\sigma_T(pp) = a1 \cdot \frac{s^{3\delta}}{(\ln s)^5}. \tag{11.8}$$

Here, δ is the excess of the intercept of the pomeron pole at $-t = 0$ over unity; this parameter is responsible for the increase in the total cross section with energy (Hagiwara et al. 2002). The analysis in Andreeva et al. (1998) shows that this function adequately describes the energy dependence of the experimental data for $\Delta\sigma_L$ and $\Delta\sigma_T$. The fitting results are shown in Fig. 11.4e by the solid line (extrapolation to 800 GeV/c is presented by the dashed line) and in Fig. 11.4f by the dashed line (extrapolation to 800 GeV/c is given by the solid line). In the measured energy range, the spin dependent cross sections continue to decrease from several millibarns near 5 GeV/c to 0.1 mb near 200 GeV/c. Such tendency could be changed if the cut in the Regge model with super-pomeron is dominant at high energies. In this case, the spin contribution to the total cross section can increase owing also to the parameter δ. Precise experiments at RHIC and LHC (the expected spin dependent cross section is ~ 1 µb) can clarify the role of the spin in the increase in the total cross section with energy.

References

Abramov, V.V., et al.: Preprint No. 96-82, IHEP, Protvino, Russia (1996)
Adams, D.L., et al.: Phys. Lett. B **264**, 462 (1991a)
Adams, D.L., et al.: Phys. Lett. B **261**, 197 (1991b)
Adams, D.L., et al.: Z. Phys. C **56**, 181 (1992)
Adams, D.L., et al.: Phys. Lett. B **336**, 269 (1994)
Adams, D.L., et al.: Phys. Lett. B **345**, 569 (1995)
Adams, D.L., et al.: Phys. Rev. Lett. **77**, 2626 (1996a)
Adams, D.L., et al.: Phys. Rev. D **53**, 4747 (1996b)
Akchurin, N., et al.: Phys. Rev. D **48**, 3026 (1993a)
Akchurin, N., Buttimore, N., Penzo, A.: In: Proceedings of the 5th Blois Workshop, Providence, Rhode Island (1993b)
Andreeva, E.A., et al.: Int. J. Mod. Phys. A **13**, 1515 (1998)
Anselmino, M., et al.: DFTT 48/94, INFNCA-TH-94-27 (1994). arXiv:hep-ph/9503290
Antille, J., et al.: Phys. Lett. B **94**, 523 (1980)
Apokin, V.D., et al.: Phys. Lett. B **243**, 461 (1990)
Artru, X.: In: Proceedings of the 5th Workshop on High Energy Spin Physics, Protvino, Russia, p. 152 (1993)
Artru, X., et al.: LYCEN/9423, TPJU 12/94 (1994)
Babcock, J., et al.: Phys. Rev. D **19**, 1483 (1979)
Bilen'kij, S.M., Ryndin, R.M.: Phys. Lett. **6**, 217 (1963)
Borisov, A.A., et al.: Sov. J. Nucl. Phys. **5**, 348 (1967)
Boros, C., Liang, Z.-T., Meng, T.-C.: Phys. Rev. Lett. **70**, 1751 (1993)
Bourrely, C., Soffer, J.: Lett. Nuovo Cimento **19**, 569 (1977)
Buttimore, N.H., Gotsman, E., Leader, E.: Phys. Rev. D **18**, 694 (1978)
Dunne, S.A.: Phys. Rev. Lett. **19**, 1299 (1967)
Fermi, E.: Nuovo Cimento **11**, 407 (1954)

Ferrari, E., Selleri, F.: Nuovo Cimento **27**, 1450 (1963)

Grosnick, D.P., et al.: Phys. Rev. D **55**, 1159 (1997)

Hagiwara, K., et al.: Phys. Rev. D **66**, 010001 (2002)

Kopeliovich, B.Z., Lapidus, L.I.: Sov. J. Nucl. Phys. **19**, 114 (1974)

Kopeliovich, B.Z., Zakharov, B.G.: In: Proceedings of the 3rd Workshop on High Energy Spin Physics, Protvino, USSR, p. 137 (1989)

Korotkiyan, V.M., Teryaev, O.V.: In: Proceedings of the 6th Workshop on High Energy Spin Physics, Protvino, USSR, p. 84 (1995)

Lapidus, L.I.: In: Proceedings of the 11th LINP Winter School, Leningrad, USSR, p. 55 (1976)

Mandelstam, S.: Proc. R. Soc. Lond. A **244**, 491 (1958)

March, R.H.: Phys. Rev. **120**, 1874 (1960)

Mcllwan, R.L., et al.: Phys. Rev. **127**, 239 (1962)

Meng, T.-C.: In: Proceedings of the 4th Workshop on High Energy Spin Physics, Protvino, USSR, p. 112 (1991)

Miettinen, H.: Preprint No. DOE/ER/05096-39, Rice University (1990)

Nurushev, S.B.: In: Proceedings of the 4th Workshop on High Energy Spin Physics, Protvino, USSR, p. 5 (1991)

Nurushev, S.B.: In: Proceedings of the 2nd Meeting Held at Zeuthen, p. 3 (1995a)

Nurushev, S.B.: In: Proceedings of the 2nd Meeting Held at Zeuthen, p. 75 (1995b)

Nurushev, S.B., Ryskin, M.G.: Yad. Fiz. **69**, 1 (2006)

Nurushev, S.B., Solovyanov, V.L.: Preprint No. R-2382, JINR, Dubna, Russia (1965)

Qiu, J., Sterman, G.: Phys. Rev. Lett. **67**, 2261 (1991)

Ramsey, G., Sivers, D.: Phys. Rev. D **43**, 2861 (1991)

Ryskin, M.G.: In: Proceedings of the 3rd Workshop on High Energy Spin Physics, Protvino, USSR, p. 151 (1989)

Saroff, S., et al.: Phys. Rev. Lett. **64**, 995 (1990)

Sivers, D.: Phys. Rev. D **43**, 261 (1991)

Soffer, J.: In: Proceedings of the Workshop on the Prospects of Spin Physics at HERA, Zeuthen, p. 370 (1995)

Szwed, J.: In: Proceedings of the 9th International Symposium on High Energy Spin Physics, Bonn, vol. 1, p. 463 (1990)

Troshin, S.M., Tyurin, N.E.: Phys. Rev. D **52**, 3862 (1995)

Chapter 12
Latest Results from the Largest Polarization Setups

The above presentation was based on review of Nurushev (1997) appearing ten years ago. Since that, large polarization setups COMPASS (CERN) and HERMES (DESY), as well as the STAR, PHENIX, BRAHMS, and $pp2pp$ setups at RHIC (BNL), have been commissioned. The first results from them have appeared partially in the final form, partially in the form of preliminary reports inaccessible to a wide audience. We present some interesting results to readers.

12.1 COMPASS Polarization Setup and the Results Obtained on It

The COMPASS setup (CERN) is one of the largest polarization setups in the world. The first engineering run on the setup whose project was proposed in 1996 was conducted in 2002 with a 160-GeV/c longitudinally polarized muon beam with a polarization of 76 %, an intensity of $2 \cdot 10^8$ μ/cycle, and a cycle duration of 4.5 s. The largest worldwide polarized target of the SMC setup (described in Sect. 6.1 "Solid polarized targets") was used, but with a different target material, namely, ^{6}LiD, which provides a larger number of polarized deuterons (the dilution factor > 0.4) than deuterated ammonium. The target polarization was 50 %.

Two runs were performed in such a configuration of the setup in 2003 and 2004 and a large statistical material has been collected. Although the COMPASS results are preliminary, it is necessary to briefly discuss them for the representation of the prospects of polarization investigations on this setup (Bressan 2004).

Figures 12.1a and 12.1b show the preliminary results of the measurement of inclusive asymmetry A_1^d in the scattering of the 160-GeV/c polarized muon beam with a polarization of 76 % on the polarized deuteron target. The asymmetry is represented as a function of the Bjorken variable x_B and is calculated from the data of two runs in 2002 and 2003. The COMPASS results are consistent with zero and are in agreement with the SMC and HERMES data also presented in Fig. 12.1a. Figure 12.1b shows that the COMPASS statistics considerably exceeds the SMC

S.B. Nurushev et al., *Introduction to Polarization Physics*,
Lecture Notes in Physics 859, DOI 10.1007/978-3-642-32163-4_12,
© Moskovski Inzhenerno-Fisitscheski Institute, Moscow, Russia 2013

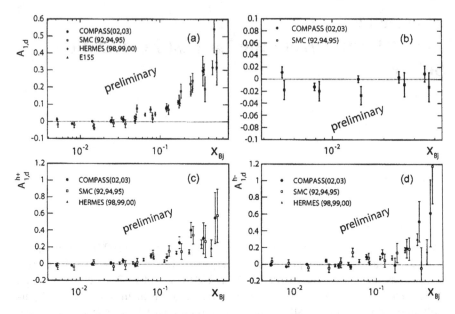

Fig. 12.1 (a) Preliminary COMPASS results for asymmetry A_1^d in comparison with the results of other experiments; (b) the same results at small x_B values in comparison with the SMC data; and asymmetry A_1^h versus x_B for semi-inclusive (c) positive and (d) negative hadrons

statistics at small x_B values. This statistics was achieved due to a higher luminosity and a better target quality factor.

Figure 12.1c shows semi-inclusive asymmetry in the production of positive hadrons. It is seen that asymmetry is close to zero at $x_B < 0.2$ and at large x_B values. Asymmetry tends to increase, on the average reaching a value of ≈ 0.3 at $x_B \approx 1$. The results of all SMC, HERMES, and COMPASS experiments presented in Fig. 12.1c coincide within the measurement accuracy. At very small x_B values, the COMPASS experiment provides the best accuracy. Semi-inclusive asymmetry for negative hadrons is presented in Fig. 12.1d. The tendency is the same as for positive hadrons, but asymmetry for negative hadrons as a whole is lower at large x_B values. Since hadrons were not identified at this stage, hadrons cannot be separated, though it is known from other sources that 80 % hadrons are pions. However, even with the identification of hadrons, it would be impossible to decompose asymmetry into the quark flavors because there are no measurements with a longitudinally polarized proton target.

Another important aim of the COMPASS experiment is the determination of the gluon polarization. This problem is solved by separating the γg fusion process. The separation of this process is carried out by detecting either open charm or hadron pairs with large transverse momenta.

Mesons with open charm are detected by reconstructing D^0 or D^* mesons through the $D^0 \rightarrow K\pi^0$ or $D^* \rightarrow D^0\pi \rightarrow K\pi^0\pi$ decays, respectively. In order to suppress the background in the case of the decay of D^0, two conditions

Fig. 12.2 Results of the
measurement of the gluon
polarization in the
COMPASS experiment
through events with large p_T
values and $Q^2 > 1$ (GeV/c)^2

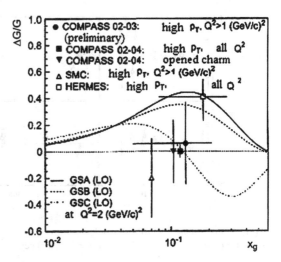

$|\cos(\theta_K^*)| < 0.5$ and $z_D = E_D/E_{\gamma^*} > 0.25$ are imposed on the K meson emission angle θ_K^* in the D^0 rest frame and the fraction of energy carried away by the D^0 meson. The second decay case is much purer from the background owing to the specificity of kinematics. The total statistics on open charm for 2002–2004 runs provides an expected estimate of $\delta(\Delta G)/G = 0.24$ for the error in the gluon polarization.

Asymmetry $A^{\gamma*d}$ in the production of hadron pairs with large p_T values in the interaction of a virtual photon with a longitudinally polarized deuteron is related to the gluon polarization as

$$A^{\gamma*d} = \frac{A^{\mu d \to hhX}}{D} \approx \left\langle \frac{\hat{a}_{LL}^{PGF}}{D} \right\rangle \left\langle \frac{\Delta G}{G} \right\rangle \left\langle \frac{\sigma^{PGF}}{\sigma_T} \right\rangle. \qquad (12.1)$$

Here, $\hat{a}_{LL}^{PGF}$ is the analyzing power of the photon–gluon fusion (PGF) subprocess, D is the depolarization factor for the virtual photon, and the ratio σ^{PGF}/σ_T is the fraction of the PGF events in the total number of events in a sample. The Monte Carlo simulation for the COMPASS experiment gives $\hat{a}_{LL}^{PGF}/D = -0.74 \pm 0.05$ and $\sigma^{PGF}/\sigma_T = 0.34 \pm 0.07$. Then, for the average gluon momentum $x_g = 0.13$, it has been found that $\Delta G/G = 0.06 \pm 0.31$ (stat.) ± 0.06 (syst.). The results of the measurement of the gluon polarization in the COMPASS experiment with the use of both methods discussed above are presented in Fig. 12.2.

The result of the HERMES experiment is also shown in this figure. Several features of the presented data can be indicated. The COMPASS data are consistent with zero within gluon polarization errors ≈ 0.27, whereas the HERMES result is about 0.4 and differs from zero by almost two standard deviations. Much larger statistics is obviously necessary for preferring one of the versions.

The following important direction of the investigation at COMPASS is aimed at determining the chirality-odd parton distribution function transverse in the nucleon spin, the so-called transversity function $\Delta_T q(x)$. To this end, it is necessary to measure semi-inclusive asymmetry in the production of hadrons in the interaction of

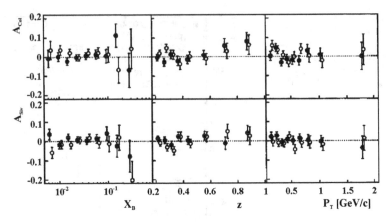

Fig. 12.3 Preliminary COMPASS results for the (*upper panels*) Collins and (*lower panels*) Sivers effects for (*open circles*) positive and (*closed circles*) negative hadrons

longitudinally polarized leptons with transversely polarized nucleons. There are two possibilities for measuring asymmetry. Asymmetry associated with the chirality-odd quark fragmentation function $\Delta D_q^h(z, p_T)$ is measured in the first case. This asymmetry is called the Collins effect after Collins who proposed such measurements (Collins 1993). In the first approximation, Collins asymmetry can be written as

$$A_{Coll} = \frac{\sum_a e_a^2 \Delta_T q_a(x, k_T^2) \Delta D_q^h(z, p_T)}{\sum_a e_a^2 q_a(x, k_T^2) D_q^h(z)}. \tag{12.2}$$

There is another mechanism leading to azimuthal asymmetry in the scattering of leptons on transversely polarized nucleons. If the distribution $\Delta_0^T q_a(x, k_T^2)$ of partons over the internal transverse momentum k_T in the initial polarized nucleon depends on the nucleon polarization direction, there is the asymmetry

$$A_{Siv} = \frac{\sum_a e_a^2 \Delta_0^T q_a(x, k_T^2) D_q^h(z)}{\sum_a e_a^2 q_a(x, k_T^2) D_q^h(z)}. \tag{12.3}$$

This asymmetry is called the Sivers effect (Sivers 1990). The measurement of these effects with the transversely polarized ^{6}LiD target at the COMPASS setup took 20 % of the general time. The preliminary measurement results for positive and negative hadrons in the colliding quark fragmentation region are presented in Fig. 12.3.

The three upper panels show the Collins effect versus the parameters x_B, z, and p_T. The measurements of the Collins effect on the polarized deuteron have been carried out for the first time at COMPASS. The three lower panels concern the Sivers effect. It is seen that both effects are consistent with zero in the measured region with statistics accumulated in 2002. The use of the total statistics accumulated in 2002 may improve the measurement accuracy by factor 2. This zero result can be interpreted either as the smallness of the Sivers and Collins effects or as a specificity of the deuteron target. The asymmetry effects from the proton and neutron probably cancel each other. To solve this problem, separate measurements of the same effects on the proton and neutron are necessary.

12.2 HERMES Polarization Setup and the Results Obtained on It

The HERMES setup was created in 1995 at the unique electron/positron–proton HERA collider at DESY (Germany). It was intended to study the nucleon spin structure with the use of a 27.5-GeV/c longitudinally or transversely polarized electron (positron) beam and an internal gas polarized target with a storage cell. The density of the polarized hydrogen atoms in the target is 10^{14} atoms/cm^2 and polarization is > 90 %. Owing to high luminosity, the experimental results have a high accuracy. Working in open geometry and having a good particle identification system, HERMES has high detection efficiency for deep inelastic scattering and semi-inclusive processes.

Figure 12.4 shows the longitudinal structure functions $g_1(x)$ for the proton, deuteron, and neutron measured at HERMES in deep inelastic scattering (inclusive process).

The data have been obtained for $Q^2 > 0.1$ (GeV)^2 and $W > 1.8$ GeV and are shown in Fig. 12.4 along with the other existing data. It is seen that the accuracy reached in the measured region $3 \cdot 10^{-3} < x < 1$ is at the world level. This function is maximal and minimal for the proton and neutron, respectively. These data provide the basis for the theoretical analysis of the nucleon spin structure.

The next series of HERMES measurements concerned semi-inclusive hadron production processes. These processes constitute an efficient tool for separating the nucleon spin in different flavors of quarks and antiquarks $\Delta q_f(x)$. The results are presented in Fig. 12.5 for the distributions of valence quarks and antiquarks separately.

It is seen that the u and d valence quarks make positive and negative contributions to the proton spin, respectively, and the $\bar{u}$ and $\bar{d}$ antiquarks make no contribution to it. There is an indication of the possible small positive contribution from sea quarks Δs. The presented curves are the theoretical calculations in the leading QCD approximation.

In addition to the structure functions of the unpolarized nucleon, $F_1(x)$, and longitudinally polarized nucleon, $g_1(x)$, the third function $h_1(x)$ called transversity is necessary for the complete description of the quark nucleon structure in the leading approximation. This function is chirality-odd and cannot be measured in the deep inelastic scattering of polarized leptons from longitudinally polarized nucleons. The measurement of transversity requires another chirality-odd function. It appears as the Collins fragmentation function in the scattering of leptons from transversely polarized nucleon targets. Another possibility of appearing azimuthal asymmetry arises with the inclusion of T-odd Sivers function $f_{1T}^\perp(x, k_T)$. This function describes correlation between the transverse polarization of the nucleon and the inter-

Fig. 12.4 Existing data on the weighted structure functions $g_1(x)$ for the proton, deuteron, and neutron; the total error is the square root of the sum of the squares of the statistical and systematic errors

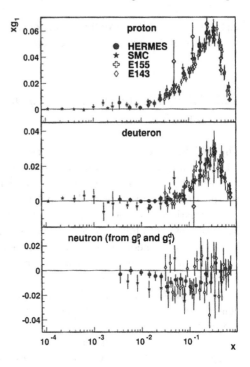

Fig. 12.5 Quark helicity distribution functions $x\Delta q(x, Q_0^2)$ calculated at $Q_0^2 = 2.5\,\text{GeV}^2$; the *lines* are the leading-QCD calculations

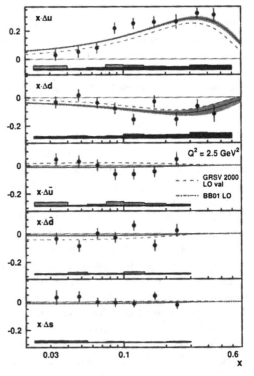

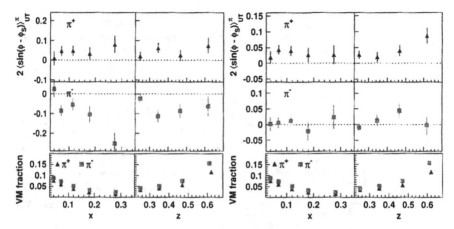

Fig. 12.6 Moments of the Compton effect of a virtual photon according to (*left panels*) Collins and (*right panels*) Sivers versus x and z; only statistical errors are presented. There is also a general normalization error of 8 %; the *lower panels* show the relative contributions from the exclusive production of vector mesons (*VM* calculations)

nal transverse momentum of a quark. A feature of the Collins effect is the azimuth-angle dependence of single-spin asymmetry in the production of hadrons in the form $\sin(\phi + \phi_s)$, whereas asymmetry in the Sivers effect has the form $\sin(\phi - \phi_s)$. Here, ϕ is the azimuth angle between the perpendiculars to the pion production plane and scattered lepton plane and ϕ_s is the angle between the pion production plane and transverse target polarization component.

The average moments of azimuthal asymmetries are presented in Fig. 12.6 as functions of x and z for the Collins and Sivers effects in the left and right panels, respectively.

It is seen that the effects are nonzero. In particular, the Collins effect for positive and negative pions is $(2.1 \pm 0.7 \text{ (stat.)})$ % and $(-3.8 \pm 0.8 \text{ (stat.)})$ %, respectively. These signs correspond to the signs in the spin distributions of valence quarks and are in agreement with the model predictions. However, the problem arises in their values. The measured moments indicate that the transverse spin distribution of d quarks, δd, is no less in the absolute value than δu; this relation is surprising, because the inequality $|\delta d| < |\delta u|$ was expected by analogy with the case of longitudinal polarization. This discrepancy has not yet been explained.

The moment for the Sivers effect is also nonzero (see the right panels in Fig. 12.6). Its average value for π^+ appeared to be $(1.7 \pm 0.4 \text{ (stat.)})$ %, which indicates a nonzero orbital angular momentum of quarks in a nucleon. However, it is not excluded that the large asymmetry could be caused by π^+ mesons from the decays of ρ mesons. The yields of ρ mesons calculated for the HERMES kinematic region are shown in the lower panels.

The next important research direction in the HERMES experiment is the so-called deeply virtual Compton scattering (DVCS) processes. One of such processes is the production of real photons in hard collisions of electrons with protons and

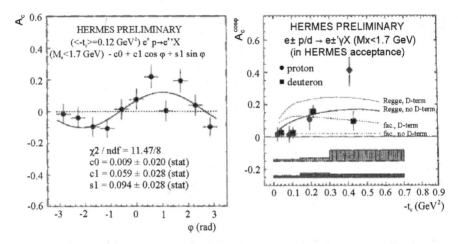

Fig. 12.7 *Left panel*: beam-charge asymmetry in the electroproduction of photons in hard collisions versus the azimuth angle ϕ; the *solid line* is the fit with the parameters given in the figure. The *right panel*: $\cos\phi$ amplitude of beam-charge asymmetry for the proton and deuteron versus $-t$; the *lines* are the calculations corresponding to various parameterizations of the generalized distribution function

deuterons. These reactions provide the most adequate method for determining the generalized parton distribution functions. The DVCS amplitudes are determined from interference between these amplitudes and the Bethe–Heitler amplitudes. This is physically due to the presence of two sources of photon radiation: quarks and charged leptons. The measurements of the azimuthal distribution of differential cross sections for various charges (spins) of the lepton beam allow the determination of the real (imaginary) parts of the DVCS amplitudes. These real and imaginary parts of the amplitudes are manifested as $\cos\phi$ and $\sin\phi$ modulations of the cross sections. The HERMES Collaboration has already published the results of the measurement of asymmetry in DVCS as a function of the beam chirality (Airapetian et al. 2001). The first results of the measurements of asymmetry A_C depending on the beam charge with hydrogen and deuterium targets are presented in Fig. 12.7.

The left panel in Fig. 12.7 shows A_C versus the azimuth angle for both beams. Since the average polarizations of the beams are different, both terms are present in the form $P_1 \cos\phi + P_2 \sin\phi$. Asymmetry is approximated by this expression, and the fit is shown in the figure. The right panel in Fig. 12.7 shows $c\cos\phi$ amplitudes determined through a two-parameter fit with corrections to the background depending on the invariant momentum transfer from a beam to the proton or deuteron. The results indicate the high sensitivity of the measured beam-charge dependent asymmetry to the parameterization of the generalized parton distribution.

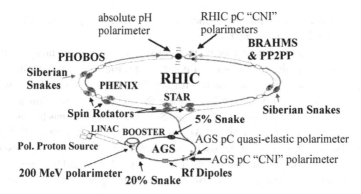

Fig. 12.8 Polarized RHIC complex

12.3 Polarized RHIC Complex and the Results Obtained on It

According to the volume of the physical program, investments, and concentration of equipment and human resources, the polarized RHIC complex has no competitors for the next decade in research of spin phenomena in hadron physics. With the implemented eRHIC program, BNL will also occupy the leading position in lepton–hadron polarization physics. For this reason, a lot of useful experience can be learnt from this group.

The scheme of the polarized RHIC complex is presented in Fig. 12.8.

The RHIC complex begins with a source of polarized negative hydrogen ions with optically pumped polarization. Its detailed description can be found in Sect. 7.1 "Polarized atomic beam sources" in the second part of the book. The polarized negative ions at the source output are accelerated by a microwave quadrupole to 750 keV and arrive at a linear accelerator that accelerates them to 200 MeV. Then, ions from the linear accelerator are transported to a booster, are stripped, and enter the booster. In the booster, protons are bunched and accelerated to 1.5 GeV. After booster polarized beam enters AGS (the Alternating Gradient Synchrotron), accelerated up to 22 GeV/c, and introduced to RHIC. Each of the proton beams is accelerated now at RHIC to 100 GeV, and each of the beams was accelerated to 250 GeV in 2006. The final desired parameters at RHIC are the c.m. proton energy $\sqrt{s} = 500$ GeV, polarization > 70 %, and a luminosity of $2 \cdot 10^{32}$ cm^{-2} s^{-1} (RHIC 1998). The following polarimeters are placed in the beam line: a Lamb shift polarimeter at the OPPIS AGS source output, a carbon polarimeter at the linear-accelerator output, and a nuclear-scattering carbon polarimeter and a Coulomb–nuclear interference carbon polarimeter at AGS. Two basic Coulomb–nuclear interference polarimeters based on proton scattering from a polarized jet target (absolute polarimeter) and from very thin carbon films are used at RHIC. Their description was given in Sect. 8.3 "Polarimetry of proton beams" in the second part of the book. The main aim of these two basic general-purpose polarimeters at RHIC is to achieve an accuracy of 5 % in the measurement of the polarization of each of proton beams.

In addition to these polarimeters, which have now certain difficulties, new polarimeters can be created. For example, it is not excluded that asymmetry found at the STAR setup in the inclusive production of π^0 mesons can be used to create a quite operative relative polarimeter.

At present, the alternating gradient synchrotron (AGS) is apparently "the narrow neck of the bottle" in the acceleration of polarized protons. A proton beam entering it with polarization > 70 % leaves it with a polarization of only 50 % (recently polarization was increased up to 60 %). Since the AGS has no rectilinear sections long enough for the installation of a full-scale Siberian snake, one has to place partial snakes. A solenoid as a 5 % partial snake has been first placed; then, a 20 % partial snake has been installed; there is also a microwave dipole for the suppression of imperfection resonances, but the problem has not yet been fundamentally solved. It is hoped that the problem can be solved by combining several partial snakes of various forces.

Figure 12.8 presents four experimental setups, PHENIX, STAR, BRAHMS, and PHOBOS, constructed to investigate interaction of heavy ions. Since 1992 when the RSC (RHIC Spin Collaboration) was organized for polarization research at RHIC, two largest groups, PHENIX and STAR, immediately joined RSC and the BRAHMS group also joined little later. The fifth experiment, $pp2pp$, was initially focused on proton–proton collisions and polarization investigations in it were included from the beginning. Two full-scale (spin rotation by 180°) groups of Siberian snakes were installed in each of two collider rings. Rotation occurs about mutually perpendicular axes lying in the beam orbit plane. In addition, each of large setups, PHENIX and STAR, is equipped with two spin rotators each for ensuring the longitudinal polarization of beams at the interaction point (IP) and for returning polarization to the initial (perpendicular to the beam orbit) position. All listed instruments have been mounted, tested, and accepted to operation at RHIC.

The first run at the collider with polarized proton beams was performed in 2002. Since then, six physical runs (one run was engineering) have been passed with the gradually improving parameters of setups and beams. A number of physical results were obtained and we present them below to the attention of readers.

12.3.1 STAR Polarization Setup

The detector (see Fig. 12.9) is a magnetic solenoid spectrometer with a magnetic field of $H = 0.5$ T, a length of 4.2 m, and an inner diameter of 2 m. The spectrometer contains track detectors based on the TPC (Time Projection Chamber) 4 m in length with an inner diameter of 30 cm and an outer diameter of 2 m. It covers an azimuth angle of 2π, detects three charged particles with high efficiency, and can identify particles with momenta lower than 1.5 GeV/c from ionization losses. Two types of electromagnetic calorimeters are placed behind the TPC. The first electromagnetic calorimeter, so-called barrel, covers the central pseudorapidity region within ± 1. The other type of electromagnetic calorimeters called the endcap calorimeter closes the setup end. A time-of-flight detector based on scintillation counters is installed in front of the barrel.

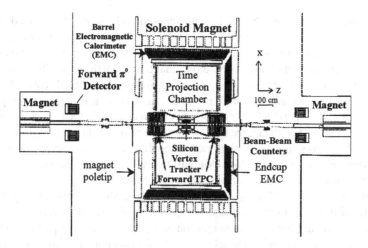

Fig. 12.9 STAR polarization setup

Fig. 12.10 Inclusive
asymmetry of π^0 mesons
measured in the beam
fragmentation region in the
STAR experiment

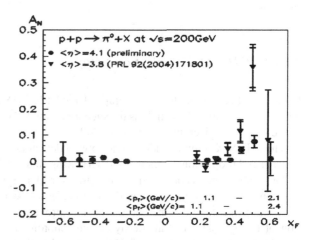

Counters monitoring collisions of two beams (beam–beam counters, BBC), and
forward pion detectors (FPD) are also shown in Fig. 12.9. The detailed description
of the STAR setup was given in Ackermann et al. (2003).

The first results on asymmetry in the inclusive production of π^0 mesons in the
beam fragmentation region have been obtained on the FPD and published in Adams
et al. (2004). These measurements were carried out at pseudorapidity $\eta = 3.8$. Similar measurements have recently been performed at $\eta = 4.1$ and both results are
presented in Fig. 12.10. It is seen that asymmetry is almost zero at negative x_F
values, deviates from zero at $x_F > 0.3$, and reaches 10–15 % near $x_F \approx 0.5$. The
measurements at the STAR setup were conducted at the c.m. energy $\sqrt{s} = 200$ GeV
(Ogawa 2004).

Similar measurements have been previously performed at $\sqrt{s} \approx 20$ GeV by the
E704 Collaboration (see Sect. 11.1). Comparison of these data shows that the inclu-

sive asymmetry of π^0 mesons in the fragmentation region of the polarized proton beam is almost independent of the initial energy. It could be concluded that asymptotic energies begins already at $\sqrt{s} \approx 20$ GeV. However, the analysis of the energy behavior of differential cross section shows that this is not the case. The differential cross section at $\sqrt{s} \approx 20$ GeV is not described in perturbative QCD (pQCD) in the measured range of kinematic variables. At the same time, it is described by pQCD at the energy $\sqrt{s} \approx 200$ GeV. If this is valid, a direct test of pQCD becomes possible for the first time in the entire history of research of single-spin asymmetry. We arrive at the depressing conclusion: pQCD predicts zero asymmetry, whereas the experiment shows a nonzero effect. We will return to this problem after the review of other results on the single-spin asymmetry obtained at RHIC.

Note another fact following from Fig. 12.10. Left–right asymmetry in the unpolarized-beam fragmentation region ($x_F < 0$ in Fig. 12.10) is zero. This could be expected because the unpolarized proton is similar to a spinless particle. The decay (fragmentation) of such a particle cannot lead to azimuthal asymmetry, because the angular distribution of its decay products in the rest frame of this particle is isotropic.

12.3.2 PHENIX Polarization Setup

Figure 12.11 shows another large setup at RHIC, PHENIX (Bruner 2002). It is based on a dipole magnet whose field is directed along the beam. The detailed description of the setup was given in Adcox et al. (2003).

The PHENIX setup includes the following detectors.

Drift and proportional chambers, RICH (Ring Image CHerenkov, Cherenkov detectors with image reconstruction) detectors for identifying charged particles, and electromagnetic calorimeters based on lead glasses, sandwiches, and scintillators are used in the central arm. These detectors provide the detection and identification of charged and neutral (decaying into photons and electrons) particles in a pseudorapidity range of $|\eta| = 0.35$ and an azimuth angle range of $\Delta\phi = 180°$. Two muon spectrometers with a total muon identification system are placed in the northern and southern directions. A muon detector covers a pseudorapidity range of $1.2 < |\eta| < 2.4$ and $\Delta\phi = 360°$. There are a vertex detector and a beam–beam counter.

In the 2002–2003 run, RHIC operated with 55 bunches each containing $5 \cdot 10^{10}$ protons. The bunch duration was ~ 1 ns and the interval between bunches was 213 ns. The polarization of protons of every bunch was set by a source and was marked by the RHIC timer. The sequence of bunches with alternating polarization ($+-, +-$, etc.) was set in one ring and the sequence of pair bunches ($++, --++, --$, etc.) was taken in the other ring. It is seen that these sequences of the polarizations of colliding protons ensures all four combinations of polarizations, namely, $++, -+, +-$, and $--$. This is equivalent to the reverse of beam polarization once in 213 ns, which is the record frequency! This procedure reduces

Fig. 12.11 PHENIX
polarization setup at RHIC

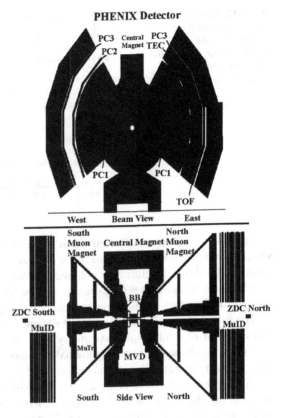

many systematic errors to minimum. The chosen combination of bunch polarizations makes it possible to determine not only double-spin, but also single-spin asymmetries, as well as false asymmetries, from the same measurements. This is an additional advantage of the polarized collider.

Transverse single-spin asymmetry measured for neutral and charged hadrons is
shown in Fig. 12.12 as a function of the momentum transfer.

Statistics for analyzing charged hadrons was $1.3 \cdot 10^7$ events collected with a
minimum bias trigger. Statistics for π^0 mesons was $1.8 \cdot 10^7$ events.

The largest uncertainty of the measured asymmetry comes from inexact information on the beam polarization at energy of 100 GeV. The beam polarization at
this energy was not directly measured. The beam polarization was measured at an
injection energy of 22 GeV using a Coulomb–nuclear interference pC polarimeter
(Jinnouchi et al. 2003a, 2003b). The analyzing power of this polarimeter was determined in Tojo et al. (2002). The polarimeter yielded a polarization of 27 % with
a relative error of 30 %. This polarization was also assigned to a beam energy of
100 GeV under the assumption that the acceleration of the polarized protons from
22 GeV to 100 GeV in RHIC was not accompanied by depolarization. This uncertainty is not taken into account in the points in Fig. 12.5.

Fig. 12.12 PHENIX results
for transverse single-spin
asymmetry in the central
region for charged hadrons
and the neutral pion at
$\sqrt{s} = 200$ GeV; the
polarization determination
error ($\sim 30\%$) is not included

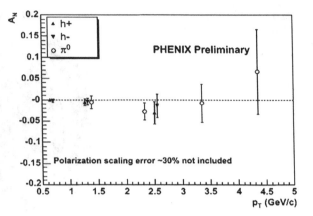

The main conclusion following from these measurements is that all inclusive single-spin asymmetries measured in the central region are equal to zero. The authors state that this conclusion is in agreement with pQCD predictions in the measured kinematic region. It is necessary to note that the same zero result was obtained ten years before these measurements in the E704 experiment (Fermilab).

The following very interesting result obtained for the first time at RHIC is the measurement of double-spin asymmetry A_{LL} in the inclusive production of π^0 mesons in collisions of two longitudinally polarized protons. The measurements were also carried out in the central region.

The parameter A_{LL} can be measured only with longitudinally polarized beams, whereas the stable polarization direction at RHIC is the vertical direction parallel to the main magnetic field of the ring. Hence, devices transforming the vertical polarization to the longitudinal direction are necessary. Two such devices named spin rotators are provided for each beam. Two devices are necessary because one rotator turns polarization to the longitudinal position before collision and the other rotator returns polarization after collision to the vertical position perpendicular to the beam orbit plane.

The correct operation of the rotators should certainly be tested. A local polarimeter is necessary for this purpose. In this sense, a discovery made by physicists when measuring the analyzing power of neutral particles in a narrow forward angular cone (Bazilevsky et al. 2003) using the ZDC detector (Zero Degree Calorimeter, a calorimeter at zero angle) (Adler et al. 2001) is good fortune. This detector is intended to measure the energy and angle of the emission of neutrons. It is located at a distance of 18 m from the interaction point of colliding beams and has a very narrow angular acceptance 0.3 mrad $< \theta_n <$ 2.5 mrad. An appreciable analyzing power of inclusively produced neutrons has been revealed in special measurements with transversely polarized beams. Neutron asymmetry in the interval $x_F = 0.3$–0.8 is 10–15 % and depends only slightly on x_F. This detector has been used as a local relative polarimeter for the adjustment of the spin rotators. This polarimeter should show zero transverse asymmetries (in the plane perpendicular to the beam) if the spin rotators are switched on and polarization becomes longitudinal. If a spin rota-

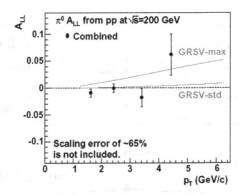

Fig. 12.13 PHENIX results for double-spin asymmetry $A_{LL}(\pi^0)$ in the inclusive production of π^0 mesons in collisions of two longitudinally polarized protons at an energy of $\sqrt{s} = 200$ GeV; the measurements are performed in the central region; an error of 65 % in the beam polarization normalization is not included in the errors shown on the plot; the curve is the next-to-leading pQCD calculation

tor is switched off, transverse asymmetry should appear. Quantitative measurements confirmed these expectations.

The first trigger on proton collisions is provided by beam–beam counters placed on both sides of the beam interaction point at a distance of ±1.44 m from this point. These counters cover a pseudorapidity range of ±(3.0–3.9) and the full azimuthal range. They detect approximately half of all interactions. A high time resolution of these counters makes it possible to reconstruct an event vertex with an accuracy of 2 cm at an interaction region of about 30 cm. The luminosity of bunches was controlled by the beam–beam counters and ZDC independently and the accuracy in the ratio of the luminosities of different bunches reached $2.5 \cdot 10^{-4}$. This corresponds to the relative error in double-spin asymmetry $\delta A_{LL} = 1.8 \cdot 10^{-3}$. The ratio of the luminosities for the selected events was unit in limits ±5 %. The algorithm developed by the collaboration was applied to reconstruct π^0 mesons.

The results of the measurements of A_{LL} are presented in Fig. 12.13.

The main conclusion is that $A_{LL}(\pi^0)$ is consistent with zero in the entire measured p_T interval. According to Fig. 12.13, the experiment is more consistent with the hypothesis of small gluon polarization than with the hypothesis of high polarization. This conclusion has been made long before these results by the E704 Collaboration.

12.3.3 The pp2pp Polarization Experiment at RHIC

Since the creation of antiproton–proton colliders has advantages over the creation of proton–proton colliders, elastic $\bar{p}p$ scattering has been studied up to an energy of 1.8 TeV, whereas investigations of pp scattering stopped at an energy of

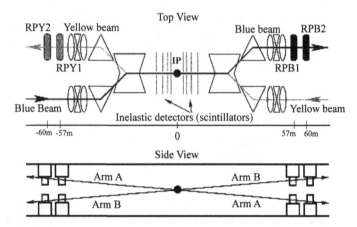

Fig. 12.14 The $pp2pp$ polarization experiment at RHIC

$\sqrt{s} = 63$ GeV after the ISR program was completed. In connection with the building of RHIC, the research program of elastic pp scattering in the energy range $\sqrt{s} = 50$–500 GeV was proposed (Guryn et al. 1995). The proposal was later supplemented by investigations of polarization effects. The setup scheme is presented in Fig. 12.14.

The idea of the experiment proposed and implemented for the first time at ISR is the use of the magnetic structure of the accelerator. Microdetectors with high spatial and time resolutions are mounted in a special device called "the Roman pot." Two such Roman pots are installed in each of two RHIC rings named blue and yellow after the colors of the rings. They are denoted as RPY1 and RPY2 (in the yellow ring) and RPB1 and RPB2 (in the blue ring) in Fig. 12.14. The places of these detectors with respect to the interaction point (IP) along the beam are chosen so that a scattered particle deviates from the beam axis by at least five sizes of the beam (5σ). The amplitude function in the detector location places should be as small as possible. In this case, the beam size is also minimal. The parameters of the beam transport matrix are chosen so that the scattering angle is unambiguously transformed to the detected particle coordinate. Therefore, determining the particle coordinate, one can unambiguously determine the particle scattering angle.

Four coordinate planes of silicon strip detectors are placed in each Roman pot. Such redefinition is necessary at least for two reasons: first, for an increase in the detector efficiency and, second, for the measurement of the efficiency of each plane. Redefinition is also necessary for the identification of elastic scattering from multiplicity.

Each plane has strips 70 µm in width and a 30-µm gap between strips. Methodical investigations show that 70 % of elastic scattering events are detected by strips and other events, by several neighboring strips (Alekseev 2004).

The selection of elastic events was accompanied by requirements of the collinearity of elastic scattering events. The high spatial and time resolutions of the detectors make it possible to significantly suppress the background of inelastic events.

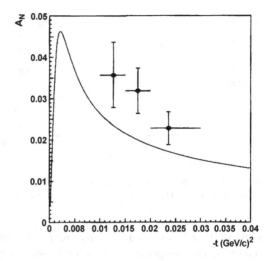

Fig. 12.15 Results of the $pp2pp$ experiment on the measurement of asymmetry in elastic pp scattering at the c.m. energy $\sqrt{s} = 200$ GeV versus t

The detailed description of the setup was given in Bültmann et al. (2004).

The first results of the measurement of polarization in elastic pp scattering at RHIC are shown in Fig. 12.15. The measurements were conducted in an interval of $|t| = 0.01$–0.03 $(GeV/c)^2$.

Figure 12.15 shows that asymmetry decreases with increasing $|t|$ in agreement with the theoretical predictions. In addition, the theoretical curve shown in the figure in which the spin-flip contribution to the nuclear amplitude is considered as negligibly small is systematically below the experimental data. The measurement accuracy should certainly be significantly improved before to insist on such a conclusion.

12.3.4 Polarization in Elastic pp and pC Scattering at RHIC in the Fixed Target Mode

Two important polarization results have been obtained on polarimeters. In both cases, polarization was measured in the Coulomb–nuclear interference region. In the first case, the experiment was performed with a new polarized jet target with the detection of recoil protons. In the second case, the measurements were conducted on a carbon target. Both these setups were described in Sect. 8.3 "Polarimetry of proton beams" in the second part of the book. Now, we briefly discuss physical results. Figure 12.16 shows the analyzing power of elastic pp scattering measured with highest existing statistical accuracy at $\sqrt{s} = 200$ GeV in the Coulomb–nuclear interference region (Okada et al. 2004).

The left panel of the figure shows that polarization measured for the first time in the expected distribution peak. The measurement accuracy is already limited by systematic errors. The solid curve represents a theoretical prediction disregarding spin flip. The experimental data are in excellent agreement with the theoretical pre-

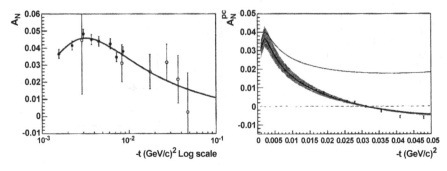

Fig. 12.16 Analyzing powers of elastic (*left panel*) pp and (*right panel*) pC scattering at initial momentum $p = 100$ GeV/c in the Coulomb–nuclear interference region

diction. If the theory also correctly predicts the energy dependence of polarization, a reliable absolute polarimeter for any proton energy would be created.

The right panel of the figure concerns the scattering of protons by carbon (Jinnouchi et al. 2004). In this case, the polarization peak region cannot be reached in measurements owing to kinematics. Nevertheless, it is seen that polarization also decreases with increasing $-t$. The shaded band presents the systematic errors.

It is surprising that, in contrast to pp scattering, the theoretical curve disregarding spin flip is much higher the experimental data. This fact has not yet been adequately explained and, consequently, deserves further study.

References

Ackermann, K.H., et al.: Nucl. Instrum. Methods Phys. Res. A **499**, 624 (2003)

Adams, D.L., et al.: Phys. Rev. Lett. **92**, 171801 (2004)

Adcox, K., et al.: Nucl. Instrum. Methods Phys. Res. A **499**, 469 (2003)

Adler, C., et al.: Nucl. Instrum. Methods Phys. Res. A **470**, 488 (2001)

Airapetian, A., et al.: Phys. Rev. Lett. **87**, 182001 (2001)

Alekseev, I.G.: In: Proceedings of the 16th International Spin Physics Symposium, Trieste, Italy, p. 511 (2004)

Bazilevsky, A., et al.: In: Proceedings of the 15th International Spin Physics Symposium. AIP Conf. Proc., vol. 675, p. 584 (2003)

Bressan, A.: In: Proceedings of the 16th International Spin Physics Symposium, Trieste, Italy, p. 48 (2004)

Bruner, N., for the PHENIX Collaboration: In: Proceedings of the 15th International Spin Physics Symposium, Upton, New York, p. 375 (2002)

Bültmann, S.L., et al.: Nucl. Instrum. Methods Phys. Res. A **535**, 415 (2004)

Collins, J.: Nucl. Phys. B **369**, 262 (1993)

Guryn, W., et al.: Proposal: Experiment to measure the total and elastic pp-cross-section at RHIC. Updated version (1995)

Jinnouchi, O., et al.: In: Proceedings of the 15th International Spin Physics Symposium. AIP Conf. Proc., vol. 675, p. 817 (2003a)

Jinnouchi, O., et al.: In: Proceedings of the 10th Workshop on High Energy Spin Physics, Dubna, Russia, p. 311 (2003b)

Jinnouchi, O., et al.: In: Proceedings of the 16th International Spin Physics Symposium, Trieste, Italy, p. 515 (2004)

Nurushev, S.B.: Int. J. Mod. Phys. A **12**, 3433 (1997)

Ogawa, A.: In: Proceedings of the 16th International Spin Physics Symposium, Trieste, Italy, p. 337 (2004)

Okada, H., et al.: In: Proceedings of the 16th International Spin Physics Symposium, Trieste, Italy, p. 507 (2004)

RHIC Collaboration: Design Manual: Polarized Proton Collider at RHIC (1998)

Sivers, D.: Phys. Rev. D **41**, 83 (1990)

Tojo, J., et al.: Phys. Rev. Lett. **89**, 052302 (2002)

Chapter 13
Results of the Experiments with Fixed Targets

13.1 Polarization in the Elastic Scattering of Hadrons

The comparative investigation of polarization phenomena in the interaction of particles and antiparticles is an important but insufficiently developed research direction. The known Pomeranchuk hypothesis states that total interaction cross sections of particles and antiparticles should be equal in the asymptotic limit in the case of unpolarized hadrons (Pomeranchuk 1958). This hypothesis is experimentally tested up to a center-of-mass energy of 63 GeV at ISR. The asymptotic limit in the sense of this hypothesis has not yet been reached. There is no any experimental test of this hypothesis for the case of the interaction of polarized particles and antiparticles. In Sect. 3.2, we mentioned the hypothesis proposed in Logunov et al. (1962) and Nambu and Iona-Lasinio (1961) of the γ_5 invariance of strong interaction in the asymptotic limit. Certain asymptotic relations between the polarization parameters in cross reaction channels are predicted in a number of theoretical works (see Sect. 3.3). Unfortunately, they have not yet been experimentally tested. There is only one experimental work on studying the polarization of particles and antiparticles in elastic scattering. It will be described below (Nurushev 1990).

In 1972–1976, physicists from IHEP, JINR, and ITEP together with physicists from Saclay prepared and conducted experiments for measuring the angular dependence of the polarization P and spin rotation parameter R in the elastic scattering of $\pi^\pm$ and $K^\pm$ mesons, protons, and antiprotons on a polarized proton target at a momentum of about 40 GeV/c at the U-70 IHEP accelerator. The layout of the experimental facility called HERA (High Energy Reaction Analysis) is shown in Fig. 13.1.

Some features of this facility are as follows:

- the beam instrumentation detects and identifies all three types of hadrons in a beam;
- the polarization in the vertical plane (with hodoscopes H5–H8; H7 and H8 are not seen in the figure, because they are under the target) and the spin rotation parameter in the horizontal plane (with hodoscopes H9–H13) are measured simultaneously (with two separate triggers);

S.B. Nurushev et al., *Introduction to Polarization Physics*,
Lecture Notes in Physics 859, DOI 10.1007/978-3-642-32163-4_13,
© Moskovski Inzhenerno-Fisitscheski Institute, Moscow, Russia 2013

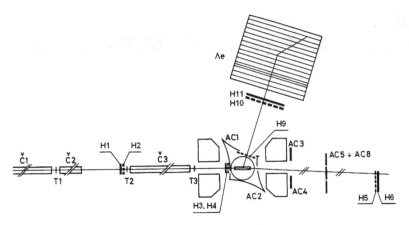

Fig. 13.1 Polarization facility HERA for the simultaneous measurement of the polarization P and spin rotation parameter R in the scattering of hadrons (π^+, K^+, p) and antihadrons (π^-, K^-, $\bar{p}$) on a polarized proton target at momentum of about 40 GeV/c

- the polarized target with a higher hydrogen content (the propanediol target with a length of 8.5 cm, a diameter of 2 cm, and a polarization of 80 %) than in the preceding lanthanum magnesium nitrate targets was used;
- a special fast matrix circuit is used for the preliminary selection of elastic scattering events according to the coplanarity and conjugation angle criteria;
- the facility operates with an intense incident beam ($\sim 3 \cdot 10^6$ particles/cycle).

The left–right asymmetry was determined by using the following three types of monitor counts:

1. Incident beam flux counts with a correction to beam load and beam drift.
2. Counts of the particles elastically scattered on a proton in the horizontal plane containing the polarization vector.
3. Counts of the particles quasielastically scattered on protons in target nuclei.

Polarization values determined from the experiment coincide within the measurement accuracy for all three normalization cases. The third monitor is the most stable monitor.

Table 13.1 presents the polarization parameters in the elastic scattering of hadrons and antihadrons measured at the U-70 IHEP accelerator during 1972–1974 beam runs. This table presents the $|t|$ measurement range in (GeV/c)^2, a dash means that the corresponding measurement has not been performed. According to the table, the complete experiment involving eight quantities (including differential cross sections) has been performed only for $\pi p \to \pi p$ scattering. For Kp scattering, one experiment, e.g., the measurement of the asymmetry in the $K^- p \to K^0 n$ reaction on a polarized target is missed. A vast field remains for further experimental investigations of nucleon–nucleon scattering, in particular, with a polarized beam.

Table 13.1 Reactions, Observables, measured $-t$ range in $(\text{GeV}/c)^2$

Reaction	$-t$ range for P	$-t$ range for R	$-t$ range $d\sigma/dt$
$\pi^- p \to \pi^- p$	[0.1–1.9]	[0.19–0.52]	[0–1]
$\pi^+ p \to \pi^+ p$	[0.08–1.00]	[0.20–0.52]	[0–1]
$\pi^- p \to \pi^0 n$	[0.1–1.2]	–	[0–2]
$K^- p \to K^- p$	[0.1–1.7]	[0.19–0.52]	[0–0.5]
$K^- p \to K^0 n$	–	–	[0–0.5]
$K^+ p \to K^+ p$	[0.08–1.00]	–	[0–0.5]
$pp \to pp$	[0.04–1.2]	– [0.2–0.52]	[0–1.5]
$\bar{p} p \to \bar{p} p$	[0.1–1.1]	–	[0–0.5]

13.1.1 Results of Polarization Measurements

The HERA polarization data (Gaidot et al. 1973, 1975) are given in Fig. 13.2. The momenta of negative and positive particles in these measurements were 40 and 45 GeV/c, respectively. The polarization behavior has the following features:

- polarizations in $\pi^+ p$ and $\pi^- p$ scatterings are almost mirror symmetric in agreement with the prediction of the asymptotic model; this symmetry can be explained by the prevailing contribution to the helicity amplitude from the ρ-pole exchange;
- polarizations in $K^+ p$ and $K^- p$ scatterings have the same signs; this is inconsistent with the asymptotic model and is in agreement with the hypothesis of strong exchange degeneracy;
- polarizations in elastic pp and $\bar{p} p$ scatterings also have the same signs and are inconsistent with both the asymptotic model and Regge models.

The dependences of the "effective" Regge trajectories α on t determined from experimental data are in good agreement with the predictions of pole models for reactions with bosons (pions and kaons) and are inconsistent with the predictions for reactions with protons and antiprotons (see Figs. 13.3, 13.4, and 13.5).

To explain the behavior of the polarization in elastic pp and $\bar{p} p$ scatterings, it was assumed that spin effects appear directly from the pomeron exchange (Pierrard et al. 1975; Aznauryan and Solov'ev 1975; Troshin and Tyurin 1976, 1981). In this case, both the angular distribution of the polarization and its energy dependence at large t values are well described; the energy dependence has the form of a weak (as the logarithm of the energy) decrease in the polarization with increasing energy. More recent measurements of the polarization in elastic pp scattering at energies up to 300 GeV at CERN and FNAL (Fidecaro et al. 1980; Kline et al. 1980) showed that this hypothesis is attractive, but has not yet been certainly confirmed. In this respect, the measurements of the polarization in elastic pp and $\bar{p} p$ scatterings at the same very high energies are very desirable.

In view of the data on the polarization in the elastic scattering of hadrons, we mention two theoretical works Kolár et al. (1976) and Solov'ev and Shchelkachev

Fig. 13.2 Polarization P versus t in the elastic scattering of (upper panels **a**, **b**, and **c**) negative and (lower panels **d**, **e**, and **f**) positive hadrons on protons at a momentum of 40 and 45 GeV/c respectively

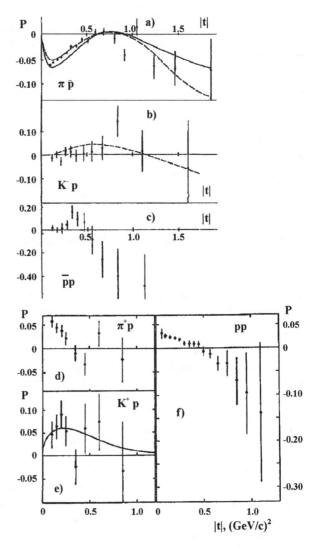

(1979). Relations between the polarization parameters at small t values are obtained in the former work using the quark interaction model (under certain assumptions). Comparison with the HERA experimental data shows that these relations are satisfied with an accuracy of about 25 % for elastic pp scattering. In the latter work, it is emphasized that, if the hadron interaction amplitudes saturate the Froissart limit, a model (of a fast increase) can be developed that satisfactorily explains both the increase in the total cross section and the slow decrease in the polarization with increasing energy. To test this model, it is particularly important to obtain data at large t values, which are yet absent.

Figure 13.6 shows the experimental dependence of the polarization for six charged hadrons elastically scattered on protons in the range $0.1 \le |t|$ (GeV/c)$^2 \le 0.3$.

Fig. 13.3 Regge trajectories $\alpha_p = 1 + 0.27t$ and $\alpha_R = 0.52 + 0.93t$ and the interference term $\alpha_R - \alpha_p$ for $\pi^- p$ scattering; the points with error bars are obtained by fitting the dependence $P(t) = A(t) \cdot S^{\alpha_R(t) - \alpha_p(t)}$ to the experimental data above 10 GeV/c

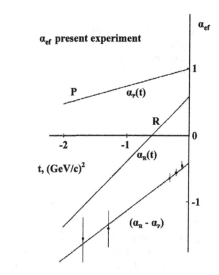

Fig. 13.4 Energy dependence of the difference between the polarization parameters for $\pi^+ p$ and $\pi^- p$; the *points* correspond to α values obtained by fitting $P(\pi^+ p) - P(\pi^- p) \approx A(t)S^{\alpha}$; the *line* corresponds to interference between the ρ meson and pomeron $\alpha = \alpha_\rho - \alpha_p$, where $\alpha_\rho = 0.52 + 0.93t$ and $\alpha_p = 1 + 0.27t$

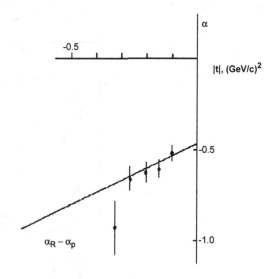

It is interesting that the polarization decrease rates with increasing energy are different for different hadrons. Therefore, the expected asymptotic regions, where polarizations should vanish, can be different for different particles. Sure more precise measurements of the energy dependence of polarization, especially for kaons and pbars, are needed.

Figure 13.6 indicates a decrease in the polarization with increasing energy for all hadrons. This is valid for small momentum transfers. There are certain indications of a weaker decrease in the polarization with increasing energy for large momentum transfers. However, this expectation has not yet been confirmed experimentally be-

Fig. 13.5 Parameter α_{eff} versus $|t|$, as determined by fitting $P(d\sigma/dt) \approx A(t)s^{\alpha_{eff}}$ to the experimental data for K^+p and pp scatterings

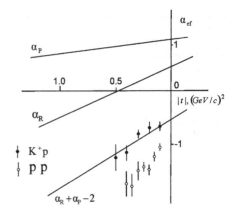

Fig. 13.6 Polarization P in the elastic scattering of hadrons on protons versus s in the range $0.1 \le |t|\ (GeV/c)^2 \le 0.3$

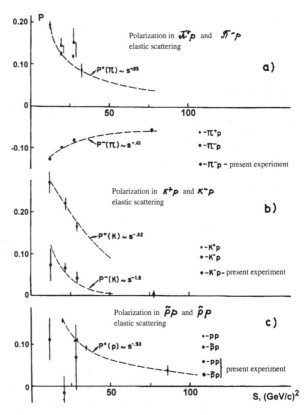

cause the corresponding experiment is difficult owing to very small cross sections for elastic processes.

Completing this brief review of the experimental data on the polarization, we note that it is very important to obtain information on spin effects in different isotopic states, for example, in the $I = 0$ state in the nucleon–nucleon interaction. This

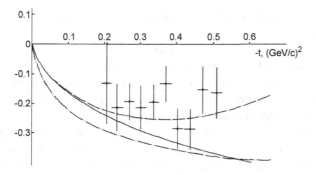

Fig. 13.7 Spin rotation parameter R in elastic $\pi^- p$ scattering at a momentum of 40 GeV/c. The *solid line* corresponds to the dependence $R = -\cos\theta_p$, where θ_p is the recoil proton emission angle in the laboratory frame. The *dashed* and *dash-dotted lines* are taken from Kline et al. (1980) and Barger and Phillips (1969)

Fig. 13.8 Spin rotation parameter R in elastic pp scattering at a momentum of 45 GeV/c. The *dotted line* is the dependence $R = -\cos\theta_p$

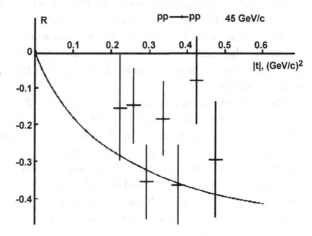

requires investigations with polarized neutrons (beam or target), which have not yet been performed at IHEP.

13.1.2 Results of the Measurement of the Spin Rotation Parameter R

The spin rotation parameter R was measured in elastic $\pi^- p$ and pp scatterings (see Figs. 13.7 and 13.8, respectively). Such measurements are important not only for the complete experiment program. The parameter R makes it possible to determine (together with the differential cross section) the degree of the conservation of the helicities in the s and t channels for πN scattering. In particular, these measurements show that helicity at a momentum of 40 GeV/c in the s channel is conserved with an accuracy of about 10 %, whereas helicity conservation is strongly violated in the t channel.

Interest in the measurement of the parameter R particularly increased in view of the idea of rotating hadronic matter proposed in Chou and Yang (1976) and Levintov (1978). The assumption that the hadronic matter current density is proportional to the electromagnetic current density (Chou and Yang 1976) provides a prediction for the angular and energy dependences of the parameter R.

Figure 13.7 provides the following conclusions. First, of three lines drawn in the figure, the solid line is closest to the experimental data. It corresponds to the model of rotating hadronic matter and the Regge model with one pomeron pole. Second, to distinguish the predictions of different models, it is necessary to perform measurements at large momentum transfers and with the accuracy an order of magnitude better than that shown in the figure. Such an experiment is very difficult because the cross sections are small and double scattering should be used. Such experiment will likely become possible at RHIC when reaching luminosity larger than 10^{32} cm^{-2} s^{-1}. It is particularly important to analyze the energy dependence of the parameter R, because the parameter R is independent of energy in both models consistent with these data. Finally, this experiment carried out more than three decades ago has not been repeated at higher energies. This fact confirms the difficulty of such an experiment.

The results of the measurement of the parameter R in pp scattering at IHEP are also in good agreement with the dependence $R = -\cos\theta_p$ (the line in Fig. 13.8) taking into account large errors in the data. Since the model of rotating hadronic matter predicts zero polarization, whereas nonzero polarization is observed in the experiment, a certain quantitative discrepancy should be expected (in ideally accurate measurements). This model should well work at high energies (higher than 100 GeV); unfortunately, such data are yet absent.

Using the data on $\pi^- p$ scattering, where the complete set of experiments is fulfilled, one can determine the quantity $\tau_0 = \sqrt{\frac{s_0}{(-t)}} \frac{|F^0_{+-}|}{|F^0_{++}|}$ specifying the relative fraction of the reduced spin-flip amplitude to the spin-nonflip amplitude in the isotopic state $I = 0$ (in the t channel). The energy dependence of τ_0 is shown in Fig. 13.9, where it is seen that τ_0 depends weakly on the energy and is about 10 %. This may mean that the pomeron can interact with changing helicity; such an interaction was not assumed before the HERA experimental results. It would be very interesting to continue these measurements at IHEP with better precisions and at very high energies.

It is interesting that the spin rotation parameter R is almost the same for $\pi^- p$ and pp scatterings and approximately follows the dependence $R_{\pi p} = R_{pp} = -\cos\theta_P$. Such dependence is obtained in the Regge model under the assumption of the prevailing contribution of the pomeron pole and factorization. In this model, the parameter R is the same for all hadrons. In the model of rotating hadronic matter, this parameter has different values for different types of particles (see Bourrely et al. 1980).

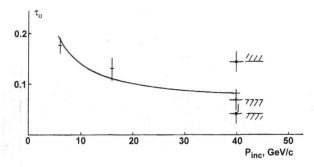

Fig. 13.9 Quantity $\tau_0 = \sqrt{\frac{s_0}{(-t)}} \frac{|F^0_{+-}|}{|F^0_{++}|}$ for $\pi^- p$ scattering versus the initial momentum

13.2 Polarization in Charge Exchange Reactions

Since 1980, the IHEP–LNP JINR–IHEP (Tbilisi University) Collaboration performed the experiments on investigating polarization effects in charge exchange reactions. To the mid-1980s, the investigations of the asymmetry in the exclusive reaction

$$\pi^- + p \uparrow \rightarrow \pi^0 + n \qquad (13.1)$$

at a momentum of 40 GeV/c were completed. The experiment was performed with a "frozen polarized proton target" made of propanediol ($C_3H_8O_2$). The target was developed at LNP JINR and the magnetic system for the build-up and keeping of the polarization was designed and manufactured at IHEP. The polarized target for this experiment is described in Chap. 2 "Polarized targets" in the second part of this book.

Reaction (13.1) is of interest because, according to the Regge pole model, this reaction proceeds through single ρ-pole exchange, and polarization should be absent as an interference effect. The first experimental data indicating large polarization effects at 5 and 12 GeV/c forced theorists to revise the foundations of the Regge model, to introduce additional poles, to take into account rescattering effects, etc. However, these experiments were conducted at low energies (below 12 GeV), whereas data at higher energies are required for the strict test of the Regge model. This circumstance motivated an experiment on the measurement of the neutron polarization in reaction (13.1) at the U-70 accelerator.

Figure 13.10 shows the layout of the PROZA (polarization in charge exchange reactions) setup, which was described in detail in Avvakumov et al. (1981).

The PROZA equipment consists of the following main units.

- The beam instrumentation consists of the scintillator counters of the total flux S1, S2, and S3; the threshold Čerenkov counters Č1, Č2, and Č3; the anticoincidence counter $A_{3,4}$ for the suppression of beam halo; and the beam hodoscopes H1 and H2. The main aims are the measurement of the flux of particles, their identification, and the measurement of beam emittance.

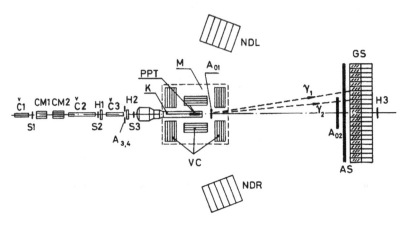

Fig. 13.10 Layout of the PROZA setup

- The polarized target with a diameter of 2 cm and a length of 20 cm ensures a proton polarization exceeding 70 % and the relaxation times > 2000 and 1000 h of positive and negative polarizations, respectively.
- The counter block VC, the protection system consisting of veto counters is used to suppress the yield of charged particles and photons constituting the background for reaction (13.1).
- The gamma spectrometer GS consists of lead glasses and ensures the determination of the energy and emission angle of photons.
- The neutron detectors NDL and NDR are the left and right units of the neutron detector and ensure the detection of neutrons for the enhancement of the kinematic selection of reaction (13.1).

The experiments were performed with the π^- meson beam with an intensity of $(0.5–3) \cdot 10^6$ particles/cycle and cycle duration of 0.5–2 s. The trigger suppressed the start of the information reception system to a level of $1.5 \cdot 10^{-5}$ of the beam intensity; i.e., the event collection rate varied in a range of 7–45 per cycle. More than million triggers were collected. The processing shows that the π^0 meson mass is reconstructed with an accuracy of $\Delta m/m = 11$ % (FWHM). Analysis shows that the background of inelastic events varies from 5 to 20 % in the measured range $0 < |t| \ (GeV/c)^2 < 2$ of the invariant momentum transfer t.

The preliminary results obtained with the use of a gas discharge gamma detector for small $|t|$ values were reported in Avvakumov et al. (1980). More accurate data were more recently obtained for large momentum transfers with a gamma detector based on Čherenkov total absorption counters (GS in Fig. 13.10). These results are shown in Fig. 13.11 (Avvakumov et al. 1982) together with the data for other exclusive charge exchange reactions obtained simultaneously with the data for main reaction (13.1) (Nurushev 1989).

The experimental data at 40 GeV/c indicate that, whereas the prediction of the quasipotential model is inconsistent with new data, a number of other complicated Regge models provide a satisfactory description of the polarization at small t values.

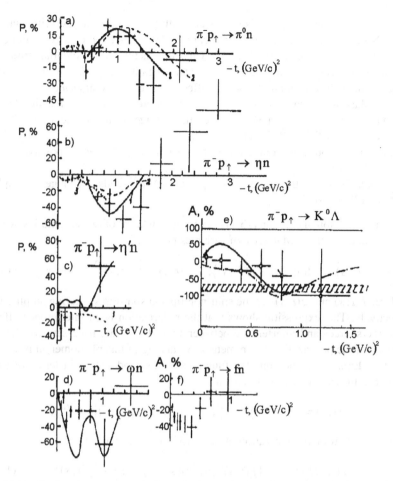

Fig. 13.11 Single-spin asymmetry in the exclusive production of neutral mesons by a 40-GeV/c π^- meson beam incident on a polarized proton target in the reactions $\pi^- + p(\uparrow) \to \pi^0 + n$ **(a)**, $\pi^- + p(\uparrow) \to \eta + n$ **(b)**, $\pi^- + p(\uparrow) \to \eta' + n$ **(c)**, $\pi^- + p(\uparrow) \to \omega + n$ **(d)**, $\pi^- + p(\uparrow) \to K^0 + \Lambda$ **(e)**, $\pi^- + p(\uparrow) \to f + n$ **(f)**

These data are also satisfactorily described in the U-matrix method (Troshin and Tyurin 1980), where the unitarity condition is taken into account. Measurements at large momentum transfers are necessary for distinguishing between different models.

The best accuracy was achieved in the measurement of the asymmetry in the $\pi^- + p(\uparrow) \to \pi^0 + n$ reaction (see Fig. 13.11a). The behavior of this asymmetry has the following features:

1. Neutron polarization $P(t)$ in the range $0 < |t|$ (GeV/c)$^2 < 0.35$ is positive and is (5.0 ± 0.7) % in average. However, the first indication of a minimum has been obtained in the t dependence of the polarization at $t = -0.22$ (GeV/c)^2. The

polarization near this point is consistent with zero within the experimental error. Note that this point is the point near which the so-called cross-over effect is observed (Antipov et al. 1973, 1976). This effect is that the differential cross sections for the elastic scattering of positive and negative pions on protons intersect each other near this point. These two effects are possibly connected.

2. The polarization near the dip in the differential cross section for reaction (13.1) changes sign and becomes negative. Its average value in the range $0.4 < |t|$ $(\text{GeV}/c)^2 < 0.6$ is $\bar{P} = -(10 \pm 6)$ %.

3. The polarization in the range $0.6 < |t|$ $(\text{GeV}/c)^2 < 1.3$ is again positive and is $\bar{P} = (22 \pm 8)$ %.

4. The polarization in the range $1.3 < |t| < 2$ $(\text{GeV}/c)^2$ tends to change sign and is $\bar{P} = -(28 \pm 28)$ %.

The polarization in reaction (13.1) is expressed in terms of two amplitudes, which are the functions of the Mandelstam variables t and s,

$$\sigma(t,s)P(t,s) = 2\,\text{Im}\big(g^*(t,s)f(t,s)\big), \quad \sigma(t,s) = \big|f(t,s)\big|^2 + \big|g(t,s)\big|^2. \quad (13.2)$$

Here, $f(t,s)$ and $g(t,s)$ are the spin-nonflip and spin-flip scattering amplitudes, respectively. This expression shows that the polarization is an interference effect: two force sources and interference between these forces are necessary for the appearance of the polarization (asymmetry). An analog of this phenomenon is known in optics. Expressing the complex functions $f(t,s)$ and $g(t,s)$ in terms of their absolute values and phases in the form

$$f(t,s) = \big|f(t,s)\big|e^{i\varphi_f(t,s)}, \qquad g(t,s) = \big|g(t,s)\big|e^{i\varphi_g(t,s)}, \qquad (13.3)$$

the expression for the polarization in Eqs. (13.2) can be represented as

$$\sigma(t,s)P(t,s) = \big|f(t,s)\big|\big|g(t,s)\big|\sin\big(\varphi_f(t,s) - \varphi_g(t,s)\big). \qquad (13.4)$$

Such an explicit representation of the polarization in terms of the absolute values and phases of the scattering amplitude allows one to understand the following statements. For the polarization to be nonzero, first, scattering amplitudes should be nonzero and, second, the difference between the phases of these amplitudes should differ from zero and multiples of 180°. On the whole, both statements are obvious, but the second statement applied to the Regge pole model provides the following conclusion: single-pole Regge exchange leads to zero polarization because both amplitudes f and g in single-pole exchange have the same phases. Therefore, for polarization to appear in the Regge model, at least two poles or a pole and a cut, etc. are necessary and this explains a frequently used statement that polarization is an interference effect. This digression directly concerns reaction (13.1). This charge exchange reaction in the t channel can involve exchange by only one ρ pole. Therefore, the appearance of the polarization in this reaction should not be expected. However, the measurements of the polarization in reaction (13.1) in an energy range of 5–12 GeV indicate a noticeable polarization effect (Bonamy et al.

1966, 1973; Hill et al. 1973). In particular, the average polarization value in a range of $0 < |t|$ (GeV/c)$^2 < 0.35$ is (16.0 ± 2.3) %, (11.4 ± 2.0) %, and (5.0 ± 0.7) % at 6, 11, and 40 GeV, respectively. Thus, these results contradict the predictions of the simple Regge pole model on the polarization.

The data on the polarization in reaction (13.1) at 40 GeV are shown in Fig. 13.11a and were widely discussed. Various Regge model modifications were used. For example, the Regge model with the ρ pole and ρ cut was used in Aleem and Saleem (1983). The experimental data were satisfactorily described. The double zero near $t = -0.5$ (GeV/c)^2 is explained by the fact that the ρ-pole trajectory has zero near this point, and interference of two terms (pole and cut) gives a double zero. Polarization in this model decreases as $1/\sqrt{s}$ with increasing energy. This prediction is also consistent with the experimental data discussed here.

Figure 13.11a presents two lines. The solid line is the prediction of the U-matrix model, which is successfully applied to describe elastic and inclusive reactions in almost all kinematic regions. Another feature of this model is that the unitarity condition leads to the appearance of spin flip in the pomeron. As a result, polarization effects weakly decreasing with increasing energy appear in this model. As seen in Fig. 13.11a, the calculated line is in agreement with the experimental result in the range under consideration (Troshin and Tyurin 1984).

The dotted line in Fig. 13.11a is the Regge model calculation with the inclusion of the odderon (Gauron et al. 1984). Interest of the authors is clear. The data on the total cross sections for pp and $\bar{p}p$ interactions at ISR indicate that they increase with energy with a theoretically maximum rate. This means that the imaginary part of the crossing-even amplitude has the form

$$F_+(s, 0) \propto s \ln^2 s. \tag{13.5}$$

The authors analyzed the data on the $\pi^+ p$, $\pi^- p$ elastic scattering, and $\pi^- p$ charge exchange reactions and concluded that the crossing-odd amplitude can also increase rapidly with energy as

$$F_-(s, 0) \propto s \left(\ln^2 s - i\pi \ln s \right). \tag{13.6}$$

At present, amplitude (13.6) is commonly attributed to the odderon exchange. The odderon corresponds to a singularity with $l = 1$ (a double pole at $t = 0$). The requirement that the odderon contribution to the amplitude F_- is smaller than the contributions of usual Regge poles (with $l \leq 1/2$) leads to the necessity of the suppression of the odderon contribution up to, for example, the Tevatron energy. As a result, it is difficult to seek the manifestation of the odderon in total cross sections. However, if the odderon can be manifested in other hadron processes, new possibilities appear. One of them is the attractable measurement of the polarization in charge exchange reaction (13.1). First, a few poles contribute to this reaction. Second, the odderon amplitude phase differs from the phases of other poles. Therefore, the odderon can interfere, for example, with the ρ pole and provide polarization. With increasing energy, the odderon contribution should increase and lead to a sharp change in the polarization as a function of t. Such an analysis was performed by the

Table 13.2 Best-fit parameters of the "maximum odderon" variant

	Parameter	ρ	ρ'	O
1	a, $(\mu b)^{1/2}$	93.8	-130.3	–
2	c, GeV^{-2}	2.53	–	–
3	b, $(\mu b)^{1/2} GeV^{-1}$	3404.6	714.3	–
4	$\alpha(0)$	0.48	0	–
5	α', GeV^{-2}	0.82	0.11	–
6	λ, GeV^{-2}	0.13	2.65	1.76
7	C, $(\mu b)^{1/2}$	–	–	-0.008
8	s_0, GeV^2	–	–	0.08

authors using the data on the polarization in reaction (13.1) including 40 GeV (the highest measured energy) data.

The authors took into account three poles and wrote the amplitudes in the form

$$A' = A'_\rho + A'_{\rho'} + A'_O, \qquad B = B_\rho + B_{\rho'}. \tag{13.7}$$

With the notation $R = \rho$ or ρ', these amplitudes can be parameterized as

$$A'_R = \left[i + \tan\left(\frac{\pi}{2}\alpha_R(t)\right)\right] a_R(t)\left[\alpha_R(t) + 1\right] e^{\lambda_R t} s^{\alpha_R(t)},$$

$$B_R = \left[i + \tan\left(\frac{\pi}{2}\alpha_R(t)\right)\right] b_R(t)\alpha_R(t)\left[\alpha_R(t) + 1\right] e^{\lambda_R t} s^{\alpha_R(t)-1}. \tag{13.8}$$

Here,

$$\alpha_\rho(t) = \alpha_\rho(1 + ct), \qquad \alpha_{\rho'}(t) = \alpha_{\rho'} = \text{const}, \qquad b_R(t) = b_R = \text{const}. \tag{13.9}$$

The following parameterization is accepted for the odderon:

$$A'_O = Cs\left(\ln^2\frac{s}{s_0} - i\pi \ln\frac{s}{s_0}\right)e^{\lambda_o t}. \tag{13.10}$$

More than 300 experimental points were included in the analysis with 13 free parameters. The fit is satisfactory with $\chi^2 = 1.6$ per point. The resulting parameter values are presented in Table 13.2. This fit is shown by the dashed line in Fig. 13.11a. This fit is also shown in Fig. 13.12 together with the data on the polarization at 5 GeV for the illustration of the conclusion by the authors that the difference between the t dependences of the polarization at 5 and 40 GeV can be explained only by interference between the odderon and ρ meson. The results of this work presented in Fig. 13.13 are impressive: polarization can increase very rapidly with energy.

In particular, polarization at $t = -0.5$ $(GeV/c)^2$ is ~ 15 % at 40 GeV and is expected to be 50 and 80 % at 100 and 200 GeV, respectively. Such an almost linear

Fig. 13.12 Odderon model applied to the description of the polarization in the $\pi^- p(\uparrow) \to \pi^0 n$ reaction at momenta 5 and 40 GeV/c

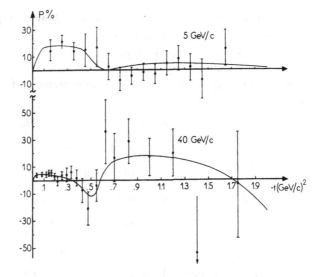

Fig. 13.13 Energy dependence of the polarization in the $\pi^- p(\uparrow) \to \pi^0 n$ reaction as predicted in the Regge pole model with the odderon

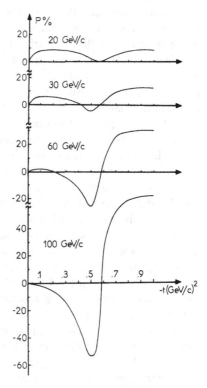

increase in polarization with energy at a given square momentum transfer t has not been predicted in any other model. Such an experiment could be performed at one of three laboratories: CERN, FNAL, and BNL.

The authors of this work state: "it has been shown that the surprising results of the polarization measurement in Protvino at 40 GeV/c in the $\pi^- p \to \pi^0 n$ reaction confirm the hypothesis that the asymptotic increase in the crossing-odd amplitude can be as fast as allowed by the general principles."

Figure 13.11 shows the results of the measurement of the asymmetry in the six reactions

$$\pi^- + p(\uparrow) \to \pi^0 + n \quad \text{(a)},$$

$$\pi^- + p(\uparrow) \to \eta + n \quad \text{(b)},$$

$$\pi^- + p(\uparrow) \to \eta' + n \quad \text{(c)},$$

$$\pi^- + p(\uparrow) \to \omega + n \quad \text{(d)},$$ (13.11)

$$\pi^- + p(\uparrow) \to K^0 + \Lambda \quad \text{(e)},$$

$$\pi^- + p(\uparrow) \to f + n \quad \text{(f)}.$$

Above, we discussed in detail reaction (13.11(a)) and now briefly discuss the features of other reactions. Before such a discussion, two remarks should be made. The first remark concerns the parameter t, the invariant momentum transfer squared.. When the masses of the initial (the target particle labeled as 2 has the rest mass m_2) and final recoil particle (labeled as 4 with mass m_4) are the same, this parameter is denoted as t and it is equal to $t = -2m_2 T_4$, where T is the kinetic energy of recoil particle. For the case where these masses are different, this parameter is denoted as t' and it equals $t' = (m_2^2 - m_4^2 - 2m_2 T_4)$. In Fig. 13.11, the notation t' is used according to the process kinematics: $t = 0$ at zero meson production angle, whereas $t' \neq 0$. This should be taken into account. The second remark concerns the spin of the produced mesons. The known equality of the polarization and asymmetry refers only to spin-0 mesons, scattered on nucleon, i.e., to pions, kaons, η and η' mesons. In the case of mesons with nonzero spins (in our case, these are the ω meson with spin 1 and f meson with spin 2), asymmetry should not be identified with polarization. After this brief digression, we continue to discuss Fig. 13.11b.

In particular, the pseudoscalar η meson has zero analyzing power in the range $0 < |t'|$ (GeV/c)^2 < 0.4, then a negative asymmetry in average of -30 % appears in the range $0.4 < |t'|$ (GeV/c)^2 < 1.4, and a tendency to the appearance of positive asymmetry is seen at $|t'| > 1.4$ (GeV/c)^2 (however, within large statistical errors). Such an asymmetry behavior was sufficiently well described in the Regge model with two poles in Aleem and Saleem (1983). The result of this work is shown by the dashed line. The solid line in this figure is the calculation in the U-matrix model (Troshin and Tyurin 1985, 1986). The dotted line in Fig. 13.11b is the calculation in the model of correlated reggeons (Arestov et al. 1984). All three models are in satisfactory agreement with the experimental data. It's obvious that the statistics should be drastically increased.

Reaction (13.11(c)) is of certain interest for theorists. As shown in Krzywicki and Tran Tranh Van (1969), polarizations in reactions involving mesons of the same multiplet are related with each other. A particular relation between the polarizations in reactions (13.11(a))–(13.11(c)) was derived in Enkovsky and Struminsky (1983) under certain assumptions on the quark–quark interaction. This relation has the form

$$P\left(\pi^0\right) + 2P(\eta) = P\left(\eta'\right). \tag{13.12}$$

Thus, the polarization $P(\eta')$ can be determined from the data on polarizations in reactions (13.11(a)) and (13.11(b)). It is shown by the solid line in Fig. 13.11c. The model and experimental data are in qualitative agreement within the statistical error, although their signs are different in the range $0 < |t|$ $(\mathrm{GeV}/c)^2 < 0.4$. The dashed line in Fig. 13.11c is the Regge model calculation from Arestov (1986). Agreement with the experimental data is quite satisfactory.

We pass to Fig. 13.11d for the ω meson. Interest in the production of vector mesons, particularly with the use of polarized targets, has long existed (Achasov and Shestakov 1983). For this reason, it is not surprising that the asymmetry in reaction (13.11(d)) of interest was predicted before obtaining experimental data at 40 GeV. This prediction is given by the solid line in Fig. 13.11d. This model involves the ρ, A_2, P poles and $A_2 A_2 + A_2 A_2 P$, $\rho\rho + \rho\rho P$, $\rho A_2 + \rho A_2 P$ cuts.

The parameters of the model were determined from the experimental data on the $\pi^- p(\uparrow) \to \omega n$ reaction at 6 GeV obtained at ZGS (Shaevitz et al. 1976) and on the $\pi^- p(\uparrow) \to \rho n$ reaction at 17.2 GeV obtained at CERN (Becker et al. 1977). As seen in Fig. 13.11d, there is no quantitative agreement between the prediction of this model and experimental data at 40 GeV.

Reaction (13.11(e)) was studied in the background regime with respect to reaction (13.11(a)) and was not optimized. Nevertheless, it is attractable because the equality $P = A$ is valid for it. This equality allows the determination of the Λ hyperon polarization in this reaction from the measurement of the left–right asymmetry of the production of K^0 on a polarized target. Figure 13.11e shows the Λ hyperon polarization as a function of t' in comparison with the predictions of three models. The solid line is the prediction of the model of weak exchange degeneracy (Arestov et al. 1983), the dash-dotted line is the eikonal calculation (Arakeliyan et al. 1983), and the shaded band corresponds to the color tube model (Anderson et al. 1982). It is seen that the first two models are in quantitative agreement with the experimental data, whereas the third model is inconsistent with the data at small t' values.

Finally, last reaction (13.11(f)) involving the f meson with spin 2 waits its theoretical interpretation.

13.3 Conclusions of Part III

The experimental program on high energy spin physics has achieved a prominent progress. The SLD experiment at the SLC accelerator demonstrates the efficiency of the use of a polarized electron beam for the accurate measurement of the parameters of the Standard Model. High-accuracy results have been obtained in the measurement of spin structure functions; they made it possible to carefully test the sum rules and to determine separate contributions to the nucleon spin from light valence and sea quarks. The same data were used to accurately determine the running coupling constant. Although many data have been collected, the problem seems to be far from the final solution, particularly concerning the polarization of gluons and sea quarks, as well as the role of the orbital angular momentum. For this reason, it is necessary to carry out new experiments in this field with better accuracy and in a wide kinematic region. Hyperon polarization, which is an important discovery in high energy spin physics, deserves more serious efforts in order to understand its dependence on energy, p_T, x_F, and flavor and the spin transfer mechanism. Single- and double-spin measurements could make the key contribution to the understanding the nucleon spin structure. A promising fact is that several spin programs have been approved at SLAC, CERN, JLab and RHIC for such investigations. Thus, spin physics has a good outlook.

13.4　Conventions for Spin Parameters in High Energy Scattering Experiments

13.4.1　Sign Convention for Particle Polarization (Basle Convention) (Nucl. Phys. 21, 696 (1960))

At the International Symposium on Polarization Phenomena of Nucleons held in Basle, the following convention was unanimously agreed upon regarding the sign of the polarization of particles taking part in any nuclear reaction:

If $\vec{k}_i$ represents the propagation vector of the incident particle and $\vec{k}_o$ the propagation vector of the outgoing particle, particles with spin pointing in the direction $\vec{k}_i \times \vec{k}_o$ are to be considered as having positive polarization.

13.4.2　Conventions for Spin Parameters (Ann Arbor Convention) (Modified Version of October 24, 1977 Report of Notation Committee (Ashkin, Leader-Chairman, Marshak, Roberts, Soffer, Thomas) Approved by Workshop on October 25, 1977)

13.4.2.1　Elastic and Pseudo-Elastic Scattering

Consider the process

$$a + b \rightarrow c + d \tag{13.13}$$

where all four particles a, b, c, and d have spin 1/2 or less. We will denote the four particles in the following ordered way:

$(a, b; c, d)$

(beam, target; scattered, recoil)

$(i, j; k, l)$,

where each index, such as l, may take values such as ↑ (spin up) or ↓ (spin down), [or → (spin along momentum)].

We will conform to the Basel convention in that the forward-scattered particle, c, goes to the left, and the normal to the scattering plane is defined by:

$$\vec{N} = \frac{\vec{P}_a \times \vec{P}_c}{|\vec{P}_a \times \vec{P}_c|} \tag{13.14}$$

(a) Fundamental Observables for Fixed Target Experiments in the Laboratory.

Each spin i, j, k, l can lie along the $\vec{N}$, $\vec{L}$, or $\vec{S}$ directions which are defined in Laboratory system (Lab) as follows:

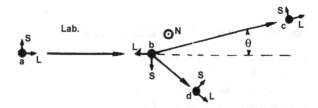

Notice that $\vec{N}$ is out of the paper for all particles, and that for each particle, $\vec{L}$ points along the particle's momentum. Note that $\vec{S}$ is defined by $\vec{S} = \vec{N} \times \vec{L}$.

(b) Fundamental Observables in the c.m. or for Colliding Beam Experiments.

Each spin $(i, j, k, l)_{c.m.}$ can lie along the $\vec{n}$, $\vec{l}$, or $\vec{s}$ directions which are defined in the center of mass system (c.m.) as follows:

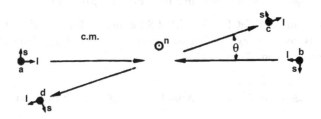

The direction of $\vec{l}$ always points along each particle's momentum. Notice that while $\vec{N}$ and $\vec{n}$ are identical, $\vec{L} \neq \vec{l}$ and $\vec{S} \neq \vec{s}$. Therefore while spin-parameters like A_{nn} are invariant in going from lab to c.m. ($A_{nn} = A_{NN}$), parameters like K_{ll} and D_{ss} are not invariant under this boost.

(c) Polarization.

"Polarization" refers to the state of a single particle or an ensemble and is short for "Degree of Polarization":

P_B = beam polarization
P_T = target polarization
P_S = scattered particle polarization
P_R = recoil particle polarization.

(d) Asymmetries Associated with One Polarized Particle.

We will use the 4-index notation to define these Asymmetries:

$(n, 0; 0, 0)$ A^a—Analyzing power for a
$(0, n; 0, 0)$ A^b—Analyzing power for b
$(0, 0; n, 0)$ P^c—Polarizing power for c
$(0, 0; 0, n)$ P^d—Polarizing power for d.

NOTE: 0 denotes unpolarized in initial state or polarization unmeasured in final state.

(e) Correlations Associated with Two Particles Polarized.[1]

$(0, 0; k, l)$ C_{kl} Final state correlation parameter
$(i, j; 0, 0)$ A_{ij} Initial state correlation parameter
$(i, 0; k, 0)$ D^a_{ik} Depolarization parameter for a
$(0, j; 0, l)$ D^b_{jl} Depolarization parameter for b
$(i, 0; 0, l)$ K^a_{il} Polarization transfer parameter for a
$(0, j; k, 0)$ K^b_{jk} Polarization transfer parameter for b

NOTE Each index i, j, k, l refers to the spin orientation along either the $\vec{n}, \vec{l}$ or $\vec{s}$ direction. Thus A_{nn}, which was previously sometimes called C_{nn} is the initial state correlation when the spins are oriented along the normal to the scattering plane.

NOTE $D^a_{ij} \neq D^b_{ij}$ for the $\vec{l}$ or $\vec{s}$ directions even for $pp \to pp$

(f) Recommend NO SPECIAL SYMBOLS for three or four polarized particle reactions. Just use 4-index notation such as $(i, j, k, 0)$ or (i, j, k, l) or else use pure spin cross sections defined below.

---

[1]The asymmetries and correlations are defined in terms of pure spin cross sections by relations such as:

$$P^c = (0, 0; n, 0) = \frac{\sigma_{00 \to \uparrow 0} - \sigma_{00 \to \downarrow 0}}{\sigma_{00 \to \uparrow 0} - \sigma_{00 \to \downarrow 0}},$$

$$A_{NN} = \frac{\sigma_{\uparrow\uparrow \to 00} + \sigma_{\downarrow\downarrow \to 00} - \sigma_{\uparrow\downarrow \to 00} - \sigma_{\downarrow\uparrow \to 00}}{\sigma_{\uparrow\uparrow \to 00} + \sigma_{\downarrow\downarrow \to 00} + \sigma_{\uparrow\downarrow \to 00} + \sigma_{\downarrow\uparrow \to 00}}.$$

(g) If an experiment measures a mixture of observables care must be taken to state exactly what has been measured.

(h) Differential Cross-sections Measurements in Pure Initial and/or Final Spin States.

Label spin direction by arrows as indicated for the different pure initial spin cross sections

n-polarization $(\frac{d\sigma}{dt})_{\uparrow\uparrow}$

l-polarization $(\frac{d\sigma}{dt})_{\rightarrow\leftarrow}$

s-polarization $(\frac{d\sigma}{dt})_{00}$

These pure 2-spin (initial) cross sections are related to pure 3-spin and 4-spin cross sections by relations such as

$$
\begin{aligned}
\left(\frac{d\sigma}{dt}\right)_{\uparrow\uparrow} &= \left(\frac{d\sigma}{dt}\right)_{\uparrow\uparrow\rightarrow 00} = \left(\frac{d\sigma}{dt}\right)_{\uparrow\uparrow\rightarrow 0\uparrow} + \left(\frac{d\sigma}{dt}\right)_{\uparrow\uparrow\rightarrow 0\downarrow} \\
&= \left(\frac{d\sigma}{dt}\right)_{\uparrow\uparrow\rightarrow\uparrow\uparrow} + \left(\frac{d\sigma}{dt}\right)_{\uparrow\uparrow\rightarrow\uparrow\downarrow} \\
&+ \left(\frac{d\sigma}{dt}\right)_{\uparrow\uparrow\rightarrow\downarrow\uparrow} + \left(\frac{d\sigma}{dt}\right)_{\uparrow\uparrow\rightarrow\downarrow\downarrow}.
\end{aligned}
\tag{13.15}
$$

13.4.2.2 One Particle Inclusive Reaction

Consider the inclusive reaction

$$
a + b \rightarrow c + X
\tag{13.16}
$$

where all three particles a, b, and c have spin $1/2$ or less. The final particle c has c.m. momentum defined by

$$
p_\perp \quad \text{and} \quad x_c = p_{||}/p_c^{max}.
\tag{13.17}
$$

All spin-parameters are defined as for the above elastic reactions. However care must be taken in defining the direction of the normal to the scattering plane

$$
\hat{N} = \frac{\vec{p}_a \times \vec{p}_c}{|\vec{p}_a \times \vec{p}_c|}
\tag{13.18}
$$

that a, b, and c are defined so that particle c goes to the left. Notice that the spin-parameters may depend on the sign of x_c.

13.4.2.3 Total Cross-Section Measurements

Arrows should indicate the spin directions of particles a and b in the process

$$a + b \rightarrow \text{anything}$$

$$\Delta\sigma_T = \sigma_{\uparrow\downarrow} - \sigma_{\uparrow\uparrow} = \sigma_{\bullet\bullet} - \sigma_{\bullet\circ}$$

$$\Delta\sigma_L = \sigma_{\rightarrow\leftarrow} + \sigma_{\rightarrow\rightarrow}$$

NOTE   First or top arrow refers to particle a.

The filled circle means the projection of polarization in the line of s from the observer; the unfilled one means the projection of polarization in the line of s to the observer.

References

Achasov, N.N., Shestakov, G.N.: In: Proceedings of the 1st International Seminar on Spin Phenomena in High Energy Physics, Protvino, USSR, p. 183 (1983)

Aleem, F., Saleem, M.: Phys. Rev. D **27**, 2068 (1983)

Anderson, B., et al.: Preprint No. 82-6, Lund University (1982)

Antipov, Yu.M., et al.: Preprint No. 73-30, IHEP, Protvino, USSR (1973)

Antipov, Yu.M., et al.: Preprint No. 76-95, IHEP, Protvino, USSR (1976)

Arakeliyan, G.G., et al.: Yad. Fiz. **38**, 1525 (1983)

Arestov, Yu.I.: Preprint No. 86-82, IHEP, Protvino, USSR (1986)

Arestov, Yu.I., et al.: Preprint No. 83-124, IHEP, Protvino, USSR (1983)

Arestov, Yu.I., et al.: Yad. Fiz. **40**, 204 (1984)

Avvakumov, I.A., et al.: Preprint Nos. 80-94 and SERP-E-112, OEF IHEP, Protvino, USSR (1980)

Avvakumov, I.A., et al.: Preprint No. 81-15, OEF IHEP, Protvino, USSR (1981)

Avvakumov, I.A., et al.: Yad. Fiz. **35**, 1465 (1982)

Aznauryan, I.G., Solov'ev, L.D.: Preprint No. 75-127, OTF IHEP, Protvino, USSR (1975)

Barger, V., Phillips, R.J.N.: Phys. Rev. **187**, 2210 (1969)

Becker, H., et al.: In: Proceedings of the 18th International Conference on High Energy Physics, Tbilisi, Georgia, vol. 1, p. 27 (1977)

Bonamy, P., et al.: Phys. Lett. **23**, 501 (1966)

Bonamy, P., et al.: Nucl. Phys. B **52**, 392 (1973)

Bourrely, C., Leader, E., Soffer, J.: Phys. Rep. **59**, 595 (1980)

Chou, T., Yang, C.N.: Nucl. Phys. **107**, 1 (1976)

Enkovsky, L.L., Struminsky, B.L.: Preprint No. ITP-83-121E, ITP, Kiev, USSR (1983)

Fidecaro, G., et al.: In: Proceedings of the International Symposium on High Energy Physics with Polarized Beams and Targets, Lausanne, Switzerland, p. 557 (1980)

Gaidot, A., et al.: Phys. Lett. **44**, 471 (1973)

Gaidot, A., et al.: Phys. Lett. **77**, 389 (1975)

Gauron, P., et al.: Phys. Rev. Lett. **52**, 1252 (1984)

Hill, D., et al.: Phys. Rev. Lett. **30**, 349 (1973)

Kline, R.V., et al.: Phys. Rev. D **22**, 553 (1980)

Kolár, P., et al.: Czechoslov. J. Phys. **26**, 1294 (1976)

Krzywicki, A., Tran Tranh Van, J.: Lett. Nuovo Cimento **12**, 249 (1969)

Levintov, I.I.: Preprint No. 144, ITEP, Moscow, USSR (1978)

Logunov, A.A., Meshcheryakov, V.A., Tavkhelidze, A.N.: Dokl. Akad. Nauk SSSR **142**, 317 (1962)

Nambu, Y., Iona-Lasinio, G.: Phys. Rev. **122**, 345 (1961)

Nurushev, S.B.: In: Proceedings of the 3rd International Symposium on Pion–Nucleon and Nucleon–Nucleon Physics, Gatchina, USSR, p. 398 (1989)

Nurushev, S.B.: In: Proceedings of the 9th International Symposium on High Energy Spin Physics, Bonn, Germany, p. 34 (1990)

Pierrard, J., et al.: Phys. Lett. **57**, 393 (1975)

Pomeranchuk, I.Ya.: Zh. Eksp. Teor. Fiz. **34**, 725 (1958)

Shaevitz, M.N., et al.: Phys. Rev. Lett. **36**, 8 (1976)

Solov'ev, L.D., Shchelkachev, A.V.: Preprint No. 79-80, OTF IHEP, Protvino, USSR (1979)

Troshin, S.M., Tyurin, N.E.: Preprint No. 76-55, OTF IHEP, Protvino, USSR (1976)

Troshin, S.M., Tyurin, N.E.: Preprint No. 80-12, OTF IHEP, Protvino, USSR (1980)

Troshin, S.M., Tyurin, N.E.: Preprint No. 81-29, OTF IHEP, Protvino, USSR (1981)

Troshin, S.M., Tyurin, N.E.: In: Proceedings of the 2nd International Seminar on Spin Phenomena in High Energy Physics, Protvino, USSR, p. 167 (1984)

Troshin, S.M., Tyurin, N.E.: In: Proceedings of the 6th International Symposium on Polarization Phenomena in Nuclear Physics, Osaka, Japan, p. 207 (1985)

Troshin, S.M., Tyurin, N.E.: Preprint No. 86-79, IHEP, Protvino, USSR (1986)

Index

S.B. Nurushev et al., *Introduction to Polarization Physics*,
Lecture Notes in Physics 859, DOI 10.1007/978-3-642-32163-4,
© Moskovski Inzhenerno-Fisitscheski Institute, Moscow, Russia 2013